ANNUAL REVIEW OF ASTRONOMY AND ASTROPHYSICS

LEO GOLDBERG, *Editor*
Harvard College Observatory

DAVID LAYZER, *Associate Editor*
Harvard College Observatory

JOHN G. PHILLIPS, *Associate Editor*
University of California, Berkeley

VOLUME 8

1970

ANNUAL REVIEWS INC.
4139 EL CAMINO WAY
PALO ALTO, CALIFORNIA, USA

EDITORIAL COMMITTEE

B. F. BURKE
L. GOLDBERG
R. P. KRAFT
G. P. KUIPER
D. E. OSTERBROCK
L. WOLTJER

Responsible for organization of Volume 8
(Editorial Committee, 1968)

R. BRACEWELL
B. F. BURKE
L. GOLDBERG
N. U. MAYALL
D. E. OSTERBROCK
F. L. WHIPPLE

ANNUAL REVIEW OF ASTRONOMY AND ASTROPHYSICS

LEO GOLDBERG, *Editor*
Harvard College Observatory

DAVID LAYZER, *Associate Editor*
Harvard College Observatory

JOHN G. PHILLIPS, *Associate Editor*
University of California, Berkeley

VOLUME 8

1970

ANNUAL REVIEWS INC.
4139 EL CAMINO WAY
PALO ALTO, CALIFORNIA, USA

ANNUAL REVIEWS INC.
PALO ALTO, CALIFORNIA, USA

©1970 BY ANNUAL REVIEWS INC.
ALL RIGHTS RESERVED

Standard Book Number 8243–0907–3
Library of Congress Catalogue Number: 63–8846

FOREIGN AGENCY
Maruzen Company, Limited
6 Tori-Nichome Nihonbashi
Tokyo

PRINTED AND BOUND IN THE UNITED STATES OF AMERICA BY
GEORGE BANTA COMPANY, INC.

PREFACE

Most of the topics and authors for the current volume of the *Annual Review of Astronomy and Astrophysics* were chosen by the Editorial Committee during its meeting in Cambridge, Massachusetts on May 3, 1968. In reading the Table of Contents, one cannot help but be impressed by the remarkably large number of new fields of astronomy that are represented. Indeed, fewer than one half of the chapters deal with relatively traditional fields of research. Such topics as Fourier Astronomical Spectroscopy, Pulsars, Optical Properties of X-ray Sources, Neutron Stars, Internal Rotation of the Sun, and Radio Recombination Lines have developed within the past 5 to 10 years and are being reviewed in this series for the first time. Whether or not this trend continues will be determined by future levels of financial support for astronomical research. It is regrettable that funds for astronomy are being sharply reduced just at the time when the potential for exciting astronomical discovery has become virtually unlimited.

It is a pleasure to thank the authors of this volume for an exceptionally fine complement of reviews. As in the past, the scientific editing of the volume has been the responsibility of the Associate Editors.

After this volume had gone to press, we learned with deep sorrow of the death of Professor Sydney Chapman, who made monumental contributions to astrophysics and geophysics. In Chapter 3 of this volume he has written one of his last published articles on auroral physics, a subject which has always been close to the center of his attention.

The heavy burden of editorial preparation of manuscripts and proofs for the printer and of correspondence with authors has been handled superbly by Joann Huddleston of the staff of Annual Reviews Inc., and we are grateful also for another outstanding printing job by the George Banta Company.

THE EDITORIAL COMMITTEE

ERRATUM

Volume 5 (1967)—Determination of Masses of Eclipsing Binary Stars, by Daniel M. Popper
  p. 102 Table I: the listing under "Radii" for Z Vul
  $\begin{matrix} 4.7 \\ 2.0 \end{matrix}$ *should read* $\begin{matrix} 4.7 \\ 4.7 \end{matrix}$

CONTENTS

THE ORIGIN OF SOLAR MAGNETIC FIELDS, *E. N. Parker*	1
THE THEORY OF STELLAR WINDS AND RELATED FLOWS, *T. E. Holzer and W. I. Axford*	31
AURORAL PHYSICS, *Sydney Chapman*	61
ATMOSPHERES OF VERY LATE-TYPE STARS, *M. S. Vardya*	87
INFORMATION-PROCESSING SYSTEMS IN RADIO ASTRONOMY AND ASTRONOMY, *B. G. Clark*	115
OPTICAL OBSERVATIONS OF EXTRASOLAR X-RAY SOURCES, *W. A. Hiltner and D. E. Mook*	139
THE COSMIC ABUNDANCE OF HELIUM, *I. J. Danziger*	161
NEUTRON STARS, *A. G. W. Cameron*	179
ASTRONOMICAL FOURIER SPECTROSCOPY, *Pierre Connes*	209
RADIOFREQUENCY RECOMBINATION LINES, *A. K. Dupree and Leo Goldberg*	231
PULSARS, *A. Hewish*	265
INTERNAL ROTATION OF THE SUN, *R. H. Dicke*	297
EXCITATION AND IONIZATION BY ELECTRON IMPACT, *Oleg Bely and Henri Van Regemorter*	329
THE NUCLEI OF GALAXIES, *G. R. Burbidge*	369
SOME RELATED ARTICLES APPEARING IN OTHER *Annual Reviews*	461
AUTHOR INDEX	463
SUBJECT INDEX	475
CUMULATIVE INDEX OF CONTRIBUTING AUTHORS, VOLUMES 4 TO 8	492
CUMULATIVE INDEX OF CHAPTER TITLES, VOLUMES 4 TO 8	493

Annual Reviews Inc. and the Editors of this publication assume no responsibility for the statements expressed by the contributors to this *Review*.

REPRINTS

The conspicuous number (2000 to 2013) aligned in the margin with the title of each review in this volume is a key for use in the ordering of reprints.

Beginning with July 1970, reprints will be available from all future Annual Reviews volumes. Reprints of most articles published in the *Annual Reviews of Biochemistry and Psychology* from 1961 to 1970 and the *Annual Reviews of Microbiology and Physiology* from 1968 to 1970 are now maintained in inventory.

Available reprints are priced at the uniform rate of $1 each postpaid. Payment must accompany orders less than $10. The following discounts will be given for large orders: $5–9, 10%; $10–24, 20%; $25 and over, 30%. All remittances are to be made payable to Annual Reviews Inc. in U.S. dollars. California orders are subject to a 5% sales tax. One-day service is given on items in stock.

For orders of 100 or more, any Annual Reviews article will be specially printed and shipped within 6 weeks. Reprints which are out of stock may also be purchased from the Institute for Scientific Information, 325 Chestnut Street, Philadelphia, Pa. 19106. Direct inquiries to the Annual Reviews Inc. reprint department.

The sale of reprints of articles published in the Reviews has been expanded in the belief that reprints as individual copies, as sets covering stated topics, and in quantity for classroom use will have a special appeal to students and teachers.

Copyright 1970. All rights reserved

THE ORIGIN OF SOLAR MAGNETIC FIELDS[1]

E. N. PARKER

Department of Physics, University of Chicago, Chicago, Illinois

1. INTRODUCTION

Since Hale (1908) first demonstrated that magnetic fields are an intimate feature of sunspots, it has become evident that magnetic fields are the basic ingredient of solar activity. If, for instance, the solar gases were nonconducting like the terrestrial atmosphere, there would be no solar magnetic fields and the most active feature of the Sun would be the hydrodynamic convection beneath the photosphere and the photospheric granulation. There would be no plages, prominences, spicules, flares, or fast particles and radio bursts. The corona, what there was of it, would give only a weak solar wind. Hence there would be little or no magnetic activity, aurora, and Van Allen radiation at Earth.

The existence of magnetic fields on the Sun, and in the astrophysical universe in general, is a direct result of (*a*) the enormous abundance of free electrical charges and (*b*) the general nonexistence of magnetic changes (monopoles). It is this curious asymmetry between electric and magnetic charge in the Universe that leads to the production of magnetic fields. The abundance of free electric charges neutralizes electric fields in the frame of reference of the gas, except for very small potential differences, of the order of the thermal energy of the gas. There are evidently no magnetic monopoles to perform the same neutralizing function for magnetic fields. So whenever there is net relative motion of electrons and ions in the gas—an electric current density **j**—there is an associated magnetic field **B**

$$4\pi j = c\nabla \times \mathbf{B}$$

The current may cause the field, or vice versa, depending upon the circumstances.

Magnetic fields can be observed in the Sun at the level of the photosphere, and at higher levels in the atmosphere wherever the density is high enough, and the temperature low enough, to permit detectable Zeeman broadening of the spectral lines. The magnetic fields are observed from their polarization effects on radio emission and from their channeling of the solar corona into visible streamers. And, finally, the solar magnetic field is extended by the solar wind into space where it is observed directly with magnetometers. Thus the fields at the photosphere and above are geneally observable. On the other hand, we know nothing of the magnetic fields

---

[1] This work was supported in part by the National Aeronautics and Space Administration under grant NASA-NsG-96-60.

beneath the photosphere beyond what extrapolation downward from the surface fields can tell us.

There are a number of recent excellent reviews of observational knowledge of solar fields (e.g. Babcock 1963, Cowling 1965, Howard 1967, Ness 1968, Sawyer 1968, Wilcox & Howard 1968 and references therein), so that the observations need not be repeated in detail here. In this review we concentrate on the origin of the magnetic fields in the Sun, insofar as the origin is currently understood.

The gross features of the solar magnetic fields are the polar fields of about 1 G within 35° of the poles, and the irregular fields at low latitudes. The north polar field was first observed (Babcock & Babcock 1955) in 1952 when it was directed outward. The south polar field was at that time directed inward, so that the polarity of the Sun was opposite that of Earth. The south polar field began to decline at the beginning of 1957 (Babcock 1959), becoming undetectable by March of that year. By June it was detectable, pointing *outward*, and by the end of the year it had increased to 1 G in this reversed sense. But by 1961 it had dwindled to undetectability, and has remained undetectable, <0.5 G. The north polar field reversed abruptly in November 1958 and has remained directed inward. The reversals took place near the peak of the sunspot activity in the respective hemispheres (Waldmeier 1960). It is speculated that the polar fields reverse in connection with the sunspot cycle. But this has not yet been confirmed by a reversal at the present peak of activity.

From latitudes of ±55° to the equator the solar magnetic field is irregular in both pattern and strength. The outstanding features are the sunspots, with fields up to 4000 G (Bray & Loughhead 1964, Zwaan 1968, Wilson 1968 and references therein, as well as recent studies by Altschuler 1966 and Altschuler, Nakagawa & Lilliequist 1968). The equatorial migration of the region where sunspots are formed is well known (e.g. Gleissberg 1953, 1968; Bappu, Grigorjev & Stepanov 1968 and references therein). It is generally believed that the sunspots, and the bipolar magnetic regions, are evidence of a general toroidal (azimuthal) field of several hundred gauss at some depth beneath the photosphere (Cowling 1953). Presumably the migration of the region of spots and bipolar regions traces the migration of the toroidal field during the solar magnetic cycle.

A particularly fascinating feature of sunspots is the small islands of reverse flux which appear in the midst of the spot field and the small umbral knots of 1500 G over scales of 10^3 km nearby (Beckers & Schröter 1968a, b; Grigorjev 1969). The general active regions of strong field ($\gtrsim$ 100 G) surrounding sunspot groups are the site of strong Ca^+ plage emission (Howard 1959, Leighton 1959, Chitre 1969), flares, and active prominences. The magnetic field is believed to cause plages by providing enhanced heating in the upper levels of the photosphere (Simon & Leighton 1963, 1964; Livingston 1968; Howe 1969). Magnetic fields apparently are the basic source of energy (up to 10^{32} ergs) for flares (Giovanelli 1947; Parker 1957b, 1963a;

Dungey 1958; Sweet 1958; Smith & Smith 1963; Petschek 1963; Jaggi 1963; Severny 1964, 1965a,b; Sturrock 1966, 1976; Sturrock & Coppi 1967; Green & Sweet 1967; Petschek & Thorne 1967; Sturrock & Smith 1968; Moreton & Severny 1968; Anzer 1968; see also Fichtel & McDonald 1967 for a review of fast particles from flares).

Prominences evidently form along the lines of force of the field, which supports their weight (Kippenhahn & Schlüter 1957, Thompson & Billings 1967, Olson & Lykoudis 1967, Harvey & Tandberg-Hanssen 1968). Prominence fields are typically 5 G, though occasionally as high as 50–80 G (Rust 1967, Malville 1968).

It was pointed out some years ago (Howard 1959, Leighton 1959, Simon & Leighton 1963) that magnetic fields ≈ 20 G are necessary and sufficient for plage emission. Hence plages occur in the active regions where the fields are typically $\gtrsim 50$ G, and in the narrow strips between the supergranules where the field is concentrated by the convection of the gas (Leighton, Noyes & Simon 1962; Simon & Leighton 1963; Parker 1963b, 1965b; Clark 1968). Spicules are also concentrated in regions of strong field (see Beckers 1963; 1968 and references therein) and, it is believed, are a direct result of the field (Parker 1964, 1965a; Wentzel & Solinger 1967).

The corona is hotter and denser over active regions than above the quiet regions, presumably because the magnetic fields enhance the transport of mechanical energy from the convective zone to the corona (Kulsrud 1955, Osterbrock 1961, Moore & Speigel 1964, Parker 1964, Kopp & Kuperus 1968, McLellan & Winterberg 1968). The plage and spicule activity associated with the strong fields also indicate this energy transport. The active regions of the corona tend to dominate the quiet corona in producing solar wind, so it is the magnetic fields of the active regions that are responsible for most of the interplanetary wind and weather, and hence for the geomagnetic and auroral effects, that typify solar activity. The magnetic fields of active regions are also responsible for the production of the fast particles observed at Earth in connection with solar activity and for shaping the various radio bursts resulting from the fast particles at the Sun.

Overall patterns in the magnetic fields in the broad equatorial band between the polar regions are ill defined. As already noted, the appearance of bipolar regions and sunspots in well-defined bands of latitude implies a toroidal field beneath. But recognition of other surface patterns at the general 1–2 G level is obscured by the patterns of the much stronger fields of active regions. The tendency for active regions to form at special longitudes throughout a given solar cycle suggests a well-defined pattern of variation with longitude deep under the surface of the Sun (see discussion and references in Sawyer 1968; Bappu et al 1968). Recently longitude patterns in the 1–2 G background fields have been uncovered by starting with the well-defined magnetic sectors of the solar field extended into space by the solar wind (Schatten, Ness & Wilcox 1968; Wilcox 1968; Wilcox & Howard 1968; Schatten, Wilcox & Ness 1969). Extrapolating from the regular sector structure, observed in space at the orbit of Earth, back to the Sun shows a general pattern of sectors in the general background field.

The expanding corona carries the lines of force of the general background solar field out through the solar system (Parker 1958a, 1963c; see reviews by Ness 1968, Wilcox 1968, Hundhausen 1968, Parker 1969) filling interplanetary space with an extended spiral field (of about 5×10^{-5} G at the orbit of Earth). The solar fields carried in the wind interact with the galactic cosmic rays, pushing them back and generally reducing the cosmic-ray density in the inner solar system (Parker 1958b, 1963c). The extended solar fields carry angular momentum away from the Sun (Brandt 1966, 1967; Weber & Davis 1967; see review by Parker 1969 and references therein) and evidently have contributed to the evolution of the rotation of the Sun. Indeed the whole problem of the nonuniform rotation of the Sun, with the equatorial period of rotation some 4 days less than the polar period, is difficult to understand without convection and the direct or indirect effects of magnetic fields (e.g. Kippenhahn 1963, Elsasser 1966, Cocke 1967, Goldreich & Schubert 1968, Schubert 1968, Nakagawa & Trehan 1968, Nakagawa & Swarztrauber 1969).

We cannot, in this brief article, explore all the many properties and consequences of the magnetic fields of the Sun. Most of the direct effects of magnetic fields are understood at best only qualitatively, and others not at all. Some effects, e.g. the precise nature of the solar flare and fast-particle acceleration, the origin of the nonuniform rotation, the precise mode of heating the active corona, the plages, and the spicules, are much debated. The problems are too complicated to be demonstrated with definitive models. Instead the present review concentrates on the origin and behavior of the general magnetic field of the Sun. We shall inquire as to why the Sun should have any magnetic fields at all, and in particular why it should have reversible polar fields, and toroidal fields which migrate from high to low latitudes. The toroidal fields (and the polar fields too, so far as one can tell) have opposite signs in the northern and southern hemispheres, and each successive migrating band of toroidal field has opposite sign from its predecessor in the same hemisphere.

2. Origin of Magnetic Fields

2.1 *General remarks.*—The basic question is the origin and present maintenance of the solar magnetic fields. As noted in the introductory remarks, the general absence of magnetic monopoles is a necessary condition for the existence of strong magnetic fields. The sufficient condition for magnetic fields appears to be a rotating conducting plasma undergoing convective motion (Parker 1955b).

Consider the well-known hydromagnetic equation (see also Mestel & Roxburgh 1962)

$$\partial \mathbf{B}/\partial t = \nabla \times (v \times \mathbf{B}) + \eta \nabla^2 \mathbf{B} \qquad \qquad 1.$$

where η is the magnetic diffusivity ($\eta = c^2/4\pi\sigma$ esu). This equation tells us

(Alfven 1950) that the magnetic field is convected with the plasma, except insofar as η permits the field to diffuse relative to the plasma. Thus, the dimensions, or scale, of the magnetic fields in nature are determined by the scale of the plasma motion. The galactic field, for instance, is stretched out by the nonuniform rotation of the Galaxy in the azimuthal direction around the Galaxy; the solar field is stretched by the solar wind out through the solar system; the fields of active regions on the Sun are extended by the supergranulation to scales of 10^5 km, etc. (see formal mathematical illustrations of these well-known effects in Parker 1963b). In every case the extension of the field is accomplished by the motions of the gases.

The fundamental question concerns the origin of the magnetic flux—the lines of force—which make up the field. On the one hand, it has been pointed out that a star, such as the Sun, must begin its life with the magnetic fields entrapped in the interstellar gas from which the star is formed (Cowling 1953, 1965). Indeed, from this point of view it is a problem to explain why the magnetic fields of a typical star are not stronger, 10^8 G or more (Mestel 1965, 1967). As we will see, there are long-term effects which would destroy and obliterate whatever primordial fields may once have been trapped in the Sun.

Both ordered and disordered gas motions have their effects on magnetic fields. The effects of random turbulence have been discussed at some length. One school of thought (Biermann & Schlüter 1951, Biermann 1953, Chandrasekhar 1955) argues that the direct coupling of the field to the fluid leads to equipartition of the energy of the fluid. Another school of thought (Batchelor 1950, Moffat 1961, Parker 1963, Pao 1963b, Saffman 1964) notes that the hydromagnetic equation 1 for **B** has the same form as the equation

$$\partial \omega/\partial t = \nabla \times (v \times \omega) + \nu \nabla^2 \omega \qquad 2.$$

for the vorticity $\omega = \nabla \times v$, and hence that the behavior of **B** in a turbulent field should be the same as ω. Hence the magnetic field should be weak in the large eddies and rise to significant levels only in the small eddies.

Kraichnan & Nagarajan (1967) have examined the complete hydromagnetic equations in detail and show that there are many terms, all of comparable magnitude, representing the dynamical interaction between the Fourier transforms of v and **B**. They point out which terms must be dropped and which retained to arrive at the differing views described above. They conclude that the question of magnetic fields in turbulent flows can be resolved only by formal calculation. Physical arguments are not sufficient.

Observations of fields on the Sun do not resolve the question of magnetic fields in turbulent flows. Typical photospheric velocities of 1 km/sec where the density is 10^{-8} g/cm^3 represent 50 ergs/cm^3, equivalent to 30 G. The general background fields of the Sun are only 1–2 G (0.04–0.1 erg/cm^3) but the fields of active regions may be 300–3000 G (4×10^3–4×10^5 ergs/cm^3). If anything, the observations suggest the irrelevance of turbulent fields, at least so far as direct production of fields on the Sun is concerned.

2.2 Hydromagnetic dynamos.

—Consider the effect of suitably coordinated fluid motions, as distinct from the random turbulence discussed above. Larmor suggested nearly 50 years ago that swirling motions of the conducting gas might generate the magnetic fields of sunspots, then recently measured by Hale. About 35 years ago Cowling (1934, 1945, 1957, 1965; Backus & Chandrasekhar 1956) showed that steady fluid motions cannot maintain a magnetic field which is confined to a finite region of space and possesses axial symmetry. The essence of Cowling's theorem is that the lines of force in each meridional plane must close on themselves, since the lines do not extend to infinity. Therefore, there must be at least one neutral point on each meridional plane, with the neighboring field lines circling the point. Hence the curl of the magnetic field is nonvanishing in the neighborhood of the neutral point, and there is an electric current flowing perpendicular to the meridional plane at the neutral point. With axial symmetry the current flows in circles centered on the axis of symmetry. It follows that the line integral of the current around each circle, $\oint d\mathbf{s} \cdot \mathbf{j}$, is nonvanishing. Since $\mathbf{B}=0$ on the circle through the neutral points, Ohm's law is $\mathbf{j}=\sigma\mathbf{E}$. Hence the line integral $\oint d\mathbf{s} \cdot \mathbf{E}$ is nonvanishing, and from Stokes' theorem the integral of $\nabla \times \mathbf{E}$ over the area enclosed by the circle is nonvanishing, and hence $\nabla \times \mathbf{E}$ itself must be nonvanishing. But $\nabla \times \mathbf{E}$ vanishes in a field that is steady in time. Hence Maxwell's equations are not compatible with the requirement that a field with axial symmetry be maintained by steady fluid motions. There is no way to provide the emf to drive the associated closed current systems. It is evident at once that the theorem is not restricted to fields with axial symmetry, but applies to any stationary field configuration in which the lines of $\nabla \times \mathbf{B}$, or $\mathbf{j}$, form closed loops (see recent discussion by Jayanthan 1968, Cowling 1968).

The next step in dynamo theory was taken by Elsasser (1945, 1946, 1947, 1950a,b) in connection with the magnetic field of Earth. Elsasser showed that there was no justification for the idea that thermoelectric effects, ferromagnetic effects, etc., were the source of the geomagnetic field. He proposed that the only tenable theory is generation by the motions in the liquid core of Earth. The liquid core is molten iron and nickel, with an electrical conductivity $\approx 10^{16}$/sec (esu). The core radius a is about half the radius $R_E=6.4\times 10^3$ km of Earth. Observations of the magnetic field at the surface of Earth over the past two centuries show that the small inhomogeneities in the field drift slowly westward and are in a continual state of evolution, some increasing, others decreasing, etc. These slow contortions of the field suggest that the liquid core is in a state of convection, with velocities $\sim 10^{-2}$ cm/sec. Presumably the convection is driven by small amounts of radioactive heating, or some differential settling of heavier material, or both (see discussion in Elsasser 1950b, 1956).

These physical considerations serve to define the problem: (*a*) what fluid motions does one expect in a rotating convecting core, and (*b*) what aspect of the fluid motion is responsible for generating the geomagnetic field which we observe at the surface of Earth, viz a dipole field with a strength of about 0.6 G at the poles at the surface $(r=R_E)$ of Earth? The dipole field is a meridional, or poloidal field, with its lines of force essentially in meridional planes. The lines are directed inward at the north geomagnetic pole and outward at the south pole.

Presumably the generation of the more complicated fields on the Sun follows from the basic principles that pertain to the relatively simple circumstances of the dipole field of Earth. So it is convenient to continue with the geomagnetic field to illustrate the basic nature of field generation. The additional effects that characterize the Sun are easily demonstrated once the ideas are clear.

The Earth is a rapidly rotating body, with the result that the Coriolis forces have a powerful effect on the convective motions. The largest effect is the nonuniform rotation of the core, with the regions closest to the axis rotating around the axis more rapidly than the more distant regions. Thus we would expect the regions near the surface of the core at low latitudes to rotate more slowly than the inner regions, as illustrated in Figure 1. The effects of this nonuniform rotation are immediately evident. They shear the poloidal dipole field $\mathbf{B}_p$ and stretch the lines of force into a toroidal field $\mathbf{B}_\phi$ around the core, as illustrated in Figure 1. The stretching of the poloidal lines of force to form a toroidal field $\mathbf{B}_\phi$ is described by Equation 1 upon introducing a nonuniform rotation $\mathbf{v}_\phi$. The shear does not affect the axisymmetric part of the poloidal field, i.e. for axial symmetry $\nabla \times (\mathbf{v}_\phi \times \mathbf{B}_p)$ has only a toroidal component (see Equation 3). Elsasser developed the mathematics for treating the interaction of the velocity with the poloidal and toroidal

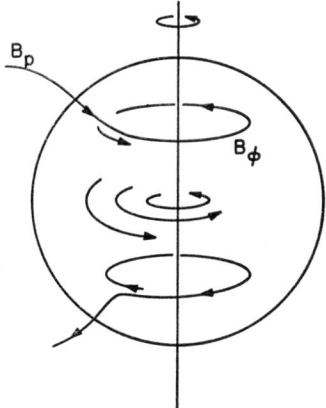

FIGURE 1. A sketch of the nonuniform rotation of the liquid core of Earth (represented by the long curved arrows in the equatorial plane) and the effect of the nonuniform rotation in producing a toroidal field from the lines of force of the dipole field.

fields in terms of the poloidal and toroidal modes of a conducting sphere. He demonstrated the inevitable strong toroidal field (perhaps 20–50 G) that must exist inside the conducting core as a consequence of the observed dipole field and the nonuniform rotation.

Cowling's theorem made it clear that the dipole field cannot be regenerated from itself, so evidently the dipole field must be generated in some way from the toroidal field, presumably as a result of convective motions. The formal approach developed by Elsasser was to expand the convective velocity field in poloidal modes, so that the term $\nabla \times (\mathbf{v} \times \mathbf{B})$ on the right-hand side of Equation 1 can be expanded in an infinite series of the modes. Equating coefficients of modes on both sides of 1 then led to an infinite sequence of algebraic equations (Elsasser 1947, 1950a,b). A thorough exploration of the method (Bullard 1949a,b, 1955; Bullard & Gellman 1954; Takeuchi & Elsasser 1954; Elsasser & Takeuchi 1955; Rikitake 1958, 1966) established that the expansion converges too slowly to be effective for treating the maintenance of a *stationary* magnetic field by stationary fluid motions. Indeed, in view of Cowling's theorem, and the evident lack of convergence of the expansions, the question was seriously raised whether there was such a thing as a stationary dynamo, in which $\eta \nabla^2 \mathbf{B}$ is offset exactly by $\nabla \times (\mathbf{v} \times \mathbf{B})$ so that $\partial \mathbf{B}/\partial t = 0$ everywhere throughout the core. After all, the geomagnetic dynamo does not have to be stationary in the core. Only the net effect at the surface of Earth must be a quasisteady dipole. If the geomagnetic field is generated by the fluid motions in the core, then the convective motions in the core must provide the mechanism for generating the poloidal field from the toroidal field.

The convective motions do not appear to be stationary. Some convective cells are growing, and others are dying out. The irregularity of the inhomogeneities in the geomagnetic field at the surface of Earth indicates that there are some 15–20 major convective cells in the core (Elsasser 1950b, 1956) and that these cells are distributed in an irregular pattern which is changing in time, much as the granules and supergranules on the Sun. We would expect that the rapid rotation of Earth causes the rising convective cells to rotate counterclockwise in the northern hemisphere and clockwise in the southern hemisphere, as illustrated on the right-hand side of Figure 2. It is then a straightforward matter to show from this that the convection does regenerate the dipole field (Parker 1955b). The rotation of each cell is initiated by the converging flow at the base of the cell and is terminated by the diverging flow at the top. The effect of such *cyclonic* convective motions is illustrated on the left side of Figure 2. The rising fluid pushes the lines of force of the toroidal field upward to form an inverted "U". The rotation of the fluid twists the inverted "U" so that it has a nonvanishing projection on the local meridional plane. The toroidal field and the cyclonic rotation of the convective cell are both of opposite sign in the southern hemisphere. Hence the projection of the resulting loop of flux onto the meridional plane has the same sense in both hemispheres. The field of the loop is southward on the

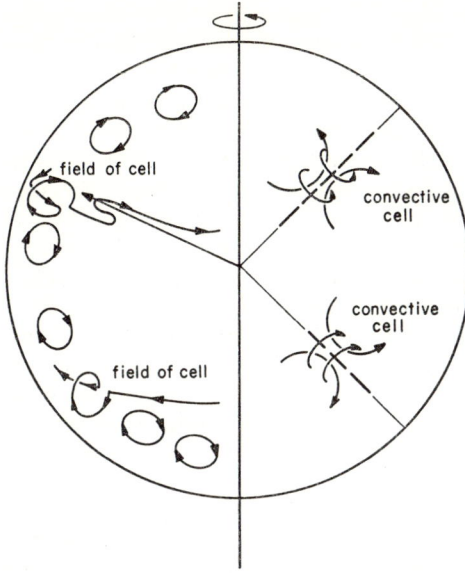

FIGURE 2. The right-hand side of the diagram is a sketch of cyclonic convective cells in the core of Earth. The arrows indicate the rising and the rotating motions associated with each cell. The left-hand side shows the deformation of the toroidal field by a rising current followed by a twist. The result is a line of force whose projection on the meridional plane is a loop of flux with the sense of circulation shown. Such loops are produced at all latitudes.

side closer to the axis of Earth and northward on the farther side. This is the same sense as the general dipole field, which is directed inward at the north pole, out at the south pole, and around to the north above the surface of Earth. Therefore, the small loops of flux generated by the individual convective cells reinforce the general circulation $\oint d\mathbf{s} \cdot \mathbf{B}$ of the dipole field. The many small loops coalesce in periods of 10^4 years to contribute a general poloidal field in the core. It is the leakage of this poloidal field (generated from the toroidal field by the cyclonic convective cells) out of the core and up through the mantle which maintains the dipole field observed at the surface of Earth (Parker 1955b).

Formal mathematical examples illustrating the twisting of the toroidal field into loops of flux in the meridional planes, and the coalescence of the loops to reinforce the dipole field have been given elsewhere (Parker 1955b, Parker & Krook 1956) and need not be repeated here. The circulation of flux in the meridional plane reinforces the dipole field, provided only that the individual rising convective cells rotate the field through an angle greater than zero, but less than π. If the rotation were greater than π, the sense of circulation in the meridional plane would be reversed and the effect would be an active degeneration of the dipole field.

Note that if the convective motions were dominated by sinking cells of heavy fluid, rather than the rising cells shown in Figure 2, the magnetic effect would be reversed and the meridional circulation of field generated by the convective motions would destroy the dipole field. There would then be no possibility for a steady regenerative dynamo, just as with cyclonic rotations in excess of π discussed in the paragraph above. Thus it is important to consider what drives the convection in the core. Convection driven by cold spots on the core surface or by the subsidence of a denser phase of liquid from the core surface would be dominated by downward-moving convective cells, with the upward motions spread out through the broader spaces between. The result would be degenerative rather than regenerative. If, on the other hand, the convection is driven by heating from below, the cells of hot fluid rising from central regions of the core (or from the surface of the small inner solid core believed to exist) find themselves separated by increasing distances. We would expect the cyclonic rotation to be concentrated in the rising cells, giving the regeneration, illustrated in Figure 2, that evidently occurs. Thus it would seem that the Earth's having an observable dipole field implies something about the causes of the convection in the core. It is an interesting and important dynamical question that needs theoretical, if not experimental, study.[2]

But to continue the general discussion of regeneration of fields by cyclonic convective motions, we must not forget that the cyclonic motions interact with the poloidal field as well. Figure 3 is a sketch of the sense of the twisted loops that result. They have opposite signs in the northern and southern hemispheres, because the cyclonic rotation, but not $\mathbf{B}_p$, has opposite signs. In the northern hemisphere the loops coalesce to produce a circulation of field around the axis of Earth which is opposite to the direction of rotation of Earth near the axis and in the same direction as the rotation far from the axis. This is a higher mode than the dipole field, and higher than the toroidal field generated by the nonuniform rotation. Hence its decay time is several times shorter than the dipole. This, and our belief that the poloidal field, from which the mode is generated, is only a fraction of the toroidal field, from which the poloidal field is generated, makes the interaction of the convective motions with the poloidal field only a secondary effect. The dominant fluid interaction with the poloidal field is the powerful nonuniform rotation, which stretches out the lines of force to form the toroidal field.

Note that the energy source of the dynamo is the convection. And the Coriolis force shapes the convection, producing the nonuniform rotation (which manufactures the toroidal field from the poloidal field) and producing the cyclonic rotation of the individual convective cells (which manufactures the poloidal field from the toroidal field). Evidently the generation of field on the Sun is likewise powered by the convection and shaped by the Coriolis forces, but with a somewhat different result (Parker 1955b, 1957a) because

[2] At the end of this section the dominant effect in a compressible atmosphere is discussed.

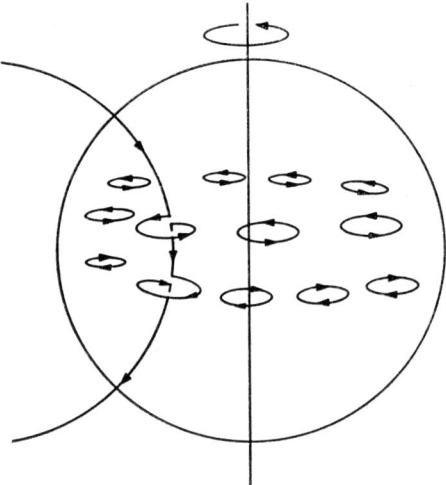

FIGURE 3. A sketch of the loops of flux produced by the interaction of the cyclonic convective cells with the poloidal field.

of the different geometry of the convective region and because of the different relative sense of the nonuniform rotation and cyclonic motions.

For both Earth and the Sun, note that the hydromagnetic equation is a linear homogeneous equation in **B**, so that the overall sense of **B** is immaterial. The Lorentz forces $(\nabla \times \mathbf{B}) \times \mathbf{B}/4\pi$ exerted on the fluid motions by the fluid motions by the field are also independent of the sense of **B**, so that if $+\mathbf{B}$ is a solution to the complete set of dynamical equations, then $-\mathbf{B}$ is too. As in any self-exciting dynamo, the field and its reverse are amplified equally well. The sign of the geomagnetic field today was determined by the sense of the field at the beginning of the present regime.[3] The magnetic field of the Sun, with its periodic reversals,[4] is equally well amplified in either sense.

Now the dynamo theory just described can be summarized by a set of dynamo equations (Parker 1955b, 1957a). Generation of the toroidal field $B_\phi(r, \theta)$ by the nonuniform rotation $v_\phi(r, \theta) = \omega(r, \theta) r \sin \theta$ follows directly from Equation 1 as

$$\left[\frac{\partial}{\partial t} - \eta\left(\nabla^2 - \frac{1}{\tilde{\omega}^2}\right)\right] B_\phi = \left(B_\theta \frac{\partial \omega}{\partial \theta} + B_r r \frac{\partial \omega}{\partial r}\right) \sin \theta \qquad 3.$$

for the simple case of axial symmetry, where $\tilde{\omega} = r \sin \theta$ is distance from the

---

[3] We refer the interested reader to the literature on the reversal of the geomagnetic field (Runcorn 1955, Wilson 1966).

[4] It is our opinion that the regular reversal of the solar field is due to causes entirely different from those of the occasional reversal of the geomagnetic field at random intervals of 10^6 years.

axis. The generation of the poloidal field from the toroidal field B_ϕ can be represented in a simple way if we describe the poloidal field in terms of a toroidal vector potential, $\mathbf{B}_p = \nabla \times \mathbf{e}_\phi A$. Then Equation 1 can be integrated at once to give

$$\left[\frac{\partial}{\partial t} - \eta\left(\nabla^2 - \frac{1}{\varpi^2}\right)\right] A = (\mathbf{v} \times \mathbf{B})_\phi \qquad 4.$$

The usual gradient of an arbitrary function that appears as an integration "constant" is zero because ϕ is a cyclic variable. The production of a poloidal loop by a cyclonic convective cell, as illustrated in Figure 2, generates a local excess of vector potential A. Thus, for instance, the local excess

$$\Delta A = \mathbf{e}_x \frac{1}{2} B_1 L \exp\left(-\frac{x^2 + y^2 + z^2}{L^2}\right) \qquad 5.$$

in some local Cartesian coordinate system with the x axis pointing in the ϕ direction, results from the localized loop of magnetic flux

$$\Delta B_y = - B_1 \frac{z}{L} \exp\left(-\frac{x^2 + y^2 + z^2}{L^2}\right)$$

$$\Delta B_z = + B_1 \frac{y}{L} \exp\left(-\frac{x^2 + y^2 + z^2}{L^2}\right)$$

around the x axis. The local concentration of A soon spreads out to merge with neighboring concentrations, to produce the overall poloidal field. The mean rate of generation of A is proportional to the product of the toroidal field B_ϕ, and the local strength of the convective motions. The decay time of the dipole field is rather longer ($\gtrsim 10^4$ years) than the short lives ($\sim 10^3$ years) of the individual convective cells, so it is sufficient in computing the dipole field to consider only the mean effective rate of production of A by the convective motions. The mean is to be taken over periods somewhat longer than the life of the individual cells, in order to accomplish a suitable smoothing of A, but short compared to any long time variations in the general level of $\mathbf{B}_p$. On this basis we replace the source term $(\mathbf{v} \times \mathbf{B})_\phi$ on the right-hand side of Equation 5 by the mean effective local production rate $\Gamma(r, \theta) B_\phi(r, \theta)$, where $\Gamma(r, \theta)$ is a measure of the strength of the local cyclonic convection. The result is

$$\left[\frac{\partial}{\partial t} - \eta\left(\nabla^2 - \frac{1}{\varpi^2}\right)\right] A = \Gamma B_\phi \qquad 6.$$

The precise value of Γ depends upon the details of the cyclonic convective motions, and must be computed for each particular case (see examples in Parker 1955b). Equations 4 and 6 together constitute the *dynamo* equations (Parker 1955b). (See also Steenbeck, Krause & Radler 1966; Steenbeck & Krause 1966; Krause & Steenbeck 1967.) Equation 6 incorporates the basic approximation that the cyclonic motions are small scale and short lived, so that $(\mathbf{v} \times \mathbf{B})_\phi$ may be replaced by the mean production rate ΓB_ϕ. In addition, we neglect the interaction of the cyclonic motions with the poloidal field for the reasons mentioned earlier. Equations 4 and 6 can be used to work out models of the terrestrial dynamo (Parker 1955b, Braginskii 1964) as well as the solar dynamo (Parker 1957a). But before we go on to further development of the dynamo equations, a number of points should be mentioned.

First of all, the original derivation of the dynamo equations (Parker 1955b, 1957) was heuristic, based on separate formal calculations of the separate aspects of the interaction of the cyclonic convective motions with the toroidal field. The whole picture, leading to Equation 6, was then assembled through physical arguments, as presented above. Thus the theory did not represent a *formal mathematical proof* from the hydromagnetic equation 1 that fluid motions can regenerate a magnetic field. There was real concern at the time that some subtle aspect of the problem was overlooked in the construction of the dynamo equations, and that regenerative dynamos were not possible, i.e. there was concern that Cowling's theorem was but a special case of a general "impossiblity" theorem for homogeneous dynamos. Cowling's theorem was not itself directly applicable, of course, because in view of the individual convective cells, the dynamo did not have axial symmetry nor was it stationary in time. But the dynamo did have axial symmetry and was stationary in time if one averaged over the individual convective cells, as was done in constructing the right-hand side of 6. There was no reason to assert that the dynamo equations were wrong, i.e. in violation of Cowling's theorem, but a *rigorous* proof of the existence of some sort of dynamo would be an important philosophical point. Consequently Backus (1958) constructed an idealized two-stage dynamo that could be handled rigorously. He avoided the poor convergence of the formal expansions of the stationary dynamo, used in earlier attempts at formal calculations, by turning on the two different fluid motions (corresponding to the nonuniform rotation and the cyclonic convective motions) alternately in short bursts, followed by extended periods of quiet. During the quiet periods the higher modes decay away, leaving only the lowest mode for the poloidal field and for the toroidal field. In this way he was able to show rigorously that there exist velocity fields in conducting fluids which regenerate magnetic fields.

At about the same time another point of interest was established. The convective dynamo proposed for Earth and for the Sun, which we have described at some length, is not stationary in time on the small scale of the individual convective cells. Only in the mean production of a poloidal field, represented by the right-hand side of 6, is there a stationary state. When the theory was first proposed, it was conjectured by this author, at least, that the nonstationary character of the small-scale motions might be an essential feature of the dynamo in avoiding Cowling's nonexistence theorem. However, Herzenberg (1958) worked out a dynamo that was completely stationary in time. The dynamo consisted of a large conducting sphere. Embedded within the large sphere were two small conducting spheres that were spun about inclined axes with constant angular velocity. Herzenberg carried out the formal mathematical treatment of his dynamo, securing rapid convergence of the formal expansions because each of the small spheres was removed many radii from the other, so that each small sphere saw only the lowest mode generated by the other. Herzenberg's treatment is then a second rigorous proof of a regenerative hydromagnetic dynamo, and is particularly interesting because it maintains a magnetic field in a completely time-independent state.

More recently, Braginskii (1964a,b, 1965) has carried out a formal derivation of the dynamo equations for a physical circumstance different from that originally treated by Parker. It will be recalled that Equation 6 was developed for the situation in which the nonuniform rotation is the dominant motion and the cyclones are small scale and short lived, but each is vigorous during its short life (to facilitate the calculations). Braginskii treated the case where again the nonuniform rotation is dominant, with a large magnetic Reynolds number $R_M = vL/\eta$, and has axial symmetry. The remaining motions must be nonsymmetric (to avoid Cowling's theorem) and he assumes that they are *weak* $O(R_M^{-1/2})$ and *slowly varying* in time, though not necessarily of small scale. Thus the formation of meridional loops of flux goes on simultaneously with their diffusion, rather than separately as in the original derivation of Equation 6. It is then possible to carry out a formal expansion in powers of $R_M^{-1/2}$, which leads to an exceedingly complex set of equations, as a consequence of the slow time variation and finite scale of the nonsymmetric motions. But the equations have the remarkable property that they reduce to the form 6 if one employs an effective vector potential $A + wB_\phi$ in place of A. Here w is a suitably chosen small quantity $O(R_M^{-1})$, so that A and wB_ϕ are of a comparable magnitude. It is interesting to see the breadth of physical circumstances that lead to the form 6.

Braginskii (1964a, 1965) gives a number of solutions of the dynamo equation appropriate for the core of Earth (see the review of Braginskii's work by Roberts 1967).

More recently, Childress (1967a,b) has developed another formal mathematical approach to the dynamo problem, based on the idea of the small- and large-scale motions and fields in the dynamo equations. The method bears some points in common with Braginskii's method, but necessarily leads to equations of a form different from the dynamo equations 3 and 6 because all the fluid velocities are included in one field, whereas 3 and 6 distinctly separate the cyclonic motions from the nonuniform rotation. A class of stationary dynamos distinct from previous dynamos is a direct result (Childress 1967c). Childress' expansion of the fields is carried out in terms of ϵ, the ratio of the small-scale to the large-scale field variations. In the Earth and the Sun this would be the scale of the cyclonic motions divided by the overall scale of the geomagnetic or solar fields. The velocity varies on the small scale and has a large magnitude $O(\epsilon^{-1/2})$. Systematic expansion of the hydromagnetic equations is then possible and Childress has given a complete mathematical development of the equations. He produces some stationary solutions as examples, for the special case that the vorticity is parallel to the velocity, but no direct application to physical circumstances pertaining to Earth or the Sun has been attempted. It is our impression that the method, together with Braginskii's equations, has great potential for further investigation of dynamos.

Dynamos based on direct solution of the hydromagnetic equation have been given by Roberts (1968, see also Roberts 1967). One of the solutions is a progressive wave, not entirely unlike the progressive fields of the Sun.

Finally, the important point has been made recently by Steenbeck, Krause & Radler (1966) that the sense of cyclonic rotation of a convective cell in an atmosphere with a large barometric density gradient may be determined more by the expansion of the rising cell, and the contraction of the sinking cell, than by the inflow and outflow pattern discussed for the core of Earth and illustrated in Figure 2 above. Figure 4 is a sketch showing the expansion and contraction of rising and sinking cells, respectively, when the convection extends over many scale heights. The horizontal motion of the expansion or contraction produces Coriolis forces, and cyclonic motion, in the directions of rotation shown. The rising and sinking cells in a stratified atmosphere rotate in opposite directions, therefore both producing the vector potential A of the poloidal field with the *same* sense. This is unlike the situation in the incompressible core of Earth, where the rising and falling cells have the same sense of rotation, producing A with opposite signs, the rising cell regenerating the poloidal field and the sinking cell degenerating the poloidal field. Note, too, that the sense of cyclonic *rotation* proposed by Steenbeck et al is opposite to the sense which we propose for the *incompressible* core of Earth.

3. Migratory Solar Dynamo

3.1 *General remarks.*—Observations indicate that the magnetic fields of the Sun are time dependent with a 22-year period. As noted above, there are two bands of toroidal field, one in each hemisphere, which become evident first at latitudes of $\pm 45°$ where the first sunspots and broader regions form. The two bands are of opposite sign and migrate to the equator, in 10–12 years, where they disappear at about the same time that the new reversed bands of field are appearing at $\pm 45°$ for the next cycle. The sunspots and bipolar regions are formed from the toroidal bands, and, indeed, are the only observable manifestation of the toroidal field.

The only unambiguous indication of poloidal fields on the Sun is the polar field, which reversed at the peak of the sunspot number in 1958–1959. As noted in Section 1, the polar fields have not yet, so far as we are aware, reversed in association with the maximum of the present sunspot activity. But it is generally expected that they will do so by the end of 1970. Away from the polar regions the poloidal fields are inextricably mingled with the strong toroidal field. In this respect conditions resemble those in the

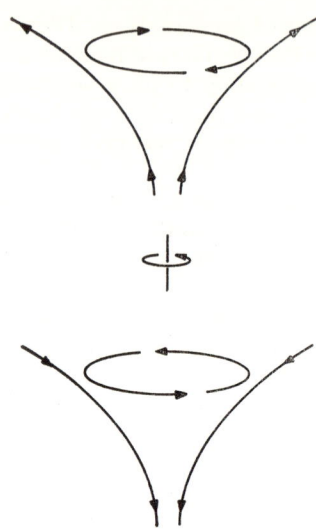

FIGURE 4. A sketch of the cyclonic rotation of rising and sinking convective cells in a compressible atmosphere rotating with the sense shown by the arrow in the center of the figure. The vertical extent of the cells is large compared to the scale height of the atmosphere, so the rotation is determined by the expansion of the rising cell and the contraction of the sinking cell.

core of Earth, where the toroidal field is somewhat stronger than the poloidal field. The field observed at the *surface* of Earth has had the toroidal field removed by the low electrical conductivity of the mantle and crust, and the higher poloidal harmonics removed by the large distance from the core at which the field is observed. The field of the Sun is much simpler too at some distance above the Sun, as shown by Schatten, Wilcox & Ness (1969).

The basic theoretical question is, what fields do we expect the solar nonuniform rotation and cyclonic convective motions to produce? There are several differences between the fluid motions in the Sun and in the core of Earth. One is that the convective zone on the Sun is a relatively thin shell, with a thickness about 10^{-1} of its radius, whereas the liquid core of Earth has a thickness of half its radius, or more. Another difference is the relatively small Coriolis force on the Sun. The period of the individual convective cell is 5×10^2 sec for a granule of some 10^3 km extent, and perhaps 5×10^4 sec for a supergranule, whereas the period of rotation of the Sun is 2×10^6 sec. The angle through which a cyclonic convective cell rotates during its lifetime is of the order of its life divided by the period of rotation of the Sun, $10^{-2} - 10^{-4}$ radians. This is in contrast to the core of Earth where the ratio is very large—3×10^5—rather than very small, compared to one. But this difference is important mainly for the dynamics of the convection and nonuniform rotation. For the dynamo it is the relative strength of the nonuniform rotation and the cyclonic motions that are important, and while both are weak on the Sun, their ratio, and hence the ratio of the poloidal and toroidal fields, on the Sun are probably not greatly different from the those of the core of Earth.

Perhaps the most important differences lie in the sign of the nonuniform rotation on the Sun and in the sign of the cyclonic motion in convective cells. On the Sun the surface equatorial regions are observed to rotate more rapidly than the poles. This has traditionally been interpreted to mean that the angular velocity of the Sun increases outward, $d\omega/dr > 0$, which is the opposite of what seems to be occurring in the core of Earth (see Figure 1). The forward tilt of sunspot fields (Maunder 1907, Minnaert 1946) and the more rapid rotation of the chromosphere (Adams 1908, Livingston 1969) would appear to confirm that $d\omega/dr > 0$ at least in the outer layers of the convective zone. On the other hand, Dicke (1964; Haurwitz 1968) has argued that the inner core of the Sun may be rotating very rapidly and $d\omega/dr < 0$. To avoid complications in the exposition that follows, we shall suppose that $d\omega/dr > 0$. We will see later that serious consideration must be given to $d\omega/dr < 0$, in which case the exposition carries over merely by reversing the helicity of the associated cyclonic convective motions.

If the vertical dimension of the convective cells (e.g. supergranulation, Leighton, Noyes & Simon 1962) is more than one scale height, we would expect the sense of rotation of the convective cells to be that suggested by Steenbeck, Krause & Radler (1966). This together with $d\omega/dr < 0$ leads to the theoretical possibility of regeneration of a stationary dipole (as in the core of Earth where $d\omega/dr < 0$ and the cyclonic rotation has the opposite sense; compare Figures 2 and 4). However, the Sun does not have a stationary dipole, but rather one that reverses periodically with the sunspot cycle. The field is basically migratory (Parker 1955b, 1957a). Hence we must consider time-dependent, rather than stationary solutions of the dynamo equations.

3.2 *Formal theory.*—The formal theory outlined in this subsection is based on the dynamo equations (Parker 1955b, Braginskii 1964). The formal theory demonstrates the necessary ingredients of the solar dynamo and provides quantitative mathematical relations between the nonuniform rotation, the cyclonic motions, and the period of the sunspot cycle. But the formal theory does *not* identify which of the motions observed on the Sun are responsible for the various aspects of the dynamo. That important question is taken up in the next subsection where the specific proposals of Babcock (1961) and Leighton (1969) are considered.

Consider the dynamo equation 6 together with 1 or 3. The solar convective zone is relatively thin compared to its radius of curvature, so the curvature has but little effect in the present formal discussion. Consider then a rectangular geometry and suppose that the nonuniform rotation is represented by the shearing velocity $\mathbf{v} = \mathbf{e}_y\, v(z)$ in the y direction, with the y axis pointing in the direction of rotational velocity of the Sun (see Figure 5). The toroidal field is then in the y direction and the positive z axis is drawn vertically upward. The x axis points southward. We suppose that conditions

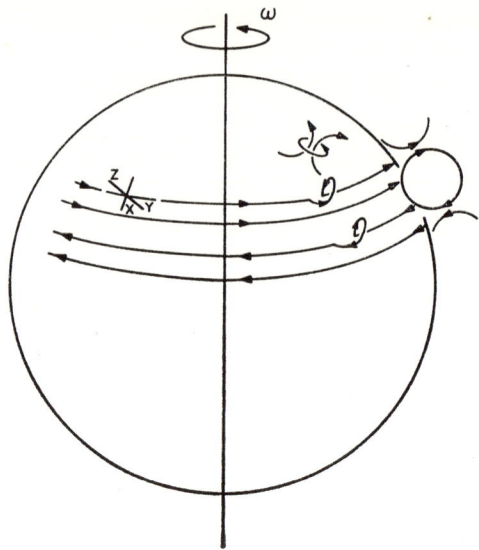

FIGURE 5. A sketch of the toroidal bands of magnetic field in the convective zone of the Sun, with hypothetical associated poloidal field (shown at the limb). The picture is drawn for the case that $d\omega/dr > 0$ and that the cyclonic convective cells have the same sense of rotation as in the core of Earth, indicated by the set of arrows in the upper right quadrant. Two loops of poloidal field produced by the convection are shown in the toroidal lines of force.

are uniform in the y direction. The lines of force of the poloidal field $\mathbf{B}_p$ are parallel to the xz plane, and the vector potential of the poloidal field is $\mathbf{e}_y$, $A(x, z, t)$ so that $B_x = -\partial A/\partial z$ and $B_z = +\partial A/\partial x$. Then the equation for the toroidal field B_y follows from 1 as

$$\left[\frac{\partial}{\partial t} - \eta\left(\frac{\partial^2}{\partial x^2} + \frac{\partial^2}{\partial z^2}\right)\right] B_y = \frac{dv}{dz} \frac{\partial A}{\partial x} \qquad 7.$$

The dynamo equation 6 reduces to

$$\left[\frac{\partial}{\partial t} - \eta\left(\frac{\partial^2}{\partial x^2} + \frac{\partial^2}{\partial z^2}\right)\right] A = \Gamma B_y \qquad 8.$$

If the strength of the cyclonic convective cells Γ and the shear $dv/dz \equiv H$ are uniform throughout the region, then solution of 7 and 8 is

$$B_y = C_1 \exp\left(\frac{t}{\tau} + ik_x x + ik_z z\right) \qquad 9.$$

$$A = C_2 \exp\left(\frac{t}{\tau} + ik_x x + ik_z z\right) \qquad 10.$$

the dispersion relation is (Parker 1955b, 1957a)

$$\frac{1}{\tau} + \eta(k_x^2 + k_y^2) = \pm (k_x \Gamma H)^{1/2} \exp i\frac{\pi}{4} \qquad 11.$$

and the ratio of the poloidal field B_z to the toroidal field B_y is

$$\frac{B_z}{B_y} = \frac{ik_x A}{B_y} = \frac{ik_x C_2}{C_1} = \left(\frac{k_x \Gamma}{H}\right)^{1/2} \exp i\frac{\pi}{4} \qquad 12.$$

Consider the dispersion relation 11. If ΓH is positive, then a necessary condition for amplification is that the $\pm$ be chosen as plus. Then for a uniform train of dynamo waves (real k_x and k_z) the fields grow exponentially at a characteristic rate given by the real part of

$$\frac{1}{\tau} = (k_x \Gamma H)^{1/2} \exp i\frac{\pi}{4} - \eta(k_x^2 + k_z^2)$$

which is positive provided that the product ΓH is large enough

$$\frac{\Gamma H}{\eta^2} > 2(k_x^2 + k_z^2)^2$$

The solution represents a wavetrain propagating with phase velocity $(k_x^3 \Gamma H/2)^{1/2}/(k_x^2+k_z^2)$ in the negative x direction. In the limit of small η the amplitude grows by a factor of $\exp \pi \simeq 23.3$ in propagating half a wavelength.

To make an estimate of the phase velocity for dynamo waves in the Sun, the nonuniform rotation amounts to some 3 or 4 days difference in the 25-day period T of rotation, so we would guess that $H \sim 1/10T \sim 4 \times 10^{-8}$/sec. The strength of the cyclones, represented by the parameter Γ, is the rate of generation of vector potential A from the toroidal field B_ϕ. With the sense of cyclonic rotation indicated in Figure 5, Γ is negative. Denote by $l = \pi/k_z$ the vertical scale of the poloidal field. Then in order of magnitude. $B_p = A/l$. Equation 8 indicates that l/Γ is the characteristic time in which an accumulation of cyclonic rotations may twist a loop of B_ϕ into a meridional plane. The characteristic time for twisting a loop is a few times T. Hence $\Gamma \sim l/T$ $\simeq 4 \times 10^3$ cm/sec if $l = 10^{10}$ cm. The wavelength L of the bands of toroidal field on the Sun is $\sim 10^{11}$ cm, so that $k_z = 6 \times 10^{-11}$ cm^{-1} and $k_x^2 \ll k_z^2$. It follows that the phase velocity is of the general order of magnitude of 2×10^2 cm/sec, which will travel 45° of latitude (5×10^{10} cm) in 2×10^8 sec or 7 years. This very rough estimate is not unlike the 11 years actually observed. Note, too, that the ratio of poloidal to toroidal field, given by 12, is $\sim (20\pi l/L)^{1/2} = 0(1)$.

Note that with the sense of nonuniform rotation suggested by the more rapid rotation of the equatorial regions ($d\omega/dr > 0$) and with the same sense of rotation of the rising cyclonic rotations as in the incompressible core of

Earth, the dynamo waves migrate from high to low latitudes (Parker 1955b, 1957a) in agreement with the observed direction of migration, i.e. $H\Gamma < 0$ so that the wave propagates in the positive x direction. The physical process of the dynamo wave is sketched briefly in Figure 5. The rotation of a rising convective cell in the northern hemisphere is sketched in the upper part of the hemisphere in Figure 5. The sense of the poloidal loops generated from the toroidal field by the cyclonic motions is indicated by twisted loops in the toroidal field. The poloidal field from which the toroidal field was generated is shown at the limb. Note that the loops reinforce the poloidal fields on the low-latitude sides, and reduce the poloidal fields on the high-latitude sides. The result is a shift of the poloidal field toward the equator. The nonuniform rotation $d\omega/dr > 0$ then generates toroidal field on the low-latitude side of the existing bands, etc. The net result is to amplify and shift the fields toward the equator.

Note that if the sense of cyclonic rotation suggested by Steenbeck prevailed, the fields would migrate away from the equator, rather than toward the equator as observed. On the other hand, if our assumption that $d\omega/dr > 0$ is wrong, and $d\omega/dr$ is actually negative in deeper layers where the field is generated, then Steenbeck's rotation would be *necessary* to account for the migration toward the equator. Leighton (1969), in fact, argues that $d\omega/dr < 0$, basing the cyclonic rotation on the sense of tilt of bipolar sunspot groups discussed in the subsection below.

Finally it should be noted that the polar fields have not entered the picture so far. The discussion has concentrated on the migrating bands of toroidal field associated with sunspots. The reversal of the polar fields in 1957–1958 at sunspot maximum suggests that they are an integral part of the solar dynamo and must be incorporated into the picture. A proper solution of the dynamo equations must include the boundary conditions (not included in the individual wave solutions considered above). Thus, in rectangular geometry suppose that $x = 0$ corresponds to the equator, where $B_y = B_z = 0$, and $x = a$ corresponds to the pole, where $B_y = \partial B_z/\partial x = 0$ (or $B_y = \partial B_z/\partial x = 0$ at $x = ta$). For a given $1/\tau + \eta k_z^2$ there are four roots of 11 for k_x, and hence four modes, whose superposition gives the complete solution satisfying the four boundary conditions. It is evident from physical considerations that the long-term solution, which neither grows nor decays exponentially, must have the real part of $1/\tau$ equal to zero, i.e. regeneration balanced by dissipation. In order of magnitude, then,

$$k_x^3 = O(\Gamma' H/\eta^2)$$

with the velocity of propagation still $\sim (\Gamma H/k_x)^{1/2}$ as when $\eta = 0$. Ordinary resistive diffusivity is much too small ($\eta \simeq 10^7$ cm^2/sec), leading to $k_x \sim 10^{-6}$ cm^{-1} (wavelengths of only 30 km), whereas the observed wavelengths are comparable to the diameter of the Sun, 10^6 km. Leighton's (1964) point, that the principal diffusivity is probably the random walk of the lines of force of B_z in the motion of the supergranules (steps of 10^9 cm in 10^5 sec), is thus a

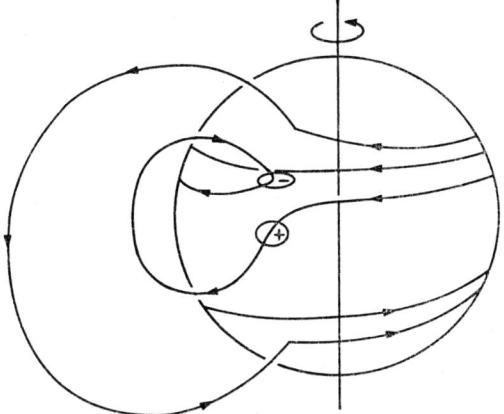

FIGURE 6. A sketch of the operation of Babcock's solar dynamo showing the toroidal field produced by $\partial\omega/\partial\theta$ and the poloidal field produced from the toroidal field by the poleward migration of following sunspots. The poloidal field produced in this way has the opposite sense to the existing poloidal field and so tends to destroy and then reverse the original polar fields.

establish that the nonuniform rotation and meridional circulation lead to the results he suggests. A particularly obvious question is whether the cycle proposed by Babcock would really progress from polar field to toroidal field, to reversed polar field which obliterates the original polar field and then establishes a pure reversed polar field to begin the next half of the cycle. Why is the toroidal field produced only from the initial pure polar field and not from the mixed polar fields that follow, giving a mixed toroidal field, etc? Babcock pointed out questions concerning the nonuniform rotation as a function of depth, and the merging rates of large-scale fields, that will have to be understood before a formal calculation can be made. An essential point of Babcock's model is that the lines of force of the poloidal field thread only the surface layers, so that the lines are drawn out by the observed $d\omega/d\theta$ to form a toroidal field with the sense indicated in Figure 6. If the lines penetrate more deeply, then it is $d\omega/dr$ which is responsible for the toroidal field, rather than $d\omega/d\theta$.

A further point in the context of the present general discussion is that the sense of shear of the poloidal field and twist of the toroidal field proposed by Babcock leaves no possibility for a *stationary* dynamo. The meridional loops destroy, rather than reinforce, the existing poloidal field. The sense of rotation of the bipolar magnetic regions is that pointed out recently by Steenbeck Krause & Radler (1966) for vertical motions in an atmosphere with a large density gradient, such as exists on the Sun.

Babcock's model has been explored by several authors (Chvojkova 1965, Tuominen 1965, Kopecky 1966, Vostry 1967).

More recently Leighton (1969) has undertaken a quantitative treatment of the solar dynamo, based on extensive numerical computations from the dynamo equations 4 and 6 with some additional terms appended to give dissipation of the toroidal field. He explores the various dynamos resulting from differing forms of nonuniform rotation, various critical fields for formation of bipolar magnetic regions, diffusion times, the rate of twisting of bipolar regions, etc. Considerable variation of magnetic activity is obtained in this way, including a tendency for the toroidal fields to have the same, rather than opposite, sign in the two hemispheres. From his extensive theoretical exploration Leighton demonstrates that a close representation of the actual solar magnetic behavior is obtained from the dynamo equations for values of the parameters which are compatible with observation. Leighton supposes that the bipolar magnetic regions, with the poleward migration of the following spot, are the cyclonic motions responsible for regenerating the poloidal field from the toroidal field, as did Babcock. But his treatment of the rest of the dynamo is quite different. With this choice of cyclonic "rotation" (which is the same sense as that proposed by Steenbeck, Krause & Radler) it follows that the angular velocity must increase inward from the surface of the Sun, $d\omega/dr < 0$, if toroidal fields are to migrate toward the equator. In this context Leighton is using the dynamo equations as a probe of magnetic and hydrodynamic conditions beneath the surface of the Sun, predicting that $d\omega/dr < 0$ if we are to understand the migration of fields in a simple way. [The inward increase of angular velocity has been suggested by Dicke (1964) on quite different grounds.]

Leighton supposed that the angular velocity $\omega(r, \theta)$ of the Sun is a function of both the polar angle θ and radial distance r. Then, $\partial B_\phi/\partial t$ follows from 3, dropping the diffusion term in the limit of high electrical conductivity. In view of the small thickness ΔR of the convective layer in the Sun, Leighton approximates r on the right-hand side by the solar radius R. And he averages over the depth of the layer, and in azimuth ϕ around the Sun. For computation Leighton assumes

$$\omega(r, \theta) = 18 \sin^2 \theta + (\alpha + \beta \sin^n \theta)(R - r)/\Delta R \text{ rad/year}$$

where the first term represents the same observed differential rotation of the surface layers employed by Babcock, and the second term represents a possible radial dependence with magnitude and latitude dependence, characterized by the parameters α, β, and n.

Leighton then represents the formation of bipolar magnetic regions and sunspots by supposing, as did Babcock, that the toroidal field erupts to the surface in a characteristic time τ if B_ϕ exceeds a critical value B_c. Since all fields are averaged over azimuth ϕ, the lifting of a rope of toroidal field to the surface does not contribute a net radial field; the preceding and following spots (separated by a mean distance a) contain equal and opposite fluxes. But if the systematic tilt through an angle γ is now introduced, so that the following spot or region appears at higher latitude than the preceding, it is readily shown that the contribution to the mean field is

$$\frac{\partial B_r}{\partial t} \simeq -\frac{\Delta RaS(B_\phi)}{2\pi R^2 \tau \sin\theta} \frac{\partial B_\phi \sin\gamma}{\partial \theta} \qquad 15.$$

at any given θ, where $S(B_\phi)$ is a step function, equal to one if $B_\phi > B_c$ and zero if $B_\phi < B_c$. Observations of sunspots suggest that $\sin\gamma \simeq \frac{1}{2}\cos\theta$.

Leighton then employs the important idea (see Subsection 3.2 above) that the radial component of the field is transported by the random walk of the supergranulation (Leighton 1964). The characteristic diffusion time T_D is, according to Leighton's estimates, about 20 years. Thus, altogether

$$\frac{\partial B_r}{\partial t} = -\frac{\Delta RaS(B_\phi)}{4\pi R^2 \tau \sin\theta} \frac{\partial (B_\phi \cos\theta)}{\partial \theta} + \frac{1}{T_d \sin\theta}\frac{\partial}{\partial \theta}\left(\sin\theta \frac{\partial B_r}{\partial \theta}\right) \qquad 16.$$

for the radial field. This equation is the radial component of the curl of the dynamo equation 6, representing the production of poloidal field from the toroidal field by rising cyclonic cells. Leighton uses the "eddy diffusivity" of the supergranulation for η. He supposes, in effect, that the production factor Γ of poloidal field in 6 is proportional to the step function $S(B_\phi)$ and to $\cot\theta$. The singularity in $\cot\theta$ at $\theta=0$ does not appear in the equations because B_ϕ, and hence $S(B_\phi)$, vanishes at $\theta=0$.

Leighton found from the numerical calculations that the strong "quantization" introduced by $S(B_\phi)$ leads to difficulty in that a weak B_ϕ dies out altogether because it does not exceed B_c, leading to no production of B_r, and hence no production of B_ϕ from 3, etc. So he introduced a small radial field B_s assumed to be generated at a very small rate from B_ϕ regardless of the magnitude of B_ϕ/B_c. For this field he supposes that the decay period T_s is 50 years and writes

$$\frac{\partial B_s}{\partial t} = -\frac{G\Delta Ra}{4\pi R^2 \tau \sin\theta} \frac{\partial (B_\phi \cos\theta)}{\partial \theta} - \frac{B_s}{T_s} \qquad 17.$$

where $G \ll 1$. This artifice is sufficient to keep the field from dying out if B_ϕ should fail to reach B_c in one cycle.

The field component B_s is computed from $B_r + B_s$ using the relation that the divergence of the total meridional field is zero.

Finally Leighton assumes that the toroidal field has a 50-year decay time T_s and, further, that the toroidal field is depleted by the eruption of flux through the photosphere to form bipolar regions and sunspots. He assumes that the loss of toroidal flux due to lifting of ropes of toroidal field up to the photosphere goes at a rate proportional to the square of the field. Hence in place of 3 he writes

$$\frac{\partial B_\phi}{\partial t} = \sin\theta\left(B_\theta \frac{\partial \omega}{\partial \theta} + B_r R \frac{\partial \omega}{\partial r}\right) - \frac{B_\phi}{T_s} - S(B_\phi)\frac{a\,|B_\phi|\,B_\phi}{2\pi R B_c \tau} \qquad 18.$$

Note, then, that in both 17 and 18 there is a dissipation term of the form $-\mathbf{B}/T_c$ with T_s assigned the large value of 50 years. This small amount of dissipation is a computational aid, without physical significance. Its presence causes the solutions to settle into their steady long-term pattern more quickly, thereby saving computer time. Computations with $T_s = \infty$ show that the convenience of $T_s = 50$ years does not introduce significant error into the asymptotic behavior of the solutions at large t.

The last term on the right-hand side of 18, representing the loss of toroidal field by magnetic buoyancy, is an effect not included in the original dynamo equations, where the only dissipation was diffusivity. The term is based on the idea that the toroidal field, of a few hundred gauss beneath the photosphere, floats upward through the photosphere and expands with the decreasing density to become unobservable. Presumably this weak field is then dispersed around the Sun and eventually freed from the gas in which it is embedded. The effect implies that the bipolar magnetic regions (produced in the intermediate stage when part of the toroidal field is above, and part below, the photosphere) disperse by a general spreading out, and increased separation, of the areas of opposite polarity as the toroidal field rises to the surface over a wider and wider region. The effect implies that the flux in each half of the bipolar magnetic regions increases monotonically until the region is so spread out as to merge with other bipolar regions at the same latitude. Detailed study of the observations of magnetic fluxes in bipolar magnetic regions would be invaluable for establishing the correctness of this important idea. The question of the eventual freeing of the toroidal field from the gas (originally from the subphotosphere) in which the field is embedded after the field rises above the photosphere needs some thought and explanation.

Leighton solves Equations 16 to 18 by numerical methods. The solutions prove to be more or less periodic, and, with suitable choices of the parameters, lead to magnetic fields with behavior remarkably similar to that of the solar fields. The calculations suggest that $\partial\omega/\partial r$, more than the $\partial\omega/\partial\theta$ used by Babcock, is responsible for generating the toroidal field. The differential rotation $\partial\omega/\partial\theta$ is fixed by observation, and Leighton finds that without $\partial\omega/\partial r$, unreasonably large values of a and γ are required to maintain the fields. With $\partial\omega/\partial r = 0$ the weakness of the nonuniform rotation $\partial\omega/\partial\theta$ must be compensated by overly strong cyclonic convective motions. To estimate $\partial\omega/\partial r$ he supposes that the interior of the Sun makes about three revolutions per year more than the surface. For this $\partial\omega/\partial r$ his calculations show that spot groups with linear dimensions of $0.2\ R_\odot$ are sufficient to maintain the dynamo. If, on the other hand, $\partial\omega/\partial r = 0$, then the linear dimensions would have to be $2\ R_\odot$, which is quite unreasonable.

With a radial dependence of ω, the fields are migratory again, as discussed in Subsection 3.2 above. The toroidal bands of field migrate toward the equator. The reader is referred to Leighton's paper for a summary of his extensive explorations of the various forms of nonuniform rotation, strength of cyclonic motions, etc. The fit to the observed behavior of the sunspot fields can be made very close with reasonable values of α, β, n, and a if suitable $\partial\omega/\partial r$ is assumed.

Altogether, then, Leighton has shown that with suitable $\partial\omega/\partial r$ the behavior of the solar magnetic fields can be understood from the dynamo equations. He obtains altogether the migratory bands of toroidal field responsible for the bipolar regions and sunspots, and the associated polar fields. His calculations suggest that the "cyclonic convective" motions can reasonably be identified with the eruption and tilt of bipolar regions through the photosphere, and that the interior of the Sun rotates more rapidly than the photosphere. It is interesting that the nonuniform rotation is the sole energy input to the solar dynamo. The cyclonic convective motions are driven by the magnetic buoyancy of the toroidal field produced by the nonuniform rotation, rather than by the thermal convective forces in the convection zone.

4. Summary and Conclusions

This review has concerned itself with the origin and behavior of the magnetic fields in the Sun, and but little with the detailed effects of the fields. For the detailed structure and behavior of sunspots, flares, spicules, plages coronal heating, etc., the reader has been referred to the literature.

The Sun has probably had a magnetic field since the beginning. The field that the Sun has today appears to be generated by the nonuniform rotation of the Sun together with cyclonic convection, in a manner describable by the dynamo equations 3 and 6. The characteristic generation time of the field is of the order of the period of the sunspot cycle, 20 years. It follows that the present field of the Sun has nothing to do with the fields in the Sun at the beginning. Indeed, it would appear, from Leighton's numerical experiments with the dynamo equations, that all memory of earlier fields is completely obliterated in 10^3 years. The field today is mainly the result of conditions and fluctuations in the past 20 years and depends only in a small way on conditions even 10^2 years ago.

Several questions and problems are outstanding. First of all, the dynamo equations, with which the generation of fields is treated, average over the cyclonic convective motions, so that their formal derivation assumes many small cyclonic convective motions at any one time. Thus there is room for improvement in the formal theory.

Babcock and Leighton have suggested that the generation of the poloidal field is associated with the emergence and north-south tilt of bipolar regions. But cyclonic motions in the supergranules have not been ruled out. We cannot overlook the point raised by Leighton that if we are to understand the solar magnetic field in terms of the emergence and north-south tilt of bipolar regions, the angular velocity of the Sun increases inward, contrary to what is indicated by the forward tilt of sunspot fields and the rate of rotation of the chromosphere, but in accord with some of the ideas proposed by Dicke.

A natural question is whether we are in a position to understand the magnetic fields of other stars (see e.g. the review by Ledoux & Renson 1966). The progress with the Sun and Earth gives some confidence in the basic idea that nonuniform rotation and cyclonic convective motions together generate the fields. But some caution is in order too, in view of the questions still outstanding for the Sun. Application to other stars must begin with the question of whether the stellar field is stationary in the star, as is the geomagnetic field in Earth, or whether it is oscillating—a migratory dynamo—as in the Sun. A stationary dynamo requires that the sense of the nonuniform rotation and the cyclonic motions be regenerative. A stationary dynamo is ruled out for the Sun or other stars if, in fact, the core rotates more rapidly and the upwelling of toroidal field tilts so that the following spot moves to higher latitudes. The dynamo can then be only migratory.

LITERATURE CITED

Adams, W. S. 1908, *Ap. J.*, **27**, 213
Alfven, H. 1950, *Cosmical Electrodynamics* (Oxford: Clarendon)
Altschuler, M. D. 1966, *Solar Phys.*, **1**, 377
Altschuler, M. D., Nakagawa, Y., Lilliequist, C. G. 1968, *Solar Phys.*, **3**, 466
Anzer, U. 1968, *Solar Phys.*, **3**, 298; **4**, 101
Babcock, H. D. 1959, *Ap. J.*, **130**, 364
Babcock, H. W. 1961, *Ap. J.*, **133**, 572
Babcock, H. W. 1963, *Ann. Rev. Astron. Ap.*, **1**, 41
Babcock, H. W., Babcock, H. D. 1955, *Ap. J.*, **121**, 349
Backus, G. E., 1958, *Ann. Phys.*, **4**, 372
Backus, G. E., Chandrasekhar, S. 1956, *Proc. Nat. Acad. Sci.*, **42**, 105
Bappu, M. K. V., Grigorjev, V. M., Stepanov, V. E. 1968, *Solar Phys.*, **4**, 409
Batchelor, G. K. 1950, *Proc. Roy. Soc. A*, **201**, 405
Beckers, J. M. 1963, *Ap. J.*, **138**, 648
Beckers, J. M. 1968, *Solar Phys.*, **3**, 258, 367
Beckers, J. M., Schröter, E. H. 1968a, *Solar Phys.*, **4**, 142; 1968b, 303
Biermann. L. 1953, *Kosmische Strahlung* (Heisenberg, H., Ed., Berlin: Springer Verlag)
Biermann, L., Schlüter, A. 1951, *Phys. Rev.*, **82**, 863
Braginskii, S. I. 1964a, *J. Exp. Theor. Phys.*, **47**, 1084, 2178
Braginskii, S. I. 1964b, *Geomag. Aeron.*, **4**, 732
Braginskii, S. I. 1965, *Soviet Phys. JETP*, **20**, 726, 1462
Brandt, J. C. 1966, *Ap. J.*, **144**, 1221; 1967, *ibid.*, **147**, 201
Bray, R. J., Loughhead, R. E. 1964, *Sunspots* (London: Chapman & Hall)
Bullard, E. 1949a, *Proc. Roy. Soc.*, **197**, 433; 1949b, **199**, 413
Bullard, E. 1955, *Proc. Cambridge Phil. Soc.*, **51**, 744
Bullard, E., Gellman, H. 1954, *Phil. Trans. Roy. Soc.*, **247**, 213
Chandrasekhar, S. 1955, *Proc. Roy. Soc. A*, **229**, 1; **233**, 322, 330
Childress, S. 1967a, *Courant Inst. Math. Sci. New York Univ. MF-53*; 1967b, *ibid, MF-54*
Childress, S. 1967c, *Geophys. Fluid Dyn.*, **1**, 165 (Woods Hole Oceanog. Inst., Ref. No. 67-54)
Chitre, S. M. 1968, *Solar Phys.*, **4**, 168
Chvojkova, E. 1965, *Bull. Astron. Inst. Czech.*, **16**, 57
Clark, A. 1968, *Solar Phys.*, **8**, 386
Cocke, W. J. 1967, *Ap. J.*, **150**, 1041
Cowling, T. G. 1934, *MNRAS*, **94**, 39; 1945, **105**, 166
Cowling, T. G. 1953, *The Sun*, 575 (Kuiper, G. P., Ed., Univ. Chicago Press)
Cowling, T. G. 1957, *Quart. J. Mech. Appl. Math.*, **10**, 129
Cowling, T. G. 1965a, *Stellar and Solar Magnetic Fields*, 405 (Lust, R., Ed., Amsterdam: North-Holland)
Cowling, T. G. 1965b, *Stars and Stellar Systems*, 425 (Aller, L., McLaughin, D. B., Eds., Univ. Chicago Press)
Cowling, T. G. 1968, *MNRAS*, **140**, 547
Dicke, R. H. 1964, *Nature*, **202**, 432
Dungey, J. W. 1958, *Cosmic Electrodynamics*, 98 (Cambridge Univ. Press)
Elsasser, K. 1966, *Z. Ap.*, **63**, 65
Elsasser, W. M. 1945, *Phys. Rev.*, **69**, 106; 1946, **70**, 202; 1947, **72**, 821; 1950a, **79**, 183
Elsasser, W. M. 1950b, *Rev. Mod. Phys.*, **22**, 1; 1956, **28**, 135
Elsasser, W. M., Takeuchi, H. 1955, *Trans.*

Am. Geophys. Union, 36, 584
Fichtel, C. E., McDonald, F. B. 1967, Ann. Rev. Astron. Ap., 5, 351
Giovanelli, R. G. 1947, MNRAS, 107, 338
Gleissberg, W. 1953, Naturwissenschaften, 40, 336
Glessiberg, W. 1968, Solar Phys., 4, 93
Goldreich, P., Schubert, G. 1968, Ap. J., 50, 571
Green, R. M., Sweet, P. A. 1967, Ap. J., 147, 1153
Grigorjev, V. M. 1969, Solar Phys., 6, 67
Hale, G. E. 1908, Ap. J., 28, 100, 315
Harvey, J. W., Tandberg-Hanssen, E., 1968, Solar Phys., 3, 316
Haurwitz, M. 1968, Ap. J., 151, 351
Herzenberg, A. 1958, Phil. Trans. Roy. Soc. A, 250, 543
Howard, R. 1959, Ap. J., 130, 193
Howard, R. 1967, Ann. Rev. Astron. Ap. 5, 1
Howe, M. S. 1969, Ap. J., 156, 27
Hundhausen, A. J. 1968, Space Sci. Rev., 8, 690
Jaggi, R. K. 1963, J. Geophys. Res., 68, 4429
Jayanthan, R. 1968, MNRAS, 138, 477
Kippenhahn, R. 1963, Ap. J., 137, 664
Kippenhahn, R., Schlüter, A. 1957, Z. Ap., 43, 36
Kopecky, M. 1966, Atti del convegna sui campi magnetici solari (Firenze) 285
Kopecky, M., Kuklin, G. V. 1966, Bull. Astron. Inst. Czech., 17, 45
Kopecky, M., Obridko, V. 1968, Solar Phys., 5, 354
Kopp, R. A., Kuperus, M. 1968, Solar Phys., 4, 212
Kraichnan, R. H., Nagarajan, S. 1967, Phys. Fluids, 10, 859
Krause, F., Steenbeck, M. 1967, Z. Naturforsch., 22, 671
Kulsrud, R. M. 1955, Ap. J., 121, 461
Ledoux, P., Renson, P. 1966, Ann. Rev. Astron. Ap., 4, 293
Leighton, R. B. 1959, Ap. J., 130, 366; 1964, 140, 1547; 1969, 156, 1
Leighton, R. B., Noyes, R. W., Simon, G. W. 1962, Ap. J., 135, 474
Livingston, W. C. 1968, Ap. J., 153, 929
Livingston, W. C. 1969, Solar Phys. (In publication)
Malville, J. M. 1968, Solar Phys., 4, 323; 5, 236
Maunder, A. S. D. 1907, MNRAS, 67, 451
McLellan, A., Winterberg, F. 1968, Solar Phys., 4, 401
Mestel, L. 1965, Quart. J. Roy. Astron. Soc., 6, 161, 265
Mestel, L. 1967, Proc. Intern. School Phys., Varenna, 185 (Sturrock, P., Ed., New York: Academic)

Mestel, L., Roxburgh, I. W. 1962, Ap. J., 136, 615
Minnaert, M. 1946, MNRAS, 106, 98
Moffat, K. 1961, J. Fluid Mech., 11, 625
Moore, D. W., Spiegel, E. A. 1964, Ap. J., 139, 48
Moreton, G. E., Severny, A. B. 1968, Solar Phys., 3, 282
Nakagawa, Y., Swarztrauber, P. 1969, Ap. J., 155, 295
Nakagawa, Y., Trehan, S. 1968, Ap. J., 151, 1111
Ness, N. F. 1968, Ann. Rev. Astron. Ap., 6, 79
Olson, C. A., Lykoudis, P. S. 1967, Ap. J., 150, 303
Osterbrock, D. E. 1961, Ap. J., 134, 347
Pao, Y. H. 1963, Phys. Fluids, 6, 632
Parker, E. N. 1955a. Ap. J., 121, 491; 1955b, 122, 293
Parker, E. N. 1957a, Proc. Nat. Acad. Sci, 43, 8
Parker, E. N. 1957b, J. Geophys. Res., 62, 509
Parker, E. N. 1958a, Ap. J., 123, 664
Parker, E. N. 1958b, Phys. Rev., 109, 1874; 110, 1445
Parker, E. N. 1963a, Ap. J. Suppl., 8, 177
Parker, E. N. 1963b, Ap. J., 138, 226
Parker, E. N. 1963c, Interplanetary Dynamical Processes (New York: Interscience)
Parker, E. N. 1964, Ap. J., 140, 1170
Parker, E. N. 1965a, Research Frontiers in Fluid Dynamics, 667 (Seeger, R. J., Templer, G., Eds., New York: Interscience)
Parker, E. N. 1965b, Solar and Stellar Magnetic Fields. IAU Symp. No. 22, 232 (Lust, R., Ed., Amsterdam: North-Holland)
Parker, E. N. 1969, Space Sci. Rev., 9, 325
Parker, E. N., Krook, M. 1956, Ap. J., 124, 214
Petschek, H. E. 1963, Proc. AAS-NASA Symp. Phys. Solar Flares, NASA-SP-50, 50 (Goddard Space Flight Ctr., Greenbelt, Md.)
Petschek, H. E., Thorne, R. M. 1967, Ap. J., 147, 1157
Rikitake, T. 1958, Proc. Cambridge Phil. Soc, 54, 89
Rikitake, T. 1966, Electromagnetism and the Earth's Interior (Elsevier)
Roberts, G. O. 1968, Proc. NATO Conf. Newcastle, 1967 (New York: Wiley)
Roberts, P. H. 1967, Geophys. Fluid Dyn. (Woods Hole Oceanog. Inst)
Runcorn, S. K. 1955, Advan. Phys. (Phil. Mag. Suppl.), 4, 244
Rust, D. M. 1967, Ap. J., 150, 313

Saffman, P. G. 1964, *J. Fluid Mech.*, **18**, 449
Sawyer, C. 1968, *Ann. Rev. Astron. Ap.*, **6**, 115
Schatten, K. H., Ness, N. F., Wilcox, J. M. 1968, *Solar Phys.*, **5**, 240
Schatten, K. H., Wilcox, J. M., Ness, N. F. 1969, *Solar Phys.*, **6**, 442
Schroeter, E. H. 1966, in *Atti del Convegna sulle Macchie Solari, 1964*, **222** (Firenze: G. Barbera Editore)
Schubert, G. 1968, *Ap. J.*, **151**, 1099
Severny, A. B. 1964, *Ann. Rev. Astron. Ap.*, **2**, 363
Severny, A. B. 1965a, *Solar and Stellar Magnetic Fields*, 358 (Lust, R., Ed., North-Holland)
Severny, A. B. 1965b, *Trans. IAU*, **12**, A, 755
Simon, G. W., Leighton, R. B. 1963, *Astron., J.*, **68**, 291
Simon, G. W., Leighton, R. B. 1964, *Ap. J.*, **140**, 1120
Smith, H. J., Smith, E. P. 1963, *Solar Flares* (New York: Macmillan)
Steenbeck, M., Krause, F. 1966, *Z. Naturforsch.*, **21**, 1285
Steenbeck, M. Krause, F., Radler, K. H. 1966, *Z. Naturforsch.*, **21**, 369
Sturrock, P. A. 1966, *Nature*, **211**, 695
Sturrock, P. A. 1967, *Plasma Ap.*, 168 (New York: Academic)
Sturrock, P. A., Coppi, B. 1967, *Ap. J.*, **143**, 3
Sturrock, P. A., Smith, S. M. 1968, *Solar Phys.*, **5**, 87
Sweet, P. A. 1958, *Electromagnetic Phenomena in Cosmical Physics, IAU Symp. No. 6*, 123 (Lehnert, B., Ed., Cambridge Univ. Press)
Takeuchi, H., Elsasser, W. M. 1954, *J. Phys. Earth*, **2**, 39
Thompson, W. I., Billings, D. E. 1967, *Ap. J.*, **149**, 269
Tuominen, J. 1965, *Observatory*, **85**, 82
Vostry, J. 1967, *Bull. Astron. Inst. Czech.*, **18**, 37
Waldmeier, M. 1960, *Z. Ap.*, **49**, 176
Weber, E. J., Davis, L., Jr. 1967, *Ap. J.*, **148**, 217
Wentzel, D. G., Solinger, A. B. 1967, *Ap. J.*, **148**, 877
Wilcox, J. M. 1968, *Space Sci. Rev.*, **8**, 258
Wilcox, J. M., Howard, R. 1968, *Solar Phys.*, **5**, 564
Wilson, P. R. 1968, *Solar Phys.*, **3**, 243, 454
Wilson, R. L. 1966, *Geophys. J.*, **10**, 413
Zwaan, C. 1968, *Ann. Rev. Astron. Ap.*, **6**, 135

Copyright 1970. All rights reserved

THE THEORY OF STELLAR WINDS[1]
AND RELATED FLOWS

T. E. Holzer

Institute for Pure and Applied Physical Sciences, Department of Applied Physics and Information Science

W. I. Axford

Institute for Pure and Applied Physical Sciences, Department of Physics

University of California San Diego, La Jolla, California

1. Introduction

In this review we discuss certain aspects of steady, radial flows that are significant in astrophysics and space physics. The literature associated with this subject has of course been dominated by work on stellar winds in general and the solar wind in particular. Parker (1969) has extensively reviewed recent theoretical work on the solar-wind problem, and additional information is available in earlier reviews by Parker (1963, 1965) and Dessler (1967). Observational aspects of the solar wind have been reviewed by Axford (1968a) and Hundhausen (1968). Rather than repeat this material we have chosen to emphasize relatively formal and well-established features of the general theory of steady, radial, spherically symmetric flow. Our purpose is to show that the results are of wide applicability, and may be used in discussing not only stellar winds, but also galactic winds, comet winds, the polar wind (and related processes in planetary ionospheres), and the classical accretion problem.

The most elementary problem of flow in the presence of a central gravitating point mass neglecting viscosity and heat conduction is discussed in Section 2. The treatment is slightly different from that usually given in that the solutions are given for positive, negative, and zero total energy in terms of the Mach number, and the usefulness of the entropy in identifying the various solutions is emphasized. In Section 3 the procedure for choosing solutions and including shocks and other gas dynamic discontinuities is described for both outflows and inflows. It is noted that the solutions can only satisfy a few boundary conditions and hence additional physical processes must be invoked to provide the arbitrariness that is required if realistic boundary conditions are to be satisfied. The effects of mass addition are discussed in Section 4; the results can be applied to problems of gas flow in galaxies with the gas originating in the stars (i.e. galactic winds). We show that it is possible to have outflow in the outer parts of a galaxy and accretion towards a "black hole" in the inner parts of the galaxy.

In Section 5 we show that it is relatively easy to cope with the problem

---

[1] This research was supported by the Advanced Research Projects Agency of the Department of Defense and was monitored by the U. S. Army Research Office—Durham under contract DA-31-124-ARO-D-257, and by NASA under contract #NGR-05-009-081, and by NASA under contract #NGR-05-009-075.

of outflow with arbitrary addition of energy. It is interesting that such a simple variation of the basic problem discussed in Section 2 allows solutions that are much more realistic, at least as far as the solar wind is concerned. The effects of heat conduction and viscosity on stellar wind flows are discussed in Section 6. These pose formidable mathematical problems but fortunately it appears that at least for the solar wind the effects of viscosity are not important. The problem including heat conduction alone is much less difficult and indeed is qualitatively rather similar to simple heat addition as described in Section 5.

In Sections 7 and 8 the problems associated with the flow of multi-component fluids are considered. The work of Hartle & Sturrock (1968) on a two-component model of the solar wind is summarized in Section 7, and it is shown that the main features of their results can be interpreted rather simply. Solar-wind models involving three and more components are described briefly. We conclude with a discussion of what is probably the most complex problem of all, the polar wind (Axford 1968b, Banks & Holzer 1968). In this case we must deal not only with a multicomponent flow, but also with sources (photoionization and charge exchange), sinks (charge exchange), and momentum loss due to frictional interactions between ions and with a stationary background of neutral gas. As the theory of the polar wind develops it will become necessary to include even more complications, such as heat conduction in the electron gas and the effects due to photoelectrons.

2. Steady, Spherically Symmetric, Radial Flow of an Ideal Gas

In the absence of viscosity and heat conduction, the Navier–Stokes equations for steady, spherically symmetric, radial flow of an ideal gas in the presence of a gravitating point mass situated at the origin are:

$$\frac{d}{dr}(\rho u r^2) = 0 \qquad 2.1$$

$$\rho u \frac{du}{dr} = -\frac{dp}{dr} - \frac{G\mathfrak{M}\rho}{r^2} \qquad 2.2$$

$$\frac{d}{dr}\left(\frac{1}{2}u^2 + \frac{\gamma}{\gamma-1}\frac{p}{\rho} - \frac{G\mathfrak{M}}{r}\right) = 0 \qquad 2.3$$

Here ρ is the mass density, u is the radial speed, and p is the pressure of the gas; γ is the ratio of specific heats, G is the gravitational constant, $\mathfrak{M}$ is the mass situated at the origin, and r is the radial coordinate. It is assumed that there are no sources or sinks of mass, momentum, or energy except at the origin. The self-gravitation of the gas has been neglected.

Equations 2.1 and 2.2 represent conservation of mass and radial momentum, respectively. Equation 2.3 represents energy conservation; it can be written in an alternative form, using 2.1 and 2.2, as follows:

$$\frac{d}{dr}(p/\rho^\gamma) = 0 \qquad 2.4$$

This implies that the flow is isentropic, since the entropy S is defined by

$$p/\rho^\gamma = R \exp(S/C_v) \qquad 2.5$$

where $R = (C_p - C_v)$ is the gas constant, and C_p, C_v are the specific heats at constant pressure and volume respectively ($C_p/C_v = \gamma$). Equation 2.1 can be integrated immediately to yield

$$\rho u r^2 = \pm \rho_0 u_0 r_0^2 \qquad 2.6$$

where the subscript (0) denotes quantities evaluated at some reference level $r = r_0, u_0 > 0$, and the choice of signs is determined by the direction of flow. In addition to the mass flow and the entropy, the equations yield a third constant of integration, which can be expressed in various ways, but most obviously (from 2.3) as the energy per unit mass (E):

$$E = \frac{1}{2}u^2 + \frac{\gamma}{\gamma-1}\frac{p}{\rho} - \frac{G\mathfrak{M}}{r} = \frac{1}{2}u_0^2 + \frac{\gamma}{\gamma-1}\frac{p_0}{\rho_0} - \frac{G\mathfrak{M}}{r_0} \qquad 2.7$$

For given values of S, E, and $\rho_0 u_0 r_0^2$, Equations 2.5, 2.6, and 2.7 represent the complete solution for p, ρ, and u as functions of r. In this form, however, the solution is inconvenient because of its highly implicit nature. It was pointed out by Bondi (1952) that an explicit solution for general γ is possible only if an auxiliary variable depending on $u^2/\rho^{\gamma-1}$ is introduced. The most appropriate form of this variable is the Mach number M, defined by

$$M = |u/c| \qquad 2.8$$

where

$$c = \sqrt{\left(\frac{\partial p}{\partial \rho}\right)_S} = \sqrt{(\gamma p/\rho)} \qquad 2.9$$

is the local value of the speed of sound. We will follow Bondi in expressing the solution in terms of M and $\xi = r/r_0$, but with a different choice of reference quantities. Dahlberg (1964) has suggested that the variables should be M^2 and $(u_0^2/c^2\xi)$, but this does not seem to be an improvement on the present treatment.

For a given choice of reference quantities, one can obtain a relationship between M^2 and ξ from 2.5, 2.6, and 2.7 by direct substitution and elimination. It is instructive however to first make the substitution in 2.1 and 2.2 using, in place of 2.3, an alternative form of 2.5:

$$\rho = \beta c^{2/(\gamma-1)} \qquad 2.10$$

where β is related to the entropy. In this manner we obtain a differential equation for M^2 as a function of ξ:

$$\frac{(M^2-1)}{2M^2} \frac{dM^2}{d\xi} = \left[1 + \left(\frac{\gamma-1}{2}\right)M^2\right]$$
$$\cdot \left[\frac{2E}{\xi} - \left(\frac{5-3\gamma}{\gamma-1}\right)\frac{G\mathfrak{M}}{2r_0\xi^2}\right] / \left[E + \frac{G\mathfrak{M}}{r_0\xi}\right] \quad 2.11$$
$$= F(M^2, \xi)$$

It is evident that in general $dM^2/d\xi = \infty$ when $M^2 = 1$; however if the condition $F(1,\xi) = 0$ holds for some real, positive value of $\xi = \xi_c$, the integral curves in the (M^2, ξ) plane have a critical point at $(1, \xi_c)$ where they intersect with finite slope. If we choose

$$E = \left(\frac{5-3\gamma}{\gamma-1}\right)\frac{G\mathfrak{M}}{4r_0}, \quad (\gamma \neq 5/3) \quad 2.12$$

then if such a critical point exists, it is situated at $\xi = \xi_c = 1$. Note that if $\gamma = 1$, there is always a critical point, but if $\gamma = 5/3$ there is no critical point.

Let us first consider the case in which γ is restricted to the open interval $(1, 5/3)$. If E is positive, a critical point exists and we can choose reference quantities which satisfy 2.12 together with

$$\gamma p_0/\rho_0 = u_0^2, \quad E = \left(\frac{1}{\gamma-1} - \frac{3}{2}\right) u_0^2, \quad \rho_0 = \beta_0 u_0^{\frac{2}{\gamma-1}} \quad 2.13$$

With these definitions 2.6, 2.7, and 2.10 yield the following relationships between M^2 and ξ:

$$\delta f(M^2) = g(\xi) \quad 2.14$$

where

$$f(M^2) = \left(\frac{1}{\gamma-1} + \frac{M^2}{2}\right) / (M^2)^{\frac{\gamma-1}{\gamma+1}} \quad 2.15$$

$$g(\xi) = \left(\frac{1}{\gamma-1} - \frac{3}{2} + \frac{2}{\xi}\right) \xi^{4\left(\frac{\gamma-1}{\gamma+1}\right)} \quad 2.16$$

$$\delta = \left(\frac{\beta_0^2}{\beta^2}\right)^{\frac{\gamma-1}{\gamma+1}} = \exp\left[\frac{2}{(\gamma+1)} \frac{S-S_0}{C_v}\right] \quad 2.17$$

If we choose S_0 to correspond to solutions which intersect the critical point, then these solutions correspond to $\delta = 1$. For a given value of γ, the complete set of solutions with $E>0$ can be displayed in the (M^2,ξ) plane; some typical solutions are shown in Figure 1. Note that solutions corresponding to $\delta<1$ lie wholly in $M^2<1$ or in $M^2>1$, while those corresponding to $\delta>1$ lie wholly in $\xi<1$ or $\xi>1$. This representation of the solutions is similar to that given by Bondi (1952), apart from slight differences in the reference quantities and the introduction of the entropy as the quantity which distinguishes the various solution curves (Axford & Newman 1966, Newman 1967). It is interesting that the functional relationship between M^2 and ξ separates in the manner shown in 2.14, since this permits a simple graphical determination of the solutions for any value of δ (Bondi 1952).

The asymptotic forms of the relationship between M^2 and ξ are as follows: $M^2 \sim \xi^{(3\gamma-5)/2} \to \infty$, and $M^2 \sim \xi^{(5-3\gamma)/(\gamma-1)} \to 0$ as $\xi \to 0$; $M^2 \sim \xi^{2(\gamma-1)} \to \infty$, and $M^2 \sim \xi^{-4} \to 0$ as $\xi \to \infty$. In the vicinity of the origin, $c^2 \sim 1/\xi$ and $u = Mc \sim \xi^{(3-2\gamma)/(\gamma-1)}$; thus $u \to 0$ as $M^2 \to 0$ and $\xi \to 0$ only if $\gamma < 3/2$. In the vicinity of the critical point $M^2 = \xi = 1$ the solutions are hyperbolas asymptotic to lines $(M^2-1) = v(\xi-1)$ where v is determined from the quadratic $v^2 - 4(\gamma-1)v - 2(5\gamma-3) = 0$. The need for special treatment of the cases $\gamma = 1$, $\gamma = 5/3$ is made very evident by these approximate relationships.

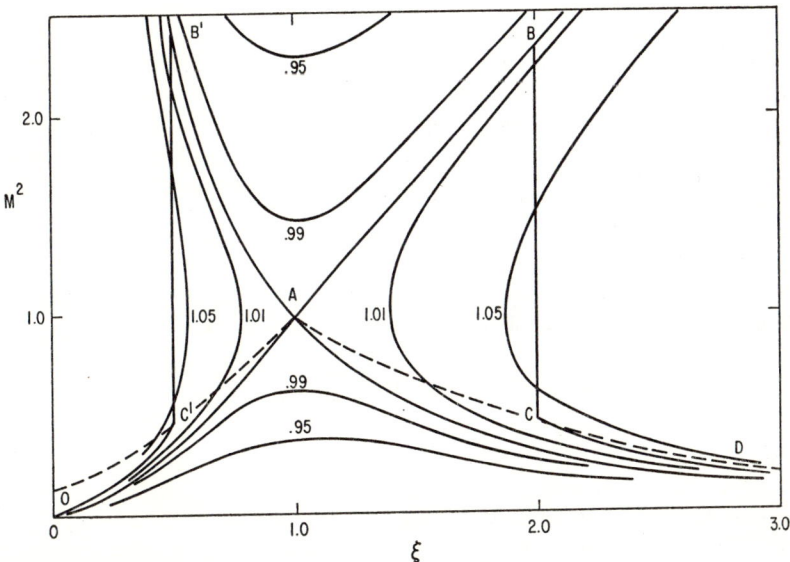

FIGURE 1. Mach number squared versus radial distance (in units of the critical point radius) for the stellar wind and accretion problems with $\gamma = 4/3$. Numbers labeling curves are values of δ (cf Equation 2.14). $OABCD$ is the complete solution for a stellar wind expanding into the interstellar medium with $p_\infty/p_* = 5.61 \times 10^{-2}$. $DAB'C'O$ is the complete solution corresponding to accretion towards a star with $p_0/p_* (r_0/r_*)^4 = 0.1797$. The dashed lines represent the square of the postshock Mach number given by Equation 3.5. (From Axford & Newman 1966.)

When the total energy per unit mass E is zero or negative the nature of the solution for M^2 as a function of ξ changes drastically. A new choice of reference quantities is necessary since 2.13 implies $E>0$ for $\gamma<5/3$. We will restrict our investigation of the solutions for $E=0$ and $E<0$ to the case $\gamma=5/3$, which is of most practical interest. In this case, the equations of motion are

$$\rho u r^2 = \pm \rho_0 u_0 r_0^2 \qquad 2.18$$

$$\rho = \beta c^3 \qquad 2.19$$

$$\frac{1}{2} u^2 + \frac{5p}{2\rho} - \frac{G\mathfrak{M}}{r} = E \qquad 2.20$$

On taking $\rho_0 = \beta_0 u_0^3$, and $\xi = r/r_0$, we obtain the following equation for M:

$$\delta(M^2+3)/M^{1/2} = (E/u_0^2)\xi + 1 \qquad 2.21$$

where $\delta = (\beta_0/\beta)^{1/2}$ and we have put $r_0 = 2G\mathfrak{M}/u_0^2$. If $E>0$ we choose $E=u_0^2$, if $E<0$ we choose $E=-u_0^2$, and if $E=0$, u_0 can be chosen arbitrarily. Typical solutions for each of these three cases are shown in Figures 2a,b,c. Note that there are no solutions which proceed smoothly from subsonic to supersonic flow, or vice versa, as there are in Figure 1. That is, there is no critical point in this case. For $E<0$ there is a limit line at $\xi=1$, beyond which no solutions with real velocity and positive temperature exist. The asymptotic forms of the relationship between M^2 and ξ when $\gamma=5/3$ and $E>0$ are: $M^2 \sim \xi^{4/3} \to \infty$ and $M^2 \sim \xi^{-4} \to 0$ as $\xi \to \infty$. In the vicinity of the point $M^2=1$, $\xi=0$, which is intersected by the curve corresponding to $\delta=1/4$, $\xi \simeq 3(M^2-1)/32$. The solutions for the case $E=0$ are all of the form $M^2=\text{const}$, and it is immediately evident from 2.18 and 2.19 that $c \propto \xi^{-1/2}$, $u \propto \xi^{-1/2}$, and $\rho \propto \xi^{-3/2}$.

The solution for the case $\gamma \to 1$ is formally obtained by integrating 2.11 with E given by 2.12, choosing $G\mathfrak{M}/2u_0^2 r_0 = 1$, and putting $\gamma=1$. Thus

$$M^2 - \log M^2 = 4 \log \xi + \frac{4}{\xi} - C \qquad 2.22$$

where C is a constant of integration. In this case it is convenient to regard γ as a polytrope index, with $p \propto \rho$ replacing a much more complicated energy equation than 2.3; accordingly the speed of sound is constant (i.e. $c^2 = p/\rho = u_0^2$). There is no constant of integration corresponding to the total energy as in the other cases we have considered, hence 2.22 represents all possible solutions. Note that 2.22 can also be obtained from 2.14 on taking the $(\gamma-1)$ root of both sides, letting $\gamma \to 1$ and using the definition $\lim_{(n \to 0)} (1+nx)^{1/n} = e^x$. The topology of these solutions is similar to that shown in Figure 1 for the case $\gamma=4/3$, $E>0$. In particular, there is a critical point $M^2=1$, $\xi=1$ that is intersected by the curves corresponding to $C=3$, and that permits transitions from subsonic to supersonic flow, and vice versa.

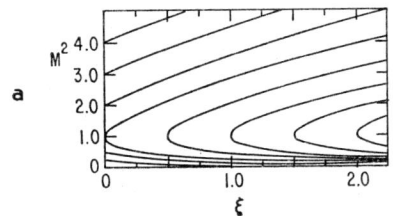

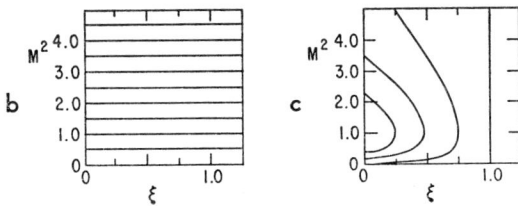

FIGURE 2. Mach number squared versus radial distance (in units of $r_0 = G\mathfrak{M}_*/2u_0^2$) for the stellar wind and accretion problems with $\gamma = 5/3$. Note that there are no solutions that proceed smoothly from subsonic to supersonic flow, or vice versa, as there are in Figure 1. (a) $E > 0$; compare these solutions with solutions for $\xi > 1$ in Figure 1. (b) $E = 0$; in $\xi > 0$, u and c both decrease with increasing ξ. (c) $E < 0$; no solutions with real u and c exist beyond $\xi = 1$.

The asymptotic forms of the relationship between M^2 and ξ in this case are: $M^2 \approx 4 \log \xi \to \infty$, $M^2 \sim \xi^{-4} \to 0$ as $\xi \to \infty$; $M^2 \approx 4/\xi \to \infty$, $M^2 \sim e^{-4/\xi} \to 0$ as $\xi \to 0$. In the vicinity of the critical point, the solution curves are hyperbolas $(M^2 - 1)^2 = 4(\xi - 1)^2 - (C - 3)$.

For the case $G\mathfrak{M} = 0$, which is relevant to the flow of gas away from a comet for example, the relationship between M^2 and ξ is easily found from 2.6, 2.7, and 2.10 to be

$$\delta\left(1 + \frac{\gamma - 1}{2} M^2\right) = \left(\frac{\gamma + 1}{2}\right)(M^2 \xi^4)^{\frac{\gamma - 1}{\gamma + 1}} \qquad 2.23$$

In this case there is no critical point and the solutions are topologically similar to those shown in Figure 2a.

3. Determination of Solutions for Given Boundary Conditions

It is evident from the results obtained in Section 2 that solutions representing both outflow and inflow with a variety of boundary conditions are possible. In this section we describe how suitable solutions can be chosen for physically interesting boundary conditions, but note that these solutions are not completely satisfactory and must be modified by allowing for additional physical processes (e.g. heat addition, thermal conduction) before they can be considered realistic. We will treat two classes of solutions, those corresponding to stellar accretion (Bondi 1952) and to stellar winds (Parker 1958). In the case of accretion ($u \leq 0$), appropriate boundary conditions are $E \geq 0$, $u \to 0$, $\rho \to \rho_\infty$, $p \to p_\infty$ as $r \to \infty$, and $p = p_* \gg p_\infty$, $M^2 \leq 1$ at $r = r_*$, where r_* can be considered to be the radius of the accreting star. In the case of stellar winds ($u \geq 0$), appropriate boundary conditions are $E \geq 0$, $p \to p_\infty$ as $r \to \infty$, $p = p_*$ and M^2 should have a moderately small value at $r = r_*$. Note that $E = 0$ corresponds in each case to $p_\infty = \rho_\infty = 0$, and that the condition $E \geq 0$ can be relaxed if heat conduction and heat addition are taken into account.

The necessity for inserting shock transitions in these problems was first pointed out by McCrea (1956). Clauser (1960) noted that the solutions are similar to those obtained in simple treatments of the convergent-divergent nozzle problem, and that in general shocks must be inserted in the flow if the boundary conditions we have imposed are to be satisfied. The Rankine-Hugoniot relationships for stationary shock transitions are:

$$[\rho u]_1^2 = 0 \qquad 3.1$$

$$[p + \rho u^2]_1^2 = 0 \qquad 3.2$$

$$\left[\frac{\gamma}{\gamma - 1}\frac{p}{\rho} + \frac{1}{2}u^2\right]_1^2 = 0 \qquad 3.3$$

$$[\delta]_1^2 \geq 0 \qquad 3.4$$

where the square brackets indicate that the difference in the arguments evaluated on the downstream side (2) and the upstream side (1) of the shock is to be taken. In terms of the Mach numbers upstream (M_1) and downstream (M_2), these equations can be combined to give

$$M_2^2 = \{(\gamma - 1)M_1^2 + 2\}/\{2\gamma M_1^2 - (\gamma - 1)\} \qquad 3.5$$

and in the notation of Section 2,

$$\delta_2 \geq \delta_1 \qquad 3.6$$

Equation 3.5 is valid for isothermal shocks ($\gamma = 1$), and in all cases, 3.6, which requires that shocks be compressive, implies

$$M_1^2 \geq 1 \geq M_2^2 \qquad 3.7$$

For $\gamma < 5/3$ and $E > 0$ the possible solutions to the accretion and stellar-wind problems are indicated in Figure 3 for $\xi_* = \xi_*' < 1$ and for $\xi_* = \xi_*'' > 1$. Where the critical point is outside the surface of the star ($\xi_* = \xi_*' < 1$), accretion solutions either must be completely subsonic ($M^2 < 1$, $\delta < 1$) or must involve the solution $\delta = 1$ (curve A) with subsonic flow in $\xi > 1$ and a region of supersonic flow in $\xi < 1$, followed by a shock transition (at $\xi = \xi_s < 1$) to the subsonic solution ($\delta > 1$) intersecting A' at ξ_s. Similarly, stellar-wind solutions for $\xi_* < 1$ either are completely subsonic, or else must involve the solution $\delta=1$ (curve B) with subsonic flow in $\xi<1$ and a region of supersonic flow in $\xi>1$, followed by a shock transition (at $\xi=\xi_s>1$) to the subsonic solution ($\delta>1$) intersecting B' at ξ_s. If the critical point lies inside the surface of the star ($\xi_*=\xi_*''\geq 1$) all accretion solutions are completely subsonic ($M^2\leq 1$, $\delta\leq\delta_c$), while stellar-wind solutions either are completely subsonic, or else must correspond to the supersonic branch of curve $C(\delta=\delta_c)$ in $\xi_*\leq\xi<\xi_s$ and (following a shock transition) to the subsonic curve ($\delta>\delta_c$) intersecting C' at ξ_s. For the case $\gamma=5/3$, accretion and stellar-wind solutions are qualitatively the same as corresponding solutions for $1\leq\gamma<5/3$ with the critical point inside the surface of the star (compare the region $\xi_*\geq 0$ in Figure 2a with the region $\xi_*\geq\xi_*''$ in Figure 3).

If $\xi_*<1$ the shock transition allows any subsonic curve intersecting the regions $\xi<1$, $M^2<M^2(A')$ for accretion and $\xi>1$ for stellar winds to contribute to the solution, so that it is possible to construct complete solutions satisfying a wide range of boundary conditions. For example, if $1<\gamma<5/3$, then

$$p = p_0\delta^{1/2}(M\xi^2)^{-2\gamma/(\gamma+1)} \qquad 3.8$$

Thus for a given $E>0$, specification of ξ_* and two appropriate boundary conditions completely determines a unique solution, which if entirely subsonic can correspond either to accretion or to a stellar wind. For wholly subsonic flow δ_* equals δ_∞ equals δ, so any two of the boundary conditions p_*, p_∞, p_0, and δ (or M_*) determine the solution. For supersonic flow the boundary conditions p_* and p_0, δ_* (or M_*) and p_0, or p_* and δ_* are formally suitable for accretion, while p_∞ and p_0, δ_∞ and p_0, or p_∞ and δ_∞ are formally suitable for stellar winds. Note that specifications of p_0 and the flux ($\rho u\xi^2$) are equivalent since $p_0=\rho u\xi^2 E^{1/2}/\gamma$. When p_∞ and p_* are used as the boundary conditions, $R=p_\infty/p_*$ must be restricted to $R\leq R(\delta_*=\delta_\infty=1)$.

If we fix p_∞ in the accretion problem, then for $\delta_*<1$, p_* decreases with increasing δ_* to a minimum value at $\delta_*=1$ and then increases with δ_* ($\delta_*>1$) up to the limiting solution intersecting A' at ξ_* ($M_*^2<1$) (see Figure 3). This behavior of p_* results from the fact that for a given p_∞ the accretion flux increases with increasing δ_∞ up to $\delta_\infty=1$ and then remains constant for all supersonic solutions (Bondi 1952). For a fixed flux in the stellar-wind problem, δ_∞ increases with decreasing p_∞ over the entire range of solutions ($M_\infty^2<1$). In other words for a given stellar-wind flux, if $p_\infty>p_1=[(5-3\gamma)/2]^{\gamma/(\gamma-1)}\rho u\xi^2 E^{1/2}/\gamma$ the flow must be entirely subsonic, while if $p_\infty<p_1$ a region of supersonic flow must exist. (This discussion can be extended with minor modifications to the cases $\xi_*>1$ and $\gamma=5/3$.)

We are somewhat restricted in the boundary conditions that can be satisfied by the solutions discussed. For example, no nonstatic solutions have $M_*=0$ if $\xi_*>0$, solutions involving supersonic flow must have $M_*=1$ if $\xi_*>1$ (or if $\gamma=5/3$), the pressure ratio p_∞/p_* cannot exceed a certain value, and so on. However, this does not mean that the solutions are not useful, but rather that it might be necessary to consider more complicated physical

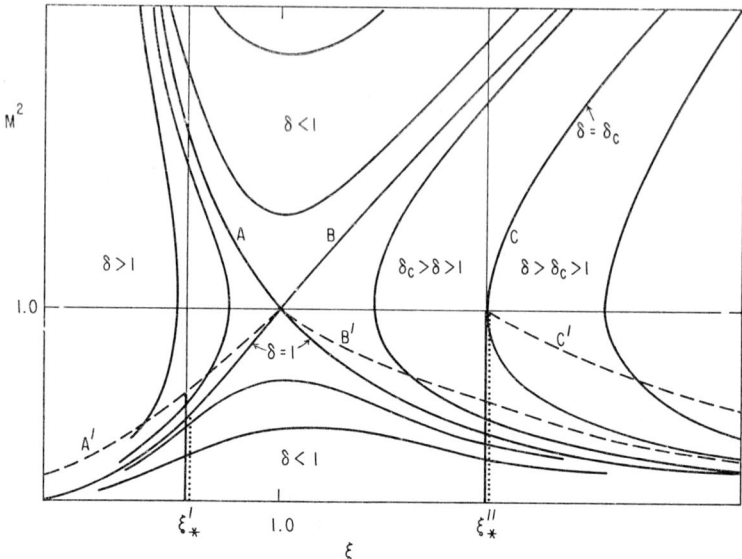

FIGURE 3. Mach number squared versus radial distance (in units of the critical point radius) for the stellar wind and accretion problems with $\gamma = 4/3$. A and B are the critical curves ($\delta = 1$), C is the solution curve ($\delta = \delta_c$) which is tangent to $\xi = \xi_*''$, and the dashed lines A', B', and C' represent the square of the postshock Mach number corresponding to the supersonic branches of A, B, and C, respectively (cf Equation 3.5). The regions $\delta > 1$, $\delta < 1$ are indicated (cf Equation 2.14). The heavy vertical line at $\xi = \xi_*'$ intersects all possible accretion solutions for the case $\xi_* = \xi_*' < 1$, while the dotted line intersects all possible stellar-wind solutions (including curve B) for the same case. The heavy vertical line at $\xi = \xi_*''$ intersects all possible accretion solutions (not including the upper branch of curve C) for the case $\xi_* = \xi_*'' > 1$, while the dotted line intersects all possible stellar-wind solutions (including the upper branch of curve C) for the same case.

processes (perhaps only locally). By introducing additional effects (especially those which increase the order of the differential equations) it is possible to satisfy a greater variety of boundary conditions. Several such effects are considered in the following sections.

Problems which involve gas-dynamic discontinuities in addition to shock waves are of some importance, especially ionization fronts in the case of accretion (Mestel 1954), and recombination fronts in the case of stellar winds (Newman & Axford 1968). The gas within the Strömgren sphere surrounding the central star can be assumed to behave isothermally, but outside the sphere adiabatic flow is probably more appropriate. As the gas traverses the ionization or recombination front bordering the Strömgren sphere it can undergo any of the four possible transitions between supersonic and subsonic flow. This permits a wide variety of flow configurations,

THEORY OF STELLAR WINDS AND RELATED FLOWS   41

usually involving one or more shock waves. The solutions are constructed by using the appropriate jump conditions for the ionization or recombination fronts, and for any shocks that might be necessary, ensuring that the Strömgren sphere has the correct size, and of course matching the boundary conditions at infinity and at the surface of the central star. In his treatment of the accretion problem Mestel (1954) rejected the possibility of strong D-type ionization fronts (e.g. Axford 1961) and thus missed a small but interesting class of solutions.

4. Steady Radial Flow Distributed Sources and Sinks: Galactic Winds

A simple generalization of the flows described in Section 2 can be achieved by permitting sources or sinks of fluid to be distributed throughout space in some arbitrarily chosen manner. This could represent for example the flow due to the release of gas by one means or another from stars in a spherically symmetric galaxy (i.e. a "galactic wind"), provided the gravitational field is chosen appropriately. The equations of motion analogous to 2.1, 2.2, and 2.3 are:

$$\frac{1}{r^2}\frac{d}{dr}(\rho u r^2) = q \qquad 4.1$$

$$\rho u \frac{du}{dr} = -\frac{dp}{dr} - \frac{G\mathfrak{M}\rho}{r^2} - qu \qquad 4.2$$

$$\frac{1}{r^2}\frac{d}{dr}\left[\rho u r^2\left(\frac{1}{2}u^2 + \frac{\gamma}{\gamma-1}\frac{p}{\rho}\right)\right] + \rho u \frac{G\mathfrak{M}}{r^2} = Q \qquad 4.3$$

where $q(r)$ and $Q(r) = [u_0^2/\gamma(\gamma-1)]q(r)$ are the mass and energy production rates per unit volume respectively, and $\mathfrak{M}(r)$ is the total gravitating mass in a sphere of radius r. We will assume that q is proportional to the number density of stars $N_*(r)$ and that the mass of the interstellar gas is negligible; thus

$$\mathfrak{M}(r) = 4\pi\mathfrak{M}_* \int_0^r N_*(r)r^2 dr + \mathfrak{M}_0 \qquad 4.4$$

where $\mathfrak{M}_*$ is the mean stellar mass, and $\mathfrak{M}_0$ is a point mass situated at $r=0$.

It is convenient to introduce a specific form for $N_*(r)$ at this stage of the analysis. For simplicity we choose $N_*(r) = N_0 e^{-r/a}$ and define

$$\alpha(\xi) = \int_0^\xi t^2 e^{-t}dt = 2 - (\xi^2 + 2\xi + 2)e^{-\xi} \qquad 4.5$$

$$I(\xi) = -\int_\infty^\xi \frac{\alpha(t)}{t^2}dt = \frac{2 - (\xi + 2)e^{-\xi}}{\xi} \qquad 4.6$$

$$W(\xi) = \int_{\infty}^{\xi} \frac{\alpha^2(t)}{t^2} dt = \frac{1}{\xi} \left[-4 + 4(\xi + 2)e^{-\xi} - \left(4 + \frac{21}{4}\xi + \frac{5}{2}\xi^2 \right.\right.$$
$$\left.\left. + \frac{1}{2}\xi^3 \right)e^{-2\xi} \right] \quad 4.7$$

where $\xi = r/a$. Integrating Equation 4.1 we obtain

$$\rho u \xi^2 = aq_0(\alpha - \alpha_0) \quad 4.8$$

where $q = q_0 e^{-\xi}$ and $-aq_0\alpha_0$ is a constant of integration representing a source or sink at the center of the galaxy according as α_0 is negative or positive. If the flow is to be outwards everywhere, then $\alpha_0 \leq 0$, with $\alpha_0 = 0$ if the central mass is not a source, since $\alpha(0) = 0$. If the flow is to be inwards everywhere and the mass flux at infinity is to vanish (i.e. there is no extragalactic medium), then we must choose $\alpha_0 = 2$ since $\alpha \to 2$ as $\xi \to \infty$. For intermediate values of α such that $0 < \alpha < 2$, the flow is inwards within a certain radius ($\xi < \xi_0$), and outwards in $\xi > \xi_0$, where ξ_0 is determined by α_0; at $\xi = \xi_0$ the radial velocity of the gas is zero.

As in Section 2, we can derive a single first-order differential equation from 4.1, 4.2, and 4.3 which involves only the Mach number and ξ:

$$\frac{(M^2 - 1)}{M^2} \frac{dM^2}{d\xi} = \left(1 + \frac{\gamma - 1}{2} M^2\right) \left\{ \frac{4}{\xi} - \frac{(1 + \gamma M^2)\xi^2 e^{-\xi}}{(\alpha - \alpha_0)} \left(1 + \frac{1}{D}\right) \right.$$
$$\left. - \gamma \frac{(\gamma + 1)\mu(\alpha + \eta)}{\xi^2 D} \right\} \quad 4.9$$

where

$$\mu = 4\pi G N_0 \mathfrak{M}_* a^2 / u_0^2 \quad 4.10$$

$$\eta = \frac{\mathfrak{M}_0}{4\pi a^3 N_0 \mathfrak{M}_*} \quad 4.11$$

$$D(\xi) = 1 - \gamma(\gamma - 1)\mu Z/(\alpha - \alpha_0) \quad 4.12$$

$$Z(\xi) = \int_{\xi_0}^{\xi} \frac{(\alpha + \eta)(\alpha - \alpha_0)}{\xi^2} d\xi = \left[W + (\alpha_0 - \eta)I + \frac{\alpha_0 \eta}{\xi} \right]_{\xi_0}^{\xi} \quad 4.13$$

Note that if $\eta \neq 0$, $D \to \infty$ as $\xi \to 0$ and that $D(\xi_0) = 1$. It is evident that critical points can exist that allow for smooth transitions from subsonic to supersonic flow; there are in general two such points if $\mathfrak{M}_0 \neq 0$ and some inflow occurs. The isothermal case, which is likely to be a good first approximation to the actual behavior of interstellar gas, can be treated by formally setting $\gamma = 1$ in 4.9:

$$\frac{(M^2 - 1)}{M^2} \frac{dM^2}{d\xi} = \frac{4}{\xi} - \frac{2(1 + M^2)\xi^2 e^{-\xi}}{\alpha - \alpha_0} - \frac{2\mu(\alpha + \eta)}{\xi^2} \quad 4.14$$

It can be shown that $c^2 = u_0^2 D/\gamma(1 + M^2(\gamma-1)/2)$; hence for a solution to exist, D must be positive and thus

$$\mu \leq \frac{1}{\gamma(\gamma - 1)} \frac{2 - \alpha_0}{Z(\infty)}. \quad 4.15$$

This condition, which can also be derived from the virial theorem, is similar to the condition $E \geq 0$ for a stellar wind to be possible that was obtained in Section 2. In effect it implies that the mean thermal speed of the gas emitted by the stars must be comparable with or larger than the mean speed of the stars themselves, if the gas is to escape from the galaxy. In realistic situations the temperature of the bulk of the gas is likely to be relatively low (perhaps about $10^4 °K$). However, in estimating the mean thermal speed it is necessary to include the contribution from cosmic rays that in general are energetically dominant. For the case $\gamma = 1$ of course there is no difficulty, and escape is always possible.

There is a wide variety of solutions to Equation 4.9, depending on the boundary conditions and the values of the parameters involved. The case of a pure galactic wind with $\gamma = 5/3$, $\eta = 0$, and $\alpha_0 = 0$ has been discussed by Burke (1968), and he has provided numerical solutions showing that a critical point can occur in this case provided u_0^2 is sufficiently large (i.e. the condition 4.15 is satisfied). Johnson (1969) has independently considered the problem using the procedure outlined here, and found solutions that represent pure accretion to the center of the galaxy (cf Spitzer 1942), as well as solutions that are mixed, containing a region of inflow ($\xi < \xi_0$) together with a region of outflow ($\xi > \xi_0$). The latter solutions are interesting in that two critical points can appear, provided $\gamma < 5/3$ and $\eta \neq 0$; an example is given in Figure 4. Note that it is more appropriate to plot M rather than M^2 in this case, and that requiring that the pressure should be continuous at $\xi = \xi_0$ is equivalent to requiring $dM/d\xi$ to be continuous.

5. Stellar Winds with Heat Addition

While the solutions described in Sections 2 and 3 provide a useful means of understanding some of the basic features of stellar-wind flows, they are far from being realistic because many physical processes which are likely to be important have been ignored. As far as the solar wind is concerned, the process that most obviously should be taken into account is heat addition associated with the dissipation of wave energy propagating upwards from the photosphere (e.g. Biermann 1946, 1948; Alfvén 1947; Osterbrock 1961; Whitaker 1963; Lighthill 1967; Barnes 1968). A simple model that illustrates the effects of heat addition rather well can be constructed by including an arbitrary distributed heat source in the energy equation 2.3, leaving the

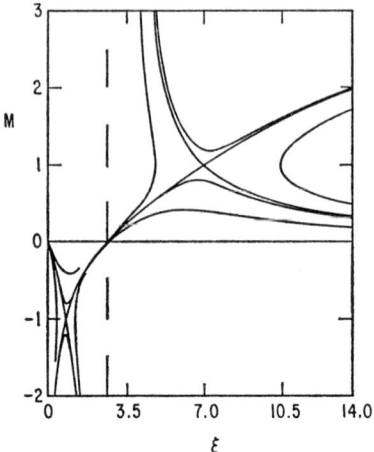

FIGURE 4. Mach number versus radial distance (in units of the scale height of stellar mass density) for the case of a galactic wind allowing for an exponential mass production function and $\gamma = 4/3$, $\mu = 1$, and $\eta = 2$. The stagnation point occurs at $\xi_0(\alpha_0) = 3$ with $M > 0$ in $\xi > \xi_0$ and $M < 0$ in $\xi < \xi_0$. A critical point exists in each of the regions $\xi > \xi_0$ and $\xi < \xi_0$ with one continuous solution intersecting both critical points.

mass and momentum conservation equations unchanged; the equations of motion are then formally equivalent to 4.1, 4.2, and 4.3 with $q=0$ and $Q(r)$ an arbitrary function. This problem has been considered by Konyukov (1967) and independently by Holzer (1968); the latter work is described in this section.

For simplicity, and because it is the case most relevant to the solar wind, we assume $\gamma = 5/3$. Taking $r_0 = r_* =$ radius of Sun, the equations of motion can be written in the form

$$\frac{(M^2 - 1)}{2M^2} \frac{dM^2}{d\xi} = \frac{(M^2 + 3)}{3H} \left[\frac{2H}{\xi} - \frac{G\mathfrak{M}}{2r_0 \xi^2}(M^2 + 3) \right.$$
$$\left. - \frac{1}{2}\frac{dH}{d\xi}(M^2 - 1) - \frac{1}{3}P(M^2 + 3) \right]$$
$$= \frac{(M^2 + 3)}{3H}\left[\frac{2E_0}{\xi} + \frac{2}{\xi}\int_1^\xi P(\xi')d\xi' \right.$$
$$\left. - \frac{5}{6}P\left(M^2 + \frac{3}{5}\right) \right] \qquad 5.1$$

where

$$P(\xi) = r_0^3 Q(\xi)\xi^2/\mathfrak{F} \qquad 5.2$$
$$\rho u r^2 = \mathfrak{F} \qquad 5.3$$

and

$$H(\xi) = E_0 + \frac{G\mathfrak{M}}{r_0\xi} + \int_1^\xi P(\xi')d\xi' \qquad 5.4$$

It is evident that E_0 is the constant of integration obtained on formally integrating 4.3, and that it is the total energy per unit mass of fluid at $r=r_0$:

$$E_0 = \frac{1}{2}u_0^2 + \frac{5}{2}\frac{p_0}{\rho_0} - \frac{G\mathfrak{M}}{r_0} \qquad 5.5$$

When $Q=0$, Equation 5.1 reduces immediately to 2.11 with $\gamma = 5/3$.

Provided $Q(\xi)$ decreases rapidly enough with increasing ξ, it is to be expected that the solutions of 5.1 are asymptotically the same as 2.21 with $E = E_0 + \int_1^\infty P(\xi')d\xi'$. However there is an important difference in the two sets of solutions in that with $Q(\xi) \neq 0$ it is possible to have a critical point and thus a transition from subsonic to supersonic flow with $\gamma = 5/3$, which is not the case when $Q(\xi) = 0$. In particular, if $E_0 \geq 0$ a critical point can exist [depending on the specific form of $P(\xi)$] unless $E_0 \geq \frac{2}{3}P(1)$ and $P(\xi)$ decreases more rapidly than $1/\xi$. If $E_0 < 0$, no critical point exists if $E_0 + \int_1^\infty P(\xi')d\xi' \leq 0$, but a unique critical point exists if $E_0 + \int_1^\infty P(\xi')d\xi' > 0$.

Some examples of computed solutions of Equation 5.1 are shown in Figures 5 and 6. Here we have assumed $Q = Q_0 e^{-\beta\xi}$, so that $P = P_0\xi^2 e^{-\beta\xi}$ with $P_0 = r^3_0 Q_0/\mathfrak{F}$, and we have chosen values of the various parameters that are reasonably representative of solar conditions. The solutions shown in Figure 5 all pass through critical points, and the Mach numbers vary continuously from $M=0$ at $\xi=1$ to $M \sim \xi^{2/3} \to \infty$ as $\xi \to \infty$. The dashed lines represent solutions corresponding to $Q=0$ (i.e. Equation 2.21) with a total energy of 1.24×10^3 eV per particle ($u_0 \approx 500$ km sec^{-1}) and various values of δ. Note that δ is not a constant in the case where heat is added to the flow since

$$p/\rho^\gamma = (p_0/\rho_0^\gamma) \exp\left\{\gamma(\gamma-1)\int_1^\xi (P/c^2)d\xi\right\} \qquad 5.6$$

and hence the entropy of the fluid increases with increasing ξ.

It is interesting that the temperature of the gas increases very abruptly from $\sim 10^4$°K to $\sim 10^6$°K within a radial distance of only $\sim 0.1 r_*$, and that there is a corresponding decrease in the density (Figure 6). Such a steep temperature gradient in a region of low-speed flow implies that heat conduction must play an important role. In fact the model is probably less realistic in this region than it seems; in particular the form assumed for the heating rate is arbitrary and may not be a good representation of a situation where there are rapid changes in the physical characteristics of the medium. However, for $\xi \geq 1.5$ the solutions are probably not unrealistic, and they provide interesting information in that it is evident that for a total energy input

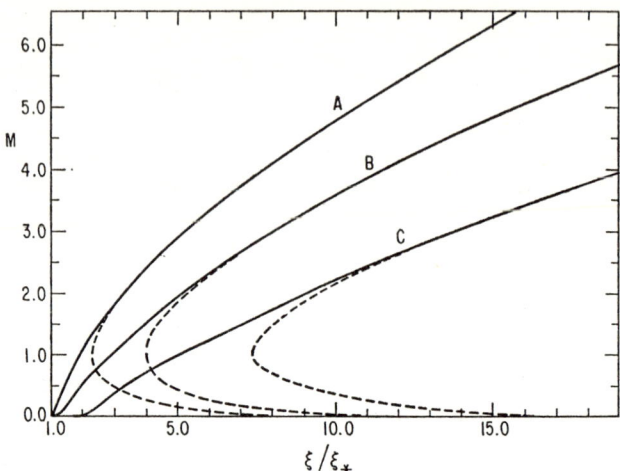

FIGURE 5. Mach number versus radial distance (in solar radii) for parameters representative of solar conditions. For curve A, $E_0 = -1.2 \times 10^{27}$ eV g^{-1}, $\mathfrak{F} = 1.6 \times 10^{11}$ g sec^{-1}, $Q_0 = 1.23 \times 10^8$ eV cm^{-3} sec^{-1}, and $\beta = 4.0$. For curve B, $E_0 = -1.2 \times 10^{27}$ eV g^{-1}, $\mathfrak{F} = 1.6 \times 10^{11}$ g sec^{-1}, $Q_0 = 5.41 \times 10^6$ eV cm^{-3} sec^{-1}, and $\beta = 2.0$. For curve C, $E_0 = -1.0 \times 10^{27}$ eV g^{-1}, $\mathfrak{F} = 1.6 \times 10^{11}$ g sec^{-1}, $Q_0 = 4.51 \times 10^5$ eV cm^{-3} sec^{-1}, and $\beta = 1.0$. The dashed curves are solutions of 2.21 for $E = 7.5 \times 10^{26}$ eV g^{-1} and various values of δ. All of the curves A, B, and C have the same asymptotic energy, $E_\infty = E_0 + \int_1^\infty P(\xi) d\xi = 7.5 \times 10^{26}$ eV g^{-1}.

comparable to what is required to produce observed solar-wind energy flows, the heating must take place throughout a fairly extended region (i.e. β is not large) for the maximum temperature produced near the Sun to be of the order of typical coronal temperatures $(1-2 \times 10^{6} {}^\circ\text{K})$.

6. THE EFFECTS OF HEAT CONDUCTION AND VISCOSITY

With viscosity and heat conduction included, the Navier–Stokes equations of motion for steady, spherically symmetric radial flow can be written

$$\rho u r^2 = \pm \rho_0 u_0 r_0^2 \qquad 6.1$$

$$\rho u \frac{du}{dr} = -\frac{dp}{dr} - \frac{G \mathfrak{M} \rho}{r^2} + \frac{4}{3} \left[\eta \left(\frac{d^2 u}{dr^2} + \frac{2}{r} \frac{du}{dr} - \frac{2u}{r^2} \right) \right.$$

$$\left. + \frac{d\eta}{dr} \left(\frac{du}{dr} - \frac{u}{r} \right) \right] \qquad 6.2$$

$$\frac{1}{2} u^2 + \frac{\gamma}{\gamma - 1} \frac{p}{\rho} - \frac{G \mathfrak{M}}{r} = E + \frac{\kappa}{\rho u} \frac{dT}{dr} + \frac{4}{3} \frac{\eta}{\rho} \left(\frac{du}{dr} - \frac{u}{r} \right) \qquad 6.3$$

THEORY OF STELLAR WINDS AND RELATED FLOWS 47

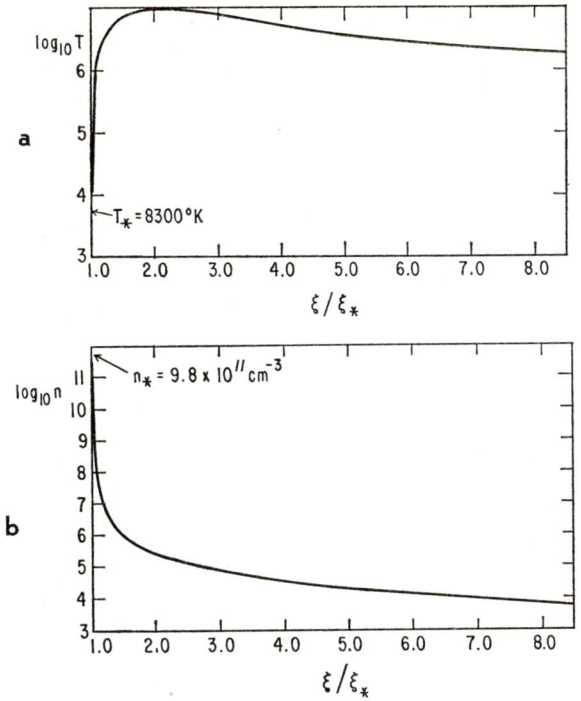

FIGURE 6. (a) Temperature (°K) versus radial distance (in solar radii) and (b) density (cm^{-3}) versus radial distance (in solar radii) corresponding to the critical solution of 5.1 with parameters appropriate to solar conditions ($E_0 = -1.2 \times 10^{27}$ eV g^{-1}, $\mathfrak{F} = 1.6 \times 10^{11}$ g sec^{-1}, $Q_0 = 3.26 \times 10^7$ eV cm^{-3} sec^{-1}, and $\beta = 2.85$).

where κ is the thermal conductivity and η the viscosity, and we have assumed there are no distributed sources or sinks of mass, momentum, and energy.

Formally it is evident that the inclusion of viscosity and heat conduction complicates the analysis enormously, since the equations are of much higher order than before and have no simple analytic solutions. However if the characteristic Reynolds number ($R_\eta = \rho_c u_c r_c / \eta_c$) and Péclet number ($R_\kappa = \rho_c u_c r_c / \kappa_c C_p$) are large compared with unity, one should not expect viscosity and heat conduction to play a significant role in determining the general features of the solutions unless there are singular regions in which gradients are locally very large. In the solutions described in Section 2 it is obvious that shock waves are the only features involving large gradients. Since the shock waves must in general be very thin, their structures are insensitive to the details of the flow elsewhere and thus they can be investigated in the usual way without taking into account the fact that they form part of the solutions of a particular class of flow problem. However, shocks occurring in the vicinity of the critical points of the solutions where the flow is transonic are weak and therefore not thin, so that their structure is affected by the overall flow. Furthermore it is to be expected that viscosity and heat conduction are important in the transonic region, even if any shocks that occur are strong and situated well away from the critical point, since the other terms in the equations of motion form groups which separately become small in this region.

A nonlinear treatment of the transonic region for both inflow and outflow problems ($E>0$, $1<\gamma<5/3$) can be carried out provided R_η and R_κ are large (Axford & Newman 1966, 1967; Newman 1967). The dependent variables in Equations 6.1, 6.2, and 6.3 are expanded in the form

$$L = L_c(1 + \epsilon L' + \epsilon^2 L'' + \cdots) \qquad 6.4$$

where $\epsilon = R_c^{-1/2} \ll 1$ and the Reynolds number is evaluated at the critical point of the solution for inviscid flow without heat conduction. The independent variable is "stretched" such that $r = r_c(1+\epsilon s)$. On substituting these forms into the equations of motion and equating to zero the coefficients of each power of ϵ, it is found that the zeroth-order equations are satisfied identically, while the first- and second-order equations can eventually be reduced to a single nonlinear equation for u':

$$\pm \left[\frac{4}{3} + \frac{(\gamma-1)}{\theta}\right] \frac{d^2 u'}{ds^2} - (\gamma-1)u' \frac{du'}{ds} - 2(\gamma-1)\frac{d}{ds}(su')$$
$$+ 2(3 - 2\gamma)s = 0 \qquad 6.5$$

where $\theta = \eta_c C_p/\kappa_c$ is the Prandtl number and the upper sign corresponds to outflow and the lower sign to inflow. This equation can be integrated twice and the solution for u' satisfying appropriate asymptotic conditions for $s \to \pm \infty$ can be written

$$u' = A_1 \nu D_{\nu-1}(as)/D_\nu(as) + A_2 s \qquad 6.6$$

where A_1, A_2, and a are known constants, ν is a parameter ($\nu \leq 0$), and $D_\nu(as)$ is the parabolic cylinder function. Examples of solutions for the case $\gamma = 4/3$ are shown in Figure 7. The straight lines intersecting the origin represent the asymptotic "critical" solutions ($\nu=0$) and the transitions between asymptotes correspond to weak shocks. As the shocks move away from the critical point (i.e. $\nu \to -0$), they become thinner as should be expected. Note that because the shocks are weak, there is no entropy change to the order we have considered (ϵ^2), hence the solutions all approach a common asymptote. It is interesting that with viscosity and heat conduction included, the "critical" solution in each case becomes one of a continuous family of solutions exhibiting no abrupt changes in character.

The problem of treating stellar-wind flows allowing for the effects of heat conduction but neglecting viscosity has been treated by a number of authors (Chamberlain 1961, 1965; Noble & Scarf 1963; Scarf & Noble 1965; Whang & Chang 1965; Parker 1964, 1965). In a notation introduced by

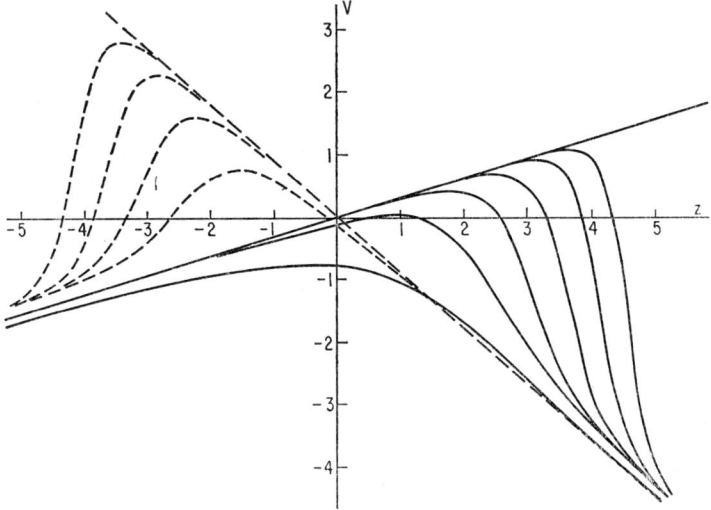

FIGURE 7. Solutions of Equation 6.5 for the structure of weak shocks in the vicinity of the critical point for stellar winds (solid lines) and accretion (dashed line) with $\gamma = 4/3$. V and Z are proportional to u' and s respectively. The more rapidly changing solutions, representing the thinnest shocks, correspond to the smallest values of ν. (From Axford & Newman 1966.)

Chamberlain, the relevant equations can be written in the form

$$\frac{1}{2}\left(1 - \frac{\tau}{\psi}\right)\frac{d\psi}{d\lambda} = 1 - \frac{2\tau}{\lambda} - \frac{d\tau}{d\lambda} \qquad 6.7$$

$$\frac{\psi}{2} - \lambda + \frac{5\tau}{2} + \frac{A}{2}\tau^{5/2}\frac{d\tau}{d\lambda} = \epsilon \qquad 6.8$$

where $\psi = (u^2/RT_0)$, $\lambda = (G\mathfrak{M}/RT_0 r)$, $\tau = T/T_0$, $\epsilon = E/kT_0$, $\rho u r^2 = C$, and $A = (2\kappa(T_0)G\mathfrak{M}/CR^2 T_0)$, with T_0 a suitably chosen reference temperature and $\kappa \propto T^{5/2}$. It is evident that to solve these equations is not easy, and that the solutions cannot be represented by a single family of curves in the (ψ, λ) plane which corresponds to the (M^2, ξ) plane in previous sections. Note that the equations have a singular point at $M^2 = 3/5$ (i.e. $\psi = \tau$) rather than at $M^2 = 1$ as is the case when $A = 0$. If one is interested in solutions for which $p \to 0$ as $r \to \infty$ (i.e. $\lambda \to 0$), one finds that the asymptotic forms of the solutions are:

$$\tau \sim c\lambda^{2/7}, \psi \sim 2(\epsilon - Ac^{7/2}/7) \qquad \text{as } \gamma \to 0 \qquad 6.9a$$

or

$$\tau \sim (35\lambda/2A)^{2/5}, \psi \sim 2\epsilon \qquad \text{as } \gamma \to 0 \qquad 6.9b$$

depending on whether or not the heat conduction term contributes to the energy flux at infinity (Whang & Chang 1965, Scarf & Noble 1965). There appear to be no solutions which yield $M^2 = \psi/\gamma\tau = \text{const}$ as $\lambda \to 0$ other than the case $\epsilon = 0$ treated by Chamberlain (1961), where $M^2 \to 0$ as $\lambda \to 0$ is a possibility. For $\epsilon > 0$ solutions are found for which $M^2 \to \infty$ as $\gamma \to 0$ and $M^2 \to 0$ as $\gamma \to \gamma_*$ (i.e. $r \to r_*$); a specific example of such a solution for the case $c = 0$ has been given by Whang & Chang (1965), using a matching procedure. Other solutions that probably correspond to $c \neq 0$, but have not been completed by extrapolation to $\lambda = 0$, have been given by Noble & Scarf (1963) and Scarf & Noble (1965).

It would be of some interest to treat the accretion and stellar-wind problems with heat conduction included from a general point of view. As far as the solar wind is concerned, however, this would be an academic exercise since it is evident that a two-component model is necessary (see Section 7) and that an extended heat source should be included.

If the effects of viscosity are to be explicitly included as well as those of heat conduction, the problem becomes extremely difficult. A first attempt by Scarf & Noble (1965) to include viscosity in a treatment of the flow of the solar wind yielded spurious solutions because the extreme delicacy of the problem was not appreciated. Whang et al (1966) have given an elegant treatment of one particular case in which $p \to 0$ as $r \to \infty$ and the energy is transported entirely as kinetic energy of bulk motion. There is of course no critical point if $\eta \neq 0$. A matching procedure is necessary to complete the solution, and as a result of the high order of the system, a wide variety of possible boundary conditions should be available. This problem has also been treated by Meyer & Schmidt (1966, 1968) and Konyukov (1969).

7. Multifluid Models of the Solar Wind

The equations discussed in Sections 2, 3, 4, and 5 can be used to describe a multicomponent plasma in which the electrons and the ions have equal isotropic temperatures, providing one takes $p = p_e + \Sigma_i p_i$, $\rho = \rho_e + \Sigma_i \rho_i$, $u = u_e = u_i$, $q = q_e + \Sigma_i q_i$, $Q = Q_e + \Sigma_i Q_i$, $\kappa = \kappa_e + \Sigma_i \kappa_i$, and $\eta = \eta_e + \Sigma_i \eta_i$, where subscripts e and i refer to electrons and ions. However, generally for a plasma composed of two or more components, the one-fluid description must be modified since it is not always reasonable to assume that the temperatures of the different components are equal when the plasma is of low density.

Consider first a two-component plasma in which the lack of a sufficiently effective electron-ion energy exchange mechanism leads to unequal but isotropic electron and ion temperatures. The equations governing the radial flow of such a plasma are:

$$\frac{1}{r^2} \frac{d}{dr}(\rho u r^2) = q \qquad 7.1$$

$$\rho u \frac{du}{dr} + \frac{d}{dr}[nk(T_e + T_i)] + \rho \frac{G\mathfrak{M}}{r^2} = -qu +$$

$$+\frac{4}{3}\left[\eta\left(\frac{d^2u}{dr^2}+\frac{2}{r}\frac{du}{dr}-\frac{2u}{r^2}\right)+\frac{d\eta}{dr}\left(\frac{du}{dr}-\frac{u}{r}\right)\right] \quad 7.2$$

$$\frac{1}{\gamma-1}nuk\frac{dT_e}{dr}=ukT_e\frac{dn}{dr}+\frac{1}{r^2}\frac{d}{dr}\left(r^2\kappa_e\frac{dT_e}{dr}\right)-\frac{1}{\gamma-1}\nu_{ie}nk(T_e-T_i)$$

$$+\frac{4}{3}\eta_e\left(\frac{du}{dr}-\frac{u}{r}\right)^2+Q_e+\frac{q_e}{m_e}\left(\frac{1}{2}m_eu^2-\frac{kT_e}{\gamma-1}\right)$$
$$7.3$$

$$\frac{1}{\gamma-1}nuk\frac{dT_i}{dr}=ukT_i\frac{dn}{dr}+\frac{1}{r^2}\frac{d}{dr}\left(r^2\kappa_i\frac{dT_i}{dr}\right)+\frac{1}{\gamma-1}\nu_{ie}nk(T_e-T_i)$$

$$+\frac{4}{3}\eta_i\left(\frac{du}{dr}-\frac{u}{r}\right)^2+Q_i+\frac{q_i}{m_i}\left(\frac{1}{2}m_iu^2-\frac{kT_i}{\gamma-1}\right) \quad 7.4$$

where $n=n_e=n_i$, $u=u_e=u_i$, $q=q_e+q_i$, $\rho=n(m_e+m_i)$, $\eta=\eta_e+\eta_i$, ν_{ie} is the reciprocal of the electron-ion energy exchange time, and Q_e (Q_i) is the rate per unit volume at which energy is added to the electron (ion) gas through processes other than electron-ion energy transfer. Equations 7.1–7.4 are obtained by writing equations for the electrons and ions separately and then combining the two continuity equations 7.1, the two momentum equations 7.2, the electron momentum and energy equations 7.3, and the ion momentum and energy equations 7.4.

The solution of 7.1–7.4 as they stand represents a formidable task. However, judicious neglect of certain terms can often considerably simplify the set of equations without seriously affecting the form of their solutions. Sturrock & Hartle (1966, Hartle & Sturrock 1968) have discussed the solar-wind problem, using 7.1–7.4 without the mass and heat addition or viscosity terms and with ν_{ie} appropriate to Coulomb collisions. As it happens, in the supersonic region of the solar wind, further simplifications can be introduced. Assuming that heat conduction dominates the electron equation, 7.3 can be written

$$T_e = T_{e0}(r_0/r)^{2/7} \quad 7.5$$

where κ_e has been taken proportional to $T_e^{5/2}$. If, on the other hand, heat conduction in the ion gas is relatively unimportant, 7.4 becomes

$$\frac{1}{T_i}\frac{dT_i}{dr}=(\gamma-1)\left[\frac{1}{n}\frac{dn}{dr}-\frac{\nu_{ie}}{u}\left(\frac{T_e}{T_i}-1\right)\right] \quad 7.6$$

The further neglect of electron-ion energy transfer leads to an adiabatic law for the ion temperature:

$$T_i = T_{i0}(n/n_0)^{\gamma-1} \quad 7.7$$

To determine the relative importance of some of the physical processes in the solar wind, the solutions of Hartle & Sturrock (1968) are compared with solutions obtained using 7.5 in place of the electron energy equation together with 7.6 and 7.7, alternately, in place of the ion energy equation (see Figure 8). Near the Sun ($\sim$3–15 solar radii), the ion temperature is determined essentially by the adiabatic law 7.7. Beyond about 10 solar radii, the combination of 7.5 and 7.6 gives a quite good approximation to the results of Hartle & Sturrock. In other words, in the region of supersonic solar-wind flow between the Sun and Earth, heat conduction determines the temperature of the electron gas (except in the innermost part of the corona), while heat transfer from the electron gas raises the ion temperature slightly above that appropriate to an adiabatic expansion. Observational results (Hundhausen 1968) often show higher ion temperatures than those predicted by Hartle & Sturrock, so that one might expect more efficient electron-ion energy transfer than that deriving solely from Coulomb collisions. Beyond the orbit of the Earth the effect of the interplanetary magnetic field (ignored in this formulation) becomes increasingly important (see Appendix) and serves to reduce the electron heat conduction so that far from the Sun the electrons and ions both should expand essentially adiabatically.

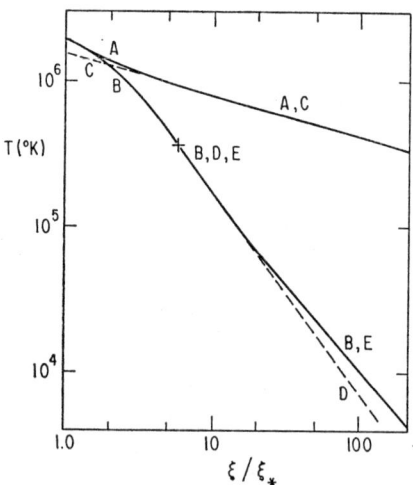

FIGURE 8. Temperature versus radial distance (in solar radii) for two-fluid models of the solar wind. Curves A and B are the results of Hartle & Sturrock (1968, Figure 2) for the electron and proton temperatures, respectively. Curve C is the electron temperature appropriate to thermal conduction 7.5. Curve D is the proton temperature appropriate to adiabatic expansion 7.7. Curve E is the proton temperature corresponding to 7.8; it coincides with B beyond $\xi/\xi_* = 7$. Curves D and E begin at the cross and extend outwards; no calculations were made for these curves in $\xi/\xi_* < 7$.

THEORY OF STELLAR WINDS AND RELATED FLOWS

In the above two-fluid formulation the electron and ion temperatures are assumed to be isotropic. However, in the vicinity of the Earth, the ion component of the solar wind exhibits a highly anisotropic temperature (Hundhausen 1968). To include the effect of ion temperature anisotropy in a fluid description, 7.2, 7.3, and 7.4 must be reformulated as follows (the continuity equation 7.1 remains unchanged):

$$\rho u \frac{du}{dr} + \frac{d}{dr}[nk(T_i^{\parallel} + T_e)] + \frac{2}{r} nk(T_i^{\parallel} - T_i^{\perp}) + \rho \frac{G\mathfrak{M}}{r^2}$$
$$= -qu + \frac{4}{3}\left[\eta\left(\frac{d^2u}{dr^2} + \frac{2}{r}\frac{du}{dr} - \frac{2u}{r^2}\right) + \frac{d\eta}{dr}\left(\frac{du}{dr} - \frac{u}{r}\right)\right] \qquad 7.8$$

$$\frac{3}{2} nuk \frac{dT_e}{dr} = ukT_e \frac{dn}{dr} + \frac{1}{r^2}\frac{d}{dr}\left(r^2 \kappa_e \frac{dT_e}{dr}\right) - \frac{1}{2}\nu_{ie}^{\parallel}(T_e - T_i^{\parallel})$$
$$- \nu_{ie}^{\perp}(T_e - T_i^{\perp}) + \frac{4}{3}\eta_e\left(\frac{du}{dr} - \frac{u}{r}\right)^2 + Q_e$$
$$+ \frac{q_e}{m_e}\left(\frac{1}{2} m_e u^2 - \frac{3}{2} kT_e\right) \qquad 7.9$$

$$\frac{1}{2} nuk \frac{dT_i^{\parallel}}{dr} = -nkT_i^{\parallel} \frac{du}{dr} + \frac{1}{r^2}\frac{d}{dr}\left(r^2 \kappa_i^{\parallel} \frac{dT_i^{\parallel}}{dr}\right) + \frac{1}{2}\nu_{ie}^{\parallel} nk(T_e - T_i^{\parallel})$$
$$- nk\nu_{ii}^{\parallel,\perp}(T_i^{\parallel} - T_i^{\perp}) - \frac{2}{3}\left\{u \frac{d\eta_i}{dr}\left(\frac{du}{dr} - \frac{u}{r}\right)\right.$$
$$+ \eta_i\left[u\frac{d^2u}{dr^2} + 3\left(\frac{du}{dr}\right)^2 - 2\frac{u}{r}\frac{du}{dr} - \frac{u^2}{r^2}\right]\right\} + Q_i^{\parallel}$$
$$+ \frac{q_i}{m_i}\left(\frac{1}{2} m_i u^2 - \frac{1}{2} kT_i^{\parallel}\right) \qquad 7.10$$

$$nuk \frac{dT_i^{\perp}}{dr} = -2nkT_i^{\perp}\frac{u}{r} + \frac{1}{r^2}\frac{d}{dr}\left(r^2 \kappa_i^{\perp} \frac{dT_i^{\perp}}{dr}\right) + \nu_{ie}^{\perp} nk(T_e - T_i^{\perp})$$
$$+ nk\nu_{ii}^{\parallel,\perp}(T_i^{\parallel} - T_i^{\perp}) + \frac{2}{3}\left\{u \frac{d\eta_i}{dr}\left(\frac{du}{dr} - \frac{u}{r}\right)\right.$$
$$+ \eta_i\left[u\frac{d^2u}{dr^2} + \left(\frac{du}{dr}\right)^2 + 2\frac{u}{r}\frac{du}{dr} - 3\frac{u^2}{r^2}\right]\right\}$$
$$+ 2Q_i^{\perp} - \frac{q_i}{m_i} kT_i^{\perp} \qquad 7.11$$

where $\gamma = 5/3$ and the basic notation is consistent with 7.1–7.4. The superscripts $\parallel$ and $\perp$ refer to the translational degrees of freedom parallel and perpendicular to the radius vector. $Q_i{}^\parallel$, $Q_i{}^\perp, \kappa_i{}^\parallel$, $\kappa_i{}^\perp$, $\nu_{ie}{}^\parallel$, and $\nu_{ie}{}^\perp$ are defined such that if $T_i{}^\parallel = T_i{}^\perp = T_i$, then $Q_i{}^\parallel = Q_i{}^\perp = \frac{1}{3}Q_i$, $\kappa_i{}^\parallel = \frac{1}{2}\kappa_i{}^\perp = \frac{1}{3}\kappa_i$, and $\nu_{ie}{}^\parallel = \nu_{ie}{}^\perp = \nu_{ie}$. Equations 7.8–7.11 are suitable for radial flow, with or without a radial magnetic field. In the presence of a nonradial magnetic field, the temperature anisotropy will be aligned with the field, and the viscosity and thermal conductivity will be modified, so that 7.8–7.11 must be reformulated. Neither the set 7.8–7.11 nor the equivalent equations for a nonradial magnetic field have as yet been applied to stellar-wind problems. Nishida (1969) has used similar equations in an attempt to find values for an effective ν_{ie} and $\nu_{ii}{}^{\parallel,\perp}$ with a point calculation at 1 au. However, because of inherent inaccuracies in the calculation and somewhat unrealistic assumptions, his results are not completely convincing.

While a complete two-fluid formulation can provide a rather detailed description of the flow properties of stellar winds, it leaves untouched the problem of wind composition when more than one type of ion is present. Extension of the full fluid equations to take account of two or more ion species is completely analogous to the generalization of the one-fluid model which led to Equations 7.1–7.4 and 7.8–7.11. It simply involves writing down continuity, momentum, and one or two energy equations for each plasma component. If the additional ions are minor constituents and do not affect the gross flow properties of the wind, then the one- or two-fluid problem can be solved first, neglecting the presence of the minor ions, and subsequently the flow of the additional ionic constituent can be described by solving separately the one-fluid equations for each minor ion species. However, if two or more types of ions do significantly affect the overall stellar-wind flow, then multifluid equations must be solved simultaneously.

Yeh (1969) has tackled the latter problem for an electron-proton-alpha particle plasma. He has used continuity and momentum equations for each constituent, neglecting viscous and frictional effects and assuming unequal but constant temperatures for the three plasma components. The results of Yeh's analysis indicate that his equations possess a critical solution that gives rise to supersonic flow for each ionic constituent. For this solution, the alpha particle-proton number density ratio decreases with increasing radial distance.

The radial expansion of an electron-proton plasma with additional minor ionic constituents has been discussed by Geiss et al (1969) and Alloucherie (1969). In this case the equation of motion for the jth ion species is

$$\frac{1}{2M_j{}^2}\frac{dM_j{}^2}{dr}(M_j{}^2 - 1) = \frac{2}{r} - \frac{G\mathfrak{M}}{c_j{}^2 r^2} - \frac{\nu_{jp}}{c_j{}^2}(M_j c_j - u) + \frac{Z_j e}{m_j c_j{}^2}\mathcal{E} - \frac{M_j{}^2 + 1}{2T_j}\frac{dT_j}{dr} \qquad 7.12$$

where $c_j = kT_j/m_j$, $M_j = u_j/c_j$, ν_{jp} represents the ion-proton momentum transfer collision frequency, Z_j is the degree of ionization, and ε is the electric field. The only new physical effect included in this formulation is dynamical friction, represented by the third term on the right-hand side of 7.12. The electric field is implicit in the two-fluid equations. Geiss et al have solved the one-fluid electron-proton problem with a polytrope law ($T_e = T_p \propto n^{\alpha-1}$), while Alloucherie has used Parker's (1963) isothermal solution and Sturrock & Hartle's (1966) solution to determine u. Both assume $T_j = T_p$ and $e\varepsilon = -(1/n_e)[d(n_e k T_e)/dr]$ in solving the ion equation of motion. The results indicate that alpha particles and many heavier ions are carried along with the solar wind with their relative velocity and density depending on the charge and mass of the ion. Again the alpha particle-proton number density ratio is found to decrease with increasing radial distance.

In earlier multifluid models (Jokipii 1964; Delache 1965, 1967; Nakada 1969) diffusion equations in the low-velocity limit (i.e. neglecting the inertia terms in the equation of motion) were used. These treatments deal primarily with the lower solar corona where large temperature gradients exist, and they are not applicable to high-flow velocity regions in the outer corona and solar wind. Results of these diffusion models indicate large enhancements of heavy elements in the lower solar corona.

Thus the principal effects predicted by the multifluid treatments, both in the lower corona and in the solar wind, are that from the top of the mixing region in the lower corona out to the coronal temperature maximum the ratios of heavy ion densities to the proton density increase, and beyond the temperature maximum these ratios decrease. In the event that the region of mixing extends out to the coronal temperature maximum, the solar-wind density ratios should be everywhere lower than those observed in the photosphere. These predictions are clearly stated by Jokipii (1964, p. 217, Figure 2).

8. THE POLAR WIND

In the high-latitude terrestrial ionosphere there occurs a multicomponent plasma flow that is considerably more complex than the multicomponent solar wind described in the last section. This flow, called the polar wind (Axford 1968b), involves particle sources and sinks as well as frictional interactions between the various ion components and between ions and a stationary neutral background (Banks & Holzer 1968). The equation of motion appropriate to the jth ion species in the polar wind is

$$\frac{1}{2M_j^2}\frac{dM_j^2}{ds}(M_j^2 - 1) = \frac{1}{A}\frac{dA}{ds} - \frac{G\mathfrak{M}}{c_j^2 r^2}\hat{r}\cdot\hat{s} - \sum_k \frac{\nu_{jk}}{c_j^2}(M_j c_j - u_k)$$

$$- \frac{T_e}{T_j}\frac{1}{n_e}\frac{dn_e}{ds} - \frac{(M_j^2 + 1)}{2T_j}\frac{dT_j}{ds} - \frac{1}{T_j}\frac{dT_e}{ds}$$

$$- \frac{A}{Q_j}\left[q_j\left(1 + M_j^2 - M_j\frac{u_n}{c_j}\right) - \ell_j\right] \quad 8.1$$

where $M_j = u_j/c_j$, $c_j^2 = kT_j/m_j$, $Q_j = n_{j0}u_{j0}A_0 + \int_{s_0}^{s} ds'(q_j - \ell_j)A$, q_j and ℓ_j are the rates of production and loss of j-type ions per unit volume per unit time, A is the cross-sectional area of a magnetic flux tube, $\hat{s}$ and $\hat{r}$ are unit vectors along the magnetic field and in the radial direction, respectively, s is the distance along a magnetic field line, u_n is the bulk flow speed of the neutral background gas, and $n_e = \Sigma_j n_j$. Comparison of 7.12 with 8.1 indicates that a source-sink term and a more complicated friction term are included in the latter equation.

We will consider here a particularly simple polar-wind model examined by Banks & Holzer (1968) to demonstrate some basic physical principles involved in the problem. Discussion of more detailed models can be found in the literature (Banks & Holzer 1969a,b). Let us assume that the topside polar ionosphere is composed of a heavy (O^+) and a light (H^+) ion component in the presence of their neutral parents (O and H), that the two ion species and the electrons have a common constant temperature, and that the neutral atmosphere is in hydrostatic equilibrium. Then 8.1 is simplified so that the right-hand side contains terms arising from expansion, gravity, friction, the polarization electric field (i.e. dn_e/ds), and sources and sinks. For a particular ion temperature and model neutral atmosphere the equations of motion 8.1 for H^+ and O^+ can be solved, yielding velocity and density profiles like those shown in Figures 9a and 9b. To understand these somewhat complex profiles we must examine the individual terms in 8.1.

The O^+ ions are produced through photoionization and are lost through O^+—H charge exchange, while the H^+ ions are produced through O^+—H charge exchange and photoionization and are lost through H^+—O charge exchange. As the ions expand outward there are frictional interactions between H^+ and O^+ (most important for H^+) and between each of the ions and the background neutral atmosphere (most important for O^+). The polarization electric field is determined by the scale height of the electron gas (i.e. the scale height of the dominant ionic constituent).

Below 1000 km the electric field is determined by the O^+ scale height (see Figure 9b) and is more than 16 times as great as a corresponding diffusive equilibrium H^+—e electric field. Consequently the H^+ ions are accelerated rapidly upward through the O^+ ions, while H^+—O^+ friction tends to retard the flow, thus inducing an H^+ pressure gradient that further enhances the acceleration. The frictional force decreases more rapidly with altitude than does the electric field, and eventually (at 2000 km in Figure 9a) the H^+ flow becomes supersonic. As in the case of the solar wind the flow of H^+ in the polar wind is analogous to the flow of gas in a convergent-divergent nozzle, with the frictional drag providing the effect of convergence (gravity plays a similar role in the case of the solar wind).

Because of the rapid acceleration of H^+, the accidentally resonant charge exchange reaction $H^+ + O \rightleftharpoons H + O^+$ produces a broad sink for O^+ ions in the vicinity of 1000 km. The presence of that sink induces a large upward flow of O^+ below 1000 km (see Figure 9a) which results in a smaller O^+ scale

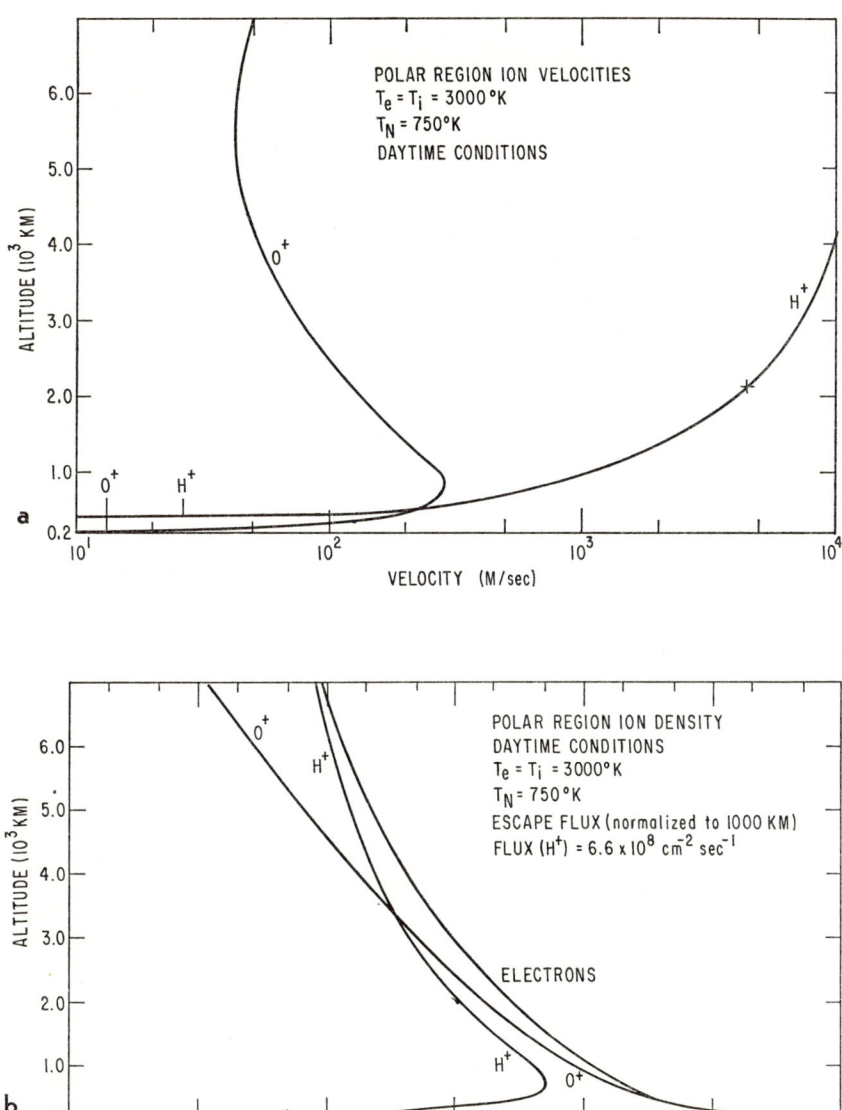

FIGURE 9. (a) Velocity (m/sec) versus altitude (km) and (b) density (cm^{-3}) versus altitude for a polar-wind model (including H$^+$, O$^+$, H, O) with parameters appropriate to daytime polar region conditions. The cross in (a) marks the location of the H$^+$ critical point. (From Banks & Holzer 1969b.)

height, a larger O^+—e electric field, and a larger H^+ acceleration. Since the O^+ flow is limited by frictional interaction with the neutral atmosphere, a dynamic equilibrium of the O^+—H^+ system is achieved.

A more realistic model of the polar wind requires the detailed consideration of energy production in the ion and electron gases, energy transfer between ions and electrons, electron thermal conduction, and the effects of photoelectrons, as well as the inclusion of minor ions in the flow. The latter is a relatively simple problem (Banks & Holzer 1969b), but treatment of the energy balance equations is quite difficult and has not yet been attempted.

APPENDIX

It has been pointed out in Section 7 that the introduction of a nonradial magnetic field in the model of Sturrock & Hartle (1966; Hartle & Sturrock 1968) can significantly affect the electron temperature at large distances from the Sun. To demonstrate this we can rewrite 7.3, allowing for the presence of a magnetic field.

$$\frac{3}{2} nuk \frac{dT_e}{dr} = ukT_e \frac{dn}{dr} + \frac{1}{r^2} \frac{d}{dr}\left(r^2 \kappa_e \cos^2\theta \frac{dT_e}{dr}\right) \qquad \text{A.1}$$

where viscosity, mass and heat addition, and electron-ion energy transfer have been ignored, and θ is the (acute) angle between the magnetic field and the radius vector. For the solar wind

$$\cos^2\theta \approx \left(1 + \frac{\Omega^2 r^2}{u^2}\right)^{-1} \qquad \text{A.2}$$

where Ω is the angular frequency of solar rotation.

Figure 10 shows four representative solutions (A, B, C, and D) of Equations A.1 and A.2, and one solution (H–S) of A.1 with $\cos^2\theta = 1$ (corresponding to the Hartle-Sturrock solution shown as curve A in Figure 8). The solutions A, B, C, and D were obtained by assuming a value for T_e at some point in $r \gg 1$ au where the electron temperature should be determined by the adiabatic law $T_e \propto r^{-4/3}$, and integrating A.1 back to $r = 1$ au.

While (H–S) is determined completely by heat conduction in $1 < r < 100$ au (that is, curves A and B in Figure 8 coincide in this region), the four curves appropriate to the case of a nonradial magnetic field all contain a region in which heat conduction is important followed by a region of adiabatic cooling with negligible heat conduction. Curve D, with $T_e = 1.2 \times 10^{5}$°K at the orbit of the Earth (cf Hundhausen 1968) is purely adiabatic at 5 au and yields an electron temperature at this distance from the Sun which is one-half that expected from heat conduction alone if $T_e = 1.2 \times 10^{5}$°K at 1 au.

It is evident from these results that a spiral magnetic field of the form A.2 affects the electron temperature in the solar wind in two ways. First,

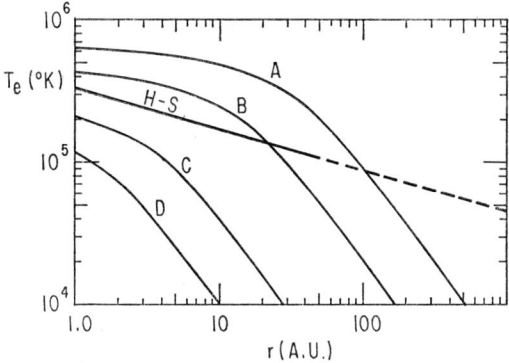

FIGURE 10. Electron temperature versus radial distance (in au) for the solar wind with no magnetic field (H–S) (from Hartle & Sturrock 1968, Figure 2) and with a spiral magnetic field (A, B, C, D) given by (A.2). All curves correspond to a flux of 3.75×10^8 cm^{-2} sec^{-1} at 1 au. For curves A, B, C, and D the flow velocity is constant ($u = 280$ km sec^{-1}), and for (H–S) the velocity increases slowly from $u = 250$ km sec^{-1} at 1 au. The dashed line is an extension of (H–S) beyond the region considered by Hartle & Sturrock using the approximate result shown as curve B in Figure 8.

it reduces the heat flux due to conduction in the electron gas, and eventually it leads to an adiabatic decrease of the electron temperature. Second, in regions dominated by heat conduction it leads to a higher electron temperature than would be expected in the absence of a magnetic field.

LITERATURE CITED

Alfvén, H. 1947, *MNRAS*, **107,** 211
Alloucherie, Y. 1969 (Submitted to *J. Geophys. Res.*)
Axford, W. I. 1961, *Phil. Trans. Roy. Soc. London A,* **253,** 301
Axford, W. I. 1968a, *Space Sci. Rev.,* **8,** 331
Axford, W. I. 1968b, *J. Geophys. Res.,* **73,** 6855
Axford, W. I., Newman, R. C. 1966, *Cornell-Sydney Univ. Astron. Cent. Rep. 34*
Axford, W. I., Newman, R. C. 1967, *Ap. J.,* **147,** 230
Banks, P. M., Holzer, T. E. 1968, *J. Geophys. Res.,* **73,** 6846; 1969a, ibid, **74,** 6304; 1969b, ibid, **74,** 6317
Barnes, A. 1968, *Ap. J.,* **154,** 751
Biermann, L. 1946, *Naturwissenschaften,* **33,** 118
Biermann, L. 1948, *Z. Ap.,* **25,** 161
Bondi, H. 1952, *MNRAS,* **112,** 195
Burke, J. A. 1968, *MNRAS,* **140,** 241
Chamberlain, J. W. 1961, *Ap. J.,* **133,** 675; 1965, ibid, **141,** 320
Clauser, T. 1960, *4th Symp. Cosmical Gas Dyn., Varenna, Italy* (August)
Dahlberg, E. 1964, *Ap. J.,* **140,** 268
Delache, P. 1965, *C. R. Acad. Sci.,* **261,** 643
Delache, P. 1967, *Ann. Ap.,* **30,** 827
Dessler, A. J. 1967, *Rev. Geophys.,* **5,** 1
Geiss, J., Hirt, P., Leutwyler, H. 1969 (Preprint, Univ. Berne)
Hartle, R. E., Sturrock, P. A. 1968, *Ap. J.,* **151,** 1155
Holzer, T. E. 1968 (Unpublished)
Hundhausen, A. J. 1968, *Space Sci. Rev.,* **8,** 690
Johnson, H. E. 1969 (Unpublished)
Jokipii, J. R. 1964, *The Solar Wind,* 215 (New York: Pergamon)
Konyukov, M. V. 1967, *Geomagn. Aeron.,* **7,** 469; 1969, ibid, **9,** 1
Lighthill, M. J. 1967, *Aerodyn. Phenomena Stellar Atmos., IAU Symp. 28,* 429 (Thomas, R. N., Ed.)
McCrea, H. C. 1956, *Ap. J.,* **124,** 437
Mestel, L. 1954, *MNRAS,* **114,** 437
Meyer, F., Schmidt, H. U. 1966, *Mitt. Astron. Ges.,* **21,** 96; 1968, ibid, **25,** 228
Nakada, M. P. 1969, *Solar Phys.,* **7,** 302
Newman, R. C., Axford, W. I. 1968, *Ap. J.,* **151,** 1145
Newman, R. C. 1967 (PhD thesis, Cornell Univ.)
Nishida, A. 1969, *J. Geophys. Res.,* **74,** 5155
Noble, L. M., Scarf, F. L. 1963, *Ap. J.,* **138,** 1169
Osterbrock, D. E. 1961, *Ap. J.,* **134,** 347
Parker, E. N. 1958, *Ap. J.,* **128,** 664
Parker, E. N. 1963, *Interplanetary Dynamical Processes* (New York: Interscience)
Parker, E. N. 1964, *Ap. J.,* **139,** 93
Parker, E. N. 1965, *Space Sci. Rev.,* **4,** 666; 1969, ibid, **9,** 325
Scarf, F. L., Noble, L. M. 1965, *Ap. J.,* **141,** 1479
Spitzer, L. 1942, *Ap. J.,* **95,** 329
Sturrock, P. A., Hartle, R. E. 1966, *Phys. Rev. Letters,* **16,** 628
Whang, Y. C., Chang, C. C. 1965, *J. Geophys. Res.,* **70,** 4175
Whang, Y. C., Liu, C. K., Chang, C. C. 1966, *Ap. J.,* **145,** 255
Whitaker, W. A. 1963, *Ap. J.,* **137,** 914
Yeh, T. 1969 (Submitted to *Planet. Space Sci.*)

AURORAL PHYSICS

Sydney Chapman

High Altitude Observatory, Boulder, Colorado

Records of the aurora go far back, as do the Chinese records of sunspots; for centuries the records of both were haphazard and scanty. One of the first to record the aurora systematically was the atomic chemist John Dalton (1793, 1828), who kept a "meteorological" journal for 57 years. Its first mention of the aurora was on March 24, 1787. This was well before Schwabe took up the regular observation of sunspots, in 1826. Since then sunspots have been increasingly well observed, but the auroral record has until recently remained mostly at the mercy of chance observation, and still more uncertain written record.

Scattered references to the aurora, generally fanciful or superstitious, and often obscure, are found in classical Greek and Latin literature, because the aurora is sometimes, at long intervals, visible from the Mediterranean. Likewise some medieval authors of northern Europe and Iceland mention the aurora, which there is much more often visible. The enlightened French scientist Gassendi (1658) seems to have been one of the first to regard the aurora as a natural phenomenon; he saw a great aurora from southern France on September 12, 1621, and gave it the name *aurora borealis*, or northern dawn. Another term commonly used for the aurora borealis is the northern lights.

Perhaps the first mention of the aurora in the periodical literature of science was by Halley (1716), who gave a description, in the *Philosophical Transactions* of the Royal Society, at its request, of the great aurora of March 16, 1716. This included a corona (Figure 1), lying somewhat south of the zenith at London. Halley, an ardent observer of Nature, then aged sixty, remarked that up to that time he had seen every kind of "meteor" *except* an aurora.

The first treatise on the aurora, by de Mairan (1733), was published by the French Academy of Sciences; it ran to a second much extended edition in 1754. It is an excellent book, still worth reading. Mairan thought that in the Southern Hemisphere lights similar to the northern lights might appear. He knew that the aurora is seen more often in higher latitudes than in France, and consequently he thought the best people to ask about the possibility would be men who had been to high southern latitudes. He consulted Frezier and Ulloa, who had rounded Cape Horn on scientific expeditions, whether they had seen such lights. Their affirmative replies do not carry conviction. The first records of indubitable southern auroras were those made in February and March 1773 by Captain Cook (1961) and others on

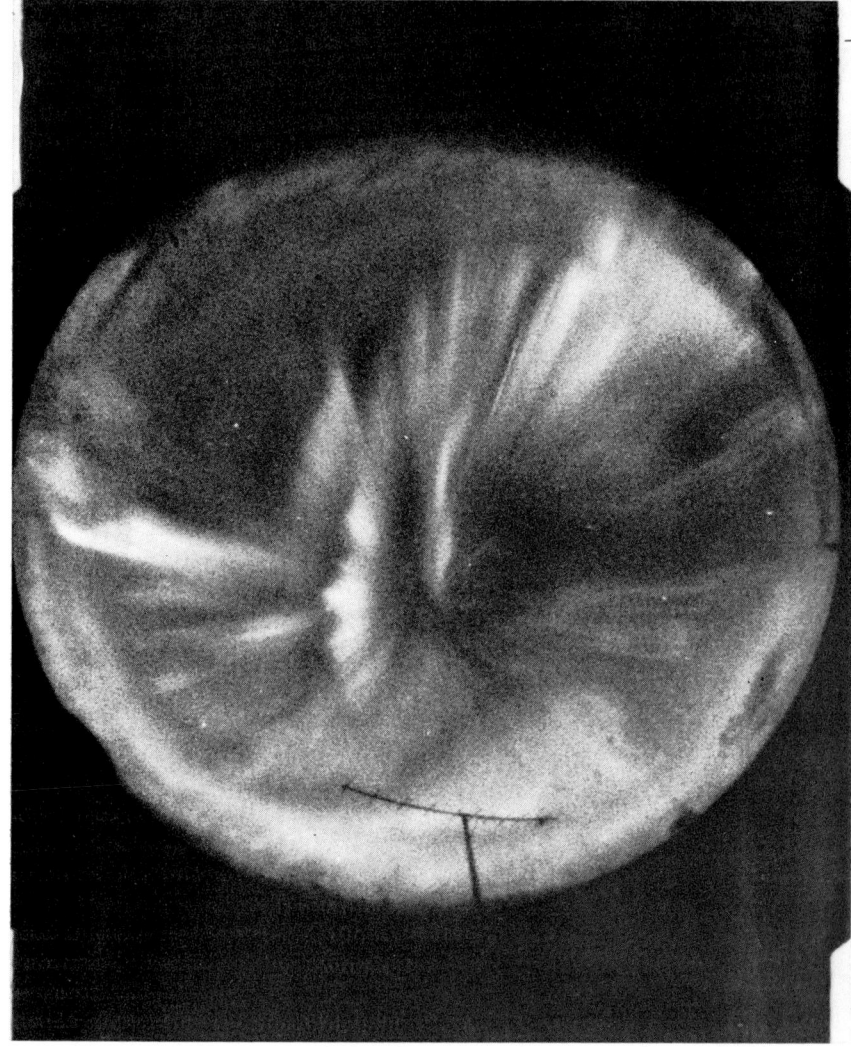

FIGURE 1. An exceptionally grand auroral corona photographed with an all-sky camera at the Geophysical Institute, College, Alaska, at 22:10:40 UT on October 10, 1969. The directions N, S, E, W, are respectively above, below, to left, to right.

the two ships (then separated after fog) of his second great voyage, during their eastward journey across the southern Indian Ocean. Cook's latitude and longitude at the time of his first sight of the southern aurora were 57.9°S and 83°E. He named it the *aurora australis* (this term has at times been inappropriately used in referring to an aurora borealis seen to the south of the observer's zenith). Such southern auroras are also called southern lights. The names aurora polaris and polar lights cover both northern and southern auroras.

Several nineteenth century writers collected the records of auroras. Up to 1872 the most extensive collection was that made by Fritz (1873). A German geographer Muncke (1833) who studied the auroral records of several arctic explorers found that although auroras are seen more often as one goes northward from mid-Europe, there is a latitude beyond which they are seen less often (and more commonly to the south than to the north). Loomis (1860) drew the first sketch of the *auroral zone* of maximum frequency of visibility of the aurora. Fritz (1874) much improved on this by his map of equilines of frequency of auroral visibility, which he called isochasms. Figure 2a is a pictorial version of his map, drawn for American readers (Gartlein 1947). The isochasms are oval curves, but are not centered on the geographical pole. Instead, as came later to be recognized, they are centered on the pole of the Earth's magnetic dipole axis, situated in north-

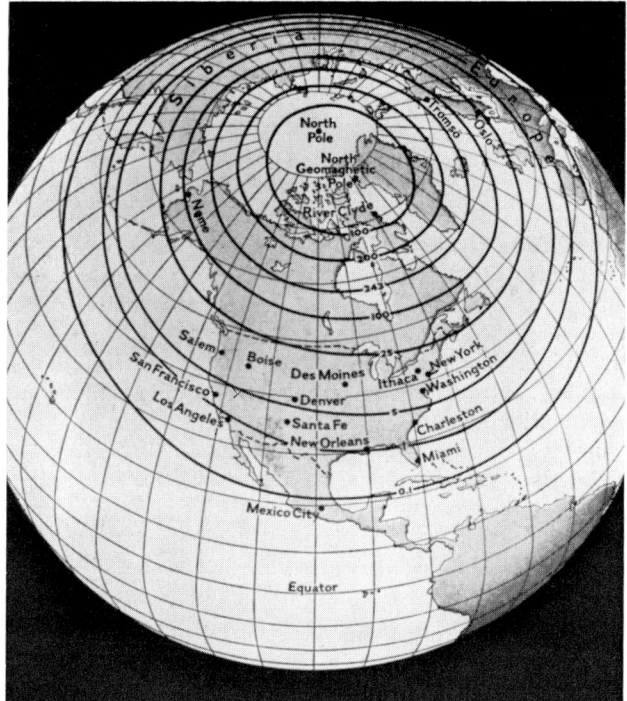

FIGURE 2a. Lines of equal frequency, averaged over many years, of auroral visibility (isochasms), drawn by Fritz (1873), here displayed on a map of the North American and polar regions. (Gartlein 1947, amended)

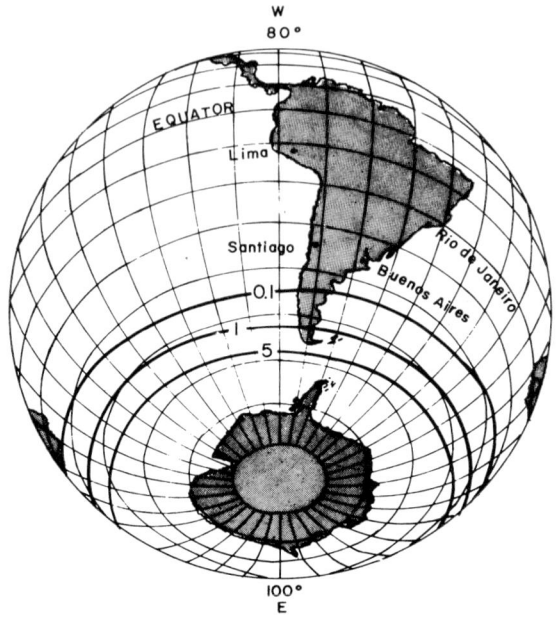

FIGURE 2b. Three of the outermost isochasms for the South American and polar regions; they are approximate, being antipodal to the corresponding northern isochasms.

west Greenland. The dipole axis is inclined by 11° to the geographical axis; its northern end is tilted towards America, giving the inhabitants of this region many more opportunities of seeing auroras than are available to Siberian viewers in the same latitude.

Conversely in south America, even at its southern extremity near Cape Horn (Figure 2b), auroras are much more rare than in similar latitudes near the antipodal meridian, somewhat west of Australia. Halley, on his pioneer magnetic survey voyage to the south Atlantic, 1698 to 1700, reached 52°S latitude without seeing any aurora; it could not be expected that he would have seen any, at that latitude and longitude, especially as the epoch was near sunspot minimum. Cape Horn is only about 4° further south.

Each of Fritz's isochasms bears a number, signifying how often per year, on the average of many years, the aurora might be seen if all nights were cloudless. The maximum isochasm passes over northern Canada, across Alaska, to the north of Siberia, and to the south of Iceland. A band a few degrees wide, centered on this isochasm, is called the auroral zone. The frequency of visibility falls off rapidly to the south of the zone, and less rapidly to the north. Vestine (1944) corrected Fritz's isochasms near the zone, but the outer isochasms have never been revised, though they are

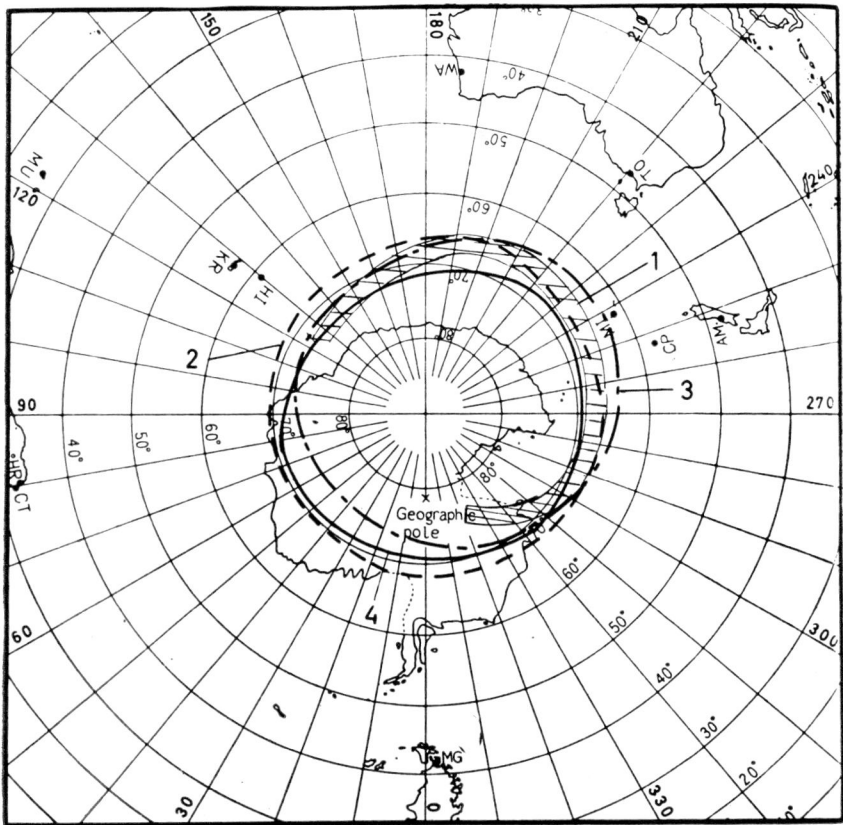

Southern auroral zones determined from observations (on a map with geomagnetic co-ordinate system). Zone 1 is according to White and Geddes[7], zone 2 to Vestine and Snyder[8], zone 3 to Bond and Jacka[15], and zone 4 to Feldstein[4].

FIGURE 3. Three maximum southern isochasms (on a map of geomagnetic dipole coordinates) inferred from observations, Nos. 2, 3, 4 being given respectively by Vestine & Snyder (1945), Bond & Jacka (1960), and Feldstein (1960); the shaded band, No. 1, shows a still earlier sketch of the southern auroral zone, drawn by White & Geddes (1939). After Hultquist (1961).

based on rather scanty data, accumulated unsystematically over the centuries.

The land distribution in the Southern Hemisphere is less favorable for the auroral record than that in the north. Scientific records of auroras, during the precolonial period, were not to be expected from the inhabitants of the southernmost parts of America, Australia, and New Zealand, though the mention of auroras can be recognized in Maori folklore. The scanty records of the aurora australis have been used to draw tentative southern isochasms, but none has yet been drawn for low southern latitudes, where too few auroras have been observed since the record there began. Four tentative sets of southern isochasms have so far been drawn. They are in fair agreement as regards shape and location, but Hultquist (1961) has shown that in certain longitudes the latitudes of their maximum isochasm may differ by up to 4° (Figure 3).

Auroras are high enough in the atmosphere to be visible at any time over a broad band of latitude, to north and south of where they are seen overhead. Equilines of frequency of *overhead* auroras are called isoaurores (Chapman 1953). Northern isoaurores were first drawn by Feldstein & Solomatina (1961). They coincide approximately with the inner isochasms, but the frequencies associated with those south of the auroral zone are less than for the isochasms. The isoaurores do not extend so far towards the equator as do the isochasms, because the auroras seen from low latitudes usually lie considerably further towards the poles, and they are very high in the atmosphere.

Auroral Relations with Geomagnetism and Sunspots

The slow secular variation of the Earth's magnetic field was discovered in 1635 by Gellibrand; it proceeds from deep within the Earth. *Transient* changes of the geomagnetic field were discovered by Graham (1724), who found that on some days the changes are quiet and regular, and on other days they are irregular and often much larger. By correspondence between Graham in London, and Celsius (1740) and Hiorter (1747) in Uppsala, it became clear that days disturbed magnetically at London were likewise disturbed at Uppsala: and moreover that the aurora is associated with such magnetic disturbance. This was the first connection established between the aurora and the Earth's magnetism. Halley, in his 1716 paper, had suggested a connection of quite another kind. He recognized that the convergence of the rays of the auroral corona (Figure 1) is only apparent, being an effect of perspective. He even suggested that they probably converge towards the observer, that is, downward, like the field lines of a magnetized sphere; he gave a diagram of these, obtained from the pattern of iron filings in the meridian plane of such a sphere. He thought that the influence of a magnet, that can act through interposed bodies like marble, is exerted by "subtle" (or penetrating) particles continually circulating through the magnet, along the field lines, in both directions. He tentatively suggested that at times certain polar regions of the atmosphere, for some reason on which he did not speculate, are sensitive to the influence of such particles, causing the aurora. His intuition that the auroral rays lie along the field lines was explicitly supported later by Wilcke (1777) at Stockholm, and confirmed recently much more accurately by Maggs & Davis (1968; vide infra.)

Halley's idea also might suggest the inference (not drawn at the time) of *conjugacy*, namely that when there is an aurora in the north there is one also in the south, at the other ends of the same field lines. Mairan's expectation that there would be auroras in the Southern Hemisphere was akin to this, but less specific. Such conjugacy has now been confirmed in large measure; it has also been observed in connection with artificial "auroras" caused by high-altitude detonation of nuclear bombs (Keys 1964). A proper test involves a knowledge of the path of the field lines above the Earth from one to the other end; the lines differ somewhat from those of a uniformly magnetized sphere or magnetic dipole, which gives only a first approximation to the field of the Earth. Only near the equinoxes are both ends of the field lines from auroral regions in darkness. Considerable technical resources have been devoted to the study of conjugacy, such as all-sky cameras and image orthicon television photography aboard two jet aircraft flying along magnetically conjugate paths over Alaska and south of New Zealand (Belon, Maggs, Davis, Mather, Glass & Hughes 1969). Hargreaves (1969) has reviewed earlier conjugacy studies of this and other kinds, which reveal some departures from conjugacy, partly related to geomagnetic disturbance. Though these deserve further attention, the substantial degree of conjugacy that prevails lessens the disadvantage of the difficulty of registering southern auroras. If the northern ones could be well observed, we should have considerable knowledge thereby of the southern ones.

Auroral observation, however, is more difficult to organize than sunspot recording; the sunspots can be seen from anywhere on the Earth during the day, when cloud is absent. The aurora is visible only over parts of the Earth, those in general the least accessible. During the International Geophysical Year (IGY: July 1, 1957 to December 31, 1958, followed by IGC, the year 1959 of International Geophysical Collaboration), auroral observation was greatly enhanced above all prior levels. The IGY auroral program and stations are described in volumes 4, 5, and 8 of the *IGY Annals* (1957/8); many results of the visual, all-sky, photometric and spectroscopic observations and their analysis are described in volumes 20 (Stoffregen 1962, ascaplots): 29 (McInnes 1962, visoplots); 38 (Akasofu & Kimball 1965, auroral morphology); 39 (Gartlein & Gartlein 1966, synoptic visual charts); 40 (Devlin et al 1966, spectrographic); and 45 (Akasofu, Kimball & Meng 1969, auroral substorms). Besides well-organized visual programs in many countries, there were about 120 all-sky cameras, divided between the Northern and Southern Hemispheres in the approximate ratio 3:1. Even so, never was so much as half the Arctic polar sky under observation at any one time; this was because of the land and sea distribution, and cloud.

Figure 4 (Akasofu 1963) shows the distribution of the clear fields of view, partly overlapping, of at least four auroral arcs, two of them of length 5000 km. Whether such arcs ever completely encircle the boreal pole is not known, and it seems almost impossible that it should be established by ground and airborne photography, valuable as the use of aircraft is in this connection (Akasofu 1968). Possibly polar-orbiting satellites might photograph the whole polar cap when this is in winter darkness, from a height enabling this great area to be covered in one picture; however, at auroral height the dark portion of the polar cap is decidedly smaller than at the ground. The world's cloud cover is continuously photographed by satellites; but the aurora changes its form and extent far more rapidly than do the clouds. Consequently reliance cannot well be placed, in seeking knowledge of the instantaneous auroral distribution and its changes, on successive pictures of the aurora from a satellite able to cover photographically only a fraction of the polar area. The project of the photography of the whole polar region from one satellite may be technically difficult, but would be free from one obstacle that hinders observation from the ground or from aircraft, the clouds. These are all well below the aurora, at least during the season of polar darkness.

The height of the aurora remained in doubt and dispute till the beginning of this century (cf Abbe 1898). Mairan much overestimated the height, others erred in the opposite direction. There were many reports of auroras being seen below the clouds, or silhouetted against mountains. Störmer (1955) substantially settled the question, by organizing, from 1910 onwards, simultaneous photography from well-separated Norwegian stations, connected by telephone, and using the same type of camera. He found the lower border to be generally located at about 100-km height, and that the auroral light could extend upwards to heights of from 150 to 1000 km. The greatest heights he found were for sunlit auroras, seen by observers on the ground in darkness, but emitting from parts of the atmosphere still in sunlight, after twilight or before dawn. He gave the height of maximum frequency of the measured points to be 100 to 110 km; some of the frequency curves show two maxima. He found very few points at a level below 90 km. At rare intervals the aurora may extend below the level (about 80 km) of noctilucent clouds (which are summer phenomena).

Very occasionally careful observers, using Störmer's technique, have reported auroras that reach down to decidedly lower levels. The first outstanding example was an intense arc with deep crimson lower border (Harang & Bauer 1932, Harang 1951) extending down to a little above 60-km height. For some years this remained a unique observation, but subsequently Canadian observers reported further cases; Currie (1955) found three that extended down to or below 70 km, and Hill (1965) found four cases (out of 1197 pairs of all-sky auroral photographs taken in the quiet sunspot years 1964/5) that extended below 70 km (one as low as 63 km). Unfortunately no spectra of these auroras were taken—it is very desirable to obtain spectra and photometric data for such exceptionally low auroras.

Points at 100-km height in the atmosphere can be seen from the ground (or photographed by all-sky cameras) within a circle of radius about 1100 km, but the outer part of an all-sky picture is compressed, and the effective radius of good record of an all-sky camera may be estimated as 500 km, as adopted in Figure 4. A satellite able to see the 100-km level of the atmosphere over a polar cap of angular radius 25° must be at least 770 km above the ground.

During the great geomagnetic disturbances, called magnetic storms, the aurora comes within sight of lower latitudes than usual, and it is on the rare occasions of the very greatest magnetic storms that it can be seen from places on the outermost of Fritz's isochasms, as was possible three times, from Cuba and Mexico, during the IGY. On February 4, 1872, it was seen even at Bombay; this is probably the greatest aurora on record (cf Chapman 1957). On such occasions many who see it may not recognize it as an aurora, and even when seen and recognized, too often little or no mention of it finds its way into scientific literature.

Whereas several institutions are equipped with cameras, photometers, and spectroscopes for the observation of auroras in the latitudes where they are commonly seen, there is very little provision for the observation by such means of the rare low-latitude auroras. Hence there is great need to arouse interest in such phenomena among astronomers and physicists who have applicable equipment. Where they are at all likely to have the good fortune to see low-latitude auroras, forethought should be taken to be ready if the opportunity comes. These events are great natural phenomena, with a strong astronomical connection with the Sun. Astronomical observatories, usually situated well away from city lights, are the institutions most likely to be able to make good observations of such auroras. They are also in many cases linked with warning services that send news of unexpected astronomical events, and hence can be alerted when great magnetic storms and worldwide auroras are in progress. It is desirable that the Astronomical Union or other astronomical agency should help to organize such observations. The Yerkes Observatory (vide infra) has set a fine example by the long-continued attention given by several members of its staff to the observation and recording of auroras—Barnard, Sullivan, Meinel, and Elvey (later Director of the Geophysical Institute of the University of Alaska, which under his leadership became an outstanding center for auroral research).

A low-latitude observatory may be able to contribute to auroral knowledge even after the visible aurora has faded away—by looking for the light of the NI excited atom (5200 Å). This light may continue to be emitted for an hour or more after the usual auroral emissions have ceased (Götz 1947).

The lowermost dipole latitude of great auroral displays shows a fair degree of correlation with the maximum value of the Dst (H) measure of magnetic disturbance (Akasofu & Chapman 1963); this measure corresponds to the mean strength of the encircling ring current that develops strongly around the Earth during a great magnetic storm. A low-latitude observatory that arranges to receive news of outstanding geomagnetic disturbance much increases its chance of making valuable auroral observations.

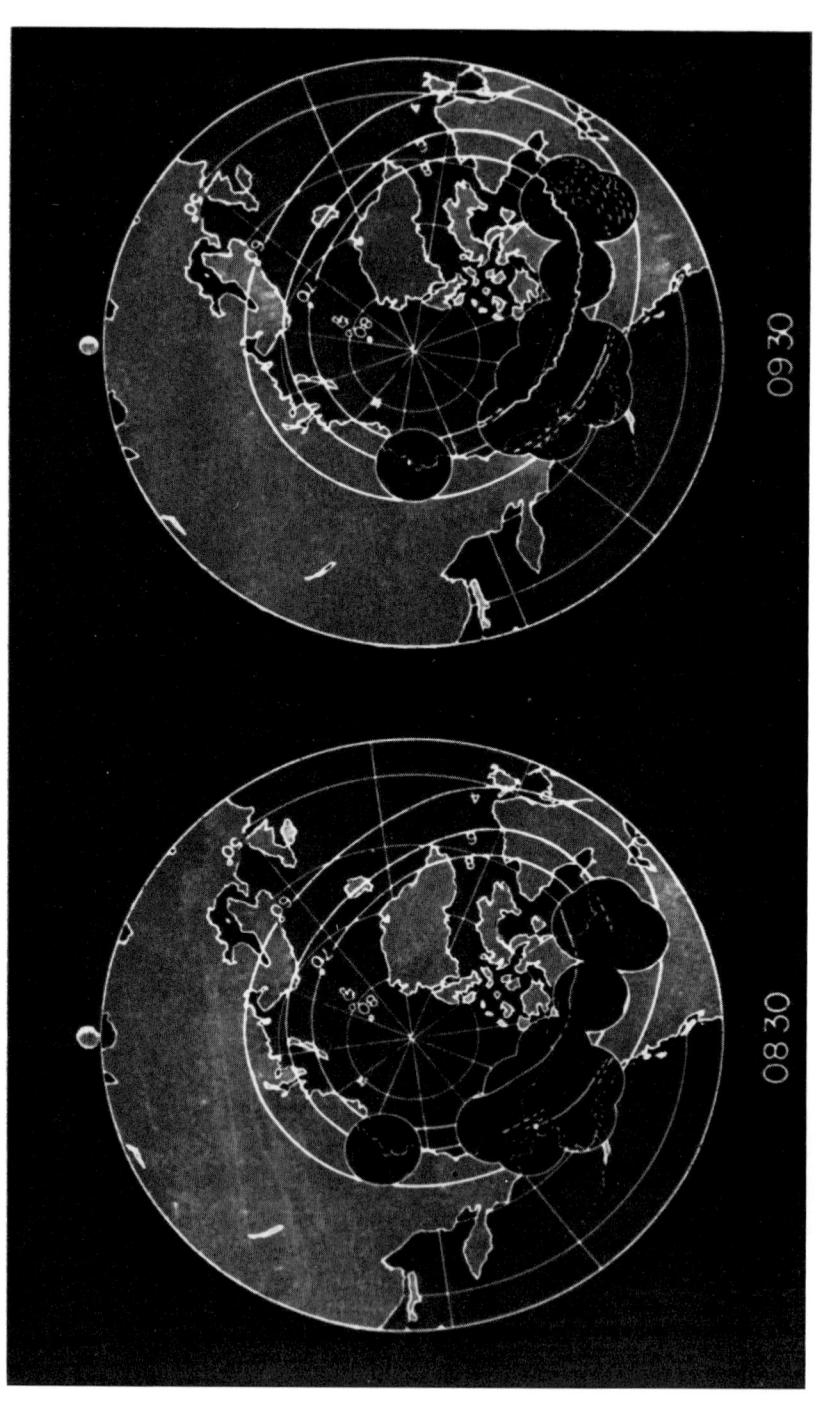

FIGURE 4. The distribution of auroral arcs at 8:30 and 9:30 UT on March 24, 1958, as shown by several all-sky cameras; the limits of the clear single or overlapping view of the cameras are shown, the radius of the area of good view being taken as 500 km. The auroral breakup of an auroral substorm occurred between the two epochs; in the second the arcs over Alaska are completely disrupted After Akasofu (1963).

When Schwabe's discovery of the 11-year sunspot cycle became widely known, in 1850, it was found within a year that the same cycle is manifested in the records of the transient magnetic variations, both on quiet and disturbed days. The relation between auroras and magnetic disturbance found in 1740 by Graham and Celsius implied that auroras should also show this influence; but for some decades this connection remained uncertain, owing to the poor quality of the auroral record. It is best manifested in the "subauroral" belts (between dipole latitudes 60° and 45°). It is clearly shown by the systematic records of the aurora made throughout three sunspot cycles at the Yerkes astronomical observatory (Meinel, Negaard & Chamberlain 1954). In auroral latitudes (above 60° dipole latitude) the aurora is so frequent that the sunspot cycle is less evident as regards frequency of visibility, though it manifests itself in other ways. In minauroral latitudes (equatorwards of 45°) the record is too scanty to give smooth frequency curves.

One of the simplest forms of auroral observation, and one of great value, in the case of the infrequent subauroral and the rare minauroral auroras, is a timed record of their appearance overhead, or of their coming within 30° of the zenith, to the north or south of it. Any such observation should be sent to the national astronomical society for its auroral section (if any), otherwise to a World Data Center.

The Auroral Spectrum

The auroral spectrum and its interpretation have been studied for a century, beginning with Ångström (1869); progress was small during its earlier half. The lines and bands, which may vary greatly in their relative strengths even during a single display, are now known and identified (Petrie & Small 1952) in great but certainly not complete detail, for the polar aurora. They are less known for the subauroral displays, and much less studied for the rarer low-altitude auroras. The spectrum was intensively observed during the IGY (Devlin et al 1966). The spectrum has been examined not only in the visible region but also, using image converters, in the infrared (Bagariazky & Fedorova 1956, Vaisberg 1959), and by rockets and satellites into the ultraviolet (Miller, Fastie & Isler 1968), which contributes a considerable share of the energy of emission. The subject is of great scope and can only be briefly summarized here. Good accounts of it have been given by Bates (1960) up to about 1959, by Chamberlain (1961) up to about 1960, and by Meinel (1966) up to about 1965; Vallance Jones (1969) has summarized more recent progress. The chief centers of auroral spectral (and other) studies include auroral or more general geophysical observatories or institutions in Alaska, Canada, Sweden, Norway, and the USSR. Important discussions of the aurora and its spectrum have been given by Krassovsky (1961, 1967, 1968) and Galperin (1963); these are references to only a few of their more recent reviews in easily accessible English translations, and enable many of their other works to be found.

One outstanding feature of the spectrum, long a mystery, is the yellow-green atomic line 5577 Å; it was finally shown by McLennan & Shrum (1924) to result from the forbidden atomic oxygen transition $^1S(0.74$ sec) to $^1D(110$ sec). These are two metastable states whose lives are indicated in brackets—the lower level is of long duration. This emission gives the most usual color of bright auroras in polar latitudes. A transition from the level ^1D to the ground-state ^3P gives the red doublet 6300–6364 Å, which is mainly emitted at heights 200 to 300 km, well above the level where most of the 5577 light is produced. Polar auroras in which this red light is prominent are said to be of type A. This light is a common feature of low-latitude auroras; it has often been misinterpreted as the sky glow of a great conflagration to the north, sometimes causing fire fighters to set out to seek and suppress it—as recorded of ancient Rome, and of Victorian London; one prompted a fire department enquiry at Göttingen as recently as 1957.

Other outstanding features of the auroral spectrum are the band systems of molecular nitrogen, both neutral (first and second positive, and Vegard-Kaplan bands) and ionized (first negative and Meinel bands). Though less conspicuous visually than the green and red light of oxygen, these emissions generally involve more energy. Some polar auroras show a purplish-red lower border (these are said to be of type B), due to an enhancement of the first positive bands. Vegard, an ardent pioneer in auroral spectral studies, was the first to attribute much of the auroral light to molecular nitrogen.

Other features of the spectrum, generally minor, are contributed by O_2 (the Kaplan-Meinel bands) and O_2^+ (first negative); there are also permitted lines of O, N, and N^+, and forbidden lines of O^+, N, and N^+. The lowest-level metastable states ^2D of O^+ and N are remarkable for their long lives, 3.6 and 26 hr respectively. Emissions by transitions from them to the ground states give doublets 3727 (O^+) and 5200 (N) that are weakly present in the spectrum.

The emissions so far described come from excited neutral or ionized molecules or atoms of the chemical elements most abundant in the atmosphere. The neutral particles N_2, O_2, and O are major constituents at the main level of emission. The ionized particles and the N atoms are minor constituents. Other minor constituents at that level are H and He, the former variously in molecular, atomic, and combined form (notably as OH). The auroral spectrum includes the α, β, γ and other lines of H (Vegard 1939, Gartlein 1950, Meinel 1951), and line 10830 Å of He (Shefov 1963). The last of these is observed in sunlit auroras and in twilight without aurora.

The hydrogen lines differ from all the other features in being notably broadened and, when viewed along the direction of the field lines, displaced to shorter wavelengths. These are interpreted as Doppler effects, of the downward and lateral motion of the emitting atoms, which are *not* previously particles of the polar atmosphere. Some Doppler broadening is shown, of course, by all the lines of the spectrum, and provides a means, along with the study of the intensity distribution in the bands, of inferring the kinetic, vibrational, and rotational temperature of the emitting gases; in some cases interferometers are used.

The Doppler displacement of the hydrogen lines gave the first proof that at least part of the auroral light is caused by the entry of particles into our atmosphere. As noted above, Halley attributed the aurora to particles issuing from within the Earth. The idea that the cause is particles coming to the Earth from outside arose late in the last century, Birkeland (1896, 1908, 1913) being its chief sponsor and propagator; he supported it by laboratory experiments. He thought that the particles are electrons coming from the Sun, which are deflected to the polar latitudes by the geomagnetic field. Störmer (1955) developed this idea mathematically, but only for the motion of charged particles each moving in the field without being influenced by accompanying charged particles. His calculations apply to cosmic rays, but not to a neutral stream of charged particles, half negative, half positive, which was the alternative hypothesis first suggested by Lindemann (1919). His conception, radically differing from that adopted by Birkeland and Störmer, is now directly confirmed by satellite observations; they indicate a continual flow, explained by Parker's theoretical studies of the solar wind (1963). But the occasional intensification of the flow from specially disturbed areas on the Sun, associated with magnetic storms and bright auroras, is not yet well understood.

Auroral Morphology

During the IGY, asca (all-sky camera) photographs were taken at most stations at 1-min intervals during dark periods in fine weather; the exposure was 55 sec followed by 5 sec to move on the film. The program produced a great volume of record, which has been extensively studied, in particular by Akasofu and his colleagues. This has thrown much light on the more systematic features of the auroral appearances and changes, enabling the course of typical *auroral substorms* to be ascertained. These events are of duration of order an hour; they occur frequently with moderate intensity, and during magnetic storms, whose duration is of order a day, several polar magnetic and accompanying auroral substorms, sometimes of great intensity, may occur. These two associated substorm events are aspects of more extensive and varied *magnetospheric* substorms (Akasofu 1964, 1965, 1968a; Davis 1966); they involve the whole ionosphere of the Earth, auroral X-ray events, proton auroras, VLF emissions, geomagnetic micropulsations, and particle flows and redistributions, partly in response to electric fields, in the magnetosphere. This is the large region, with dimensions of order ten Earth radii on the Earth's sunward side (and far greater on the opposite side, in what is called the tail of the magnetosphere), within which the geomagnetic field is confined by the flow of neutral ionized gas from the Sun.

"The characteristics of auroral displays depend greatly on both universal time and local time, so that if observed at a particular point on the earth it is very difficult to distinguish between those that depend on universal time and those that depend on local time" (Akasofu 1968). The description of an auroral substorm in terms of these two time variables is too complex to be detailed here. The onset is a universal time phenomenon, but the characteristics seen at different longitudes following the onset are by no means similar. Between auroral substorms there may be extensive quiet auroras; when the substorm begins (the auroral "breakup"), the arcs become rayed and thinner; they multiply and move rapidly, some to the north, others to the south; waves travel along them, with great curtains and folding in rayed forms (Figure 1). Some time after this expansion the activity decreases—the auroral arcs return to a narrower band of latitude and become quiet again; the last phase may consist of diffuse pulsating patches of auroral light. The inferences drawn from the ascafilms have been confirmed and extended by observation of the aurora from an aircraft that in polar latitudes can remain on or near a constant local-time meridian (Akasofu 1968).

Auroral photographic techniques have steadily improved in recent years, and one post-IGY development is outstanding, the image orthicon television system. This can make successful exposures in 1/60 sec, recording at 24 or 30 per sec (Davis 1967, 1969); it has been used in jet aircraft as well as on the ground. The field of view in most cases was $12° \times 16°$, so that the full grand spectacle of a great aurora is not shown; but the system has been most useful in the study of conjugacy by simultaneous aircraft flights in the two hemispheres, and in revealing auroral changes too rapid to be seen even by the eye (the IGY ascafilms were of course still less percipient). Under favorable operating conditions a linear element of low brightness and angular width $0.03°$ can be recorded. The spectral response resembles but is rather wider than that of the eye.

One most striking feature of the aurora on many occasions is the confinement of the visible light to thin sheets, which lie along the geomagnetic field lines (Figure 5). When a rayed arc passes across the radiant point (the magnetic "zenith"), the ray structure momentarily disappears. An ellipse containing 2/3 of 30 measures of the radiant, made at College, Alaska, by the TV technique (Maggs & Davis 1968), has a north-south minor axis of $1°$, and a $2°$ major axis, showing how close is the alignment along the field lines (which may be slightly modified during magnetic disturbance). The thickness of various auroral sheets was measured on many occasions during 15 hr of observation time in 1966 and 1967. Errors of order 50 to 100 m may occur, partly because the picture is not taken exactly when the sheet lies

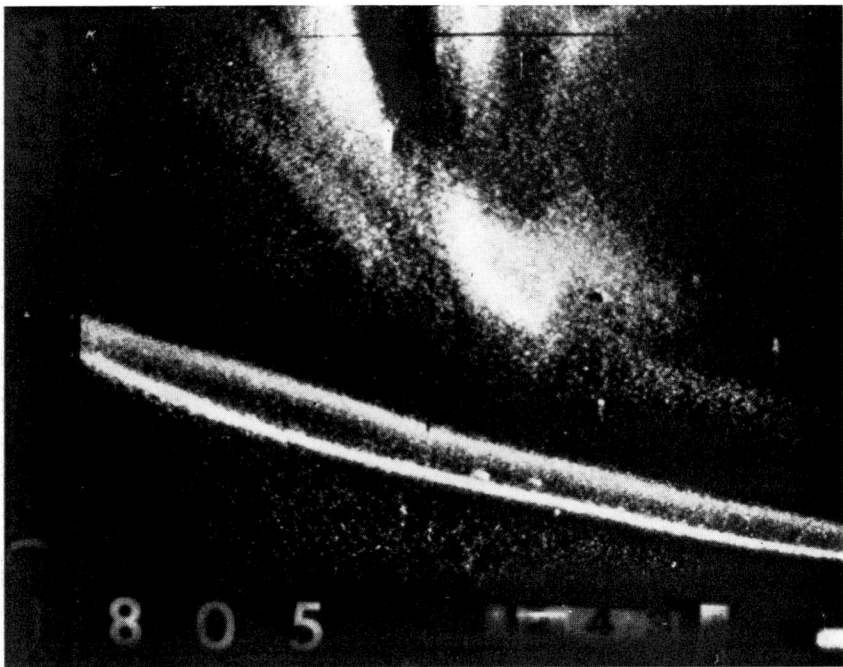

FIGURE 5. A close pair of auroral arcs photographed nearly edge on, by image orthicon TV camera. The lower arc is measured to have a thickness of 350 m; it is unusually regular. The forms shown in the upper part are seen sideways. After Maggs & Davis (1968).

along the radiant direction. It was found that the brighter the arc, the thinner; the median thickness among 581 measures was 230 m; the range was from the lower (instrumental) limit of 70 m up to 4440 m. Both thinner and thicker sheets certainly occur. Such edge-on photographs of the auroral sheets also often reveal a fine pleating of the sheet (Figure 6; Akasofu 1963a). Seen broadside on (or obliquely), the pleats would show as rays, owing to the triple amount of light from the pleats.

Another remarkable and very frequent feature of auroral morphology is the occurrence of multiple arcs, sometimes as many as seven or eight, which are approximately parallel (transverse to the magnetic meridian) and often move in unison, especially in quiet periods, from north to south or vice versa. Their north-to-south spacing may be 40 km. But as in Figure 5, what may appear as a single arc when seen from the side may be revealed as a close double arc when seen edge on.

Despite the incomplete coverage of the polar skies by all-sky cameras,

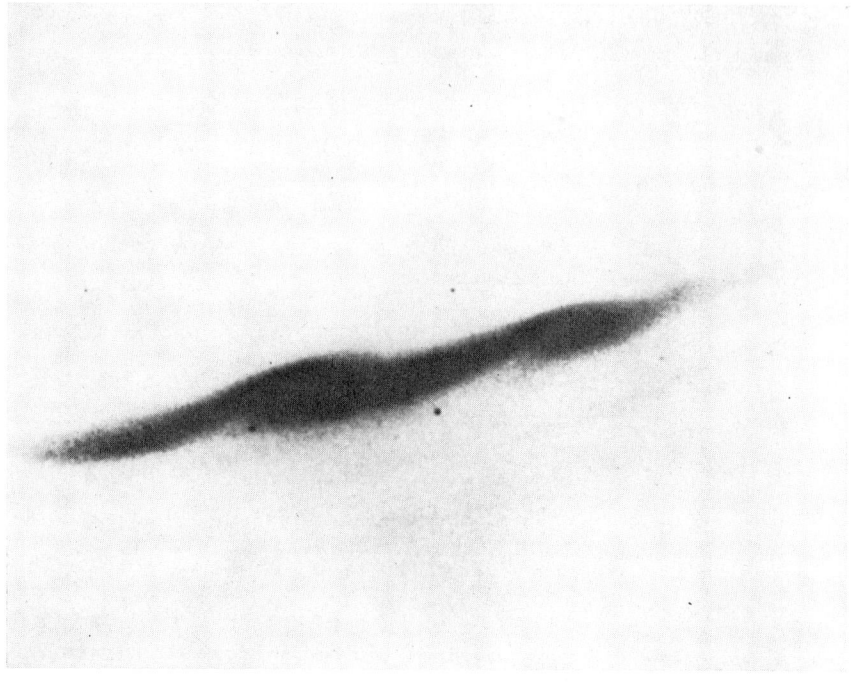

FIGURE 6. An auroral rayed arc seen edge on, showing pleating of the auroral sheet. Three stars of Ursa Major are shown. After Akasofu (1963a).

Feldstein (1960; see also Feldstein & Starkow 1968, Khorosheva 1962) inferred the general nature of the synoptic pattern of the auroral form and location at any instant. The arcs tend to lie in a narrow oval belt encircling the dipole pole, but not centered on it. The center is displaced by about 3° along the midnight meridian, away from the sunward direction. The radius of the oval is about 20°, so that near the midnight meridian the belt, called the *auroral oval*, coincides with the auroral zone; elsewhere it lies within the zone. To a first approximation the oval is fixed relative to the Sun, and the Earth rotates daily under the oval. The zone is a statistical result of the daily passage of each part of it across the midnight meridian, where it coincides with the average position of the oval, where the auroras are often most intense. As the oval is eccentric relative to the dipole pole, a station may be within the oval at midnight but outside it at noon. Auroral substorms, and polar magnetic disturbance, are best described with reference to position relative to the oval rather than to the zone. The oval encloses the area of polar-cap absorption, caused by 1–10 MeV protons, which produce a polar-glow aurora (Sandford 1962).

At times of great magnetic disturbance the radius of the oval is enlarged; this corresponds to the visibility of the aurora from subauroral or even minauroral latitudes (Akasofu & Chapman 1962).

Mishin & Popov (1969) contest the continuity of particle penetration along the Feldstein oval synoptic pattern of the aurora. Using magnetic and ionospheric as well as auroral data, they infer the occurrence of two quasi-circular auroral activity zones. Akasofu (1968a), who has taken part in long-continued exploration of the auroral arctic region by jet aircraft, following the auroral bands, maintains that they confirm the oval pattern accurately and completely.

Another important (and highly variable) feature of auroral morphology is the brightness distribution or emission rate as a function of height, or of distance along the field lines. Early studies of this kind were made by Vegard (1930), who considered different spectral components of the light, and their ratio at different heights. Harang (1946, 1946a) made densitometer measurements of his auroral photographs, and inferred the variation of the scale height of the atmosphere along the rays. More recently, as during the IGY, photometers and scanning photometers, some provided with filters to isolate different spectral regions, were applied to such studies. Belon, Romick & Rees (1966) reported on the height distribution of the light of wavelength $3914(N_2^+)$, from scanning photometer records from two Alaskan stations 226 km apart, Fort Yukon and College. Romick & Belon (1967), using their data from the same two stations, went beyond the thin-sheet approximation, and by a complicated method of analysis derived contours of the *volume* emission rates for the 3914 and 5577 emissions (Figure 7; note that the scales of abscissae are not the same for the two sets of contours). The respective peak rates are 2.3×10^4 and 4.4×10^4 photons/cm^3 sec.

The unit of sky brightness used in connection with the aurora (and also for the airglow) is the Rayleigh, namely 10^6 photons/cm^2 (column) sec (cf Chamberlain 1961, p. 570). Auroras are classed by visual observers according to an International Brightness Coefficient (Chamberlain 1961, p. 571).

The detailed processes of these and the other auroral optical emissions are too complex to be discussed here. Kaplan (1932) produced in the laboratory an afterglow discharge spectrum in nitrogen that he named the auroral afterglow, because of its resemblance to the auroral spectrum. This is dominated by nitrogen contributions, though they are not the most conspicuous to the eye. Even the complex nitrogen spectra obtained under controlled laboratory conditions are still under debate (Oldenberg 1969). Secondary electrons produced by the primary ones are the main agents in producing the spectrum. Green & Barth (1965) have discussed the nitrogen emission resulting from bombardment of the atmosphere by 30-keV electrons. They concluded that 40% of the energy should appear in the far-ultraviolet spec-

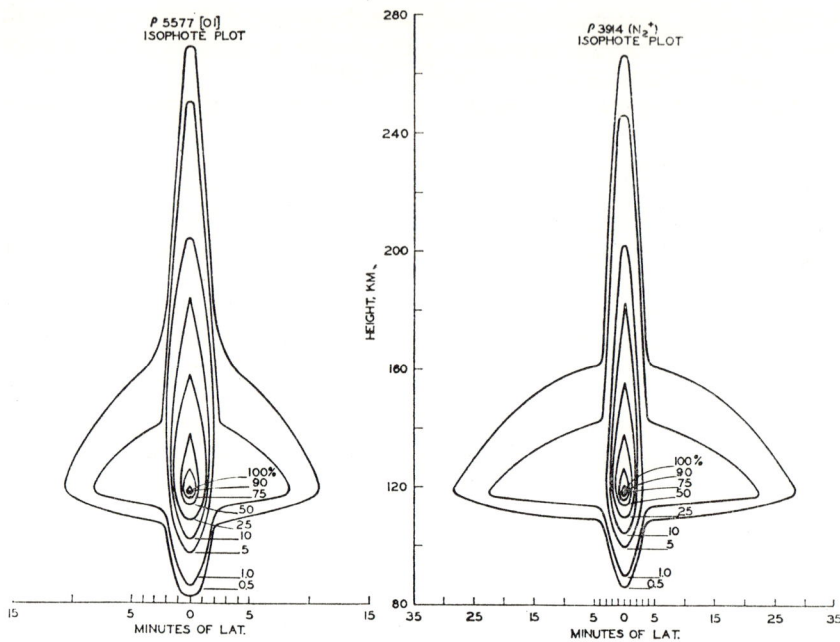

FIGURE 7. The volume distribution of emission of two components of the auroral spectrum, in the meridian plane normal to the auroral sheet; note the difference in the two scales of latitude. The isophotes show percentages of the maximum emission rate. After Romick & Belon (1967). The difference between the distributions implies that the 5577 emission is not necessarily due to the excitation of oxygen atoms by secondary electrons ejected from nitrogen molecules. The precise cause of the 5577 emission is still a difficult problem.

trum, and that another 40% goes to ionization, which later contributes a recombination spectrum.

Besides the auroral light emission, auroras at some times and places emit radio waves and possibly also Cerenkov radiations (Chamberlain 1961). Infrasonic waves from auroras have been studied by Wilson (1969, 1969a), using two microphone arrays at College and Palmer, Alaska; he associates such waves with electrojets (electric currents of limited cross section) in supersonic motion, flowing along moving auroral arcs. The ionosphere, ionized during the day by solar wave radiation, is ionized at night (as also, in certain regions, by day) by entering particles; some of these produce X rays that travel down to balloon levels, where they have been observed by balloon-borne instruments (Winckler et al 1958, 1959; Anderson 1965; Brown 1966).

THE PARTICLES THAT CAUSE THE AURORA

In 1904 Maunder interpreted the 27-day recurrence tendency shown by magnetic disturbance and auroras as indicating emission (often long continued) of limited streams of gas from active regions on the Sun. For about half a century thereafter it was generally accepted that such solar streams cause magnetic disturbance and auroras, but in 1950 it still seemed necessary to defend this idea (Chapman 1950) against the alternative proposal (Hulburt 1929) that the cause is solar ultraviolet light. Meinel's discovery (1950), soon afterward, of the Doppler displacement of the hydrogen lines in the auroral spectrum, strengthened the particle hypothesis.

But the simple theory advocated by Birkeland and Störmer, that the aurora is caused by particles from the Sun, guided polewards by the geomagnetic field, faced a serious difficulty. Chapman & Ferraro (1930) had inferred that the impact on the Earth of a solar stream of neutral ionized gas, as proposed by Lindemann, would be prevented by the geomagnetic field, and that this field would be confined and compressed within a cavity in the solar stream (this cavity is now called the magnetosphere; Gold 1959). Chapman & Ferraro (1940) also showed that unless the solar stream is of such low density as to be unable to have any appreciable effect on the Earth, the mutual attraction of its positively and negatively charged particles makes their paths entirely different from those calculated by Störmer, who took account only of the force exerted by the magnetic field. But though concluding that the entry of particles of the stream into the cavity would be severely restricted, Chapman & Ferraro (1931) believed that during magnetic storms an electric ring current grows inside the cavity, in some way they were unable to explain. Their conception of this ring current as being of toroidal form was later corrected by Singer (1957), using ideas of Alfvén (1940, 1950) as to the paths of particles trapped within the cavity. The satellite discoveries (Van Allen & Frank 1959, Vernov et al 1960) of particles trapped within the cavity, forming the Van Allen belts, soon confirmed these conclusions.

In 1955 Van Allen and his colleagues (Meredith et al 1955) found evidence of particle entry into the auroral atmosphere, additional to and different from that of Meinel; by rockoon observations they had detected X rays (Van Allen & Kasper 1956), attributed to the close impact of energetic electrons with the nuclei of atmospheric atoms. Thus the atmosphere receives from outside both protons and electrons. Since 1955 the X rays have been extensively studied by balloon-borne recorders (Anderson 1958). The flux and energy spectra of electrons and protons have been measured at higher levels of the atmosphere by rockets, and further out by low-altitude satellites at various local times during and between magnetospheric substorms. Many of the results have been summarized by O'Brien (1967);

Akasofu (1968a) gives some later references. The details are too voluminous for inclusion here.

Even before 1955 the nature of the particles responsible for the different components of the auroral spectrum had been much discussed. It is now clear that there are three types of auroral emission, classified according to the exciting cause, which may be electrons, protons, or the heating of the atmosphere (Seaton 1955). The first two kinds, caused by electrons and protons entering from outside, may be called *energetic* auroras (Cole 1969), and the third kind may be called *thermal*. The oxygen red lines, of light from great heights, have on occasion shown Doppler broadening indicating, according to Krassovsky (1959, 1961) and Mulyarchik & Shcheglov (1963), temperatures up to 3500°K, which the electron temperature may exceed.

These types may appear singly, or two or more together. They may have different ranges of height or latitude and longitude (or local time).

The electrons are the most visually effective cause. Their energy can be inferred from the extent of their penetration of the atmosphere. The ordinary lower limit, about 100-km height, corresponds to energy of a few kiloelectron volts (keV). For the rare cases of penetration down to 80, 70, or even 60 km, the energy may range up to a few hundred keV. As the electrons descend along the field lines, they collide with atmospheric particles (mainly N_2, O, and O_2), which they may ionize, dissociate and/or excite; also the electrons are scattered, changing their pitch angles. The average energy loss to the electron, per ion-electron pair produced by the impact, is about 35 eV, so that a 10-keV electron may ionize about 300 particles. Thus the sheet of gas that emits the light of an auroral arc will be highly ionized and electrically conducting. Often the arcs move (usually, in the evening, from north to south); thus ionization of the air may be spread over a band much wider than the thin auroral sheet, though the ionization decays after the entry of particles has passed away to another region.

The thinness and movement of the auroral sheets produce notable horizontal gradients of electron density and electric conductivity; Ungstrup (1969) has discussed the influence of such gradients on the power of whistlers and very low-frequency emission coming from the magnetosphere, to penetrate to the ground. He finds that for a certain range of the ratio of the horizontal scale length of the electron density to the atmospheric scale height, auroral arcs may greatly facilitate such penetration.

The intensified nocturnal ionization associated with auroras was first observed at Tromsö during the second International Polar Year 1930/1 (Appleton et al 1937). It was extensively observed in several ways in polar (and other) regions during the IGY (the *IGY Annals*, 3, 8, 13–15, 17–19, 23 report the ionospheric program and results of the IGY). Often the D region of the ionosphere is affected when the auroral particles penetrate below 90 km; this is well recorded by riometers (Little 1957), which register the degree of absorption of extraterrestrial radio waves (e.g. of 30 Mc/sec). Radio waves are also reflected when they impinge nearly perpendicularly on the auroral sheets (Booker 1960); such radio echoes are received from regions in sunlight by day, thus somewhat mitigating the difficulty of observing visual auroras in daylight. Doppler shift and spread of such echoes has been studied, and gives information about the motion of the electron concentrations associated with the aurora. At College, Alaska, Nichols (1957) observed such auroral echoes, using low-power continuous-wave transmissions, one directed 30°W of magnetic N, the other 30°E, and obtained Doppler shifts of opposite sign on the two antennas; his measures indicated motion along an east-west line, $\sim$700 m/sec, generally eastward before midnight and westward after midnight. Such speeds cannot be due to wind, and are interpreted as indicating the motion of secondary electrons in the auroral sheet, moving along the sheet, and hence carrying an electric current in the opposite direction. Owing to the anisotropy of the electric conductivity engendered by the presence of the geomagnetic field, a westward current would be produced by a component electric field directed nearly equatorward (Akasofu 1960).

The proton precipitation occurs in general over a band broader than that of the electron precipitation; this band is not far outside the border of the auroral oval. The proton aurora is subvisual; its photons are those produced by the entering particles, in the repeated intervals during their descent in which they have temporarily acquired an electron, and thus become able to radiate.

The heating in the case of the thermal aurora has been variously attributed to atomic resistance to electric current flow in the ionosphere, to electric fields, to conduction from the magnetosphere (Cole 1969), and to absorption of hydromagnetic waves from the magnetosphere (Krassovsky 1968). Thermal auroras include the stable red arcs observed in middle latitudes (Barbier 1958, 1960; Roach & Marovich 1959).

THE EVENTS PRECEDING THE ENTRY OF AURORAL PARTICLES

According to Alfvén's theory of the motion of trapped particles in the magnetosphere, they travel northwards and southwards between mirror points, but some have mirror points low enough for them to collide in the ionosphere and become part of it; these may be considered auroral particles. Other auroral particles not in the trapping region may enter the ionosphere along field lines that extend out into the tail of the magnetosphere. The auroral conjugacy suggests that these must come from near the median plane of the tail, where about equal numbers are moving northward and southward.

The electrons that cause the auroral arcs, as they descend to the ionosphere, must constitute thin sheets, and while the arcs are quiet these sheets must be stable. At the auroral breakup, when the arcs become rayed and pleated and thinner, some instability is indicated, apparently of the kind studied in the laboratory by Webster (1957).

Some remaining fundamental questions concerning the events prior to the entry of the auroral particles into the atmosphere involve problems of magnetospheric and general plasma physics beyond the scope of this article. Do the auroral particles come from the Sun, as Birkeland, Störmer and others supposed, or are they particles of the Earth's outer atmosphere, which extends far out into the magnetospheric cavity, and in some way influenced and energized by the changes in the solar wind and its magnetic field? The energy of the electrons in the solar wind is of order 25 eV (Hundhausen 1968); if any of them become auroral electrons, they must be energized in the cavity.

To what extent particles from the solar wind enter the cavity is not known. Axford & Hines (1961; see also Nishida 1966 and Brice 1967) suggested that some of them diffuse into the cavity, especially along its sides, and set up circulatory motion therein. Axford (1967; see also Dungey 1968, Piddington 1968) proposed that solar particles accumulate in the tail and inflate it increasingly until it becomes unstable, with quick release of its excess energy in the form of a magnetospheric storm or substorm. Akasofu (1964a) proposed that the solar wind includes a varying proportion of uncharged particles, whose entry into the cavity is unimpeded, bringing solar energy into it.

The selective process that restricts the entry of auroral electrons to thin sheets, often multiple, is not understood. Akasofu & Chapman (1961) proposed that these electrons come from regions closely bordering on neutral lines of the geomagnetic field; they speculated that these might lie in the ring current of the trapping region, but this is now seen to be unlikely (Parker 1962); neutral lines will, however, occur near the median plane of the tail region (Coppi, Laval & Pellat 1966).

Akasofu has estimated the energy of a substorm as $\sim 10^{22}$ ergs, and for a duration $\sim 3 \times 10^3$ sec the average rate of energy supply needed is about 3×10^{18} ergs/sec. The kinetic energy flux in the solar wind has a variation from time to time of more than tenfold; taking a value for it of 10^{-1} erg/cm^2 sec, this, over a cross-sectional area 10^{20} cm^2, comparable with that of the cavity, gives a supply rate 10^{19} ergs/sec. But much of this must be diverted round and pass with the solar wind away from and beyond the Earth.

Schuster (1911), in a discussion of theories of magnetic storms, concluded that the energy must come from that of the Earth's rotation (cf Krassovsky 1968), but it is not clear how this energy can be drawn upon.

Dungey (1968) attaches importance to interconnection of the magnetic field of the solar wind and the field within the cavity (cf Petschek 1964, Piddington 1968, Sweet 1963), proposing that this may provide a channel for the flow of solar particles and energy into the cavity. Electric fields and magnetospheric instabilities have been considered by various authors (Fejer 1964, Kern 1962, Speiser 1968, Swift 1968, Taylor & Hones 1965), in discussing the origin and nature of magnetospheric storms.

Further observational discoveries concerning the solar wind and the magnetosphere (Ness 1968) will doubtless give guidance in the search for the true chain of events and causes that produce auroras, but it seems clear that many difficult theoretical questions will also be involved.

LITERATURE CITED

Abbe, C. 1898, *Terr. Magn. Atmos. Elec.*, **3**, 5, 53, 149

Akasofu, S.-I. 1960, *J. Atmos. Terr. Phys.*, **19**, 10

Akasofu, S.-I. 1963. *J. Geophys. Res.*, **68**, 1667 (see p. 1670)

Akasofu, S.-I. 1963a, *J. Atmos. Terr. Phys.*, **25**, 163

Akasofu, S.-I. 1964, *Planet. Space Sci.*, **12**, 273

Akasofu, S.-I. 1964a, *Planet. Space Sci.*, **12**, 801, 905

Akasofu, S.-I. 1965, *Space Sci. Rev.*, **4**, 498

Akasofu, S.-I. 1968, *Planet. Space Sci.*, **16**, 1365

Akasofu, S.-I. 1968a, *Polar and Magnetospheric Substorms* (Dordrecht: Reidel)

Akasofu, S.-I., Chapman, S. 1961, *Phil. Trans. Roy. Soc.*, **253**, 359

Akasofu, S.-I., Chapman, S. 1962, *J. Atmos. Terr. Phys.*, **24**, 785

Akasofu, S.-I., Chapman, S. 1963, *J. Atmos. Terr. Phys.*, **25**, 9

Akasofu, S.-I. Chapman, S., Meinel, A. B., 1966, *Handb. Phys.*, **49**(1), 1

Akasofu, S.-I., Kimball, D. S. 1965, *IGY Ann.*, **38**

Akasofu, S.-I., Kimball, D. S., Meng, C.-I. 1969, *IGY Ann.*, 45

Alfvén, H. 1940, *Ark. Mat. Astron. Fys.*, **27A**, 122

Alfvén, H. 1950, *Cosmical Electrodynamics* (Oxford)

Anderson, K. A. 1958, *Phys. Rev.*, **111**, 1397

Anderson, K. A. 1965, *Auroral Phenomena* (Walt, M., Ed., Stanford Univ. Press)

Angström, A. J., 1869, *Pogg. Ann.*, **137**, 161

Appleton, E. V., Naismith, R., Ingram, L. J. 1937, *Phil. Trans. Roy. Soc. A*, **236**, 191

Axford, W. I. 1967, *Aurora and Airglow*, 499 (McCormac. B. M., Ed., New York: Reinhold)

Axford, W. I., Hines, C. O. 1961, *Can. J. Phys.*, **39**, 1433

Bagariazky, B. A., Fedorova, N. I. 1956, *The Airglow and the Aurorae*, 174 (Armstrong, E. B., Dalgarno, A., Eds., Pergamon)

Barbier, D. 1958, *Ann. Geophys.*, **14**, 334

Barbier, D. 1960, *Ann. Geophys.*, **16**, 544

Bates, D. R. 1960, *Physics of the Upper Atmosphere*, Chap. 7 (Ratcliffe. J. A. Ed., New York: Academic)

Belon, A. E., Romick, G. J., Rees, M. H. 1966, *Planet. Space Sci.*, **14**, 597

Belon, A. E., Maggs, J. E., Davis, T. N., Mather, K. B., Glass, N. W., Hughes, G. F. 1969, *J. Geophys. Res.*, **74**(1), 1

Birkeland, K. 1896, *Arch. Sci. Nature Genève*, **1**, 497

Birkeland, K. 1908, 1913, *Norwegian aurora polaris expedition*, 1902–3, Parts 1 and 2 (Christiania, Oslo: Aschehoug, 801 pp.)

Bond, F. R., Jacka, F. 1960, *Aust. J. Phys.*, **13**, 611

Booker, H. G. 1960, *Physics of the Upper Atmosphere*, Chap. 8, 366 (Ratcliffe, J. A., Ed., New York: Academic)

Brice, N. M. 1967, *J. Geophys. Res.*, **72**, 5193

Brown, R. R. 1966, *Space Sci. Rev.*, **5**, 301

Celsius, A. 1740, *Svensk. Vet. Akad. Handl.*, 296

Chamberlain, J. W. 1961, *Physics of the Aurora and Airglow* (New York: Academic, 704 pp.)

Chapman, S. 1950, *J. Geophys. Res.*, **55**, 361

Chapman, S. 1953, *Proc. Indian Acad. Sci.*, **37**, 175

Chapman, S. 1957, *Bull. Nat. Inst. Sci. India*, **9**, 180 (1872 aurora)

Chapman, S., Ferraro, V. C. A. 1930, *Nature*, **126**, 129

Chapman, S., Ferraro, V. C. A. 1931, *Terr. Magn. Atmos. Elec.*, **36, 37, 38**

Chapman, S., Ferraro, V. C. A. 1940, *Terr. Magn. Atmos. Elec.*, **45**, 245

Cole, K. D. 1969, *IAGA Symp.*, Madrid

Cook, J. 1961, *The Journals of Captain James Cook*, **2**, 95 (Cambridge)

Coppi, B., Laval, G., Pellat, R. 1966, *Phys. Rev. Letters*, **16**, 1207

Currie, B. W. 1955, *Can. J. Phys.*, **33**, 773

Dalton, J. 1793, *Phil. Trans. Roy. Soc.*, **53**, 144

Dalton, J. 1828, *Phil. Trans. Roy. Soc. A*, **118**, 291

Davis, T. N. 1966, *Space Sci. Rev.*, **6**, 222

Davis, T. N. 1967, *Aurora and Airglow*, 133 (Amsterdam: Reinhold)

Davis, T. N. 1969, *Atmospheric Emissions* (McCormac, B. M., Omholt, A. Eds., New York: Van Nostrand, Reinhold)

Devlin, J. J., Oliver, N. J., Carrigan, A. 1966, *Auroral Spectrographic Data*, IGY Ann., 40

Dungey, J. W. 1968, *Earth's Particles and Fields*, 385 (McCormac, B. M., Ed.)

Fejer, J. A. 1964, *J. Geophys. Res.*, **69**, 123

Feldstein, Y. I. 1960, *Aurorae and Airglow*, No. 4, 61 (Moscow: Publ. House Acad. Sci.)

Feldstein, Y. I., Solomatina, E. K. 1961, *Results of IGY Researches, Aurorae and Airglow*, **7**, Sec. 4, IGY Program, 51–59

(Moscow: Publ. House Acad. Sci.)
Feldstein, Y. I., Starkov, G. V. 1968, *Planet. Space Sci.*, **16**, 129
Fritz, H. 1873, *Verzeichniss beobachteter Polarlichter* (Vienna: Gerold's Sohn)
Fritz, H. 1874, *Petermann's Geogr. Mitt.*, **20**, 347
Galperin, Yu.I. 1963, *Planet. Space Sci.*, **10**, 187
Gartlein, C. W. 1947, *Nat. Geogr. Mag.*, **92**, 673
Gartlein, C. W. 1950, *Trans. Am. Geophys. Union*, **31**, 18
Gartlein, C. W., Gartlein, H. E. 1966, *Northern Hemisphere Synoptic Charts . . . IGY Ann.*, **39**
Gassendi, P. 1658, *Opera omnia* (Lyons, France)
Gold, T. 1959, *J. Geophys. Res.*, **64**, 1219
Götz, F. W. P. 1947, *Prisma, Schweiz. Monatschr.*, **8**, 2
Götz, F. W. P. 1947, *Experientia*, **3**(5), 185
Graham, G. 1724, *Phil. Trans. Roy. Soc.*, **33**, 96
Green, A. E. S., Barth, C. A. 1965, *J. Geophys. Res.*, **70**, 1083
Halley, E. 1716, *Phil. Trans. Roy. Soc.*, **29**, 46
Harang, L. 1946, *Geofys. Publ.*, **16** (13)
Harang, L. 1946a, *Terr. Magn. Atmos. Elec.*, **51**, 381
Harang, L. 1951, *The Aurora*, 53 (London: Chapman & Hall, 166 pp.)
Harang, L., Bauer, W. 1932, *Gerlands Beitr. Geophys.*, **37**, 109
Hargreaves, K. K. 1969, *Atmospheric Emissions* (McCormac, B. M., Omholt, A. Eds., New York: Van Nostrand, Reinhold)
Hill, J. E. 1965, *Can. J. Phys.*, **43**, 1917
Hiorter, O. P. 1747, *Svensk. Vet. Akad. Handl.*, 27
Hulburt, E. O. 1929, *Phys. Rev.*, **34**, 344, 1167; **36**, 1560
Hultquist, B. 1961, *Planet. Space Sci*, **8**, 142
Hundhausen, A. J. 1968, *Space Sci. Rev.*, **8** (5/6), Figures 21/2
Kaplan, J. 1932, *Phys. Rev.*, **42**, 807
Kern, J. W. 1962, *J. Geophys. Res.*, **67**, 2649
Keys, J. G. 1964, *J. Atmos. Terr. Phys.*, **26**, 979
Khorosheva, O. V. 1962, *Geomag. Aeron.*, **2**, 696
Krassovsky, V. I. 1959, *Planet. Space Sci.*, **1**, 14
Krassovsky, V. I. 1961, *Planet. Space Sci.*, **8**, 125
Krassovsky, V. I. 1967, *Solar Terrestrial Physics*, Chap. 8 (King. J. W., Newman, W. S., Eds., New York: Academic)
Krassovsky, V. I. 1968, *Planet. Space Sci.*, **16**, 47
Lindemann, F. A. 1919, *Phil. Mag.*, **38**, 669
Little, C. G. 1957, *IGY Ann.*, **3**, 207
Loomis, E. 1860, *Am. J. Sci. Arts*, **30**, 89
Maggs, J. E., Davis, T. N. 1968, *Planet. Space Sci.*, **16**, 205
Mairan, J. J. d'O. de 1733, 1754, *Traité physique historique de l' aurore boréale* (Acad. Sci. Paris, 570 pp.)
McCormac, B. M., Ed. 1967, *Aurora and Airglow* (Amsterdam: Reinhold, 698 pp.)
McCormac, B. M., Omholt, A., Eds. 1969, *Atmospheric Emissions* (New York: Van Nostrand, Reinhold)
McInnes, B. 1962, *IGY Visoplots, IGY Ann.*, 29
McLennan, J. C., Shrum, G. R. 1924, *Proc. Roy. Soc. A*, 106
Meinel, A. B. 1951, *Ap. J.*, **113**, 50
Meinel, A. B. 1966, *Handb. Phys.*, **49** (1) 100; part of Akasofu, Chapman & Meinel
Meinel, A. B., Negaard, J. B., Chamberlain, J. W., 1954, *J. Geophys. Res.*, **59**, 407
Meredith, L. H., Gottlieb, M. B., Van Allen, J. A. 1955, *Phys. Rev.*, **97**, 201
Miller, R. E., Fastie, W. G., Isler, R. C. 1968, *J. Geophys. Res.*, **73**, 3353
Mishin, V. M., Popov, G. V. 1969, *IAGA Madrid Assembly*
Mulyarchik, T. M., Shcheglov, P. V. 1963, *Planet. Space Sci.*, **10**, 215
Muncke, G. W., 1833 *Gehler's Physikalische Wörterbuch*, 2nd ed. 1823–1845, **7**(1), 113
Ness, N. F. 1968, *Ann. Rev. Astron. Ap.*, 79
Nichols, B. 1957, *J. Atmos. Terr. Phys.*, **11**, 292
Nishida, A. 1966, *J. Geophys. Res.*, **71**, 5669
O'Brien, B. J. 1967, *Solar-Terrestrial Physics*, 169 (King, J. W., Newman, W. S., Eds., New York: Academic)
Oldenberg, O. 1969, *Phys. Sci. Res. Pap. 390, AFCRL–69–0311* (Bedford, Mass.)
Parker, E. N. 1962, *Space Sci. Rev.*, **1**, 62
Parker, E. N. 1963, *Interplanetary Dynamical Processes* (New York:Wiley)
Petrie, W., Small, R. 1952, *Ap. J.*, **116**, 433
Petschek, H. E. 1964, *AAS-NASA Symp. Phys. Solar Flares, NASA SP-50*, 425 (Hess, W. N., Ed.)
Piddington, J. H. 1968, *Earth's Particles and Fields*, 417 (McCormac, B. M. Ed., New York: Reinhold)
Roach, F. E., Marovich, E. 1959, *J. Res. NBS*, **63D**, 297
Romick, G. J., Belon, A. E. 1967, *Planet.*

Space Sci., **15**, 1695
Sandford, B. P. 1962, *J. Atmos. Terr. Phys.*, **24**, 155
Schwabe, S. H. 1844, *Astron. Nachr.*, **21**
Seaton, M. J. 1955, *The Aurora and Airglow* (Dalgarno, A., Armstrong, E. B. Eds., Pergamon)
Schuster, A. 1911, *Proc. Roy. Sci.*, **85**, 44,
Shefov, N. N. 1963, *Planet. Space Sci.*, **10**, 73
Singer, S. F. 1957, *Trans. Am. Geophys. Union*, **38**, 175
Speiser, T. W. 1968, *J. Geophys. Res.*, **73**, 1112
Stoffregen, W. 1962, *IGY Ascaplots, IGY Ann.*, **20**
Störmer, C. 1955, *The Polar Aurora* (Oxford)
Sweet, P. A. 1963, *AAS-NASA Symp. Solar Flares, NASA-SP-50*, 409 (Hess, W. N., Ed.)
Swift, D. W. 1968, *Planet. Space Sci.*, **16**, 329
Taylor, H. E., Hones E. W., Jr., 1965, *J. Geophys. Res.*, **70**, 3605
Ungstrup, E. 1969, *Danish Space Res. Inst. Pap.*
Vaisberg, O. L. 1959, *Aurora and Airglow, Rep. IYG Comm.*, **10**, 454 (Moscow: Acad. Sci.)
Vallance Jones, A. 1969 (See McCormac & Omholt, p. 47)
Van Allen, J. A., Kasper, J. E. 1956, *Bull. Am. Phys. Soc.*, **1**, 230
Van Allen, J. A., Frank, L. A. 1959, *Nature*, **183**, 430
Vegard, L. 1930, *Z. Geophys.*, **6**, 42
Vegard, L. 1939, *Nature*, **144**, 1089
Vernov, S. N., Chudakov, A. E., Valukov, P. V., Logachev, Yu.I., Nikolaev, A. G. 1960, *Artificial Earth Satellites*, 5(24) (Moscow; translation in: 1961, *ARS J.*, **31**, 967)
Vestine, E. H. 1944, *Terr. Magn. Atmos. Elec.*, **49**, 77
Vestine, E. H., Snyder, E. J. 1945, *Terr. Magn. Atmos. Elec.*, **50**, 105
Webster, H. F. 1957, *J. Appl. Phys.*, **28**, 1388
White, F. G. W., Geddes, M. 1939, *Terr. Magn. Atmos. Elec.*, **44**, 367
Wilcke, J. C. 1777, *Svensk. Vet. Akad. Handl.*, 273
Wilson, C. R. 1969, *Planet. Space Sci.*, **17**, 1817
Wilson, C. R. 1969a, *J. Geophys. Res.*, **74**, (7), 1812
Winckler, J. R., Peterson, L., Arnoldy, R., Hoffman, R. 1958, *Phys. Rev.*, **110**, 1221
Winckler, J. R., Peterson, L., Arnoldy, R., Hoffman, R. 1959, *J. Geophys. Res.*, **64**, 597

ATMOSPHERES OF VERY LATE-TYPE STARS

M. S. Vardya

Tata Institute of Fundamental Research, Bombay, India

INTRODUCTION

The study of late-type stars is important in many respects. It is in this part of the Hertzsprung-Russell diagram that a collapsing interstellar cloud becomes an identifiable star. And it is to this part that the star returns at an advanced stage of evolution before we lose track of its future movements. The atmospheres of these late-type stars are important not only because they frequently show rather peculiar elemental abundances, which may reflect the effect of nucleosynthesis during the advanced stages of evolution, but also because the structure of these stars depends very sensitively on the boundary conditions provided by the atmosphere. In this review, we will restrict ourselves mainly to stars of M, S, and C spectral types, though some of the discussion will be more general. It has not been possible to discuss several interesting topics in this brief review. The selection of topics and the emphasis laid on it may reflect, in part, the author's bias and his awareness that it has been elegantly discussed in detail elsewhere.

GROSS PROPERTIES

Spectral Classification of M, S, and C Stars

Spectra of very late-type stars have three different branches, M, S, and C (or R and N). For the same temperature, the differences between these branches have been attributed to difference in elemental abundance ratios, specially of oxygen to carbon. A detailed discussion has been given by Keenan (1963).

Type M.—In the original Henry Draper classification, M stars were classified as Ma, Mb, and Mc. This was replaced by M0, M1, M2, M3, M4, M5, and M6 in the Mount Wilson classification, based mainly on the strengths of TiO bands, which was later extended to M variable stars by Joy (1942) and Merrill (1941). The MKK classification (Morgan, Keenan & Kellman 1943) went up to M3 only; it was extended to M8 by Sharpless (1956), though there are stars that have been classified as late as M10 (cf Wing, Spinrad & Kuhi 1967).

Classification between K5 and M0 is made by the ratio of Ca $\lambda4226$/Fe $\lambda4144$, $\lambda4325$; one has to be careful because Ca $\lambda4226$ is sensitive to luminosity also. Strong TiO bands at $\lambda\lambda$ 4950, 5446, 6153, and 7050, have also been used to assign spectral types as early as K5 (Keenan 1970). Beyond M2, bands of TiO are used. From M5 to M8, it is possible to classify very accurately because of the appearance of several faint bands. The appearance of VO bands defines the M7 subclass.

Deutsch, Wilson & Keenan (1969) have recently discussed the effect of stellar population on the spectral classification of K2-M6 giants.

The line ratios of Sr II $\lambda 4077$/Fe $\lambda 4063$; Hδ/Fe $\lambda 4077$; Ti $\lambda 4496$/Fe $\lambda 4495$; Fe II, etc $\lambda 7712$/Ni $\lambda 7714$; and Fe $\lambda 8514$/Ti $\lambda 8435$ have been used, among others, as luminosity criteria (Keenan 1963; Yamashita 1966, 1967). Sr II lines $\lambda\lambda 10327$ and 10914 appear to be luminosity sensitive (Spinrad & Wing 1969). The bands of CaH and MgH have been used to distinguish between dwarf and giant stars (Öhman 1934, 1936; Spinrad & Wood 1965; Treanor & McCarthy 1966), and the bands of CN and CO to separate classes I, II, and III (Keenan 1963, Spinrad et al 1969).

Type S.—Pure type M to pure type S forms a continuous sequence. A star is classified as an S star only if ZrO bands are visible at moderate dispersion. The atomic lines of Zr, La, Y, Sr, and Ba and the bands of their oxides strengthen as one goes from M to S stars (Keenan 1963). The bands of CN are rather weak in normal M giants but are enhanced in M supergiants and S stars (Spinrad et al 1969). The bands of LaO are also used for spectral classification.

As the strength of ZrO bands depends on the temperature and the abundance of the Zr group of metals relative to the Ti group, Keenan (1954) defined the temperature class by the sum of the band strengths of ZrO and TiO and the abundance class by the relative intensities of these two sets of bands. There are a few borderline stars which do not fit the temperature sequence, perhaps because of variation in the abundance of carbon.

There are no dwarf S stars and no satisfactory luminosity classification exists for these stars.

Type C or R and N.—The main characteristics of carbon stars are the bands of C_2, CN and the Sanford bands at $\lambda 4979$ and $\lambda 4686$, which are perhaps due to SiC_2. In the HD classification, they were divided into R and N type. As it was not a temperature-decreasing sequence, Keenan & Morgan (1941) introduced the C classification (see also Fujita, Yamashita, Kamijo, Tsuji & Utsumi 1965; Yamashita 1967). There is a group of high-velocity carbon stars, called CH stars, which show very strong CH features, besides moderately strong C_2 and CN bands and a strong resonance line of Ba II.

The criteria used for temperature class depend on the absolute intensity of Na D lines, the relative intensity of the continuous spectra near $\lambda 5190$, $\lambda 5670$, and $\lambda 6150$, and the relative intensity of the vibrational bands of the Swan system of C_2 at $\lambda 5685$ and $\lambda 5585$. For carbon abundance, the absolute intensity of the Swan bands is used. The infrared CO bands appear weaker in carbon stars compared to M stars of the same temperature, perhaps because of overlapping bands of CN (Wing & Spinrad 1970).

There appear to be no dwarf carbon stars. It has not been possible to determine any luminosity indicator for carbon stars.

Effective Temperatures, Luminosities, Masses, and Radii

Johnson (1965) has obtained bolometric corrections (BC), effective temperatures (T_e), and bolometric magnitudes (M_{bol}) for M dwarf stars, using $UBVRIJKL$ photometry, as given in Table 1. These should be fairly reliable.

Johnson (1964) has measured $UBVRIJKLMN$ colors for M0-M5 giants and Mendoza V. & Johnson (1965) $BVIJKL$ colors for M6-M9 giants. They have derived, as given in Table 1, BC and T_e for M giant stars. As these values are based on a very few stars, they are not of the same quality as for dwarf stars. Blanco (1965) has derived absolute visual magnitudes (M_v) for M0-M7 giants, using which we have derived the M_{bol} as given in the table; the values with a colon in front are from somewhat uncertain values of M_v.

Smak (1964) has obtained, using UBV measurements, mean T_e, BC, and M_{bol} for M2-M7 variables using absolute visual magnitudes as given by Osvalds & Risley (1961).

For luminosity class II from M0-M5, Johnson (1964) has given BC and T_e based on $UBVRIJKLMN$ photometry as given in Table 1. However, stars measured are too few and the results may be affected by interstellar reddening as well as by molecular absorption. No reliable values of absolute magnitude are available though Keenan (1960) has given $M_v \simeq -2.3$ for these stars.

There are no separate measurements for S stars. Temperaturewise, S3, S5, S7, and S10 stars may be approximately equivalent to M3, M6, M8, and M10 of luminosity class III (Keenan 1960).

Mendoza V. et al (1965) have given BC and T_e for carbon stars, based on $UBVRIJKLN$ photometry. The stars observed are too few and for each

TABLE 1. Bolometric corrections, effective temperatures, and bolometric magnitudes

Spectral class	Bolometric correction			$T_e(°K)$			M_{bol}		
	V	III	II	V	III	II	V	III	II
M0	−1.20	−1.29	−1.30	3920	3680	3680	7.55	−1.59	
M1	−1.48	−1.37	−1.38	3680	3600	3600	8.04	−1.87	
M2	−1.76	−1.43	−1.44	3500	3600	3600	8.52	−2.23	
M3	−2.03	−1.74	−1.75	3360	3370	3370	9.00	−2.84	
M4	−2.31	−2.51	−2.52	3230	3060	3060	9.50	−3.51	
M5	−2.62	−3.33	−3.34	3120	2800	2800	9.95	−4.2:	
M6	−2.97	−4.8		2960	2550		10.65	−5.7:	
M7	−3.6	−7.7		2720	2150		11.30	−8.6:	
M8	−4.2	−8.6		(2660)	1900		12.05		
M9		−10.8			1650				

subclass, scatter appears to be larger than in M giants. This scatter may be due in part to abundance differences and in part to luminosity differences. They have combined these BC with the values of M_v given by Vandervort (1958) to obtain M_{bol}. Gordon (1968) has determined improved values of M_v, and using BC of Mendoza V. et al (1965), of M_{bol}, for carbon stars; she finds that R stars have luminosities like K giants but N stars are distinctly brighter than M stars and hence may be of luminosity class II. Temperature-wise, C_2, C_3, C_4, C_5, C_6, C_7, C_8 and C_9 correspond approximately to M0.5, M2, M3.5, M5, M6, M7, M8 and M9, respectively (Fujita et al 1965).

Knowing T_e and M_{bol} one can obtain the radius of a star. Any inaccuracy in these values will be reflected in the derived radii. There are hardly any direct measurements of the radii in this spectral range.

Thanks to the work of Limber (1958) and van de Kamp (1969), our knowledge about the masses of M dwarfs is fairly good. For stars fainter than $M_{bol} = 7.5$, one has the following approximate mass-luminosity relation (Harris, Strand & Worley 1963)

$$M_{bol} = 5.2 - 6.9 \log \mathfrak{M}$$

However, for the same spectral subclass, differing values of masses are found in the literature. Part of this scatter may be due to inaccurate spectral classification and part due to difference in stellar population.

No reliable values of masses are available for M giant and supergiant, S, and C stars. For Mira variables, it has been assumed that their masses are around 10-20$\mathfrak{M}_\odot$; however, Fernie & Brooker (1961) have given reasons why it should be $\sim 1 \mathfrak{M}_\odot$. For carbon stars, Morris & Wyller (1967) have assigned mass values of $\sim$2-3$\mathfrak{M}_\odot$ for early carbon stars and $\sim$10 $\mathfrak{M}_\odot$ for late ones from stellar evolution calculations of Iben (1964); these are indirect estimates and hence are of low weight.

ELEMENTAL ABUNDANCES

The determination of surface elemental abundances of cool stars is beset with several difficulties, e.g., uncertainty in fixing the continuum, line blending, ignorance of sources of opacity, and lack of good model atmospheres. Of late, a few analyses and estimates have become available.

M stars.—In going from M to S to C stars, the elemental abundance ratio of O/C decreases and the abundance of heavy elements increases (cf Fujita 1965). Spinrad & Vardya (1966) obtained, using infrared molecular bands, O/C = 1.05 for M giants and 1.04 for a mild S star, and Greenstein & Oinas (1968) have found this ratio very close to unity for one K dwarf star (ρ' Cancri) and somewhat less than unity for another K dwarf star (70 Oph A) compared to a value of $\sim$1.7 for the Sun (Lambert 1968). If the value of <1 for 70 Oph A is confirmed, this may be the first case of a dwarf carbon star. Note that the O/C ratio is very close to unity for galactic cosmic rays of energies >100 MeV/nucleon (Rosenvinge, Webber & Ormes 1969).

Merrill, Deutsch & Keenan (1962) have found weakening of certain atomic lines in Mira variables compared to nonvariable M giant stars. This they have interpreted as due to metal deficiency by a factor of ~100, but Feast (1963) has contested this simplified conclusion.

Lithium has not been found in K and M dwarfs. The abundance of Li decreases as one goes from G to K type stars, as expected, but it increases, on the average, as one proceeds from K to M giants (Conti & Wallerstein 1969). This increase has yet to be explained.

Boesgaard (1968) has found that Mg^{25} and Mg^{26} isotopes may be slightly enhanced over their terrestrial values in K and M giant and supergiant stars.

No detailed abundance analysis is available for M-type stars. Some K stars have, however, been analyzed in detail (see e.g. Pagel 1964, Pagel & Powell 1966, Griffin & Griffin 1967, Peat & Pemberton 1968; cf Helfer & Wallerstein 1968, Greene 1969). The abundances thus obtained are similar to solar values except for high-velocity stars, which are metal deficient.

S stars.—S stars are characterized by strong bands of ZrO, implying an enhanced abundance of Zr relative to M stars, and lines of Tc are observed in the spectra of S stars. No detailed abundance analysis is available for these stars.

S stars are similar in element abundances to Ba II stars, except for Tc which is not observed in Ba II stars. Several detailed elemental analyses are available for Ba II stars (Danziger 1965b, Warner 1965, Cowley 1968). These stars are characterized by a high abundance of *s*-process elements.

Tsuji (1968) has found Zr/Ti about 30 times the solar value, and $C^{12}/C^{13} \geq 10$ in S stars. Schadee & Davis (1968) have found Zr^{93} in a nonvariable mild S star to be overabundant, but only slight evidence of Tc compared to variable S stars (Davis 1968). If this is confirmed, it supports Danziger's (1965b) suggestion that Ba II stars evolve to S stars. In this connection, the possibility that many S stars have Nb (Merrill 1948, Davis & Keenan 1969) is exciting.

C stars.—Carbon stars have C/O ≥1. There is a sequence of carbon stars so far as hydrogen abundance is concerned; some have near normal abundance and some are highly deficient in H. In many of the carbon stars N is enhanced, perhaps He also, and *s*-process elements. The high-velocity CH stars are metal-deficient carbon stars (Wallerstein & Greenstein 1964).

Most of the carbon stars analyzed are hot stars (cf Danziger 1965a, Warner 1967, Hunger & Klinglesmith 1969). Recently, Fujita & Tsuji (1965) have analyzed, in detail, Y CVn, a C5,4 star with T_e~2800°K. They have found that *s*-process elements Sc, Y, La, Nd, etc, are enhanced

100 to 1000 times the solar value. Lithium is also overabundant, with Li/Na ratio with respect to the Sun being $\sim 4 \times 10^3$.

Lithium is present in most of the low-velocity carbon stars but not in the high-velocity ones (Torres-Peimbert & Wallerstein 1966). This has yet to be explained satisfactorily.

Sanford (1942) identified lines at λ5184.2 Å and λ6210.9 Å with CaCl in the spectrum of U Cygni; the presence of chlorine in cool stars has yet to be confirmed.

The C^{12}/C^{13} ratio in carbon stars is far smaller than the terrestrial value of 90 and in some cases it is close to the theoretically predicted equilibrium value of 4 from the CNO cycle (cf Climenhaga 1966; Fujita, Tsuji & Maehara 1966).

Gordon (1968) has suggested, on the basis of the observed intensities of spectral lines, that N stars are perhaps a logical lower-temperature extension of Ba II stars. In fact, Ba II, S, and C stars are similar so far as the enhancement of s-process elements is concerned (cf Utsumi 1967).

Emission Lines

There are a large number of dwarf M stars which show emission lines of H, Ca II, He, Fe I, Fe II, Ti II, [S II], etc (Joy 1960).

The long-period variable stars also show emission lines of H, Ca II, [Fe II], AlH and many unidentified lines (Merrill 1960). These emission lines show cyclical changes in intensity with variability. Even in N-type spectra, bright emission lines of M and S types have been observed.

The presence of these lines indicates the existence of a chromosphere and of physical conditions, in the outer layers of the star, in which deviations from local thermodynamic equilibrium (LTE) exist.

Magnetic Field

There are hardly half a dozen cool stars for which magnetic field has been established (Ledoux & Renson 1966). The measured magnetic fields are less than 2000 G. Some of these stars show abundance anomalies (Babcock 1958) and variation in magnetic-field strengths. The statistics are so meager that it is difficult to draw any general conclusion. No measurements are available for the magnetic field in M dwarf and carbon stars.

A large number of M dwarf stars are flare stars. Whether the flare activities of these stars are of similar origin to that of the Sun is not known. Wilson (1969) has found variation in the Ca II H-K line fluxes in the dwarf stars 61 Cygni A (K5V) and B (K7V), which indicate solar activity type of cycle. Perhaps this may be true for M dwarf stars also. A determination of the magnetic field will help in understanding the origin of the activity.

Space Velocity

The dMe stars have considerably lower space motion relative to the Sun and also lower velocity dispersion than dM stars (Delhaye 1965). This had led to the belief that dMe stars belong to Population I and dM stars to Population II. However, dMe stars, which show H emission along with Ca emission, appear kinematically to be closer to dM stars than to dMe stars which show only Ca-emission features (Gliese 1958). Does this indicate that dM stars evolve via dMe (H-emission) stars from dMe (Ca-emission) stars? Applying various other population criteria may help to resolve this problem.

We will not discuss the kinematic properties of other cool stars. For a detailed discussion see *Galactic Structure*, edited by Blaauw & Schmidt (1965).

Rotation

Rotational velocities decrease as one goes from B to A stars, and around G stars they become negligible; only a low upper limit of 12 km/sec has been set for cool stars. This may be due to the outer convective zone in these stars (cf Wilson 1966). If the ideas relating to spindown (Dicke 1964) are correct, the cool stars may be rotating with fairly high velocities at deep layers.

Variability

The red region of the H-R diagram abounds in intrinsic variable stars. They can be broadly divided into two broad groups: 1. eruptive variable stars and 2. pulsating variable stars, with each group having several subgroups (cf Ledoux & Walraven 1958, Joy 1960, Merrill 1960). Here we will just enumerate the various types with a few of their defining features.

Eruptive variable stars.—
1. Many emission lines (H, Ca II, He, Fe II, Fe I, Ti II, S II, etc) of T Tauri or chromospheric type: These are found from dGe to dMe.
2. Few emission lines (H, Ca II, and He I only) of GW Aurigae type: These are found from dG5e to dM2e.
3. Haro's flash stars: They show outbursts of light of duration $\sim$100 min. H and Ca II emission lines are weak or absent. The spectral type is between dK6 to dM6.
4. UV Ceti-type flare stars: They usually show very strong and narrow emission lines. During flare, emission lines of H are wider and stronger. They are closeby M dwarf stars (dM3e–dM6e).

Pulsating variable stars.—
1. Irregular variables: They show slow variation in luminosity with no periodicities and are mostly M, S, and C spectral-type giants and supergiants.
2. Semiregular variables: They have well-defined periods, which show appreciable irregularities. They are mostly giants and supergiants of class

M but some belong to G and K classes. The period ranges between 50 to 300 days.

3. Mira-type variables: They have periods between 70 to 800 days with amplitude of light variation between 3 to 10 mag. They show emission lines near maximum light, specially of H. They belong to M, S, and C spectral types. The period and luminosity follow an inverse correlation.

Polarization

Polarization has been measured in M, S, and C type variable stars by Appenzeller & O'Dell (1967), Dyck (1968), Kruszewski, Gehrels & Serkowski (1968), Serkowski & Kruszewski (1969), and others. Most of these measurements extend from 0.33–1.0 μ. The polarization and position angle of the electric vector are wavelength dependent and show large fluctuations with time. The wavelength dependence of polarization also changes with time. In M-type semiregular and Mira variables, the polarization increases steeply in the ultraviolet; in carbon stars, a flat wavelength dependence in the blue-yellow region is found. There may be some correlation between changes in polarization with light variations; this may be due to intrinsic changes in the circumstellar envelope of the star. Dyck & Johnson (1969) have found a correlation between the intensity of the Ca II K2 peak and polarization in non-Mira-type variable M giant and supergiant stars in the sense that stars with stronger K2 peak show smaller variation in polarization; they have tried to interpret this relation by associating Ca II emission with chromospheric heating and polarization with molecules or particles.

In a few stars, polarization is as large as 13.5 percent and may change, with time, from 2.2 to 13.5 percent, in the ultraviolet. Ordinary red giants may have no or very little intrinsic polarization (Zappala 1967). Carbon stars, on the average, may have a slightly higher polarization than other variable stars, though the statistics are rather poor at the moment (Zappala 1967, Kruszewski et al 1968).

The very infrared stars in Cygnus and Taurus have a polarization of $\sim$5 percent at $\lambda \sim 1.62$ μ and $\sim$12 percent at $\lambda \sim 0.5$ μ, respectively. Note that the Taurus object is a Mira-type variable, whereas the Cygnus star has shown no variability (cf Wing et al 1967).

The polarization may be due to Rayleigh scattering, which gives a far higher degree of polarization than the classical value of 5.5 percent if limb darkening is taken into account (Harrington 1969). Aligned graphite platelets for cool stars (Donn, Wickramasinghe, Hudson & Stecher 1968) and SiO_2 particles (Hoyle & Wickramasinghe 1969) might be an alternative to explain polarization.

The wavelength dependence of the position angles can be interpreted, following Treanor (1963), in terms of the starlight passing through more than one layer with different mean grain sizes, and will have different orientations, depending on the strength and configuration of the stellar

THEORETICAL MODELS OF THE ATMOSPHERE
BASIC PROBLEM

A model atmosphere for a star of given mass, radius, luminosity, and relative elemental abundances, or in the plane-parallel case, of given T_e, g (the acceleration due to gravity), and relative elemental abundances, provides the law of variation of the physical variables, like the total pressure P, temperature T, and other quantities, as a function of a suitably chosen depth parameter. The basic problem is to match the predicted emergent spectrum of radiation from such a model with observations. Such a matching can be crude, covering only the continuum spectrum or very detailed, covering all the observed features. This requires an iterative procedure. For a detailed account refer to Pecker (1965).

The temperature stratification depends on the mode of transport of energy. If radiative equilibrium prevails, the transfer of energy is given for the plane-parallel atmosphere, by the equation:

$$(1/c)\, \partial I_\nu/\partial t + \mu\, \partial I_\nu/\partial x = -\kappa_\nu I_\nu + \epsilon_\nu \qquad 1.$$

Here

$I_\nu d\nu$ (erg/cm²/sec/sr) = specific intensity of radiation
μ = cosine of the angle between the direction of the radiation and the normal to the atmosphere
κ_ν (cm⁻¹) = coefficient of absorption
$\epsilon_\nu d\nu$ (erg/cm³/sec/sr) = coefficient of spontaneous emission
c (cm/sec) = velocity of light

and x and t are space and time coordinates. Unless one is interested in phenomena of duration of the order of the characteristic time $1/(c\kappa_\nu)$, one can neglect the time dependence, in which case, Equation 1 reduces to:

$$\mu\, dI_\nu/d\tau_\nu = I_\nu - S_\nu \qquad 2.$$

where $S_\nu = \epsilon_\nu/\kappa_\nu$ is the source function and the optical depth τ_ν is given by

$$d\tau_\nu = -\kappa_\nu\, dx$$

The equation of radiative transfer retains the form 2 even when scattering is present and line formation is incorporated.

If the atmosphere does not contain any sink or source, as is usually the case, the above equation is solved under the condition that the total flux, integrated over all wavelengths, remains constant at each layer of the atmosphere. If part of the energy is carried by convection, then the sum of the radiative and convective fluxes should remain constant.

The pressure stratification, on the assumption of hydrostatic equilibrium, is given by:

$$dP/d\tau = g\rho/\kappa \qquad 3.$$

Here P is the total pressure and includes the partial gas, radiation, turbulent, magnetic etc pressures, g is the inertial acceleration due to gravity, and ρ is the mass density. In general, one uses for κ, opacity at a standard wavelength or some kind of mean value, and τ is defined accordingly.

It is appropriate now to consider the equation of state, sources of opacity, and convection before discussing the construction of model atmospheres.

Equation of State

The main constituents in the equation of state of cool stars are H_2, H, H^+, He, and e^- except when elemental abundances are very abnormal. In the computation of opacity, however, one needs to consider a large number of constituents, some of which, though small in concentration, may contribute directly to opacity, like H^-, or may affect the opacity indirectly by increasing or decreasing the concentration of an active opacity source. One has to consider, therefore, atoms, ions (both positive and negative), and molecules of a large number of elements. For very cool stars, even liquid and solid phases need to be considered. We will assume here, unless noted otherwise, that matter is in a gaseous phase with no interparticle interaction and is in thermodynamic equilibrium.

The electrons are donated chiefly by the low ionizing metals, like Na and K. At very high densities, the electrons may be depleted by the formation of negative ions (Vardya 1966a, Carson 1969). It is important to consider all the important molecules formed out of the elements considered; otherwise the electron pressure, thus obtained, will be an upper limit (Vardya & Kandel 1967).

If there are n elements, then for given relative elemental abundances, temperature, and pressure, one has to solve n simultaneous equations to obtain the various thermodynamic quantities. For pressure, one generally takes either p_e, the electron pressure, or P_H, the fictitious partial pressure of the nuclei of hydrogen; the method of approach is slightly different in the two cases (Vardya 1966f).

The equilibrium constants are given by the law of mass action, with appropriate values of the ionization, detachment or dissociation energy, partition functions, and reduced mass. Generally one assumes that the partition function is a function of temperature only. However, at high densities, one may have to consider pressure ionization (cf Brush 1967) or pressure dissociation (Vardya 1965b, 1966e; Grossman 1969; Vardya, Giannone & Virgopia 1969) as well. For negative ions (nonmolecular) one takes for the partition function the statistical weight of the ground state of the neutral atom of the element with the same number of electrons. In the case of molecular negative ions, its structure is assumed to be similar to that of the parent molecule. The internal partition function of a molecule is usually assumed to be a product of electronic, vibrational, rotational, and nuclear partition functions, though strictly this is not correct. Except for H_2, one can ignore nuclear partition function.

The first ionization energies of the elements of astrophysical interest are known with sufficient accuracy (cf Moore 1949, 1952, 1958). Vardya (1967a) has tabulated the detachment energies and partition functions of a large number of negative ions, which may be important in cool stars; some of these detachment energies, e.g., that for C_2^- and CN^-, are not well known, and H_2O^- may not even exist (Vardya 1970a).

For diatomic molecules, the main source of basic spectroscopic data is Herzberg (1950), though this needs to be updated in individual cases and uncertainties do persist, even for some important molecules like CN (cf Gaydon 1968). For tri- and polyatomic molecules, however, the spectroscopic constants are very inadequately known. Equilibrium constants can be computed using these spectroscopic data though several extensive tables are now available (JANAF 1960; McBridge, Heimel, Ehlers & Gordon 1963; Tsuji 1964; Tatum 1966).

Condensation.—It is possible that in cool stars, especially in their circumstellar envelopes, condensation takes place. Merrill (1916) tried to explain the pulsation of long-period variables by the condensation of carbon soot. Recently, Kamijo (1963, 1966), Lord (1965), Donn et al (1968), Wickramasinghe (1968), and Gilman (1969) have considered the composition and formation of condensates. It has not been possible to treat condensation in a satisfactory way in astrophysics. There is no satisfactory theory of nucleation; some progress can be made by using the theory of homogeneous nucleation. The vapor pressure data, which determines where condensation can take place, are rather unsatisfactory for many of the compounds of interest. If condensation takes place, it will affect the abundances and hence opacity, which in turn will affect the total pressure at a given level of the star.

Thermodynamic derivatives.—The treatment of convection involves the knowledge of a few derivatives (like specific heat and adiabatic gradient) of the basic thermodynamic variables. For simple cases, one can write explicit expressions to determine these derivatives but in complicated cases it is simplest to calculate the various thermodynamical quantities by direct numerical differentiation (Vardya 1965a).

Sources of Opacity

Our understanding of the sources of opacity in cool stars is far from satisfactory and we should be prepared for surprises as the recent work of

Wing & Spinrad (1970) has shown. An earlier review (Vardya 1966c) on the subject needs considerable updating, which we will attempt here.

Pure continuous absorption sources.—
Bound-free and free-free absorption by H^-: It is an important source of opacity, specially in cool dwarf stars. The bound-free absorption cross sections for H^- (Geltman 1962; Doughty, Fraser & McEachran 1966; Krogdahl & Miller 1967) are known to better than 5 percent except near the threshold (John 1966b).

The free-free absorption coefficients for H^-, as computed by Geltman (1965), John (1966a), Doughty & Fraser (1966), and Dalgarno & Lane (1966) agree fairly well, and may be accurate to 10 per cent.

Weinberg & Berry (1966) have considered the forbidden transitions induced when H^- is in the field of H^+. In the Sun, this may amount to ~ 10 percent of the total absorption (Myerscough 1968a). It may be important in stars cooler than the Sun. It appears, however, that the formulation and numerical results as given by Weinberg et al (1966) need revision (Tarafdar & Vardya 1969a) before their importance can be assessed.

The effect of interparticle interaction on the bound-free absorption coefficient of H^-, under the Debye-Hückel approximation, shifts the maximum of the coefficient as well as the threshold wavelength redward (Tarafdar & Vardya 1970). This may be important in high-density cool dwarfs. However, the effect of H and H_2 as perturbers, when considered, may be even more important.

Free-free absorption by H_2^-: Wildt (1942, 1957) suggested the possibility of the importance of bound- and free-free absorption by H_2^- in cool stars. It appears that H_2^- may not be a stable ion (Somerville 1966, Taylor 1967). If this is confirmed, then bound-free transitions need not be considered.

The free-free absorption is important in cool stars because of the very high abundance of H_2 in these stars. The cross sections of Dalgarno et al (1966), who have used experimental phase shifts, are of the order of free-free cross sections for H^-; these cross sections are an improvement over the ones given by Somerville (1964), though it is difficult to estimate their accuracy.

Free-free absorption by He^-: The only bound state of He^- is a metastable level about 19 eV above the ground state of He; hence we need to consider only the free-free transitions.

The first calculations of the free-free absorption coefficient for He^- by Somerville (1965) have been improved by McDowell, Williamson & Myerscough (1966). For $\lambda \geq 1\,\mu$, the values given by John (1968) are better than those of McDowell et al. These results may be uncertain by about 20 percent. The values computed by Dalgarno et al (1966), using experimental elastic scattering cross sections, are close to, though somewhat smaller than, the values given by McDowell et al. These results can be further refined by using the phase shifts given by LaBahn & Callaway (1966). The contribution of He^- opacity will be less than 20 percent in solar-composition dwarf stars, but will be large in the hydrogen-deficient and helium-enriched stars.

Bound-free and free-free absorption by C^-: The photodetachment cross sections for C^-, given by Cooper & Martin (1962), Henry (1966), Robinson & Geltman (1967), and Myerscough & McDowell (1964) for dipole velocity formalism, do not show the correct energy variation near the threshold when compared with experimental results (Seman & Branscomb 1962); the dipole length results of Myerscough et al accord reasonably well with experimental results. The free-free cross sections have been computed by Myerscough & McDowell (1966). The bound- and free-free cross sections for C^- are reliable to no more than 50 percent.

In carbon stars with surface temperatures greater than 2500°K, bound-free absorption may provide as much as 10 percent of the opacity, though this contribution will be increased in hydrogen-deficient carbon stars (Vardya 1967a, Myerscough 1968b). The free-free absorption will be important only in hydrogen-deficient stars, if at all.

Bound-free absorption by Cl^- and F^-: The bound-free continuum of Cl^- and F^-, operative shortward of $\lambda = 0.3 \mu$, may not be negligible (Vardya 1966d). The bound-free cross sections for Cl^- and F^- have been computed by Cooper et al (1962) and Robinson et al (1967). Berry, David & Mackie (1965) have measured the photodetachment cross sections for Cl^- and Berry & Reimann (1963) for F^-. The cross sections for Cl^- are two to three times larger than that for F^-. Unless F is more abundant than Cl, F^- will be far less important than Cl^- as a source of opacity.

Bound-free absorption by S^- and Si^-: Robinson et al (1967) have estimated the bound-free cross sections for S^- and Si^-. They may contribute a few percent to the total opacity in the spectral region $\lambda \lesssim 0.8 \mu$.

Bound-free absorption by H_2O^-: In very cool stars having solar composition, the abundance of H_2O is very large. Hence the possibility exists that H_2O^-, if it is a stable ion, may contribute to opacity. However, the electron affinity of H_2O is not known (Vardya 1970a).

Bound-free absorption by C_2^- and CN^-: The negative ions C_2^- and CN^- are fairly abundant in carbon stars (Vardya 1967a). The bound-free cross sections are not known. These ions may contribute to the blackout in carbon stars, shortward of $\lambda 4500$ Å.

Bound-free absorption by SH^-, CH^-, and SiH^-: The bound-free absorption by SH^-, operative at $\lambda < 0.535 \mu$ may be significant in M dwarf stars (Vardya 1967a) and may explain the Lindblad (1935) depression in these stars. The absorption cross sections, however, are not known.

Bound-free absorption by CH^- and SiH^-, having photodetachment energies of 1.61 and 1.46 eV (Cade 1967), may also be important in cool stars, and needs to be investigated.

Bound-free and free-free absorption by H and C: The bound-free and free-free cross sections for atomic hydrogen are well known (Vardya 1964). For $T < 5000°K$, it is not very important.

In carbon stars, especially the early-type hydrogen-deficient ones, bound- and free-free absorption by atomic carbon may be important (cf Myerscough 1968b). Absorption coefficients, using the quantum defect method, have been given by Peach (1967) and Wilson & Nicolet (1967).

Free-bound absorption by quasi-H_2: An absorption continuum is produced when a quasimolecule of two colliding hydrogen atoms in the $1s\ ^2S$ state absorbs a light quantum, leaving the hydrogen molecule, thus formed, in the unstable $1s\sigma 2s\sigma^3 \sum_g{}^+$ state. Wildt (1949) found it unimportant in solar-composition late-type stars; this has been confirmed by more refined work (Solomon 1964, Soshnikov 1964, Doyle 1968a).

Following the suggestion by Varsavsky (1966), Vardya (1967b) and Doyle (1968b) have shown that in metal-deficient stars, the main contribution of quasi-H_2 is in the ultraviolet region and is not important for $T \lesssim 3200°K$.

Photoionization of molecules: The ionization potential of most of the astrophysically interesting molecules is more than 10 eV. Hence, the photoionization continua will be important only in the far ultraviolet (cf Ditchburn & Öpik 1962).

Photodissociation of molecules: This should be an important source of opacity in cool stars. However, it has yet to be investigated, in detail, for conditions obtaining in these stars. Recently Goon & Auman (1970) have suggested that photodissociation of NaCl, operative for $\lambda \leq 0.37\mu$, is worth considering.

Pure continuous scattering sources.—

Electron scattering: The free electron or Thomson scattering may contribute a few percent to the total opacity in the far-infrared region in the outermost layers of late-type giant and supergiant stars (cf Vardya 1964).

Rayleigh scattering by H and H_2: The Rayleigh scattering due to H and H_2 is important in the atmospheres of giants and supergiants and its relative contribution increases with decrease in the relative abundances of the metals, the electron donors. The Rayleigh scattering cross section for atomic hydrogen has been given by Dalgarno & Kingston (1960) and for molecular hydrogen by Dalgarno & Williams (1962), which are good for wavelengths larger than Lyman α.

Rayleigh scattering by He, C, and N: The Rayleigh scattering contribution of He and heavier elements to opacity is very small in stars with solar elemental abundances. However, in carbon stars, specially in hydrogen-deficient ones, the contribution of some of these less abundant elements may not be negligible. At very long wavelengths, the Rayleigh scattering cross sections for H, He, C, N, and O are approximately in the ratio of 1: 0.1: 7.9: 4.1: 1.4 (Tarafdar & Vardya 1969b). Considering the various possible peculiar abundances in stars, it appears that oxygen can be neglected in all cases.

*Bound-bound transitions.—*Cool stars are havens for molecules. The bound-bound rotation, rotation-vibration, and electronic transitions of the various molecules are effective agents in blocking the radiation from various parts of the spectrum.

Pressure-induced opacity of H_2: Though abundant, allowed electric dipole rotation and rotation-vibration lines of H_2 and N_2 are absent in their spectra because they have no permanent dipole moment. Very weak quadrupole lines of H_2 have been observed (Spinrad 1966). However, a collision of the type H_2-H_2 or H_2-He induces a dipole moment, which gives rise to complex spectra. Linsky (1969) has computed pressure-induced monochromatic opacity for collisions of the type H_2-H_2 and H_2-He considering translational, rotational, and vibrational transitions. For $T \lesssim 2500°K$, it should be an important source of opacity in the spectral range 1–10 μ, not only in dwarf stars but in giants as well. One should investigate H_2-H type collisions also.

In cool hydrogen-deficient and nitrogen-enriched stars, one needs to consider the opacity of N_2 induced by the collisions of types N_2-He and N_2-H_2 also.

H_2O: Water vapor is an important constituent of cool stars with solar-type elemental abundances. Tsuji (1966a,b) has considered the H_2O rotation-vibration and rotation lines and has combined the numerous lines, assuming partial-overlapping as well as just-overlapping approximation. Auman (1967) has carried out a very detailed analysis of the rotation-vibration lines of H_2O, considering positions and strengths of more than two million lines, and has given harmonic means at intervals of 100 cm^{-1}.

The lines, especially the pure rotation ones, will be affected as a result of pressure broadening by collisions of the type H_2O-H_2 and H_2O-H; this effect should be more important for dwarf than for giant stars. and needs to be considered along the lines of H_2O-N_2 collisions computed by Benedict & Kaplan (1959).

CO: Tsuji (1966a) has considered the rotation-vibration and pure rotation lines of CO in an approximate way. Kunde (1968) has computed, in detail, the CO absorption coefficient, considering fundamental, first- and second-overtone rotation-vibration bands, for temperatures ranging from 175° to 3500°K.

CN: Wing et al (1970) have found CN to be "the most important of all bound-bound opacity sources in K giants, K and M supergiants, and carbon stars" in the infrared. Detailed opacities are being computed by S. D. Price (Wing 1970).

TiO, MgH, CaH, and SiH: Tsuji (1969) has considered the electronic transitions of TiO, MgH, CaH, and SiH. The oscillator strengths of these transitions are, however, not well established. The electronic transitions of C_2, VO, AlO, and ZrO also need to be looked into.

Rotation-vibration and rotation bands of other molecules: The rotation and rotation-vibration bands of SiO, OH, NO, CN, and C_3 may be important in the atmospheres of cool stars. Tsuji (1966b) has computed, in an approximate way, the rotation and rotation-vibration absorption for OH. The detailed absorption coefficients for SiO and other above-mentioned molecules are not yet available.

Microturbulence and line opacity: The microturbulence, through line broadening coupled with an increase in equivalent width, contributes indirectly to opacity. This introduces an unknown parameter, which can be determined only after a detailed analysis. However, one can estimate it fairly well by using "curve of linewidth correlation" (Van den Heuvel 1963).

Opacity due to solid particles.—The extinction due to grains depends on the refractive index, composition, size distribution, mean size, and orientation, and can be computed using Mie's theory (Gaustad 1963, Krascella 1965, Main & Bauer 1966).

Blanketing.—The main blanketing in the atmospheres of cool stars comes from the molecular lines. A proper appraisal requires not only the bound-bound transition probability, but also the shape and position of the line, as well as knowing whether the line is formed by scattering or absorption. Normally, it is assumed that the lines are not shifted from their unperturbed position. For the shape of the line, one has to consider the natural, Doppler (including turbulent), and collisional broadening. Combination of these broadenings gives the conventional Voigt profile, which is modified if the velocity of the radiating atom or molecule is also considered (Edmonds 1968).

As the number of lines involved is very large, simplified methods need to be evolved to consider the line spectra (cf Böhm 1966). The necessity to consider such a large number of lines arises because in the harmonic mean of the opacity, the main contribution comes from the transparent windows, where a large number of weak lines may be present, rather than from opaque portions of the spectrum (Auman 1967). In taking a mean, a good amount of this detailed information will be lost. This requires a judicious decision about the strength of the weakest lines one needs to consider. No satisfactory criterion for such a decision exists.

A large number of idealized statistical models exist for considering line blanketing (cf Vardya 1966g, Golden 1967, Kyle 1967). Strom & Kurucz (1966) and Mihalas (1967) have given a statistical procedure in which the detailed opacity variation is transformed into an equivalent relation and is perhaps more representative of the real situation short of considering each line in detail. In this procedure, one divides the spectrum into a number of frequency intervals and determines what fraction of each of these frequency intervals is occupied by a given value of opacity. Thus the opacity versus frequency curve is transformed into a weight parameter versus opacity curve, which is a far smoother curve. In this way, the labor in studying the transport of energy is considerably reduced, though the details of the emergent line spectrum are lost.

Convection

In the outer layers of cool stars, the transport of energy is not only by radiation but also by convection. The convective instability sets in when (cf Cox 1968)

$$\nabla_{ad} \equiv (d \ln T/d \ln P)_{ad} < (d \ln T/d \ln P)_{rad} \equiv \nabla_{rad} \qquad 4.$$

Here ∇_{ad} and ∇_{rad} are the adiabatic and radiative gradients, respectively. Such a condition may arise because of ionization or dissociation of an abundant element, or high opacity, or very large temperature gradients. Note that the above criterion is modified in the presence of a magnetic field (cf Kovetz 1967).

The fulfillment of the above instability condition does not ensure that convection is efficient. In fact, there exists an extensive superdiabatic zone in cool stars, especially giants and supergiants, where the adiabatic gradient is larger than the true gradient. At present no satisfactory treatment exists for this superadiabatic zone. The most widely used formalism (Böhm-Vitense 1958) is, at best, a crude approximation. It is based on the Boussinesq approximation, with several arbitrary parameters (cf Henyey, Vardya & Bodenheimer 1965), especially the mixing length. Even if we were able to specify the mixing length, this formalism should not be considered as anything but a stopgap arrangement.

Various refinements have been introduced into the work of Mrs. Böhm-Vitense. Shaviv & Chitre (1968) explicitly introduced the aerodynamic drag in the formalism. Spiegel (1963) has tried to generalize the mixing-length theory to large mixing lengths; in this treatment, mixing length is no longer a local quantity. Ulrich (1969) has attempted to improve the nonlocal mixing-length theory and has considered the effect of convective overshoot on the overlying layers in radiative equilibrium.

If the radiative core is rotating faster than the surface layers (Dicke 1964), the convection in the outer layers will be affected; this may modify the surface abundance of Li and of other elements (cf Spiegel 1968).

Construction of Model Atmospheres

The computation of a model atmosphere generally involves iteration. An initial guess of the (T,τ) relation leads to a (P,τ) relation via the equation of hydrostatic equilibrium. If the integrated flux is not constant at different values of τ, a correction to the temperature is applied and the process repeated till relative errors in both the flux and the flux derivative are smaller than a given tolerance. A large number of procedures have been proposed for temperature corrections for rapid convergence in the construction of nongray atmospheres (cf Pecker 1965, Mihalas 1967, Peterson 1969). Most of these methods are for atmospheres in radiative equilibrium.

Mihalas (1965) extended the method of Avrett and Krook to include convection. This method has been used by Auman (1966, 1969), with some modification, to construct models for cool stars. Auman (1966, 1969), Tsuji

(1966b, 1969), and Carbon, Gingerich & Latham (1969) have included molecular line opacities in the models. The problem of convergence of the iterative solution still presents certain difficulties (cf Auman 1969) when convection is considered, bound-bound opacity is incorporated, and distinction is made between pure absorption and scattering. The method of integral equation, first suggested by Mrs. Böhm-Vitense, offers hope (Peterson 1969), though its superiority over other methods has yet to be established as far as cool stars are concerned.

Main Features of the Computed Models

Table 2 lists some of the low-temperature, nongray models computed so far. All these models have been computed on the assumptions of plane-parallel homogeneous layers, hydrostatic equilibrium, local thermodynamic equilibrium, and no condensation. Except for Auman (1966, 1969), the others have not incorporated convection. Solar elemental abundances have been assumed except for a few cases in which the effects of the variable C/H ratio (Auman 1969) and of metal deficiency (Gingerich, Latham, Linsky & Kumar 1966, Auman 1969, Carbon et al 1969) have been considered. In all of these models H, H^-, He^-, H_2^-, H_2^+, and metals as sources of continuous absorption and H, H_2, and e^- as sources of scattering have been considered. Many of these models also include molecular line blanketing, as indicated in the table. So far, no model has been computed with chemical composition and opacity relevant to cool S or C spectral-type stars.

We will now discuss some of the main conclusions from these models.

Convection.—The convective instability sets in at a very shallow optical depth in M-type stars as a result of dissociation of molecular hydrogen (Vardya 1960, 1966f; Gingerich et al 1966; Auman 1969; Berg, Hershey & Kumar 1969; Carbon et al 1969). For example, for a star of $T_e = 3000°K$, log $g = 1.0$ or 5.0, the convective instability sets in at $\tau_{1.17\mu} = 0.015$ (Auman 1969). In low-density giant and supergiant atmospheres, the convective zone may separate into H_2-dissociation and H-ionization zones, though not in dwarfs. The flux carried by convection, at a given optical depth, is far larger in dwarf stars than in giant stars; at $\tau_{1.17\mu} = 1$, for example, about half the flux is carried by convection in a dwarf star of $T_e = 3000°K$, whereas it is negligible in a giant star of the same temperature at that optical depth. The superdiabatic zone is far thinner in dwarf stars than in giant stars; this makes giant and supergiant stars more susceptible to uncertainties in the treatment of the superadiabatic zone than the dwarf stars.

Density inversion occurs in many of the models computed. In the models computed by Gingerich et al (1966) and Carbon et al (1969), it may be due to neglect of convection (Vardya et al 1967). One should expect density inversion in the models of Tsuji (1966b, 1969) since he neglected convection; however, he has not tabulated values of ρ. Auman (1969) has tabulated ρ for only one model and does find that a density inversion occurs even

TABLE 2. Grid of nongray model atmospheres

$T_e(°K)$	log g	Abundances	Convection	Molecular bound-bound opacity	References
4000	4.7; 2.0; 1.0	Solar	Yes	H_2O	Auman (1969)
4000	2.0	0.1 and 0.01 times solar metals	Yes	H_2O	Auman (1969)
4000	4.5	Solar	No	None	Berg et al (1969)
3500	1.5; 0.5	Solar	Yes	H_2O	Auman (1969)
3500	5.0	Solar	No	None; H_2O	Carbon et al (1969)
3000	4.8; 1.0	Solar	No	H_2O, CO	Tsuji (1966)
3000	5.0; 1.0; 0.0	Solar	Yes	H_2O	Auman (1969)
3000	1.0	Solar except C/H	Yes	H_2O	Auman (1969)
3000	5.0; 1.0	Solar	Yes	H_2O-straight	Auman (1969)
3000	4.8	Solar	No	H_2O, CO, TiO, MgH, CaH, SiH, Pres. Ind. H_2	Tsuji (1969)
2520	3.0	Solar	Yes	H_2O	Auman (1969)
2500	5.0; 3.0; 1.0	Solar	No	None	Gingerich et al (1966)
2500	1.0	0.01 times solar metals	No	None	Gingerich et al (1966)
2500	0.0; −1.0	Solar	Yes	H_2O	Auman (1969)
2500	5.0	Solar	No	None; H_2O	Carbon et al (1969)
2500	5.0	0.01 times solar metals	No	H_2O; H_2O, Pres. Ind. H_2	Carbon et al (1969)
2000	−1.0; −2.0	Solar	Yes	H_2O	Auman (1969)
1500	5.0	Solar	No	H_2O; H_2O, Pres. Ind. H_2	Carbon et al (1969)

though he has considered convection. However, Auman has assumed a mixing length proportional to the gas pressure scale height rather than the total pressure scale height; note that this may lead to negative *effective* gravitational acceleration, with attending difficulties (Böhm-Vitense 1958, Cox 1968). The density inversion may be a real phenomenon, but the present evidence does not rule out its being due to inadequate treatment of the superadiabatic zone (Vardya 1970b).

T–τ relation.—The variation in T–τ relation is large over the range of T_e and g, for $T_e < 4000°K$; hence scaled models are not reliable at the lower end of the temperature scale. This variation is due both to the variation in the opacity and to convective efficiency.

For a heavy H_2O-blanketed model with $T_e = 2500°K$ and log $g = 0.0$, the ratio of surface temperature to effective temperature is 0.63 compared to ~0.81 for a nonblanketed model. Thus the line blanketing normally steepens the temperature gradient near the surface (cf Unno 1962). The lowering of the surface temperature appears to increase with increase in g value for a given value of T_e.

If enough flux is carried by convection in the layers with $\tau \lesssim 1$, the convective flux will be released in the overlying regions as radiation or as a mass loss carrying energy with it (Vardya et al 1967, Auman 1969). This will decrease the temperature gradient in the outermost layers of the atmosphere. Convection in the transparent region thus counteracts the lowering of the surface temperature due to blanketing. This effect should be more important in dwarf stars than in giant stars, because very little flux is carried by convection in $\tau \lesssim 1$ layers of giant stars.

The temperature stratification in the outer transparent radiative zone will be altered if overshooting of convective elements from the underlying convective zone occurs. Ulrich (1969) has found that, in the Sun, convective overshoot reduces the radiative flux by 40 percent at $\tau = 2.5$ and by 10 percent at $\tau = 0.2$. This effect should be important in cool stars and needs to be investigated.

Gingerich et al (1966) are the only ones who have found a temperature inversion very close to the surface. Their models are unrealistic in many respects (cf Vardya et al 1967) and it is very likely that this is not a genuine phenomenon in these stars. In fact, the inversion disappears when H_2O opacity is included in some of these temperature-inverted models (Linsky 1966, Carbon et al 1969).

Emitted flux.—The variation of flux with frequency at the surface depends on T_e and g, if relative elemental abundances are the same. At $T_e = 4000°K$, the frequency dependence of the flux does not differ very much in going from dwarf to supergiant stars and it shows a peak at $\lambda \sim 1.6$ μ, corresponding to the minimum of H^- opacity. The sharpness of this peak is somewhat reduced for the dwarf stars as a result of water vapor opacity becoming important (Auman 1969).

At $T_e = 3000°K$, the opacity due to H_2O dominates the infrared spectrum in all computed models. There is still a peak at $\lambda \sim 1.6$ μ but there is one also at $\lambda \sim 2.4$ μ, the peaks being higher for supergiants than for dwarfs. These peaks are due to gaps between the H_2O bands. The continuous opacity, mainly due to H^-, is small, which allows one to probe fairly deeply into the atmosphere at these wavelengths.

At still lower temperatures, the infrared opacity is very much dominated by H_2O and by MgH and TiO in the visible and near-infrared regions, and the computed surface fluxes reflect these features (Auman 1969, Tsuji 1969).

Variation of elemental abundances.—The concentration of H_2O is very sensitive to the elemental abundance ratio of C/O. Auman (1969) computed two models for $T_e = 3000°K$ and $\log g = 1.0$, one with 1.5×10^{-4}, the other with 7.2×10^{-4} for carbon abundance relative to hydrogen. Between these two models, the concentration of H_2O changes by a factor of ~ 4. However, the change in the rotation-vibration features of H_2O in the flux is rather small.

A reduction in the metal abundance decreases the temperature and the opacity and increases the surface density and the efficiency of the convection, at a given optical depth, and the upper boundary of the convective zone is shifted towards the surface (cf Auman 1969, Krishna Swamy 1969).

VALIDITY CHECKS ON THE ASSUMPTIONS

As stated earlier, the theoretical models of the atmosphere are, perforce, based on several assumptions. It is desirable, therefore, to discuss the validity and consequences of some of these assumptions.

Assumption of plane-parallel homogeneous atmosphere.—Auman (1969) has shown that the plane-parallel approximation is valid for dwarf stars but the errors in flux for cool giants and supergiants can be rather large because of the neglect of the curvature. For $T_e = 2000°K$, $\log g = -2.0$, as an example, the error in flux is ~ 20 percent.

The assumption of homogeneity is bound to be violated, perhaps drastically in giants and supergiants, because of convection in the H_2-dissociation and H-ionization zones. It is difficult to estimate the extent of the deviation without a satisfactory theory of convection. Further, in the outermost layers in radiative equilibrium, overlying the convective layers, inhomogeneity may arise because of the fluctuations in the lower layers and because of overshooting of convective elements in the radiative zone.

Hydrostatic equilibrium.—The assumption of hydrostatic equilibrium is very reasonable. Cox (1968) has shown that a departure of ~ 1 percent from hydrostatic equilibrium would lead to ~ 10 percent change in the radius of a star like the Sun in ~ 1 hour. Such changes have not been observed. However, deviations are very likely for giants and supergiants.

In the convective zone, one assumes pressure equilibrium between a convective element and its surroundings. This holds only when the convective velocity is smaller than the velocity of sound in the medium; otherwise shock formation will result in disrupting the equilibrium.

Convection.—As stated earlier, the structure of the outer layers, as well as of the star as a whole, is highly dependent on the treatment of convection, and a better treatment than the present crude formulation of the superadiabatic zone may alter the current ideas of the structure of the cool atmospheres drastically.

One assumes that convection does not change the dissociation/ionization equilibrium. In a supergiant star, where the densities are low and convective velocities high, it is very likely that deviation from this equilibrium occurs. This needs to be investigated, to start with, along the lines of Eddington (1941). This may explain, at least in part, the peculiar abundances observed in cool giant and supergiant stars.

The turbulent velocities obtained by theoretical computations are far smaller than the microturbulent velocities obtained from the curve of growth. A proper understanding should explain this difference, if they represent different phenomena, or the discrepancy should disappear, within observational error, when a better treatment of convection becomes available.

Local thermodynamic equilibrium.—Emission lines are observed in some of the dwarf as well as giant and supergiant stars, the lines of AlH in χ Cygni type stars are due to an "inverse predissociation" effect, and the infrared helium line at $\lambda 10830$ Å observed in some of the cool stars is a 19.7-eV line. All these are indicators of deviation from LTE.

Tsuji (1964) has considered the attainment of thermal equilibrium; he has concluded that even if the radiation field is not in thermal equilibrium, the molecular concentration will not deviate from equilibrium, because biomolecular reactions are predominant, even if collisions are unimportant.

Auman (1969) has found, following Pande (1968), that the rotation-vibration levels of H_2O are governed by collisional rather than radiative transitions in the atmospheres of M dwarf stars, though deviations may occur in giant and supergiant stars. He has, however, not considered bimolecular reactions in arriving at this conclusion.

The problem is difficult. The needed reaction rates and collisional cross sections are, in general, not available. Unless the complete problem is carried out, it is difficult to estimate the deviations from equilibrium. It is likely that the continuum picture is satisfactory so far as the LTE description is concerned, with deviation occurring in the description of certain lines or bands.

Molecular and other sources of opacities.—Most of the major sources of molecular opacities have perhaps been discussed though many of them have not yet been incorporated in the models computed. Detailed opacity calculations for many of the molecules, like CN, SiO, C_2, C_3, CH, and HCN, have not yet been carried out. Part of the difficulty lies in the unavailability of reliable transition probabilities. For example, the two estimates that are available for the (0,0) band of MgH are 0.008 and 0.002 (Tsuji 1969).

Doppler and turbulence broadening are the only mechanisms of line broadening that have been used in the computation. It is assumed that the turbulent velocity is the same in the entire line-forming region; this is an approximation and the existence of a turbulent velocity gradient will affect the flux as well as the determination of the chemical abundance.

Pressure broadening, which has not been treated so far, is bound to be important for pure rotation bands and in the dwarf stars, even for rotation-vibration bands.

As it is impossible to consider all the lines individually, one has to divide the spectrum into a number of segments and take some kind of mean over each segment. What kind of mean should one take, a harmonic or a straight mean? There is no sure way to decide. The models are affected, specially the ones for dwarf stars, when one switches from one kind of mean to another (Auman 1969).

In considering the molecular lines, it has been assumed that they are formed by pure absorption. If the lines are formed by scattering, the radiation will become decoupled from the local radiation field and large changes in the structure of the atmosphere will ensue. Chances of this happening increase as one goes from dwarf to supergiant stars, where the frequency of radiative transitions may be higher than that of collisional transitions.

Elemental abundances.—The models so far computed are based on solar elemental abundances. If the ratio of O/C is not 1.7 (solar value) but about 1.05 (Spinrad et al 1966), H_2O will be far less abundant. The concentration of other sources of opacities, like CN, will also be affected. And the structure of the atmosphere may be altered greatly. Similarly, the atmosphere will be affected if the relative abundances of other elements are changed.

Dissociation energies.—The knowledge of the dissociation energy and spectroscopic constants for the astrophysically important diatomic molecules is now fairly good except for a few cases. However, the same cannot be said for the tri- and polyatomic molecules.

SUGGESTIONS FOR FUTURE WORK

We have all along indicated the areas where further work will throw light on our understanding of the atmospheres of cool stars. Summarizing, we would like to stress the need for:

1. A high-dispersion study of a few representative very late-type stars.
2. Infrared observations for more cool stars and at larger wavelengths than have been made so far.
3. Ultraviolet spectra.
4. Luminosity classification for S and C stars.
5. Accurate determination of masses, radii, and effective temperatures of M giant and supergiant and S and C stars.
6. Polarimetric measurements of a large number of cool variable and nonvariable stars.
7. Accurate reaction rates and collision cross sections for conditions relevant to cool stars.
8. A satisfactory treatment of nucleation and condensation.
9. Detailed opacities for CN, C_2, C_3, SiO, HCN, etc. (This will require determination of accurate transition probabilities for many of these systems.)
10. A satisfactory theory of convection.
11. Models for S and C stars.
12. An estimate of the abundance of helium using HeI $\lambda 10830$ line, the

line-free infrared continuum (Vardya 1966b), and the effect of helium on the line formation.

13. The determination of the relative abundances of other elements and of isotope ratios.

14. Computation of synthetic spectra and their comparison with observations.

ACKNOWLEDGMENTS

It is indeed a pleasure to thank Drs. S. M. Chitre and G. A. Shah and Mr. S. P. Tarafdar for their help and suggestions in preparing this review, and Drs. J. R. Auman, P. C. Keenan, and R. F. Wing for their valuable comments on the manuscript.

This article is dedicated to Dr. R. Wildt, who inspired the author's interest in the field of cool stars and to the late Dr. L. G. Henyey, who encouraged the author to pursue this study during his stay at Berkeley.

LITERATURE CITED

Appenzeller, I., O'Dell, C. R. 1967, *Ap. J.*, **149**, L5
Auman, J. R. 1966, *Colloquium on Late-Type Stars*, 313 (Hack, M., Ed., Trieste)
Auman, J. R. 1967, *Ap. J. Suppl.*, **14**, 171
Auman, J. R. 1969, *Ap. J.*, **157**, 799
Babcock, H. W. 1958, *Ap. J. Suppl.*, **3**, 141
Benedict, W. S., Kaplan, L. D. 1959, *J. Chem. Phys.*, **30**, 388
Berg, R. A., Hershey, J. L., Kumar, S. S. 1969, *Low Luminosity Stars*, 493 (Kumar, S. S., Ed., New York: Gordon & Breach)
Berry, R. S., David, C. W., Mackie, J. C. 1965, *J. Chem. Phys.*, **42**, 1541
Berry, R. S., Reimann, C. W. 1963, *J. Chem. Phys.*, **38**, 1540
Blaauw, A., Schmidt, M., Eds. 1965, *Galactic Structure* (Univ. Chicago Press)
Blanco, V. M. 1965, *Galactic Structure*, Chap. 12 (Blaauw, A., Schmidt, M., Eds., Univ. Chicago Press)
Boesgaard, A. M. 1968, *Ap. J.*, **154**, 185
Böhm, K. H., Ed. 1966, *J. Quant. Spectrosc. Radiat. Transfer*, **6**, 534
Böhm-Vitense, E. 1958, *Z. Ap.*, **46**, 108
Brush, S. G. 1967, *Progress in High Temperature Physics and Chemistry*, **1**, 1 (Rouse, C. A., Ed., London: Pergamon)
Cade, P. E. 1967, *Proc. Phys. Soc.*, **91**, 842
Carbon, D., Gingerich, O. J., Latham, D. W. 1969, *Low Luminosity Stars*, 435 (Kumar, S. S., Ed., New York: Gordon & Breach)
Carson, T. R. 1969, *MNRAS*, **142**, 409
Climenhaga, J. 1966, *Colloquium on Late-Type Stars*, 54 (Hack, M., Ed., Trieste)
Conti, P. S., Wallerstein, G. 1969, *Ann. Rev. Astron. Ap.*, **7**, 99
Cooper, J. W., Martin, J. B. 1962, *Phys. Rev.*, **126**, 1482
Cowley, C. R. 1968, *Ap. J.*, **153**, 169
Cox, J. P. 1968, *Principles of Stellar Structure*, **1**, Chap. 14 (New York: Gordon & Breach)
Dalgarno, A., Kingston, A. E. 1960, *Proc. Roy. Soc. A*, **259**, 424
Dalgarno, A., Lane, N. F. 1966, *Ap. J.*, **145**, 623
Dalgarno, A., Williams, D. A. 1962, *Ap. J.*, **136**, 690
Danziger, I. J. 1965a, *MNRAS*, **130**, 199; 1965b, ibid, **131**, 51
Davis, D. N. 1968, *Ap. J.*, **152**, L13
Davis, D. N., Keenan, P. C. 1969, *Publ. Astron. Soc. Pac.*, **81**, 230
Delhaye, J. 1965, *Galactic Structure*, Chap. 4 (Blaauw, A., Schmidt, M., Eds., Univ. Chicago Press)
Deutsch, A. J., Wilson, O. C., Keenan, P. C. 1969, *Ap. J.*, **156**, 107
Dicke, R. H. 1964, *Nature*, **202**, 432
Ditchburn, R. W., Öpik, U. 1962, *Atomic and Molecular Processes*, 79 (Bates, D. R., Ed., London: Academic)
Donn, B., Wickramasinghe, N. C., Hudson, J. P., Stecher, T. P. 1968, *Ap. J.*, **153**, 451
Doughty, N. A., Fraser, P. A. 1966, *MNRAS*, **132**, 267
Doughty, N. A., Fraser, P. A., McEachran, R. P. 1966, ibid., **132**, 255
Doyle, R. O. 1968a, *J. Quant. Spectrosc. Radiat. Transfer*, **8**, 1555
Doyle, R. O. 1968b, *Ap. J.*, **153**, 987
Dyck, H. M. 1968, *Astron. J.*, **73**, 688
Dyck, H. M., Johnson, H. R. 1969, *Ap. J.*, **156**, 389
Eddington, A. S. 1941, *MNRAS*, **101**, 177
Edmonds, F. N. 1968, *J. Quant. Spectrosc. Radiat. Transfer*, **8**, 1447
Feast, M. W. 1963, *Ap. J.*, **137**, 342
Fernie, J. D., Brooker, A. A. 1961, *Ap. J.*, **133**, 1088
Fujita, Y. 1965, *Vistas in Astronomy*, **7**, 71 (Beer, A., Ed., Oxford: Pergamon)
Fujita, Y., Tsuji, T. 1965, *Publ. Dominion Ap. Obs.*, **12**, 339
Fujita, Y., Tsuji, T., Maehara, H. 1966, *Colloquium on Late-Type Stars*, 75 (Hack, M., Ed., Trieste)
Fujita, Y., Yamashita, Y., Kamijo, F., Tsuji, T., Utsumi, K. 1965, *Publ. Dominion Ap. Obs. Victoria*, **12**, 293
Gaustad, J. E. 1963, *Ap. J.*, **138**, 1050
Gaydon, A. G. 1968, *Dissociation Energies* (3rd ed., London: Chapman & Hall)
Geltman, S. 1962, *Ap. J.*, **136**, 935
Geltman, S. 1965, *Ap. J.*, **141**, 376
Gilman, R. C. 1969, *Ap. J.*, **155**, L185
Gingerich, O., Latham, D. W., Linsky, J., Kumar, S. S. 1966, *Colloquium on Late-Type Stars*, 291 (Hack, M., Ed., Trieste)
Gliese, W. 1958, *Z. Ap.*, **45**, 293
Golden, S. A. 1967, *J. Quant. Spectrosc. Radiat. Transfer*, **7**, 225
Goon, G., Auman, J. R. 1970, *Ap. J.* (In press)
Gordon, C. P. 1968, *Publ. Astron. Soc. Pac.*, **80**, 597
Greene, T. F. 1969, *Ap. J.*, **157**, 737
Greenstein, J. L., Oinas, V. 1968, *Ap. J.*, **153**, L91

Griffin, R., Griffin, R. 1967, *MNRAS*, **137**, 253
Grossman, A. S. 1969, *Low Luminosity Stars*, 247 (Kumar, S. S., Ed., New York: Gordon & Breach)
Harrington, J. P. 1969, *Ap. Letters*, **3**, 165
Harris, D. L., Strand, K. A., Worley, C. E., 1963, *Basic Astronomical Data*, Chap. 15 (Strand, K. A., Ed., Univ. Chicago Press)
Helfer, H. L., Wallerstein, G. 1968, *Ap. J. Suppl.*, **16**, 1
Henry, R. J. W. 1966, *J. Chem. Phys.*, **44**, 4357
Henyey, L. G., Vardya, M. S., Bodenheimer, P. 1965, *Ap. J.*, **142**, 841
Herzberg, G. 1950, *Molecular Spectra and Molecular Structure I. Spectra of Diatomic Molecules* (New York: Van Nostrand)
Hoyle, F., Wickramasinghe, N. C. 1969, *Nature*, **223**, 459
Hunger, K., Klinglesmith, D. 1969, *Ap. J.*, **157**, 721
Iben, I., Jr. 1964, *Ap. J.*, **140**, 1631
JANAF Thermochemical Tables, 1960, and later (Compiled by Dow Chemical Co., Midland, Michigan)
Johnson, H. L. 1964, *Bull. Tonantzintla Tacubaya Obs.*, **3**, 305
Johnson, H. L. 1965, *Ap. J.*, **141**, 170
John, T. L. 1966a, *MNRAS*, **131**, 315; 1966b, ibid, **133**, 447; 1968, ibid, **138**, 137
Joy, A. H. 1942, *Ap. J.*, **96**, 344
Joy, A. H. 1960. *Stellar Atmospheres*, Chap. 18 (Greenstein, J. L., Ed., Univ. Chicago Press)
Kamijo, F. 1963, *Publ. Astron. Soc. Jap.*, **15**, 440
Kamijo, F. 1966, *Colloquium on Late-Type Stars*, 252 (Hack, M., Ed., Trieste)
Keenan, P. C. 1954, *Ap. J.*, **120**, 484
Keenan, P. C. 1960, *Stellar Atmospheres*, Chap. 14 (Greenstein, J. L., Ed., Univ. Chicago Press)
Keenan, P. C. 1963, *Basic Astronomical Data*, Chap. 8 (Strand, K. Aa., Ed., Univ. Chicago Press)
Keenan, P. C. 1970 (Private communication)
Keenan, P. C., Morgan, W. W. 1941, *Ap. J.*, **94**, 501
Kovetz, A. 1967, *MNRAS*, **137**, 169
Krascella, N. L. 1965, *J. Quant. Spectrosc. Radiat. Transfer*, **5**, 245
Krishna Swamy, K. S. 1969, *Ap. Space Sci.*, **3**, 552
Krogdahl, W. S., Miller, J. E., 1967, *Ap. J.*, **150**, 273
Kruszewski, A., Gehrels, T., Serkowski, K. 1968, *Astron. J.*, **73**, 677
Kunde, V. G. 1968, *Ap. J.*, **153**, 435
Kyle, T. G. 1967, *Ap. J.*, **148**, 845
LaBahn, R. W., Callaway, J. 1966, *Phys. Rev.*, **147**, 28
Lambert, D. L. 1968, *MNRAS*, **138**, 143
Ledoux, P., Renson, P. 1966, *Ann. Rev. Astron. Ap.*, **4**, 293
Ledoux, P., Walraven, T. 1958, *Handbuch der Physik*, **51**, 353 (Flügge, S. Ed., Berlin: Springer-Verlag)
Limber, N. 1958, *Ap. J.*, **127**, 387
Lindblad, B. 1935, *Stockholm Obs. Ann.*, **12**, no. 2
Linsky, J. L. 1966, *Astron. J.*, **71**, 863
Linsky, J. L. 1969, *Ap. J.*, **156**, 989
Lord, H. C. 1965, *Icarus*, **4**, 279
Main, R. P., Bauer, E. 1966, *J. Quant. Spectrosc. Radiat. Transfer*, **6**, 1
McBridge, B. J., Heimel, S., Ehlers, J. G., Gordon, S. 1963, *NASA SP-3001*
McDowell, M. R. C., Williamson, J. H., Myerscough, V. P. 1966, *Ap. J.*, **144**, 827
Mendoza, V., E. E., Johnson, H. L. 1965, *Ap. J.*, **141**, 161
Merrill, P. W. 1916, *Publ. Michigan Obs.*, **2**, 70
Merrill, P. W. 1941, *Ap. J.*, **94**, 171; 1948, ibid, **107**, 303
Merrill, P. W. 1960, *Stellar Atmospheres*, Chap. 13 (Greenstein, J. L., Ed., Univ. Chicago Press)
Merrill, P. W., Deutsch, A. J., Keenan, P. C. 1962, *Ap. J.*, **136**, 21
Mihalas, D. 1965, *Ap. J.*, **141**, 564
Mihalas, D. 1967, *Meth. Comput. Phys.*, **7**, 1
Moore, C. E. 1949, *Atomic Energy Levels*, **1**, NBS Circ. 467
Moore, C. E. 1952, *Atomic Energy Levels*, **2**, NBS Circ. 467
Moore, C. E. 1958, *Atomic Energy Levels*, **3**, NBS Circ. 467
Morgan, W. W., Keenan, P. C., Kellman, E. 1943, *An Atlas of Stellar Spectra* (Univ. Chicago Press)
Morris, S., Wyller, A. A. 1967, *Ap. J.*, **150**, 877
Myerscough, V. P., McDowell, M. R. C. 1964, *MNRAS*, **128**, 287; 1966, ibid, **132**, 457
Myerscough, V. P. 1968a, *Ap. J.*, **152**, 1115; 1968b, ibid, **153**, 421
Öhman, Y. 1934, *Ap. J.*, **80**, 405
Öhman, Y. 1936, *Stockholm Obs. Ann.*, **12**, Nos. 3 & 8
Osvalds, V., Risley, A. M. 1961, *Publ. Leander McCormick Obs.*, **11**, 147
Pagel, B. E. J. 1964, *Roy. Obs. Bull. No. 87*
Pagel, B. E. J., Powell, A. L. T. 1966,

Roy. Obs. Bull. No. 124
Pande, M. C. 1968, Sov. Astron. AJ, 11, 592
Peach, G. 1967, Mem. RAS, 71, 29
Peat, D. W., Pemberton, A. C. 1968, MNRAS, 140, 21
Pecker, J. C. 1965, Ann. Rev. Astron. Ap., 3, 135
Peterson, D. M. 1969, Smithsonian Spec. Rep. No. 293
Robinson, E. J., Geltman, S. 1967, Phys. Rev., 153, 4
Rosenvinge, T. T. von, Webber, W. R., Ormes, J. F. 1969, Ap. Space Sci., 3, 4
Sanford, R. F. 1942, Publ. Astron. Soc. Pac., 54, 158
Schadee, A., Davis, D. N. 1968, Ap. J., 152, 169
Seman, M. L., Branscomb, L. M. 1962, Phys. Rev., 125, 1602
Serkowski, K., Kruszewski, A. 1969, Ap. J., 155, L15
Sharpless, S. 1956, Ap. J., 124, 342
Shaviv, G., Chitre, S. M. 1968, MNRAS, 140, 61
Smak, J. 1964, Ap. J. Suppl., 9, 141
Solomon, P. M. 1964, Ap. J., 139, 999
Somerville, W. B. 1964, Ap. J., 139, 192; 1965, ibid, 141, 811
Somerville, W. B. 1966, Proc. Phys. Soc., 89, 185
Soshnikov, V. N. 1964, Optics Spectrosc., 17, 186
Spiegel, E. A. 1963, Ap. J., 138, 216
Spiegel, E. A. 1968, Highlights of Astronomy, 261 (Perek, L., Ed., Dordrecht, Holland: Reidel)
Spinrad, H. 1966, Ap. J., 145, 195
Spinrad, H., Vardya, M. S. 1966, Ap. J., 146, 399
Spinrad, H., Wing, R. F. 1969, Ann. Rev. Astron. Ap., 7, 249
Spinrad, H., Wood, D. B. 1965, Ap. J., 141, 109
Strom, S. E., Kurucz, R. L. 1966, J. Quant. Spectrosc. Radiat. Transfer, 6, 591
Tarafdar, S. P., Vardya, M. S. 1969a, Proc. 3rd Harvard-Smithsonian Conf. Stellar Atmospheres, 143 (Gingerich, O. J., Ed., Cambridge, Mass: MIT Press)
Tarafdar, S. P., Vardya, M. S. 1969b, MNRAS, 145, 171
Tarafdar, S. P., Vardya, M. S. 1970, J. Quant. Spectrosc. Radiat. Transfer (In press)
Tatum, J. B. 1966, Publ. Dominion Ap. Obs. Victoria, 13, 1
Taylor, H. S. 1967, Proc. Phys. Soc., 90, 877

Torres-Peimbert, S., Wallerstein, G. 1966, Ap. J., 146, 724
Treanor, P. J. 1963, Astron. J., 68, 185
Treanor, P. J., McCarthy, M. F. 1966, Colloquium on Late-Type Stars, 109 (Hack, M., Ed., Trieste)
Tsuji, T. 1964, Ann. Tokyo Astron. Obs., Ser. 2, 9, 1
Tsuji, T. 1966a, Publ. Astron. Soc. Jap., 18, 127
Tsuji, T. 1966b, Colloquium on Late-Type Stars, 260 (Hack, M., Ed., Trieste)
Tsuji, T. 1968, Astron. J., 73, S120
Tsuji, T. 1969, Low Luminosity Stars, 457 (Kumar, S. S., Ed., New York: Gordon & Breach)
Ulrich, R. K. 1969 (Preprint)
Unno, W. 1962, Publ. Astron. Soc. Jap., 14, 153
Utsumi, K. 1967, Publ. Astron. Soc. Jap., 19, 342
Van de Kamp, P. 1969, Publ. Astron. Soc. Pac., 81, 5
Van den Heuvel, E. P. J. 1963, Bull. Astron. Inst. Neth., 17, 148
Vandervort, G. L. 1958, Astron. J., 63, 477
Vardya, M. S. 1960, Publ. Astron. Soc. Pac., 72, 364
Vardya, M. S. 1964, Ap. J. Suppl., 8, 277
Vardya, M. S. 1965a. MNRAS, 129, 205; 1965b, ibid, 129, 345
Vardya, M. S. 1966a, Observatory, 86, 32; 1966b, ibid, 86, 162
Vardya, M. S. 1966c, Colloquium on Late-Type Stars, 242 (Hack, M., Ed., Trieste)
Vardya, M. S. 1966d, MNRAS, 132, 475; 1966e, ibid, 134, 183; 1966f, ibid, 134, 347
Vardya, M. S. 1966g, J. Quant. Spectrosc. Radiat. Transfer, 6, 539
Vardya, M. S. 1967a, Mem. RAS, 71, 249
Vardya, M. S. 1967b, Observatory, 87, 247
Vardya, M. S. 1970a, ibid (In press)
Vardya, M. S. 1970b, ibid (In press)
Vardya, M. S., Giannone, P., Virgopia, N. 1969, Low Luminosity Stars, 517 (Kumar, S. S., Ed., New York: Gordon & Breach)
Vardya, M. S., Kandel, R. 1967, Ann. Ap., 30, 111
Varsavsky, C. M. 1966, Space Sci. Rev., 5, 419
Wallerstein, G., Greenstein, J. L. 1964, Ap. J., 139, 1163
Warner, B. 1965, MNRAS, 129, 237; 1967, ibid, 137, 119
Weinberg, M., Berry, R. S. 1966, Phys. Rev., 144, 75

Wickramasinghe, N. C. 1968, *MNRAS*, **140**, 273
Wildt, R. 1942, *Observatory*, **64**, 195
Wildt, R. 1949, *Relations entre les phenomènes solaires et géophysiques*, 7 (Paris: Ed. Rev. Opt. Théorique Instr.)
Wildt, R. 1957, *Mém. 8° Soc. Roy. Sci. Liège, 4th Ser.*, **18**, 319
Wilson, K. H., Nicolet, W. E. 1967, *J. Quant. Spectrosc. Radiat. Transfer*, **7**, 891
Wilson, O. C. 1966, *Science*, **151**, 1487
Wilson, O. C. 1969, *Low Luminosity Stars*, 103 (Kumar, S. S., Ed., New York: Gordon & Breach)
Wing, R. F. 1970 (Private communication)
Wing, R. F., Spinrad, H. 1970, *Ap. J.* (In press)
Wing, R. F., Spinrad, H., Kuhi, L. V. 1967, *Ap. J.*, **147**, 117
Yamashita, Y. 1966, *Colloquium on Late-Type Stars*, 94 (Hack, M., Ed., Trieste)
Yamashita, Y. 1967, *Publ. Dominion Ap. Obs. Victoria*, **13**, 47
Zappala, R. R. 1967, *Ap. J.*, **148**, L81

Copyright 1970. All rights reserved

INFORMATION-PROCESSING SYSTEMS IN RADIO ASTRONOMY AND ASTRONOMY

2004

B. G. CLARK

National Radio Astronomy Observatory,[1] Green Bank, West Virginia

INTRODUCTION

Until a few years ago, the application of information theory to the process of gathering astronomical data proceeded in an ad hoc fashion, if at all. Each experiment was designed to display the quantity with which it was concerned in the simplest, most direct fashion. More recently, the great complexity and cost of large optical and radio telescopes has made it mandatory that they be so designed to serve as many purposes as possible with high efficiency. This is facilitated if the maximum amount of information coming from the sky is preserved to a rather late point in the chain of processing the information. Some sophistication must now be used to efficiently extract the wanted data from a general format.

The earliest astronomical observations were made visually, where one can measure only the positions and brightnesses of the stars (and the colors of the brightest) and the brightness distributions of the extended objects. Of the extended objects, only the Sun, Moon, and planets are of much interest visually in a small telescope, and these were adequately represented by sketched maps from the time of the earliest telescopic observations. Perhaps the most successful quantitative adjunct of measurements of this type was the introduction, by Wolf, of the relative sunspot number in 1848.

The most comprehensive and useful series of visual measurements of stars is the remarkable effort of Argelander (1859) and his assistants, who, in the space of 11 years beginning in 1852, made, reduced, and published approximately 700,000 observations of 324,198 stars, which comprise the BD Catalogue.

Visual observations are now primarily used only in the measurement of close double stars (see the review by Van den Bos 1956). The rate of accumulation of data by visual observers is sufficiently slow that it may easily be analyzed by the traditional graphical and arithmetic methods, although in practice it is desirable even here to program the reductions to lessen the labor required to achieve an impartial solution.

---

[1] Operated by Associated Universities, Inc., under contract with the National Science Foundation.

The introduction of the photographic plate brought about a revolution in astronomical observations. The information content of a photographic plate is enormous. It may range from about 200 resolution elements for an unwidened small-dispersion spectrogram to 10^8 resolution elements for a large plate taken with the 48-inch Schmidt telescope. An observer may spend many weeks extracting the information from a plate whose exposure required only a few hours. When properly treated, the photographic plate is also a compact and permanent means of storing this information.

Special-purpose devices have been constructed to extract various items of information from photographic plates in convenient numerical or graphical form. Among these may be listed the measuring engines for measuring the position of a small image (e.g., Vasilevskis 1960, Strand 1966), the iris photometer for measuring its size, the microdensitometer for integrating densities in one-dimension, and the isodensitometer for plotting contour diagrams (e.g., de Vaucouleurs 1956). Each of these devices requires more or less assistance from a human operator in selecting the appropriate data for processing. In addition, some simple information processing may be done photographically, such as the plate-subtraction technique (Zwicky 1964).

Photoelectric observations have a still different appearance from an information-processing point of view. The output rate of information is vastly lower—the result of an hour's observations may be ten numbers of eight bits each, instead of the 10^6–10^8 bits of information in a complex direct photograph. Because these numbers are so few but of such high accuracy, general-purpose digital computers were very quickly programed (Arp 1959, Schulte 1960) to apply the various small corrections needed to preserve the intrinsically high accuracy of the measurements.

Photoelectric imaging devices are, from the point of view of image processing, simply equivalent to direct photography, except that signal and sky noises are both amplified and the exposure times are shorter.

Radio astronomy has developed from the beginning along the lines followed by photoelectric photometry. In the beginning of radio astronomy, only a single receiver was installed on each radio telescope, and a few seconds or a few minutes integration time was needed to detect the signals of interest, so that the output of the radio telescope was again a few numbers per hour. Digitization of the output was clearly profitable, and was first attempted at a rather early stage in the development of the science (Heeschen 1961).

The construction of radio astronomy receivers with many output channels is a development of only the last few years. These receivers are, for instance, array receivers where many beams are synthesized, or spectral-line receivers in which many narrow-frequency channels are recorded. When these receivers came into use and quickly demonstrated that their output was too formidable to be reduced by pencil and paper alone, the natural tendency was to increase the use of general-purpose digital computers in the reduction. For the most part, only solar radio astronomy, in which large information rates developed at a very early date, has utilized the enormous data capacity of photographic film, for recording dynamic spectra of solar bursts.

General Principles of Information Processing

It is very difficult to extract general principles or definitions that apply to all the vast variety of methods and devices used to make observations in astronomy. However, a few problem areas can be discussed.

Any information-processing system can be divided into three general areas. First, there is some sort of real-time reduction device, which is attached to the telescope. In optical astronomy this may simply be optics, forming an image, or it may include a filter, or it may be a slit and spectroscope, or any of several other devices. In radio astronomy, it is likely to include a low-noise front-end amplifier, usually with some sort of switching system so that the noise from the sky is compared with some stable noise source to eliminate the gain instability of the preamplifier as a source of uncertainty. The preamplifier is likely to be followed by a conversion to some convenient intermediate frequency, a set of filters determining the frequency observed, and a set of detectors. The problem of constructing noise-free front ends for radio astronomy, or of low-noise photoelectric devices for optical and infrared astronomy is extensive and specialized—far beyond the scope of this article. The only necessary comment is that care should be taken that information of interest is not lost before the recording process. For example, interplanetary scintillation information is lost by the use of too long a time constant in recording radio data.

At some stage in any information-processing system, the partially processed information is recorded on a more or less permanent recording medium. For density and compactness in storing large quantities of information, the photographic plate is without peer. A hundred years ago, when the entire choice in recording media was between the photographic plate and pencil and paper, the choice was clear, and astronomy began its long career of photographic observations. Now, a strong competitor for information storage has arisen, in the form of the computer-readable media, usually magnetic-tape recordings. The cost per bit of information is much higher on magnetic tape—about 10 cents per million bits—than on photographic plate—about 2 cents per million bits for astronomical plates. However, this may be somewhat offset by the possibility of compressing the data before recording.

Since modern astronomical instruments are very expensive and the operating costs very high, the equipment preceding and including the recording process should be as reliable, and as infrequently changed as possible, so that expensive telescope time is not expended for equipment development. In order that this equipment may satisfy the demands of various experiments, very much more information than any one experiment can utilize must be recorded. Latter stages of the information processing may work from the storage medium, and, since they do not destroy it, it is required only that the processing be fast enough that an infinitely growing backlog of unprocessed data does not form.

The third part of information processing consists of taking the data from the intermediate storage and processing it further, to reduce it eventually to the comparatively few numbers which will appear in a journal article. This phase is so specialized to specific experiments that it cannot to any extent be discussed in general. This final reduction can consist of anything from describing a photograph to comparing a complicated spectrum tracing with a model-atmosphere theory. When the amount of information to be compressed, or data to be published, is very large, this final stage of processing is now usually implemented in some sort of software data-reduction program.

An important intermediate step in this third stage of data processing is the problem of data display. What has been said above has as its basis the assumption that the investigator knows all of the quantities of interest—the quantities he will eventually publish in his paper—and that the problem is to gather all of the information about his chosen quantities, together with appropriate weightings, to get the best estimates of his numbers. Actually, this is only part of the problem. Before making his observation, the observer may not know enough about the object he is studying to formulate an intelligent hypothesis. He may need to conduct a preliminary study of the data he has in hand before he can decide just which quantities he wishes to determine. To ascertain the content and reliability of his data, he wishes to look at hundreds or thousands of data points rather than the relatively few he will eventually publish. The problem is to make enormous quantities of information available in an intelligible form with a minimum of study. In this application, the direct photograph is a strikingly good example. The human brain is well adapted to preceiving forms and shades, and a good direct photograph may furnish all the material needed to originate hypotheses about data that can conveniently be exhibited as pictures. This is a large class, as much effort has lately been expended on giving computers the capability of producing picture representations of any variables present in their memory (Harmon & Knowlton 1969).

The complexity of a data-display problem depends on its dimensionality; that is, the number of independent variables. The display of a quantity as a function of one independent variable, for instance a spectrum (either optical or radio) of one particular point in the sky, is a straightforward problem. Many computers come equipped with a graph plotter, and if not, it may be able to draw a presentable graph on its printer.

Two-dimensional displays (e.g., the conventional optical direct photograph and its radio equivalent, the brightness map) are slightly more difficult, but still reasonable. One has essentially three choices—the density-modulated map (cf Schroeder 1969), which most resembles a direct photograph, the contour plot (e.g., Figure 1), and the graphs of slices through the true three-dimensional object (e.g., Figure 2), which may optionally be arranged to form an isometric projection of the three-dimensional graph.

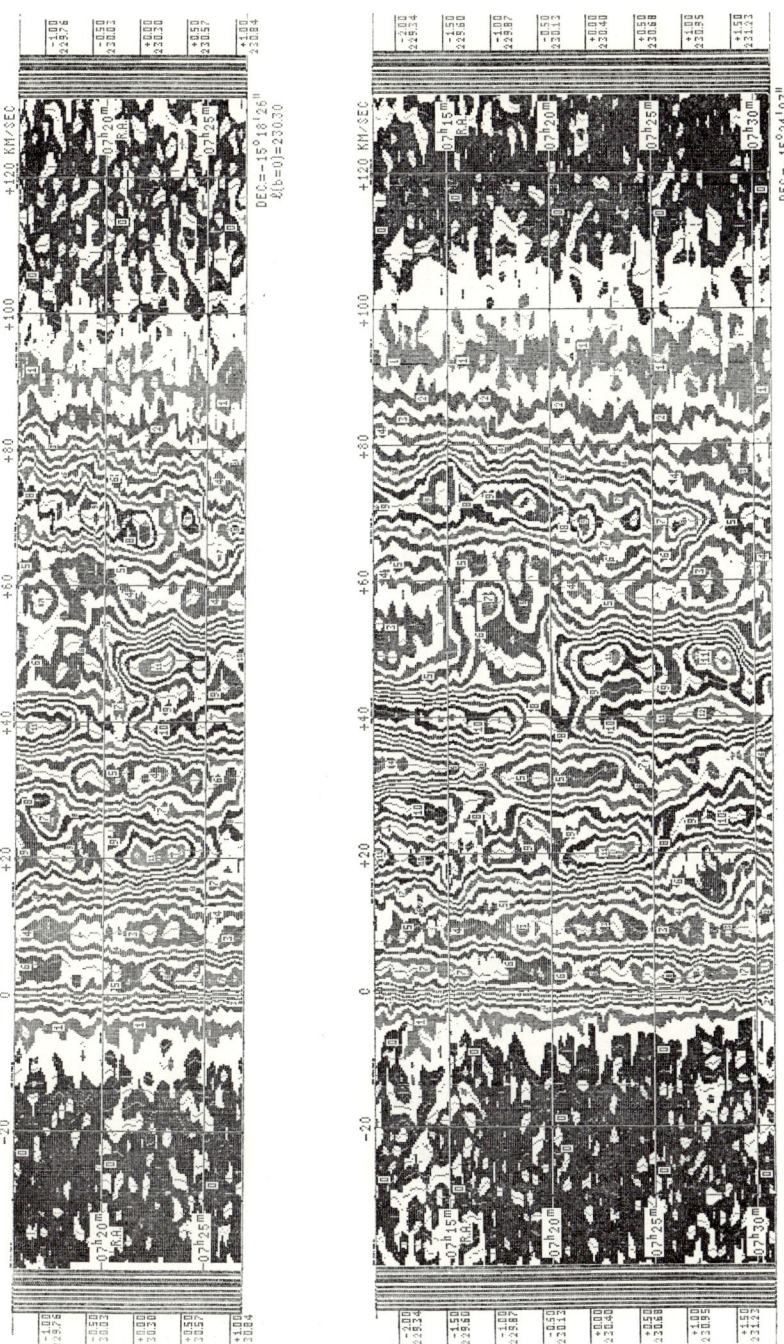

FIGURE 1. A sample map from the *Maryland-Green Bank Galactic 21-cm Line Survey* (Westerhout 1969). The two strips are contour diagrams of brightness as a function of velocity and right ascension for a given declination. Galactic coordinates are indicated along the side.

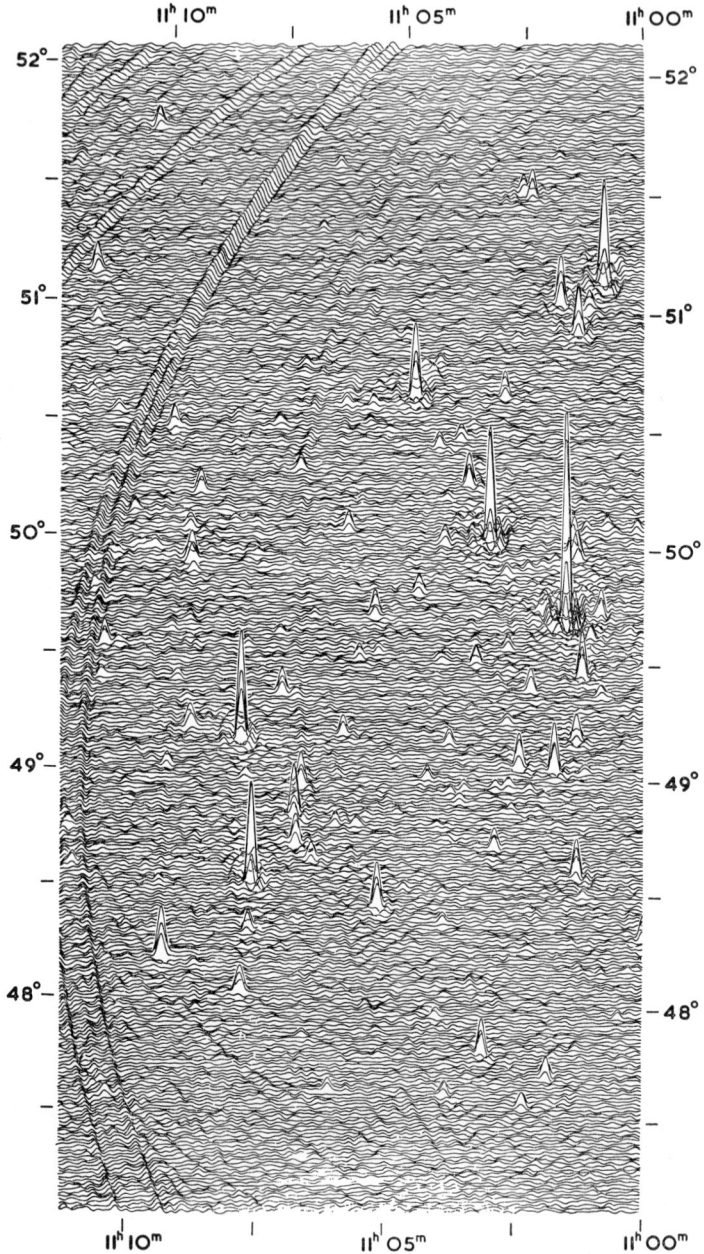

FIGURE 2. The 5C2 survey. Adjacent scans of the sky brightness are displaced 52″ in declination. The curved ridges are instrumental sidelobes. The strong source at $10^h55^m18^s$, 49°56′ has a flux of 0.66×10^{-26} W/m^{-2} Hz^{-1} at 408 MHz. (By permission, Blackwell Publications, Ltd.)

INFORMATION PROCESSING IN ASTRONOMY 121

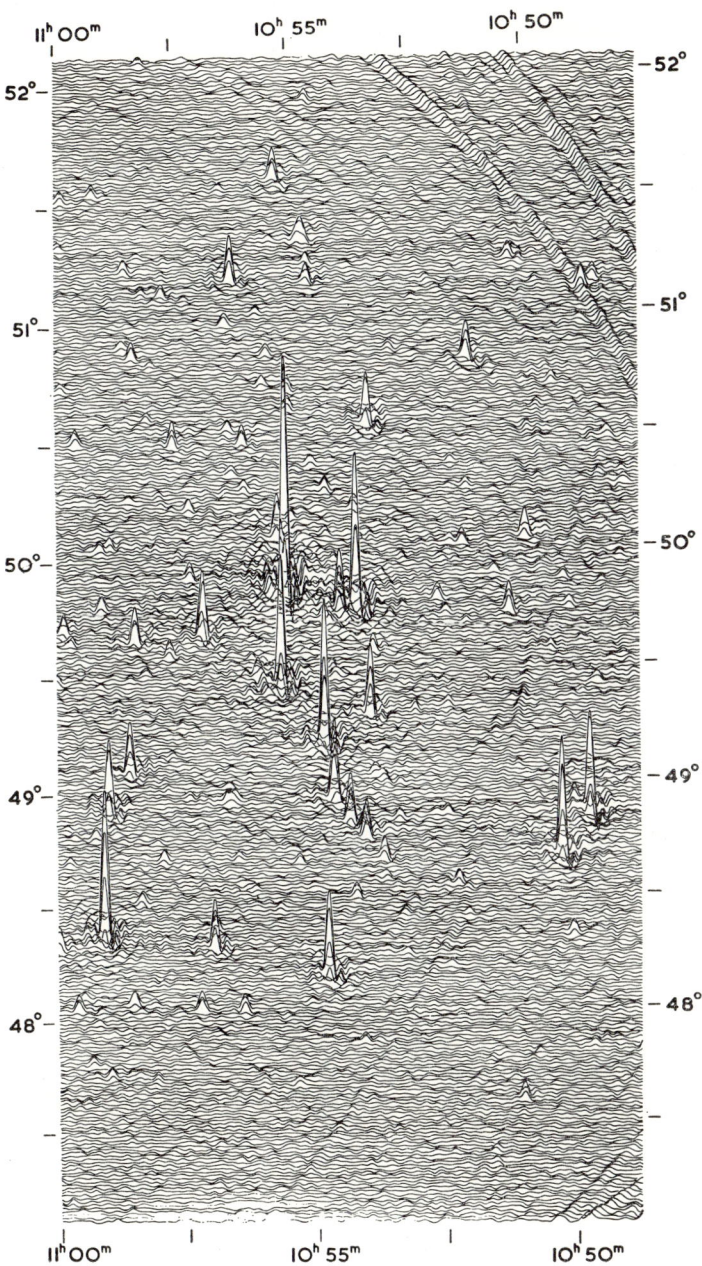

The really difficult case occurs when an observed quantity is to be displayed as a function of three independent variables. This case occurs in spectral-line observations of extended objects, in which it is desirable to plot specific intensity as a function of frequency and of two space coordinates. An early attempt at displaying this function was, for the 21-cm hydrogen line observations (Müller & Westerhout 1957), simply to plot the spectrum at many discrete points. Clever pictorials have also been used (Dieter 1965). This sort of display, while marginally practical for the 694 points of Müller & Westerhout, clearly becomes impractical for a large number of points. The approach in this case has usually been to reduce the case to two independent variables and to display a separate contour or density plot for each value of the third independent variable (Braes 1963, Heiles 1967, Westerhout 1969; see Figure 1). If desired, these plots may then be strung together to produce a movie, thus utilizing time as the last dimension. It may also be informative to make plots of some function of the spectrum against the two spatial coordinates, for instance, the frequency of the peak, or the integrated emission (e.g., Roberts 1966). An integral along one of the two space coordinates may also be useful, as in the famous map of neutral hydrogen in the Galaxy given by Westerhout (1957), which is really a plot of intensity against frequency and galactic longitude, with intensity integrated over galactic latitude. None of these methods of presenting this basically four-dimensional function is completely satisfactory. For various applications one may be markedly better than the others, and utilizing several different displays may be profitable for private use, if not for publication. It may be that no really satisfactory solution of this data-display problem exists.

ONLINE INFORMATION-PROCESSING DEVICES

Several classes of devices are directly attached to the telescope to perform some processing before data recording. The simplest of these display directly the one quantity of interest. This is the usual output for photoelectric observations, where a signal proportional to the brightness of the star in the diaphragm is directly produced by the photocell. Much more complicated quantities can be computed with simple electrical or mechanical devices for a similar direct display. Perhaps the classic example of this sort of analog processing is the solar magnetograph (Babcock 1953). The magnetograph determines the wavelength of a selected spectral line by comparing the intensity at selected points on the slope on either side of the line. The longitudinal component of the magnetic field is calculated by subtracting the wavelengths of a magneto-sensitive line as measured in left- and right-circular polarization. A small mechanical analog computer makes adjustments to the apparatus to remove the effects of the general solar rotation. The magnetic-field data are neatly displayed on a cathode-ray oscilloscope, ready for a photograph directly displaying the magnetic-field intensities.

Elaborate devices constructed for very special purposes are somewhat less desirable on very large and expensive telescopes. Many different types must be constructed, each taking up valuable telescope time to check out on actual observations. If possible, it is more profitable to record data in a more general format, and to perform the final, specialized reductions with pencil and paper, or with software, neither of which requires telescope time. This was early done in photoelectric photometry (Arp 1959) and in continuum radio astronomy (Heeschen 1961), and has been continued in both fields (e.g., Oke 1969, Dixon & Kraus 1968). The early work usually involved recording the data on punched paper tape, which could then be read into a general-purpose computer. This was done because paper tape is the least expensive of the machine-readable media to implement. However, paper-tape punches tend to be slow, noisy, and unreliable. Where the data rates are at all high, and elsewhere where at all feasible, experimenters have turned to recording on magnetic tape.

We now mention a most remarkable development in the economics of electronic design. Integrated-circuit logical elements have become so inexpensive that most of the cost of a digital circuit is design and labor cost, rather than component cost. This being so, complex devices that are sold by the hundreds may be less expensive than much simpler devices designed for the project at hand. Thus, a special-purpose device to couple an experiment to a digital-tape recorder may actually cost more than a general-purpose computer with a digital-tape drive interfaced. It has become economic for an expensive telescope to come equipped with a small general-purpose computer dedicated to the special purpose of serving as an interface between its experiments and a digital magnetic-tape recorder.

Utilizing a general-purpose computer as a device coupling the magnetic-tape recorder to the experiment carries a great bonus of data-processing capability, which may conveniently be used for moderately complex processing before recording. In a few experiments, however, the data-processing load becomes so great that the small online computer may not have sufficient computing power to handle the necessary reductions in real time. In this case, it may be economical to construct a special-purpose digital device. The high one-time design fee may be offset by the fact that, for a specific problem, a highly specialized device can operate several times as fast as a general-purpose computer. If the load of computations is sufficiently great, the special-purpose device may replace several general-purpose digital computers. An example of a system of this sort is the image-forming radioheliograph at Culgoora, Australia (Wild 1967, Beard et al 1967), which will be described in detail below. In this array, a small special-purpose digital-analog hybrid computer forms and tracks the array beam. The computer not only tracks the beam at the solar rate, but scans a number of simultaneous beams over the field of veiw, and corrects the natural beam of the circular array for its sidelobe effects.

It is possible that in the future, the computing demands on more instruments may also become so high that other implementations of special-purpose computers will be made. In fact, the digital autocorrelation-function receivers, discussed in detail below, may also be considered as very special-purpose online data-processing computers.

EXAMPLES OF MODERN INFORMATION-PROCESSING SYSTEMS

In the following paragraphs, several information-processing problems are considered and a group of solutions to each by various observatories is described. These examples are heavily weighted toward radio astronomy, because only in isolated fields of optical astronomy have important observations been made utilizing either sophisticated information-processing techniques or extensive data-processing computer systems.

Photoelectric photometry.—Since the early days of photoelectric photometry, it has been realized that the nasty details of determining and making the small corrections to the observed magnitudes and colors may profitably be left to the computer. Since then it has been tacitly assumed that computers will handle the details, leaving the astronomer free to comment only on the information processing (e.g., Hardie 1959) or on the results.

Most early photoelectric photometers had, as primary readout, a strip-chart recorder, whose output was then read and punched on punched cards for computer reduction. However, an increasing number of photometers are coupled to paper-tape punchers or magnetic-tape recorders, and record the data directly in a machine-readable format. At least one multichannel photoelectric system, that at the Hale telescope of Palomar Observatory, is being equipped with an online computer to prepare data for output on punched cards (Oke 1969). This multichannel photometer has 33 channels, variable in width from about 360 Å to about 20 Å. The channel spacing is about 360 Å in the red and 150 Å in the blue. The signal from the object being investigated is recorded only half of the time. The remainder of the time the light from a patch of sky 40″ distant is recorded to remove the effects of sky brightness fluctuations. The sensitivity of the instrument is limited only by the photon statistics of the light from the night sky, and, with a few hours integration time on the Hale telescope, is expected to produce a spectrum, with a bandpass of 80 Å in the red and 40 Å in the blue, of an object of magnitude 22. The limiting magnitude for a spectrum scanner in the same time would be two magnitudes brighter.

Plate-overlap astrometry.—It is perhaps surprising that one of the more interesting developments in data processing in astronomy comes from one of the oldest fields of astronomy, that of astrometry. To review briefly, fundamental star catalogues are compiled using meridian circles and are checked by a variety of instruments of similar high precision, such as astrolabes. The fundamental catalogue stars are then used as reference points for a photographic transfer of the coordinates to fainter and more numerous stars. In the traditional technique, a photographic plate is taken that contains several of the reference stars from a fundamental catalogue. The positions of the reference star images and also the unknown star images are then measured with a two-dimensional measuring engine. It is then assumed that the plate positions are some fairly simple function of the star positions in the sky. Unfortunately, because of the nonlinear response of the photographic plate, this may be a function of magnitude of the star; because of the chromatic aberrations of the lenses of the optical system, this may also be a function of the color of the star. That is, the measured positions (x, y) on the plate are a function of position, color, and magnitude of the star,

$$x = f(\alpha, \delta, m, B - V)$$
$$y = g(\alpha, \delta, m, B - V)$$

1.

The functions f and g are different for each plate. It is usually assumed that these functions may be described by some fairly small number of parameters, describing plate centers, the distortions of the optics, and the color and magnitude terms. These parameters are determined by the method of least squares from the measured plate coordinates of the reference stars. The difficulty arises in the determination of the color and magnitude terms, since the reference stars have been chosen for ease of observation with a transit instrument and are therefore rather uniform. They are also rather few, and so the statistical vagaries of their distribution may leave them too poorly distributed for accurate determinations of the distortion terms.

It has been pointed out by Eichhorn (1960) that the plate parameters may be much more accurately determined if an area of the sky is photographed twice over, with the centers of the second series of plates falling on the corners of the first series. Each star then appears on two plates. A general least-squares solution for the positions of all of the unknowns and the plate constants of all of the plates is then made in a straightforward manner. Stars of various brightnesses and colors fall at different places on the two plates on which they appear, and thus are affected in different fashion by the distortion, magnitude, and color terms of the plate constants. By using stars with a wide range of characteristics as secondary standards, it is possible to determine the plate constants with much higher accuracy than is possible from the measurements on the reference stars alone. This secondary reference system is implemented implicitly and automatically by the use of least-squares adjustment on all the available data (Eichhorn et al 1967).

The story of plate-overlap astrometry has a general moral, which we shall now draw. The impact of the computer on many areas of science has been that the necessity for clever approximations has been removed. In the past, magnitude and color corrections to catalogues have been improved by comparison of catalogues taken with different instruments, or at different times, and the small plate constants derived by a sophisticated analysis of the residuals. The straightforward application of least-squares analysis to all of the available data is really a much less clever and sophisticated procedure than what might have been done, in terms of deriving sets of secondary standards, etc. However, it is undoubtedly a better way to do things. It is merely computationally intractable. If, as in the case of plate-overlap astrometry, the problem is difficult even for straightforward programing, the proper solution is to first exhaust the avenues of clever programing (Googe 1967) rather than to proceed at once to making physically significant approximations.

Although plate-overlap astrometry only extracts from the data a few percent more accuracy than would be obtained with more traditional reductions, the additional computing complexity is a sufficiently small part of the whole process of obtaining accurate stellar positions that the improvement the method offers justifies the additional computing time and complexity.

The Minnesota proper-motion survey machine.—Because the photographic plate is such an efficient and versatile form of display, astronomy has been reluctant to turn to other information-storage media. And, indeed, in some cases, the photographic plate or its reproduction is the best and most efficient form of conveying information. A case in point is the publication of the National Geographic Society-Palomar Observatory Sky Survey, which surely ranks as one of the great achievements of observational astronomy, as it has supplied so much material for astronomers all over the world.

On the other hand, for many purposes it is desirable to have extensive photographic material converted to numerical data of one form or another. The example of an enormous data-handling program in extracting information from photographic plates is the Astrographic Catalogue. This catalogue has not fulfilled its promise, for several reasons. First is the delay in completion of the project, although in astrometry a project whose completion requires 60 years is not as disastrous as it would be in many areas of astronomy. Second, adequate positional controls were not included from the beginning, partly because the concept of what constitutes adequate positional control has changed, and partly because the Astrographic Catalogue did not fall heir to the serendipity that has made the Palomar-National Geographic survey so useful without accurate positions. Third, the Astrographic Catalogue has been published in printed form, in something over 100 bound volumes, rather than being published on some sort of machine-readable medium, as any collection of numbers of such length should be.

Were a project of the size of the Astrographic Catalogue to be begun today, it would utilize very different techniques. An example of a highly automated system for extracting positional information from photographic plates is the proper-motion survey machine now being constructed at the University of Minnesota under the direction of Willem J. Luyten (Newcomb 1968). This automatic plate scanner scans a 16-mm-wide strip of a photographic plate with a raster scan of a point of light. The raster is generated by the motion of a lead screw along the length of the strip and the motion of a rotating prism, which causes the point of light to trace across the strip at a rate of 1200 times per second. The opacity of the plate is monitored by a photocell. A stellar image is crossed by the flying spot several times as the plate is carried past by the lead screw. The star is recognized by the detection of several successive dark areas at the same position on several successive scans. The magnitude is estimated by the size of the image in both directions.

Since this machine is specifically constructed to detect stars of high proper motion, it is constructed to scan two plates at the same time, and record data for stellar images on both plates, taken at two different epochs. The stellar-image recognition logic is implemented partly in special-purpose devices, which also record the data on magnetic tape, and partly in a large general-purpose computer.

The output of this system will be a machine-generated list of the stars of high proper motion, their estimated magnitudes, and their positions. The technique has obvious applications to other two-plate experiments, such as searching for variable stars, flare stars, or stars of unusual colors. It is even possible to envision using a similar machine to compile a list of all stars to magnitude 20, or so. This project would take a few thousand hours of time on the plate scanner, plus an undisclosed amount of computer time. The cost of the magnetic tape would be only a few times greater than the cost of a copy of the Sky Survey prints.

Autocorrelation-function spectrometers.—The measurement of power spectra of electromagnetic radiation is such an important part of astronomical instrumentation that it is little wonder that a great deal of ingenuity should have been devoted to it. The techniques we shall deal with now are based on the fact that the autocorrelation function is the Fourier transform of the power spectrum of the function from which it is derived. This fact, which follows directly from the Fourier transform theorem and the convolution theorem, is known to electrical engineers as the Wiener-Khintchine theorem.

In infrared astronomy, the autocorrelation-function receiver has an intrinsic advantage (J. & P. Connes 1966, Hunten 1968). The infrared autocorrelation-function spectrometer obtains the autocorrelation function from a Fabry-Perot or Michelson interferometer. The field function at time τ, $E(t)$, is added to the same function delayed by τ in an arm of the interferometer, $E(t-\tau)$, and detected. The result is $E^2(t)+2E(t)E(t-\tau)+E^2(t-\tau)$. The square terms are constant as τ is changed, and are removed in the analysis.

The source of noise in an infrared spectrometer is the detecting element itself. Therefore, the noise output from the receiver is independent of the input bandwidth, in contrast to most applications in optical or radio astronomy where the output noise is proportional to the square root of the bandwidth. In this case, the noise is the same whether we are sampling one spectral channel, as with a monochromotor spectrometer, or all channels, as with the autocorrelation-function spectrometer. Clearly, on a given monochromatic signal, the Fourier spectrometer, which looks at this signal all of the time during the observation, will have a higher signal-to-noise ratio than a scanning-monochromator spectrometer, which sees the signal only when it is tuned to exactly the right wavelength. In general, for highest sensitivity, the bandwidth of a receiver should be limited by operations performed after the noisiest element of the system.

In radio astronomy there is no a priori reason to build a receiver that measures the autocorrelation function rather than the power spectrum. However, the recent development of inexpensive digital logic has made the autocorrelation-function receiver economically competitive with the conventional spectrometers. In radio astronomy receivers, the autocorrelation function is not measured directly, but is estimated from a sampled, digital representation of the signal. The digitization is usually carried to only one bit. The relation between one-bit representations of the noiselike signals of radio astronomy has been investigated by several authors, notably Van Vleck (1943) and Weinreb (1963). This relation is so important that a description of the most elementary form is given below.

Consider a noiselike signal $f(t)$, sampled at uniform intervals. At each sample, we have recorded only the sign; that is, we have a set of data a_i

$$a_i = \begin{cases} 1 & \text{if } f(i\Delta t) > 0 \\ -1 & \text{if } f(i\Delta t) < 0 \end{cases} \qquad 2.$$

and a similar set of data b_i relating to the function $g(t)$. The functions f and g are assumed to be random variables whose joint probability distribution is the binormal distribution,

$$\text{Probability} \begin{cases} x < f(t) < x + dx \text{ and} \\ y < g(t) < y + dy \end{cases}$$

$$= A \exp - \left[\frac{x^2 + y^2 - 2\rho xy}{2(1 - \rho^2)\sigma^2} \right] dx \, dy \qquad 3.$$

where A is the normalizing factor

$$A = \frac{1}{2\pi\sigma^2} (1 - \rho^2)^{-1/2} \qquad 4.$$

The coefficient ρ is the correlation coefficient of the functions f and g,

$$\rho = \frac{1}{\sigma^2} \langle f(t) \, g(t) \rangle \qquad 5.$$

as may readily be seen by evaluating the defining integral of $\langle fg \rangle$.

$$\langle fg \rangle = \int\!\!\int_{-\infty}^{\infty} A \, xy \exp - [(x^2 - 2\rho xy + y^2)(2\sigma^2(1 - \rho^2))^{-1}] dy \, dx \qquad 6.$$

The quantity of interest now is the expectation of the digitized version of f and g, $\langle a_i b_i \rangle$. This expectation is simply the expectation that a_i and b_i have the same sign, less the expectation that they have different signs. Let us evaluate the probability that a_i and b_i are both positive,

$$p_{++} = \int_0^\infty \int_0^\infty A \exp - [(x^2 - 2\rho xy + y^2)(2\sigma^2(1-\rho^2))^{-1}] dy\, dx \qquad 7.$$

We now make the substitution

$$v = (y - \rho x)(1 - \rho^2)^{-1/2} \qquad 8.$$

which restores the integrand to circular symmetry. We then have

$$p_{++} = A \int_0^\infty \int_{-\rho(1-\rho^2)^{-1/2} x}^\infty (1 - \rho^2)^{1/2} \exp - [x^2 + v^2]/2\sigma^2] dv\, dx \qquad 9.$$

At this point, the usual conversion from (x, v) rectangular coordinates to (r, θ) polar coordinates results in the simple integral

$$p_{++} = A(1 - \rho^2)^{1/2} \int_0^\infty \int_{-\arcsin \rho}^{\pi/2} \exp - [r^2/2\sigma^2] r d\theta\, dr \qquad 10.$$

$$= \frac{1}{2\pi}\left(\frac{\pi}{2} + \arcsin \rho\right) \qquad 11.$$

The binormal distribution clearly has the symmetry relations

$$p_{--} = p_{++},\ p_{+-} = p_{-+} = \tfrac{1}{2} - p_{++} \qquad 12.$$

Using these relations, it clearly follows that

$$\langle a_i b_i \rangle = \frac{2}{\pi} \arcsin \rho \qquad 13.$$

For small correlation coefficients,

$$\langle a_i b_i \rangle \simeq \frac{2}{\pi} \rho \qquad 14.$$

We now see that the arc sine of the correlation coefficient, whose appearance is so puzzling to the uninitiated, has appeared because, in evaluating the integral over a quadrant of the binormal distribution function, we change to a coordinate system in which the distribution function becomes the familiar circular Gaussian, and the quadrant is altered to a sector of opening angle involving the arc sine of ρ.

If the functions f and g have flat power spectra from zero to some frequency ω_0, and the samples a_i and b_i are taken at the Nyquist frequency

$\Delta t = \pi/\omega_0$, then the various a_i are uncorrelated, and it may readily be shown that the signal-to-noise ratio on the spectrum computed in this fashion is lower by the factor $2/\pi$ than the signal-to-noise ratio calculated by undigitized samples, which, by the Fourier sampling theorem, is the same as that on the output of a matched-bandpass filter.

One technical advantage of the autocorrelation-function receiver is that the one-bit digitization is the equivalent of an extremely good automatic-level control, which would otherwise be extremely difficult to make, so the receiver has very high stability for long integrations.

Various authors have investigated extensions of this simple theory. For instance, with infinite digitization, no increase in signal-to-noise ratio is achieved by increasing the sample rate beyond the Nyquist rate. With the one-bit digitization, this is no longer true. Several authors have independently calculated the signal-to-noise ratio for other sampling rates and for an unsampled one-bit digitized signal (Ekre 1963, Burns & Yao 1969). Various authors have also investigated the two-bit digitization (Cooper 1969) and higher digitizations (Cole 1968). It appears that implementing a multilevel scheme either seriously reduces the maximum sampling rate that the receiver will handle or requires so much additional hardware that the competition with a multichannel-filter receiver is uneconomical.

The economics of construction of the autocorrelation-function receivers favor them only in the situation that a large variety of bandwidths is needed, which can readily be obtained by varying the clock of the autocorrelation-function receiver. A single set of filters with the same number of channels can be built for less than the cost of an autocorrelation-function receiver. However, such receivers have at the moment a considerable popularity in radio astronomy. They have been constructed at the National Radio Astronomy Observatory (e.g., Gordon et al 1969), the Haystack Microwave Facility of Lincoln Laboratory, and the Nuffield Radio Observatory of Jodrell Bank (Davies et al 1969), and are under construction at the Dominion Astrophysical Observatory and the Parkes Field Station of CSIRO in Australia. A receiver constructed at the NRAO is now in use at the Owens Valley Observatory.

In order to effectively handle the enormous data output capacity of a large multichannel line receiver, elaborate data-handling systems have been constructed at several observatories. The 100-channel filter receiver at the Hat Creek Observatory of the University of California has gone through two generations of data-handling systems. In the first, the output of the filter bank was stored in a pulse-height analyzer, and accumulated for the duration of the integration. The spectrum was then punched on a standard five-character-per-second card punch. This system is convenient chiefly in looking at the weaker lines in a comparatively small number of objects, so that the lengthy and laborious process of punching the spectrum on cards must be done less often. It is rather less convenient for observing the strong 21-cm hydrogen line at a large number of points in the sky. Consequently, a more convenient system has been constructed, using an online digital computer for giving format to the data for output on a magnetic tape.

A similar multichannel receiver data system has been put into use by the Onsala Space Observatory of the Chalmers University of Technology (Ellder et al 1969, Rydbeck & Kollberg 1968). This receiver system also employs an online computer to prepare data for both online displays—a CRT display for the spectrum—and for punching on punched cards for later processing in a general-purpose digital computer.

A variety of line-observation systems have also been made at the NRAO. Perhaps the most interesting, because of the enormous amount of data actually contained in the published graphs, is the system used by Westerhout (1969) to make the Maryland-Green Bank hydrogen-line survey. The example reproduced in Figure 1 is a contour diagram showing brightness temperature as a function of frequency and of one of the two spatial coordinates (right ascension) for various values of the other spatial coordinate (declination). The software system following the actual autocorrelation-function spectrometer has been used to perform very many functions. First, the Fourier transform was taken to convert the autocorrelation function into the power spectrum, and a rough display of all the data made so that the observer can view the quality of his data. This display has the same basic form as the published contour plots, but is a plot of spectrum against time, and is thus not necessarily in the orderly arrangement displayed in the published catalogue. The programing system then has the ability to edit out bad data and to apply appropriate calibration information to convert the measured spectrum into units of antenna temperature. The final step is to rearrange data into the appropriate form to produce the contour maps in the published catalogue. A provision is also being made in the software to remove the effects of the error beam of the 300-ft telescope, an extended sidelobe pattern about 6° diam concentric with the main beam. Removing the error beam thus consists of combining spectra from about 6° around and subtracting this average spectrum from the observed spectrum at a point.

Interferometer and array data-processing systems.—Interferometers at their most spectacular involve their use as a large-field-of-view aperture-synthesis instrument (Swenson 1969). Using an instrument in this fashion requires a large degree of automation in the data reduction to handle the many data points that are combined in a Fourier transform to produce the scan of the region under investigation by the synthetic aperture. Since this usually involves combining the observations from several days, clearly data storage and access are an important part of the data-handling system.

The most successful aperture-synthesis interferometer has been the One Mile Telescope of Cambridge University (Ryle et al 1965). In this telescope, three interferometer elements are present, two of which are stationary, and the third can be moved along the extended line connecting them. Thus, the interferometer produces three interferometer outputs on the first configuration, and two new ones thereafter. The telescopes lie on an east-west line, so that the paths of the interferometers in the transform plane (spatial-frequency plane, with the two spatial frequencies being u and v) are ellipses centered on the (0, 0) point, with axial ratio $\sin \delta$. If there are N such ellipses all observed for 12 hr, with major axis a_i (in wavelengths), it may readily be shown that the beam pattern of the synthesized aperture, in the principal solution (defined by Bracewell & Roberts 1954), is

$$B(\Delta\alpha, \Delta\delta) = \sum_{i=1}^{N} J_0(2\pi a_i r) \qquad 15.$$

where r is the normalized radius

$$r = [(\Delta\alpha \cos \delta)^2 + (\Delta\delta \sin \delta)^2]^{1/2} \qquad 16.$$

It is further apparent that if N is reasonably large and if the various a_i have the form

$$a_i = ia/N \qquad 17.$$

then we may expand the various J_0 in the asymptotic approximation and from the resultant trigonometric series determine the location of the first large distant sidelobe to be

$$r = \frac{1}{2\pi} \frac{N}{a} \qquad 18.$$

and its amplitude to be

$$B_{\max} \approx \frac{1}{\sqrt{\pi N}} \qquad 19.$$

In practice, when the one-mile telescope is used to observe a strong discrete source, it is simply used at a sufficient number of configurations that this large ring sidelobe lies outside the extent of the discrete source.

The one-mile telescope is also used to observe areas of the sky containing no strong radio sources (Kenderdine et al 1966, Pooley & Kenderdine 1968). In this case, to reach the faint flux limits desired, the telescope must not only accumulate many days of integration time, but must also occupy many configurations in order that the several weak sources expected to be in an arbitrarily selected beam area should not have their appearances distorted by each other's sidelobes. An example of this sort of synthesis program is shown in Figure 2, taken from Pooley & Kenderdine (1968). Ryle & Hewish (1960) have argued that this sort of synthesis, if carried to the limit that the interferometer spacings are taken at intervals of the dish diameter, is nearly as sensitive as a scan of the same field by a 1-mile filled-aperture telescope in the same time. The filled-aperture telescope with only a single beam uses the same total integration time to scan the field, rather than to accumulate

Fourier components over the whole field. The 1-mile telescope used in this mode is about as sensitive as a 750-ft filled aperture (Ryle 1962).

Sophisticated interferometer data systems have also been constructed at Owens Valley Observatory and at the NRAO. The Owens Valley system developed from an interferometer data-recording system that merely permitted coherent integration for a length of time determined by the stability of a mechanical analog computer, which predicted the fringe function fitted to the observed data (Morris et al 1963). In its more sophisticated form (Fomalont et al 1967), the system employs the mechanical analog computer to reduce the natural fringe rate to a reasonable value, usually one per minute, so that the fringe function may be sampled less frequently, and recorded on magnetic tape by a special-purpose interface. The behavior of the mechanical analog computer is also monitored and recorded on tape, so that its imperfections can be removed by later analysis with a large general-purpose digital computer.

The NRAO interferometer data system includes an online digital computer, which samples the fringes at their natural rate and fits them by a best-fit fringe function. An extensive program library has been compiled to make handling the copious data coming from a three-element interferometer recording four polarization products (corresponding to the four Stokes parameters; see Seielstad et al 1964, Conway & Kronberg 1969) as automatic as possible. This interferometer also operates as a synthesis system (Hogg et al 1969).

A rather different approach to forming a beam from a multielement antenna array is used in the Culgoora solar array (Wild 1967). This array is a circular ring with 96 elements spaced about its rim. Rather than preserving Fourier components for a long period of time for later beam formation, it forms a beam instantaneously by combining the IF signals from all of the elements and detecting. It is easily shown that a uniform level ring of radius a has, if evenly phased, a beam pattern

$$B(z) = J_0^2(2\pi\ az)$$

where z is the zenith distance. The beam may be steered to other parts of the sky by appropriate phasing, though it becomes elongated by the factor sec z. The function J_0^2 has undesirably high near-sidelobes. Wild (1965) has shown that beams with forms described by Bessel functions of higher order, J_2, J_3, etc, may be formed by appropriately phasing the ring. A beam of shape $J_n^2(2\pi\ az)$ is formed by phasing the kth antennas by $2\pi nk/N$ radians (antennas being numbered counterclockwise around the ring). A sidelobe-free, well-behaved beam can be formed by suitable combinations of J_0^2 and J_n^2.

The field of view of the Culgoora array is such that about 3000 independent beams can be formed. The construction of this many beam-forming equipments was too large a project, so, in practice, 48 beams are formed, and steered to 60 locations, each within the field of view, within 1 sec. The interval of time allotted for one beam position is further subdivided, so that part of the time is allocated for J_0^2 beams, part for J_1^2, etc, which are averaged together to give a good final beam shape. McLean et al (1967) have described a technique for obtaining the full number of beams instantaneously, using an optical analog computer.

Continuum data-handling systems.—As was stated before, the recording of the output of a continuum receiver is one of the early tasks that obviously should be performed as soon as the volume of data has exceeded the amount that the observer is willing to handle by himself. The introduction of computer processing has introduced a number of benefits to the observer: the greatest is a repeatable procedure that can be accurately analyzed, so that the observer may know just what the errors in his data are.

Various procedures are applied to continuum data. The most usual is a filtering to remove the low-frequency components from the radiometer output ("removing baseline drifts") because it is usually felt that these are more probably due to slow instabilities in the receiver than to actual large-scale structure in the sky. If there is structure of a scale much larger than their beam, most observers are willing to leave it to someone with a smaller telescope. The baseline removal may be done by filtering out the low-frequency components with some kind of digital filter (e.g., Dixon & Kraus 1968), or by fitting a polynomial to the source-free regions of the observation.

The radiometer output may be fitted to various functions. Fitting Gaussians to observed data has always been popular among line observers (Kaper et al 1966), who have attempted to interpret them in terms of thermal Doppler-broadened emission lines. The same has been done with continuum records (von Hoerner 1967), where the Gaussian is interpreted as an approximation to the beam size or source brightness distribution, whichever is larger. Dixon & Kraus (1968) take an interesting approach, in which the record is approximated by a sum of beam shapes, the approximation being done by a highly nonlinear procedure. That is, whenever a significant positive deviation occurs, a beam shape is subtracted, centered at the most positive deviation. This procedure is repeated until no significant positive deviations from the (filtered) baseline occur. For complex sources, the subtracted beam shapes are added together and examined by eye to determine the true shape of the source. This procedure is, of course, guaranteed to yield a reconstructed profile that is a possible output of the antenna, but no analysis yet indicates the extent to which the reconstructed output resembles the actual source. Nonetheless, the technique has an interesting advantage: it uses the fact that, in any real, physical radio source, the brightness is always positive.

This extra bit of information has long tempted information theorists, because it is something beyond the information given by the linear filters envisioned by Bracewell & Roberts (1954). Even in the linear theory, resolution may be increased somewhat beyond the apparent beam size of the antenna, but the price one pays in signal-to-noise ratio has made the procedures impractical (Bracewell 1958). The fact that the true brightness is positive everywhere actually adds information to the problem, and one might therefore expect to extract more information from an analysis using this fact. In practice, inserting this condition in the analysis appears to result in a remarkably large increase in the usable resolution.

The most straightforward way to incorporate this condition is simply to do a nonlinear least-squares fitting of a positive model to the observed data, as has often been done in the case of interferometer data (e.g., Berge 1966, Wade 1966). A model-fitting approach may be especially suitable for interferometer data, because one may weigh amplitudes more heavily than phases, which are not so well determined. This is possible because (Bates 1969) a satisfactory recovery of the brightness is possible with only a small fraction at the existing phase information.

Biraud (1968) has proposed a much more complex model-fitting procedure, which involves noting that since the brightness is positive, it may be expressed as the square of a real, analytic function. A reasonable model is fitted to the Fourier transform of this function. Criteria are proposed to indicate when the procedure can be extended to higher spatial frequencies, increasing the resolution of the observations. It is claimed that resolution may be increased by fairly large factors without incurring the decay in signal-to-noise ratio experienced by the conventional analytic continuation methods (supergaining).

A method of interpolating unevenly spaced interferometer data also using the fact of positive brightness has been described by Högbom (1969). This technique has also been used by Zisk (1968) and appears to have originated with Max (1965). The data in hand are scattered samples of the Fourier transform of the true brightness distribution. The inverse transform is applied to these scattered samples, and the reconstructed brightness distribution is examined for negative values. Whenever these occur, they are replaced by zero. The Fourier transform is taken. This will now differ, in the sampled locations, from the measured values. The transform at these points is replaced by the measured value, and the procedure is then repeated until it converges with an all-positive brightness distribution and a Fourier transform having the measured values at the sampled points.

Pulsar-search techniques.—With the discovery of pulsars by Hewish et al (1968), a new class of information-processing problems was introduced into radio astronomy. These have been briefly reviewed by Burns & Clark (1969). Given the characteristics of the pulses which are sought, it is necessary to devise an optimum linear filter to detect them. Two problems arise. First, the optimum linear filter is difficult to devise. The filter is matched to the pulsar repetition rate, and to the position of the pulse within the interval. Therefore, to search a given frequency range with a given integration period, the number of channels that must be synthesized is the integration period times the frequency range divided by the duty cycle of the pulsar. Further, the bandwidth must be narrow, or else a third dimension, that of dispersion measure, must be introduced into the search parameters. This number of filter channels is expensive to implement in either software or hardware, even in the most efficient implementation (Staelin 1969).

The second problem is that the optimum linear filter may not be the optimum way to detect the pulses. For a very erratic pulsar, the pulsar may be better detected by noticing the occasional very strong pulse than by the linear filter. Indeed, most known pulsars have been detected by noticing the pulses on a chart recorder. Various nonlinear procedures have been suggested also, because they, though not as efficient as the linear filter, require much less processing, and a fairly sensitive pulsar receiver can be built with little expense. Perhaps the most productive is to multiply two adjacent frequency channels after detection, with an appropriate delay, to detect the dispersed pulse.

For the most part, information processing in astronomy is a matter of simply deciding what you want, and sitting down and building the system to produce it, in either software or hardware. The main problems are problems of economical design. However, in a few cases, astronomy has been strongly influenced by discoveries about information processing, as in the infrared Fourier spectroscopy, or even by computational techniques, as in the Fast Fourier Algorithm (Cooley & Tukey 1965).

LITERATURE CITED

Argelander, F. W. A. 1859, *Astron. Beob Sternwarte Bonn*, **3,** 1
Arp, H. C. 1959, *Ap. J.*, **129,** 507
Babcock, H. W. 1953, *Ap. J.*, **118,** 387
Bates, R. H. T. 1969, *MNRAS*, **142,** 413
Beard, M. Morimoto, M. Hedges, P. 1967. *Proc. Radio Electron. Eng. Aust.*, **28,** 345
Berge, G. L. 1966, *Ap. J.*, **146,** 767
Biraud, Y. 1968, *Astron. Ap.*, **1,** 124
Bracewell, R. N. 1958, *Proc. IRE*, **46,** 106
Bracewell, R. N., Roberts, J. A. 1954, *Aust. J. Phys.*, **7,** 615
Braes, L. L. E. 1963, *Bull. Astron. Inst. Neth.*, **17,** 132
Burns, W. R., Clark, B. G. 1969, *Astron. Ap.*, **2,** 280
Burns, W. R. Yao, S. S. 1969, *Radio Sci.*, **4,** 431
Cole, T. 1968, *Aust. J. Phys.*, **21,** 273
Connes, J., Connes, P. 1966, *J. Opt. Soc. Am.*, **56,** 896
Conway, R. G., Kronberg, P. P. 1969, *MNRAS*, **142,** 11
Cooley, J. W., Tukey, J. W. 1965, *Math. Comput.*, **19,** 29
Cooper, B. F. C. 1969. *Electron. Lett.* (In press)
Davies, R. D., Ponsonby, J. E. B., Pointon, L., de Jager, G. 1969, *Nature*, **222,** 933
de Vaucouleurs, G. 1956, *Occas. Notes Roy. Astron. Soc.*, **3,** 126
Dieter, N. H. 1965, *Astron. J.*, **70,** 552
Dixon. R. S., Kraus, J. D. 1968, *Astron. J.*, **73,** 381
Eichhorn, H. K. 1960, *Astron. Nachr.*, **285,** 233
Eichhorn, H. K., Googe, W. D., Gatewood, G. 1967, *Astron. J.*, **72,** 626
Ekre, H. 1963, *IEEE Trans. Inform. Theory*, **9,** 18
Elldér, J., Rönnäng, B., Winnberg, A. 1969, *Nature*, **222,** 67
Fomalont, E. B., Wyndham, J. D., Bartlett, J. F. 1967, *Astron. J.*, **72,** 445
Googe, W. D. 1967, *Astron. J.*, **72,** 623
Gordon, C. P., Gordon, K. J., Shalloway, A. M. 1969, *Nature*, **222,** 129

Hardie, R. 1959, *Ap. J.*, **130,** 663
Harmon, L. D., Knowlton, K. C. 1969, *Science*, **164,** 19
Heeschen, D. S. 1961, *Ap. J.*, **133,** 322
Heiles, C. 1967, *Astron. J.*, **72,** 1040
Hewish, A., Bell, S. J., Pilkington, J. D. H., Scott, P. F., Collins, R. A. 1968, *Nature*, **217,** 709
Högbom, J. 1969, *Rep. URSI 16th Gen. Assembly, Ottawa, Aug.*
Hogg, D. E., Macdonald, G. H., Conway, R. G., Wade, C. M. 1969, *Astron. J.* (In press)
Hunten, D. M. 1968, *Science*, **162,** 313
Kaper, H. G., Smits, D. W., Schwarz, U., Takakubo, K., van Woerden, H. 1966, *Bull. Astron. Inst. Neth.*, **18,** 465
Kenderdine, S., Ryle, M., Pooley, G. G. 1966, *MNRAS*, **134,** 189
McLean, D. J., Lambert, L. B., Arm, M. 1967, *Proc. Inst. Radio Electron. Eng. Aust.*, **28,** 375
Max, Joel 1965, *Rep. 5th Midwest Conf. Inform. Process.*
Morris, D., Clark, B. G., Wilson, R. W. 1963, *Ap. J.*, **138,** 889
Müller, C. A., Westerhout, G. 1957, *Bull. Astron. Inst. Neth.*, **13,** 151
Newcomb, J. S. 1968, *Intern. Rep., Control Data Corp., Minneapolis, Minn.*
Oke, J. B. 1969, *Publ. Astron. Soc. Pac.*, **81,** 11
Pooley, G. G., Kenderdine, S. 1968, *MNRAS*, **139,** 529
Roberts, M. S. 1966, *Ap. J.*, **144,** 639
Rydbeck, O. E. H., Kollberg, E. 1968, *IEEE Trans. Microwave Theory Tech.*, **16,** 799
Ryle, M. 1962, *Nature*, **194,** 517
Ryle, M., Elsmore, B., Neville, A. C. 1965, *Nature*, **207,** 1024
Ryle, M., Hewish, A. 1960, *MNRAS*, **120,** 220
Schroeder, M. R. 1969, *Commun. Assoc. Comput. Mach.*, **12,** 95
Schulte, D. H. 1960, *Astron. J.*, **65,** 507

Seielstad, G. A., Morris, D., Radhakrishnan, V. 1964, *Ap. J.*, **140**, 53
Staelin, D. H. 1969, *Proc. IEEE*, **57**, 724
Strand, K. Aa. 1966, *Astron. J.*, **71**, 873
Swenson, G. W. 1969, *Ann. Rev. Astron. Ap.*, **7**, 353
Van den Bos, W. H. 1956, *Vistas Astron.*, 1035 (Beer, A., Ed., Pergamon)
Van Vleck, J. H. 1943, *Internal Rep., Harvard Univ. Radio Res. Lab.* Republ. as Van Vleck, J. H., Middleton, D. 1966, *Proc. IEEE*, **54**, 2
Vasilevskis, S. 1960, *Astron. J.*, **65**, 208
von Hoerner, S. 1967, *Ap. J.*, **147**, 467
Wade, C. M. 1966, *Phys. Rev. Lett.*, **17**, 1061
Weinreb, S. 1963, *MIT Tech. Rep., No. 412, Dept. Electron. Eng.*
Westerhout, G. 1957, *Bull. Astron. Inst. Neth.*, **13**, 201
Westerhout, G. 1969, *Maryland-Green Bank Galactic 21-cm Line Survey* (2nd ed., Univ. Maryland)
Wild, J. P. 1965, *Proc. Roy. Soc. London A*, **286**, 497
Wild, J. P. 1967, *Proc. Inst. Radio Electron Eng. Aust.*, **28**, 279
Zisk, S. 1968 (Private communication)
Zwicky, F. 1964, *Ap. J.*, **139**, 1394

OPTICAL OBSERVATIONS OF EXTRASOLAR X-RAY SOURCES

W. A. HILTNER AND D. E. MOOK

Yerkes Observatory, University of Chicago
Williams Bay, Wisconsin

INTRODUCTION

This review of optical observations of X-ray sources will be concerned primarily with those sources for which there are well-established optical counterparts. A brief discussion of the methods of optical identification will be followed by remarks on the individual objects. The entire field of X-ray astronomy has been reviewed in these pages by Morrison (1967) and elsewhere (e.g. Oda 1970 and references cited therein).

OPTICAL IDENTIFICATION OF X-RAY SOURCES

Introduction.—Present astronomical X-ray experiments yield source positions accurate, at best, to about 1 min of arc and this is likely to remain true for the forseeable future. In general, several optical objects will lie within the X-ray position error box, the number depending upon the faintness of the objects one chooses to consider and the region of the sky. The problem of identification is then one of choosing criteria, in addition to location in or near the error box, to decide which object is the optical counterpart of an X-ray source (if, indeed, a counterpart exists). Currently accepted criteria include:

(*a*) Time correlations between X-ray and optical flux variations (if variations are present at all).

(*b*) Agreement between the observed optical properties of a candidate (colors, brightness) and those predicted from an extrapolation of the X-ray data to optical energies on the basis of a simple physical theory.

(*c*) If the X-ray source is extended, similar spatial distribution of the optical and X-ray sources.

(*d*) The optical similarity of a candidate to sources previously identified.

(*e*) The unusual optical nature of an object.

In the following two sections we will briefly discuss the X-ray and optical techniques currently used in applying these criteria.

Accurate X-ray positions.—The ability to identify X-ray sources optically depends primarily upon a good position from the X-ray observations. This will restrict the area of the sky to be searched so that the number of optical candidates is sufficiently small to make practical the application of one or more of the above criteria. As an example of the difficulty encountered in searching a large area, Figure 1 shows the region of the sky around GX3+1; the search area defined by the X-ray position error box is shown along with a tentative identification by Blanco et al (1968b).

Techniques of X-ray astronomy have been reviewed in these pages by Giacconi et al (1968). Those experiments employing slat collimators give positions with a precision of from 15 min of arc to several degrees (see Bradt et al 1968b). The precision depends upon the initial goal of the experiment, the strength of the source, whether there is overlapping of sources, and the quality of information on the payload orientation. Positions of much higher precision can be obtained with the modulation collimator suggested by Oda (1965; for a review see Bradt et al 1968a). Precisions of approximately 1 min of arc can be achieved (Gursky et al 1966b). Recently a modification has been introduced (Schnopper et al 1968) in which the modulation collimator is rotated. It is then possible to find the positions of several sources that may fall within view of the detector. Again the precision is approximately 1 min of arc.

A different approach to obtaining accurate X-ray positions involves the use of lunar occultations of X-ray sources. More precisely, where the X-ray observations are made from rockets, this technique is most useful for confirming a suspected identification since the X-ray position must be known with sufficiently high precision to predict the time of occultation. This limitation will lessen with the advent of X-ray satellites, but, of course, the method will always be restricted to that band of the sky covered by the Moon. This technique was beautifully applied by Bowyer et al (1964) in the definitive optical identification of the Crab Nebula as the X-ray source Tau X-1.

Methods of optical identification.—Given an X-ray position, one or more of the criteria listed above must be applied to discriminate the optical object associated with the X-ray source from the other objects in the field. The ease with which this can be done will depend upon the size of the error box, the crowding of objects in the field around the X-ray position, and the faintness of the objects that one must consider. These problems are enhanced since most X-ray sources are concentrated to the galactic plane, most strongly near the galactic center.

Of the five criteria listed, (*b*) and to a certain extent (*e*) do not necessarily require detailed observations of individual objects; when the number of candidates within the search area becomes large these criteria may be applied over a fairly large region of the sky by photographic survey techniques. Criterion (*b*) has been applied by assuming some physical model for the radiation mechanism based on the X-ray spectrum. Two primary processes have been considered, synchrotron radiation and thermal bremsstrahlung. Although the spectral forms predicted by these processes are not necessarily

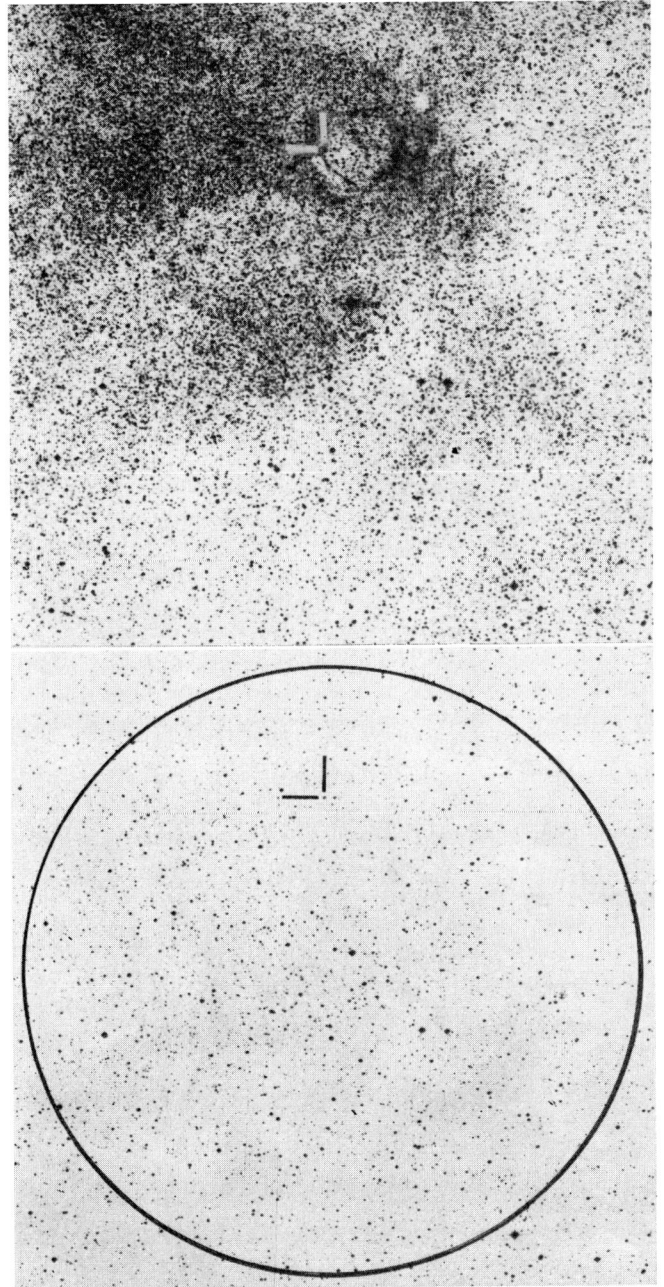

FIGURE 1. Blue (left) and red (right) photographs of the region containing GX3+1 from Blanco et al (1968b). The error circle is 17' in radius. A tentative optical identification is indicated. Copyright National Geographic Society-Palomar Sky Survey.

unique in complex systems, simple models have aided identification in the past. The model permits extrapolation of the observed X-ray spectrum to the optical region and yields approximate colors and brightness, including, of course, some correction for interstellar extinction. To compute a brightness, some angular diameter of the source must be assumed. It is customary to assume a stellar source in present search programs.

With some indication of the expected magnitude and color one can search the area either by direct photography, if the number of stars in the field is large, or by photoelectric photometry of individual objects, if the number of stars is sufficiently small to make this technique practical. The direct photographs are multiple exposures of the same field taken through two or more filters. The images resulting from exposure through the different filters are separated from one another by a small distance on the plate so that rapid evaluation of the color is possible by visual inspection. Another procedure is to substitute objective prism spectra for the multicolor photography. Whether or not this approach is practical will depend on the expected magnitude. Similar survey programs can also be used to apply criterion (e) by searching the plates for objects with unusual characteristics. The identification of Sco X-1, discussed later, exemplified these techniques.

Survey techniques may reduce the number of candidates so that one can proceed with more accurate photoelectric photometry and spectroscopy of individual objects. Application of the other criteria for identification may then proceed. With regard to criterion (d), thus far only two sources have been firmly identified: the Crab Nebula, showing a power-law X-ray spectrum from 2-20 keV, and Sco X-1, showing an exponential X-ray spectrum from 2-20 keV. The published identification of Cyg X-2, discussed below, is based in part upon its optical similarity to Sco X-1.

Fulfillment of criterion (a) gives the most compelling argument for the validity of an identification. On the other hand, one cannot necessarily reject an identification on the basis of a failure to discern any time correlation between the X-ray and optical flux variations. Sources might well exist in which there is little or no coupling between the X-ray and optical components of the spectrum. Here the task of confirming a tentative identification will be much more difficult in principle, and one will have to rely on the other criteria to reduce doubt.

REMARKS ON INDIVIDUAL SOURCES
Tau X-1 (The Crab Nebula)

Tau X-1 was the first extrasolar X-ray source to be identified optically. The coincidence between the X-ray source position and that of the Crab Nebula and the uniqueness of this object made it seem likely that the two are one and the same. The lunar-occultation technique mentioned earlier was used by Bowyer et al (1964) to determine that the X-ray source is centered on the Crab Nebula and that it has an angular extent of 1 to 2 min of arc, thereby confirming the identification. These results were later confirmed by means of a modulation collimator experiment (Oda et al 1967).

Subsequently, a radio pulsar, NP 0532, was discovered in the direction of the Crab Nebula; this was followed by the discovery that Baade's (1942) "south preceeding" star in the Crab, displaying a featureless spectrum, is an optical pulsar. That this same object is associated with the X-ray emission from the Crab was later demonstrated when X-ray pulses (amounting to $\sim 9\%$ of the total X-ray emission) were detected in synchronism with the radio and optical pulses. The review by Oda (1970) discusses these and subsequent observations of the Crab pulsar; more recent optical observations are given by Warner et al (1969), Oke (1969), Neugebauer et al (1969a), Wampler et al (1969), Cocke et al (1969), and references cited in these papers. It should be emphasized that research in this area is extremely active at this time, and there is lack of consensus on features of the pulsed emission. There is an extensive literature on the optical properties of the Crab Nebula itself. We refer to the reviews cited in the introduction for discussions within the context of X-ray astronomy. Also it would appear, from the limited information currently at hand concerning the spectra of the X-ray sources, that few are characterized by a power-law spectrum like Tau X-1, whereas several show exponential spectra like Sco X-1.

Sco X-1

The optical identification.—Historically, Sco X-1 was the first extrasolar celestial X-ray source to be discovered (Giacconi et al 1962) and except for transient phenomena in other sources (Chodil et al 1968b, Conner et al 1969, Evans et al 1970), it remains the strongest yet discovered. Oda (1968) has written an excellent summary of the events leading to the optical identification of this source. Early optical searches were based on magnitudes and colors derived from an extrapolation of the X-ray spectrum (assumed to be due to thermal bremsstrahlung from an optically thin plasma) to optical wavelengths. These searches were hampered by large uncertainties in the position and angular extent of the X-ray source. Gursky et al (1966a) determined 20" as an upper limit to the angular size, which suggested that the optical counterpart might be bright enough to be visible. The workers at Tokyo (Sandage et al 1966, Oda 1968) isolated a possible optical counterpart of Sco X-1 from two-color plates on the basis of its predicted ultraviolet excess. This proposed candidate was found to lie in one of two accurate X-ray positions obtained with a modulation collimator (Gursky et al 1966b). Johnson & Stephenson (1966) noticed the same object as a possible old nova and suggested its association with Sco X-1, but they could not be certain of the identification because the object lay so far outside the then-current X-ray position error box. Recent simultaneous observations of Sco X-1 in the optical and X-ray bands (see below, Figure 8) have further strengthened this identification.

Thus, the optical identification of Sco X-1 can be said to have resulted from the extrapolation of the observed X-ray spectrum to the optical region using physical theory and from an accurate X-ray position. Fortunately, Sco X-1 is far enough from the Galactic plane ($b^{II} \sim 24$) that the number of stars within the surrounding field is relatively small and its isolation was possible prior to the availability of an accurate X-ray position. Still, the initial identification was not beyond doubt until further observations established the unusual nature of the optical object, and especially the correlation between the X-ray and optical fluxes.

Optical properties of Sco X-1.—

A. The spectrum: The spectrum is the least well studied of the optical properties of Sco X-1. Most of our knowledge depends on the 10 spectrograms discussed by Sandage et al (1966) and by Westphal et al (1968). The spectrum of Sco X-1 has some similarity to that of an old nova (see e.g. Kraft 1964). This suggestion is based on the presence and character of emission lines, notably He II 4686, hydrogen, and the C III-N III blend near λ4650. It is similar to V 603 Aql (1918) and U Sge except that the latter has broader hydrogen lines. Representative spectra of Sco X-1 taken at several different magnitudes are shown in Figure 2. In addition to the emission lines

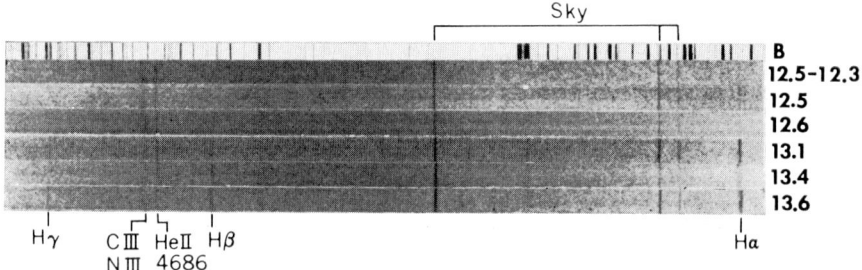

FIGURE 2. Spectra of Sco X-1 at various magnitudes from Hiltner et al (1970). The dispersion is 120 Å/mm and the exposure times are ~10 min. Exposure of the first spectrum began at $B \approx 12.5$ when the object commenced a flare of amplitude 0.2 mag. The exposure was terminated somewhat after the flare peak. The second spectrum was made immediately preceding the flare.

noted above, Westphal et al (1968) have reported emission features due to He I, Fe II, and O II. Also Johnson & Golson (1968) have reported Hβ in absorption.

The intensities of the emission lines relative to the continuum are seen to vary. Westphal et al (1968) have suggested, on the basis of their limited data, that the emission lines do not change in absolute intensity, and consequently are decoupled from the continuum, which does vary. Visual inspection of a series of spectrograms available to the authors does not confirm this conclusion, however. Also, an inspection of Figure 4 in Sandage et al (1966) strongly suggests that He II 4686 varies less relative to the continuum than the lower excitation hydrogen Balmer lines.

Radial-velocity data are also based primarily on the ten spectrograms discussed above (Westphal et al 1968). These authors report velocity changes of about 100 km/sec over an interval of 3 hr on two different nights; the velocity measured for He II 4686 changed in a direction opposite to that of the hydrogen lines. Also systematic changes of velocity from night to night were reported. These changes in radial velocity have been interpreted as evidence for duplicity of the system, a feature consistent with the known binary nature of most old novae (see e.g. Kraft 1963). However, not all of the published radial-velocity observations (and some recent unpublished ones by Hiltner et al 1970) show He II 4686 and hydrogen varying out of phase. It is premature to call Sco X-1 a binary since the most fundamental quantity, the period, has not been determined either from radial-velocity observations or, if the system is eclipsing, from photometry.

The need for more spectroscopic observations particularly at high time and spectral resolution cannot be overemphasized, not only for radial velocities and line intensities but also for line profiles. Hα is variable in profile and may at times appear to be double.

B. *Photometric properties*: Figure 3 shows some representative light curves for Sco X-1 from Hiltner & Mook (1967). The great variety in the character of the light curves from night to night, with no evident pattern, is typical. Figure 4 illustrates the long-term variability of the system.

The light variations may be divided into four components:
1. Night-to-night variations of up to $\sim 1^m$.
2. Smooth changes of $\sim 0^m.5$ in $\sim 1^h$.
3. Flickering with amplitude $\sim 0^m.02$ at a time scale of minutes.
4. Flares with typical amplitudes of $\sim 0^m.2$ and rise times as short as 90^s.

Westphal et al (1968), in a high-time-resolution study of the brightness variations of Sco X-1, have shown that the flickering is superimposed on all of the other types of variations listed above, including the flaring. Hiltner & Mook (1967) noted that Sco X-1 flares only when it is brighter than $B \sim 12^m.6$; in nearly 260 hr of monitoring over a 3-year period, they (1970) have not yet seen this statement contradicted. Flares typically occur in groups, often by threes, with no recognizable regularity. X-ray flares have been detected by Lewin et al (1968b) and by Hudson et al (1970).

Figure 5 shows brightness histograms for Sco X-1 from monitoring done by Hiltner & Mook (1970) over the three observing seasons since the optical identification became available. Arguments by these authors suggest that the peaks represent real "preferred" brightness states of the system. If these are reliable samples of the photometric state of Sco X-1 in each of the 3 years, then an evolution of the system is indicated on a time scale of months. Attention is called to the shift in magnitude. In 1967 the object was brighter than $B = 12^m.5$ for 45% of the time while in 1968 and 1969 this quantity was 13 and 12% respectively. Changes in the faintest observed magnitude are also notable. This type of analysis could prove to be of utmost significance for the determination and study of long-term changes in the system.

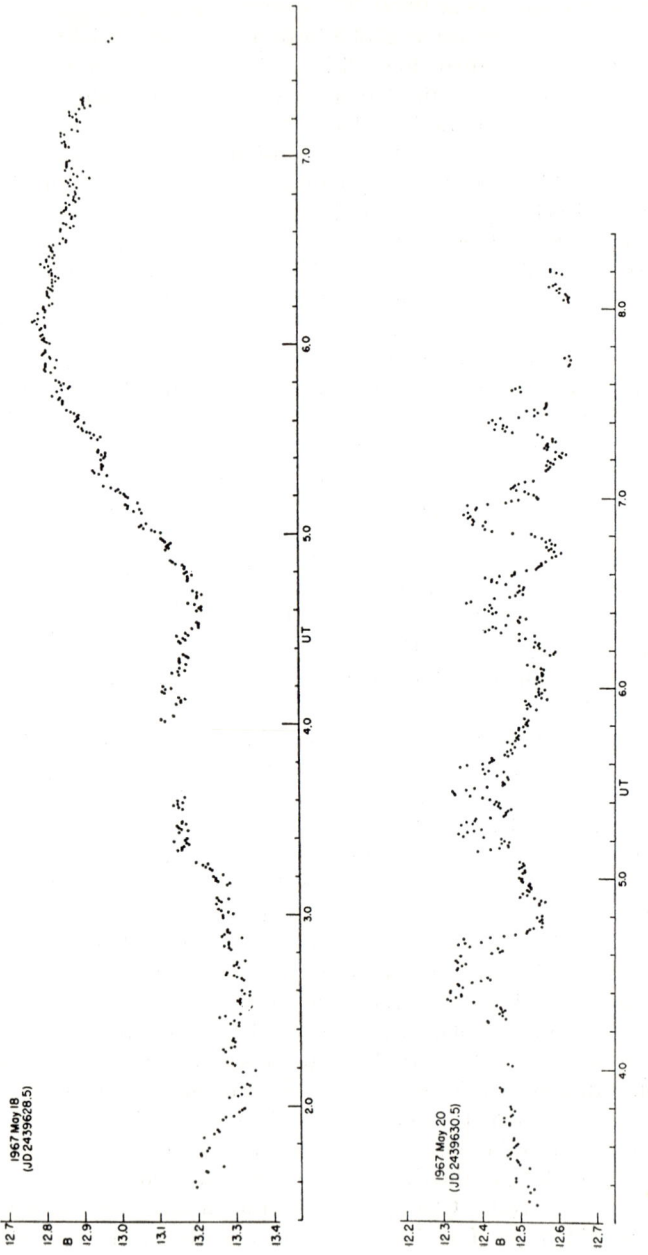

FIGURE 3. Representative light curves for Sco X-1 from Hiltner & Mook (1967).

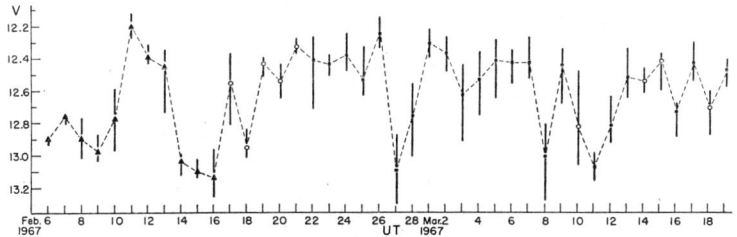

FIGURE 4. Average visual magnitude of Sco X-1 for 42 nights of monitoring by Mook (1967). Bars through the data points are not error bars but represent the total range in V over which Sco X-1 varied on that night.

Figure 6 is a color-magnitude locus for Sco X-1 determined from simultaneous B and V monitoring of the system by Mook (1970a) during seven nights in 1968. All data are consistent with the linear relation shown in the figure (solid line) to within the photometric accuracy (about $0^m.03$). The dashed line in Figure 6 is the least-square fit to the 42 nightly averages of V and $B-V$ obtained by Mook (1967) in 1967; no significant change in the color-magnitude relation is indicated from 1967 to 1968, in contrast to the brightness histograms for these 2 years.

C. *Periodicities in the light variations*: There is no evidence for any periodic component in the brightness variations of Sco X-1. Periods have been reported by Rao et al (1969b) and by van Genderen (1969); however, these periods do not agree with one another, and because the proposed periods are large compared with the total length of continuous data available for analysis, no period has been established. The difficulty is that if any periodic components do exist, they are not appreciably larger in amplitude than the irregular variations, so that phasing of noncontiguous series of data is difficult. Time series analysis by Hiltner & Mook (1967) and Lasker & Hesser (1969) shows no evidence for periodic activity over the range of periods from 0.2^s to 1^h. Extension of this analysis to longer periods will require continuous monitoring for lengths of time in excess of 9^h, the practical limit presently available. The power spectra obtained by Lasker & Hesser (1969) show an inverse square frequency dependence for periods from 45^s to 120^s.

D. *Infrared photometry*: Neugebauer et al (1969b) carried out scanner observations of Sco X-1 from .33 to 1.08 μ and infrared photometry at 1.65 and 2.20 μ. Their data were averaged to obtain a mean spectrum for Sco X-1; see Figure 7. At longer wavelengths, the spectrum approaches that of a blackbody. This suggests that strong self-absorption may be taking place in the source. Indeed, these authors are able to explain their observations by extrapolating the observed X-ray spectrum to the visual and infrared wavelengths by means of a self-absorption bremsstrahlung model discussed below.

E. *Radio observations*: Radio observations of Sco X-1 have been reported

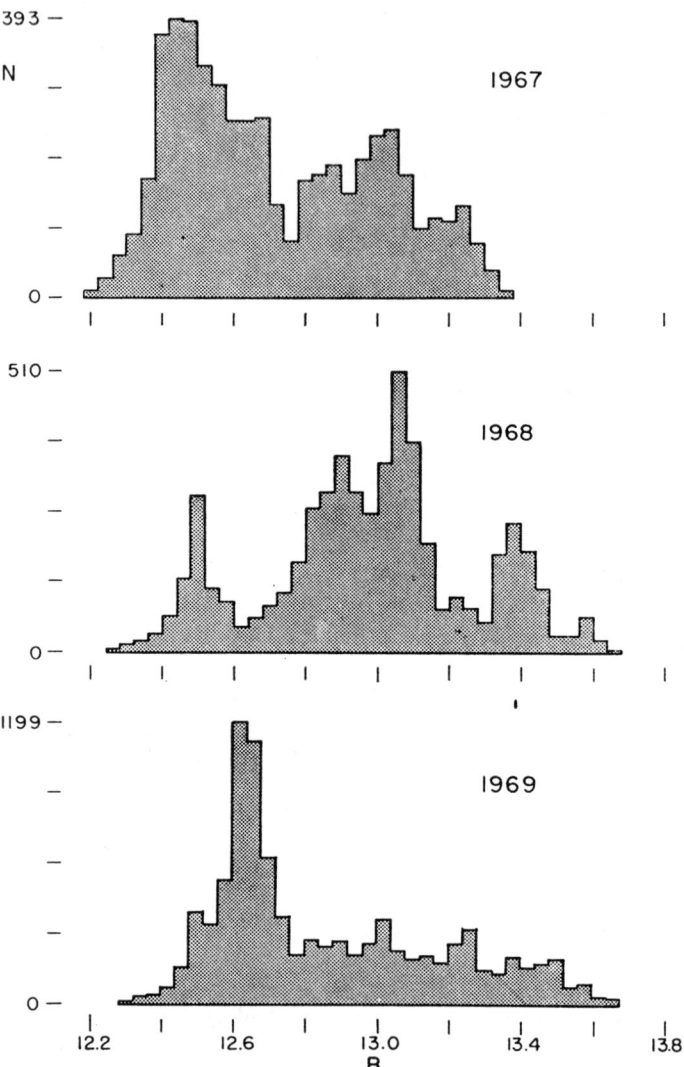

FIGURE 5. Brightness histograms of Sco X-1 from Hiltner & Mook (1970). The number of times (30-sec integrations) the object was observed to be within a given magnitude interval (of width .04 mag) is plotted versus the B magnitude. Note that the scales of the vertical axes are different in each plot.

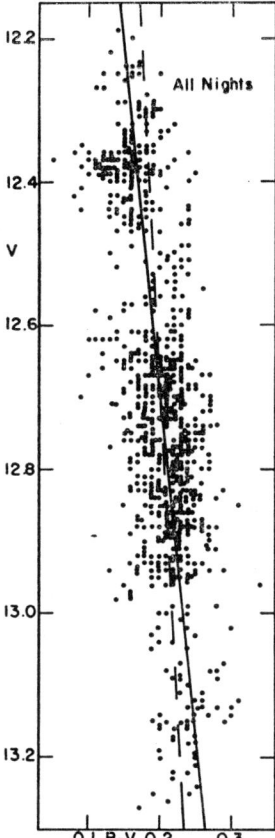

Figure 6. The color-magnitude locus for Sco X-1 from simultaneous B and V monitoring by Mook (1970a). The solid line is a least-square fit to the data; the dashed line shows the least-square fit to the 42 nightly V and $B-V$ averages by Mook (1967).

by Andrew & Purton (1968) and by Ables (1969). The latter study indicates variability in the source with a time scale of 1^h. No spectral information is yet available; however, the observed flux levels are well below a simple bremsstrahlung extrapolation of the X-ray spectrum as shown in Figure 7.

F. *Ultraviolet observations*: Preliminary ultraviolet filter photometry from the OAO by Bless et al (1969) indicates a flat spectrum (corrected for interstellar extinction) from 3300 to 1380 Å, which is consistent with a thermal bremsstahlung source.

Distance estimates for Sco X-1.—Some distance estimates published for

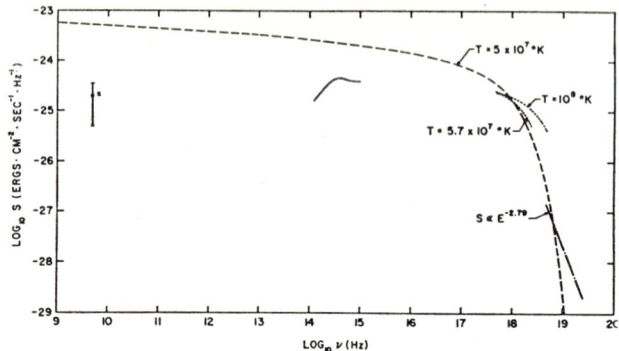

FIGURE 7. The spectrum of Sco X-1 from Riegler & Ramaty (1969). Radio data are from Andrew & Purton (1968) (cross) and Ables (1969) (circle; the bar indicates the range of observed fluxes); the solid line in the visual and infrared region is from Neugebauer et al (1969b); the dotted lines correspond to the low-energy X-ray observations of Gorenstein et al (1968); the dot-dash line fits the high-energy data of Buselli et al (1968); the dashed line is the bremsstrahlung spectrum from an isothermal, optically thin gas at $T = 5 \times 10^{7}$°K, normalized to typical X-ray spectra.

Sco X-1 are listed in Table 1. These estimates are based on three types of data:
  A. Properties of the interstellar medium.
  B. Knowledge of the intrinsic brightness of Sco X-1 on the assumption that it is an old nova.
  C. Proper-motion data.

A. Use of the interstellar medium:

STRENGTH OF THE CA II LINES. Wallerstein (1967) has shown that the radial velocity of the interstellar Ca II absorption features in spectra of Sco X-1 is consistent with those of five other stars nearby in the sky; this argues that the lines are due to interstellar absorption and not to material intrinsic to the system itself. He also shows that the equivalent width of the K line (∼280 mÅ) in the spectrum of Sco X-1 is larger than in any of these nearby stars including stars in the Sco-Cen association. A firm lower limit on the distance to Sco X-1 of 270 pc is obtained. Westphal et al (1968) use mean correlations of K-line strength with distance in the Galaxy to estimate $d \sim 500$–600 pc, but these estimates ignore any local structure of the interstellar medium in the line of sight to Sco X-1. If interstellar line strengths become available for more distant stars near Sco X-1 on the sky, this method could yield a reliable distance estimate.

Westphal et al (1968) also use the ratio of the line strengths H/K to estimate the distance by Münch's (1966) method; they find, depending on the density of interstellar hydrogen assumed, ∼240 pc for $N_H = 0.4$ cm^{-3} and ∼425 pc for $N_H = 0.3$ cm^{-3}, values that are very uncertain.

INTERSTELLAR ABSORPTION OF X RAYS. The absorption cross section of

TABLE 1. Distance estimates for Sco X-1

Method	Distance	Comments
Ca II K-line equivalent width	>270 pc	Wallerstein (1967); a lower limit
Ca II K-line equivalent width and mean equivalent width-distance relation for the Galaxy	500–600 pc	Westphal et al (1968)
Ca II H/K ratio and assumed cosmic abundances of Ca and H	240 pc for $N_H \sim 0.4$ 425 pc for $N_H \sim 0.3$	Westphal et al (1968)
Attenuation of low-energy (<1 keV) X rays	<25 pc for $N_H \sim 1.0$ <50 pc for $N_H \sim 0.5$	Fritz et al (1968); the X-ray data have been questioned
Observed reddening of the bremsstrahlung spectrum assumed due to interstellar extinction	200 pc	Oda (1968)
Assuming Sco X-1 is an old nova	$230 < d < 1000$ pc	Westphal et al (1968)
Proper-motion data	170 pc–200 pc	Sofia et al (1969); Gatewood & Sofia (1968); there is uncertainty in the interpretation of these results (Wallerstein 1969)

the interstellar medium has been computed by Felten & Gould (1966) and by Bell & Kingston (1967) for X rays, assuming "normal" cosmic abundances. An upper limit of $\tau < 0.5$ to the optical depth of the interstellar medium to the distance of Sco X-1 at 0.2 keV has been set by Fritz et al (1968). For $N_H = 0.5$ cm^{-3} this means that d is less than 50 pc, in sharp contrast to the lower limit set by the K-line equivalent width. There seems to be some disagreement with the X-ray measurement at 0.2 keV, however. Grader et al (1970) find evidence for substantial absorption in the spectrum of Sco X-1 at low energies that would increase the distance estimate considerably. Another alternative is to assume that N_H is ~ 0.1 cm^{-3}, a value that is low compared to 21-cm data but not unreasonable if clumping of the interstellar medium is present (Bowyer & Field 1969).

INTERSTELLAR REDDENING. Oda (1968) estimates the distance by assuming the intrinsic optical spectrum to be due to pure bremsstrahlung from an optically thin plasma for which $B - V \sim 0.1$. Since $B - V$ is observed to be ~ 0.2 for Sco X-1, a value of $E_{B-V} \sim 0.1$ for interstellar reddening is indicated.

Observations of two nearby stars suggest that this reddening will take place at $d \sim 200$ pc (note that any self-absorption in the source, as suggested by the simultaneous X-ray and optical observations discussed below, will tend to increase the required value of E_{B-V} and hence the distance determined by this method).

INTERSTELLAR POLARIZATION. Hiltner et al (1967) have measured the polarization of the light from Sco X-1 to be 0.015 mag. If it is assumed that this polarization is not intrinsic to the system, a distance can be estimated by comparison with the interstellar polarization of field stars near to Sco X-1 on the sky for which distances can be determined spectroscopically. A preliminary study by Mook (1970b) has shown that this technique may be practical.

B. *Assuming Sco X-1 is an old nova*: Sandage et al (1966), Westphal et al (1968), and Johnson & Stephenson (1966) argue that Sco X-1 may be identified as an old nova from its spectroscopic and photometric properties. This leads to 230 pc $<d<$ 1000 pc.

C. *Proper-motion data*: Sofia et al (1969) determined the proper motions of Sco X-1 and 82 other stars in a surrounding area. They found three stars in their sample which showed proper motions similar to that of Sco X-1 and hence could be considered a co-moving group with the X-ray source. The similarity of proper motion of this group to that of the Sco-Cen association at a distance of 170 to 200 pc suggests membership of these stars and of Sco X-1 in this association. However, Wallerstein (1969) points out two difficulties with this interpretation. First, stars in the Sco-Cen association over a wide region of the sky including Sco X-1, as well as the four co-moving stars listed by Gatewood & Sofia (1968) for which interstellar line strengths are available, show much weaker K lines than Sco X-1; this indicates that Sco X-1 must lie beyond the Sco-Cen association.

Second, Wallerstein questions the interpretation of the proper-motion data as presented by Sofia et al (1969). He suggests that these data do not confirm the existence of a group of stars co-moving with Sco X-1. In view of the uncertainty in the interpretation of these proper motions, and the arguments based on the interstellar line strengths, any relationship of Sco X-1 to the Sco-Cen association is obscure.

Simultaneous optical and X-ray observations of Sco X-1.—Knowledge of the spectrum of Sco X-1 and its variability over as wide a wavelength range as possible is vital for the formulation and testing of models for the system. In view of the substantial flux variability found in every observed wavelength region, this information can only be obtained by simultaneous observations in the wavelength bands of interest.

Since 1967 several programs of simultaneous optical and X-ray observation have been carried out. The major difficulty with this type of observation is the very limited observing time available to the rocket experiments.

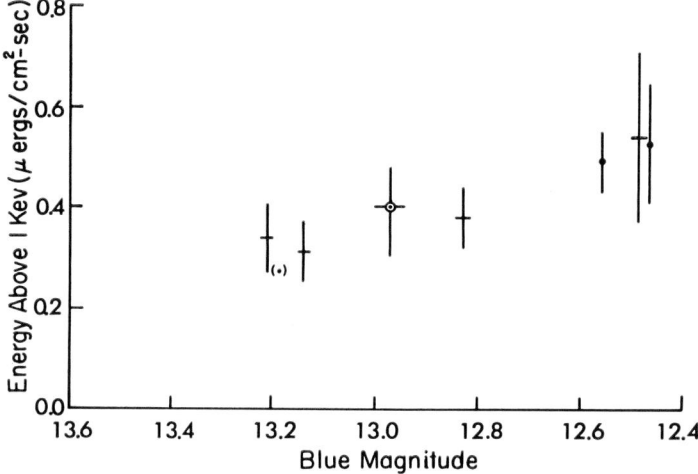

FIGURE 8. Results of simultaneous X-ray and optical observations of Sco X-1 from Burginyon et al (1970).

Balloons (for high-energy measurements) and especially satellites offer the prospect of extended periods of simultaneous monitoring.

Figure 8 is a summary by Burginyon et al (1970) of the simultaneous X-ray and optical observations made to date by the Lawrence Radiation Laboratory and several optical observatories. The monotonic relationship that is indicated between optical brightness and X-ray luminosity lends support to the optical identification and offers yet another empirical relation useful for testing theoretical models.

The work of Chodil et al (1968a) and Mark et al (1969b) has shown that the optical flux of Sco X-1 does not agree with a simple extrapolation of an optically thin bremsstrahlung spectrum from the X-ray region to lower frequencies. In fact, the discrepancy differs from night to night so that one cannot attribute it to a constant interstellar extinction. These authors suggest that variable free-free self-absorption is taking place within the source in addition to a constant interstellar extinction, and they are able to bring the X-ray and optical observations into agreement in this way. Interestingly, the infrared flux measurements mentioned earlier can also be explained by this type of model using the same order-of-magnitude parameters, although as yet no simultaneous infrared and optical or X-ray observations have been made to check this result in detail.

Hudson et al (1970) have recently reported the simultaneous X-ray and optical observation of two flares from Sco X-1. The X-ray and optical fluxes increased simultaneously (within the timing accuracy) and, as in previous simultaneous observations, variable self-absorption within the system is indicated.

Cyg X-2

The X-ray position of Cyg X-2, a source with an X-ray intensity about 20 times less than Sco X-1, is known to an accuracy of about 0.05 square degrees (Giacconi et al 1967b). From an extrapolation of the X-ray spectrum Giacconi et al (1967a) predicted that the optical counterpart should appear at $V \sim 15.1$, $B - V \approx +0.1$, and $U - B \approx -0.9$, without correction for interstellar extinction. From the search of a region 4° on a side, three candidates were found, one near the X-ray position. Its magnitude and colors were estimated to be $V \approx 15.5$, $B - V \approx +0.4$, and $U - B \approx -0.4$ (subsequent photoelectric measurements of magnitudes and colors gave $V = 14.7$, variable, $B - V = +0.44$, and $U - B = -0.22$; Kristian et al 1967). These values agree sufficiently well with those predicted that this object was considered a prime candidate. Furthermore, it was found to be variable, as is Sco X-1; a

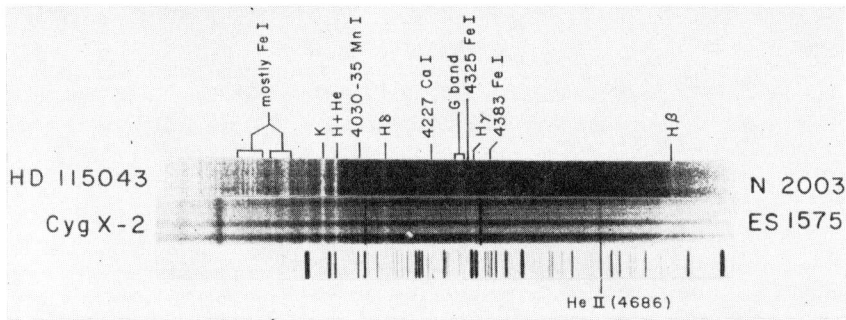

FIGURE 9. Spectrum of the object identified as Cyg X-2, and HD 115043 (G IV) from Kraft & Demoulin (1967). The two spectra were obtained with different optical systems and were not precisely registered over the whole range. Metallic lines in the spectrum of Cyg X-2 are noted. The higher members of the Balmer series are clearly contaminated by the strong lines of Fe I present in a G-type star shortward of the K line.

study by Kristian et al (1967) shows fluctuations of ~ 0.04 mag on a time scale of minutes, and slower variations of ~ 0.1 mag. Possible color changes have also been detected. Aside from the work by Kristian et al (1967), this object seems to have been neglected by photometrists.

Considerably more telescope time has been used for spectroscopic studies of the object (see Figure 9). Spectrograms by Lynds (1967) showed Ca II and hydrogen in absorption and He II 4686 in emission. The absorption lines gave radial velocities on the order of -400 km/sec while He II 4686 in emission was near -100 km/sec. A spectrogram obtained about 1 month later by Burbidge et al (1967) gave a reversal of velocities, -410 km/sec for He II 4686 emission and -96 km/sec for the absorption. Binary motion was the obvious explanation although subsequent studies have raised doubts about this interpretation. Spectrograms by Kraft & Demoulin (1967) showed

OPTICAL OBSERVATIONS OF X-RAY SOURCES   155

that, in addition to the absorption lines of Ca II and hydrogen, there were numerous sharp metallic absorption lines suggesting a type-G spectrum, probably a subdwarf (see Figure 9). Also the He II 4686 emission and Ca I absorption varied together in radial velocity with a time scale ~5.7 hr. In addition, Ca II, Hβ, and Hγ absorption did not vary 180° out of phase with respect to He II. In fact, judging from the published figures, the He II emission-velocity curves were a reasonably good representation of the absorption curves.

The most recent work by Kraft & Miller (1969) has further demonstrated the complexity of this object. They reported that the velocity curves of He II emission and Ca II absorption tended to be mirror images, but were not compatible with binary motion, both because of the shape of the curves and because of the lack of a common systematic velocity. There are velocity changes ~100 km/sec for both absorption and emission (He II 4686) lines. This similarity to Sco X-1, in addition to the variable magnitude, helps to strengthen but does not establish the identification of this object with the X-ray source.

OTHER OPTICAL IDENTIFICATIONS

Table 2 is a list of five other proposed optical counterparts of X-ray sources. Three are of particular interest to us. GX3+1 has an X-ray in-

TABLE 2. Proposed optical identifications[a]

X-ray source	X-ray position 1950		Error square degrees	Optical identification	References
	$12^h30^m.7$	+12°30′	6.5	M87	Friedman & Byram (1967) Bradt et al (1967)
				LMC	Mark et al (1969a)
Cen X-2	13^h24^m	−62°	20	NGC 5189 Central star WX Cen	Feast (1967) Blanco (1967) Eggen et al (1968)
Cen X-4	14^h56^m	−32°15′	13	S 5003	Eggen & Rodgers (1969)
GX3+1	$17^h43^m.4$	−26°06′	0.25	13.4 mag star	Blanco et al (1968b)

[a] *Note added in proof*: Gorenstein et al (Gorenstein, P., Kellogg, E. M., Gursky, H. 1970, *Ap. J.*, **160**) have reported positions of two X-ray sources in Cassiopeia with a precision of about one-tenth degree. These positions are consistent with those of the supernova remnants Cas A and SN 1572 (Tycho's Supernova).

tensity about 1.5% that of Sco X-1. It has been tentatively identified with a 13.4-mag star with a $B-V = +1.0$ and $U-B = -0.9$ (Blanco et al 1968b), 10' from the most probable X-ray position of GX3+1 (see Figure 1). Spectroscopically the star belongs to an unusual group of objects with strong, broad emission lines of He II 4686 and O VI 3811 and 3834 (Blanco et al 1968b). Its spectrum has been described in some detail by Freeman et al (1968) who conclude that it is an extreme Wolf-Rayet star at a distance of about 5 kpc. Little is known about any changes in brightness. The identification must remain tentative until a more accurate X-ray position is available and more extensive optical observations are made.

A recent rocket flight with a rotating modulation collimator indicates a position with a precision of 1 min of arc which is removed from this optical candidate by one-half degree (Schnopper, H. W., Bradt, H. V., Rappaport, S., Boughan, E., Burnett, B., Doxsey, R., Mayer, W., and Watt, F. 1970, private communication).

Both Cen X-2 and Cen X-4 are known for their extreme changes in X-ray intensity. Both were novalike. Cen X-2 faded from an intensity about equal to that of Sco X-1 to invisibility in 5 months or less (Chodil et al 1968b). It was reported again on November 3 and 7, 1968 (Rao et al 1969a) from balloon observations. The behavior of Cen X-4 was even more spectacular (Evans et al 1970). On the discovery date, the object had an intensity 30% *greater* than that of Sco X-1; 80 days later it was invisible (<0.5% of original intensity). Although the X-ray positions of these two sources are very poorly known, attempts have been made to identify them. Several identifications have been suggested for Cen X-2. The peculiar nebula NGC 5189 was suggested by Feast (1967), and the central star of the nebula by Blanco (1967). The spectrum of the central star shows strong He II 4686, O VI 3811 and 3834 emission (Blanco et al 1968a); its variability is unknown. The more recent X-ray position by Lewin et al (1968a) places NGC 5189 outside the error box. This prompted Eggen et al (1968) to suggest WX Cen as the optical counterpart to Cen X-2. The spectrum of WX Cen (Eggen et al 1968) shows numerous emission lines; hydrogen and He II are especially strong. Lasker & Hesser (1969) have observed WX Cen for rapid fluctuations similar to those observed in Sco X-1; no periodic components were detected with amplitude greater than 0.003 mag from 0.2 to 520 sec. However, the power spectrum is different from that for Sco X-1. The optical identification of Cen X-2 must remain in doubt.

The optical identification of Cen X-4 with the known variable star S5003 Cen (Eggen & Rodgers 1969) is as precarious as that of Cen X-2, first, because of the inaccurate X-ray position, and second, because of the variability of the X radiation. S5003 Cen was observed to be constant in brightness for a period of 1 month while the X radiation decreased by approximately a factor of 10. Since there appears to be coupling between the optical and X radiation in Sco X-1 (see above) this lack of optical variability in S5003 Cen raises doubts about the optical identification of Cen X-4; the same considerations apply to Cen X-2. Furthermore, the absence of He II emission in S5003 sets it apart from other optical identifications, both secure and tentative.

The Large Magellanic Cloud

Mark et al (1969a) have reported an extended X-ray source 12° wide at the position of the Large Magellanic Cloud. The intensity corresponds to an emission rate of about 4×10^{38} ergs/sec. On the basis of the number of known X-ray sources and their estimated distances, Friedman et al (1967) have estimated the X-ray power output of the Milky Way Galaxy to be approximately 7×10^{39} ergs/sec in the 1 to 10 keV range. These two power levels are more or less compatible in view of the inaccuracy of the observed and estimated fluxes and the factor of 10 smaller mass of the Large Magellanic Cloud.

M87 (Vir A)

A source near M87 has been reported by two groups of observers (Friedman & Byram 1967, Bradt et al 1967). The error box has an area of 6.4 square degrees. However, since few sources are known to exist at high galactic latitudes, and since M87 is a strong radio source and a peculiar galaxy, the identification is generally accepted. The measured X-ray flux is such that the total power radiated is estimated to be approximately 10^{43} ergs/sec—10^4 to 10^5 times greater than that estimated for the Galaxy (Friedman et al 1967). The optical radiation from M87 is $\sim 2 \times 10^{42}$ ergs/sec.

SUMMARY

There are two aspects of optical observations in X-ray astronomy: the identification of the optical counterparts of X-ray sources and the determination of the properties of these systems at optical energies. Of the current optical identifications only two are firm and one is fairly certain. However, with additional rocket flights of experiments using modulation collimators and with the advent of X-ray satellites capable of giving positions to approximately 1 min of arc, we can anticipate that the number of secure identifications will increase manyfold during the next few years. The burden of obtaining the data necessary for greater understanding of these sources will then fall on the optical as well as the X-ray observer. Indeed, the paucity of spectroscopic and photometric data on Sco X-1 and Cyg X-2 respectively suggests that the optical observers are already falling behind!

If one is permitted to compute the anticipated optical magnitudes from the observed X-ray intensities, most of the optical counterparts are likely to be fainter than 15 mag. Consequently, relatively large telescopes will be needed, not only for spectroscopic observations, but also for photometry. In addition it appears as if many X-ray sources are variable; for these, simultaneous observations of the X-ray and optical (or infrared and radio) wavelengths can provide not only definitive tests of optical identifications, but, even more important, information critical to the testing of proposed models for the systems. Proper optical coverage for support of present and future X-ray observations is a challenge and an opportunity for the optical astronomer.

The authors' work on the optical counterparts of X-ray sources has been supported by the National Science Foundation and the National Aeronautical and Space Agency.

LITERATURE CITED

Ables, J. 1969, *Ap. J.*, **155**, L27
Andrew, B., Purton, C. 1968, *Nature*, **218**, 855
Baade, W. 1942, *Ap. J.*, **96**, 188
Bell K., Kingston, A. 1967, *MNRAS*, **136**, 241
Blanco, V. 1967, *IAU Circ. 2035*
Blanco, V., Kunkel, W., Hiltner, W., Chodil, G., Mark, H., Rodrigues, R., Seward, F., Swift, C. 1968a, *Ap. J.*, **152**, L135
Blanco, V., Kunkel, W., Hiltner, W., Lynga, G., Bradt, H., Clark, G., Naranan, S., Rappaport, S., Spada, G. 1968b, *Ap. J.*, **152**, 1015
Bless, R., Code, A., Houck, T., Lillie, C. 1969 (Private communication)
Bowyer, S., Byram, E., Chubb, T., Friedman, H. 1964, *Science*, **146**, 912
Bowyer, S., Field, G. 1969, *Nature*, **223**, 573
Bradt, H., Garmire, G., Oda, M., Spada, G., Sreekantan, B., Gorenstein, P., Gursky, H. 1968a, *Space Sci. Rev.*, **8**, 471
Bradt, H., Mayer, W., Naranan, S., Rappaport, S., Spada, G. 1967, *Ap. J.*, **150**, L199
Bradt, H., Naranan, S., Rappaport, S., Spada, G. 1968b, *Ap. J.*, **152**, 1005
Burbidge, E., Lynds, C., Stockton, A. 1967, *Ap. J.*, **150**, L95
Burginyon, G., Grader, R., Hill, R., Price, R., Rodrigues, R., Seward, F., Swift, C., Hiltner, W., Mannery, E. 1970 (Private communication)
Buselli, G., Clancy, M., Davison, P., Edwards, P., McCracken, K., Thomas, R. 1968, *Nature*, **219**, 1124
Chodil, G., Mark, H., Rodrigues, R., Seward, F., Swift, C., Turiel, I., Hiltner, W., Wallerstein, G., Mannery, E. 1968a, *Ap. J.*, **154**, 645
Chodil, G., Mark, H., Rodrigues, R., Swift, C. 1968b, *Ap. J.*, **152**, L45
Cocke, W., Disney, M., Gehrels, T. 1969, *Nature*, **223**, 576
Conner, J., Evans, W., Belian, R. 1969, *Ap. J.*, **157**, L157
Eggen, O., Freeman, K., Sandage, A. 1968, *Ap. J.*, **154**, L27
Eggen, O., Rodgers, A. 1969, *Ap. J.*, **158**, L111
Evans, W., Belian, R., Conner, J. 1970, *Ap. J.*, **159**, L57
Feast, M. 1967, *Nature*, **215**, 1158
Felten, J., Gould, R. 1966, *Phys. Rev. Lett.*, **17**, 401
Freeman, K., Rodgers, A., Lynga, G. 1968, *Nature*, **219**, 251
Friedman, H., Byram, E. 1967, *Science*, **158**, 257
Friedman, H., Byram, E., Chubb, T. 1967, *Science*, **156**, 374
Fritz, G., Meekins, J., Henry, R., Byram, E., Friedman, H. 1968, *Ap. J.*, **153**, L199
Gatewood, G., Sofia, S. 1968, *Ap. J.*, **154**, L69
Giacconi, R., Gorenstein, P., Gursky, H., Usher, P., Waters, J., Sandage, A., Osmer, P., Peach, J. 1967a, *Ap. J.* **148**, L129
Giacconi, R., Gorenstein, P., Gursky, H., Waters, J. 1967b, *Ap. J.*, **148**, L119
Giacconi, R., Gursky, H., Paolini, F., Rossi, B. 1962, *Phys. Rev. Lett.*, **9**, 439
Giacconi, R., Gursky, H., Van Speybroeck, L. 1968, *Ann. Rev. Astron. Ap.*, **6**, 373
Gorenstein, P., Gursky, H., Garmire, G. 1968, *Ap. J.*, **153**, 885
Grader, R., Hill, R., Seward, F., Hiltner, W. 1970, *Ap. J.*, **159**, 201
Gursky, H., Giacconi, R., Gorenstein, P., Waters, J., Oda, M., Bradt, H., Garmire, G., Sreekantan, B. 1966a, *Ap. J.*, **144**, 1249
Gursky, H., Giacconi, R., Gorenstein, P., Waters, J., Oda, M., Bradt, H., Garmire, G., Sreekantan, B. 1966b, *Ap. J.*, **146**, 310
Hiltner, W., Mook, D. 1967, *Ap. J.*, **150**, 851
Hiltner, W., Mook, D. 1970 (Private communication)
Hiltner, W., Mook, D., Ludden, D., Graham, D. 1967, *Ap. J.*, **148**, L47
Hiltner, W., Mook, D., Lynds, C. 1970 (Private communication)
Hudson, H., Peterson, L., Schwartz, D. 1970, *Ap. J.*, **159**, L51
Johnson, H., Golson, J. 1968, *Ap. J.*, **153**, 307
Johnson, H., Stephenson, C. 1966, *Ap. J.*, **146**, 602
Kraft, R. 1963, *Advan. Astron. Ap.*, **2**, 43
Kraft, R. 1964, *Ap. J.*, **139**, 457
Kraft, R., Demoulin, M. 1967, *Ap. J.*, **150**, L183
Kraft, R., Miller, J. 1969, *Ap. J.*, **155**, L159
Kristian, J., Sandage, A., Westphal, J. 1967, *Ap. J.*, **150**, L99
Lasker, B., Hesser, J. 1969, *AAS Meet.*, *December 1969, New York*
Lewin, W., Clark, G., Smith, W. 1968a, *Ap. J.*, **152**, L49
Lewin, W., Clark, G. Smith, W. 1968b, *Ap. J.*, **152**, L55
Lynds, C. 1967, *Ap. J.*, **149**, L41
Mark, H., Price, R., Rodrigues, R., Se-

ward, F., Swift, C. 1969a, *Ap. J.*, **155**, L143

Mark, H., Price, R., Rodrigues, R., Seward, F., Swift, C., Hiltner, W. 1969b, *Ap. J.*, **156**, L67

Mook, D. 1967. *Ap J.*' **150**, L25

Mook, D. 1970a (Private communication)

Mook, D. 1970b (Private communication)

Morrison, P. 1967, *Ann. Rev. Astron. Ap.*, **5**, 325

Münch, G. 1966, *Publ. Astron. Soc. Pacific*, **78**, 305

Neugebauer, G., Becklin, E., Kristian, J., Leighton, R., Snellen, G., Westphal, J. 1969a, *Ap. J.*, **156**, L115

Neugebauer, G., Oke, J., Becklin, E., Garmire, G. 1969b, *Ap. J.*, **155**, 1

Oda, M. 1965, *Appl. Opt.*, **4**, 143

Oda, M. 1968, *Space Sci. Rev.*, **8**, 507

Oda, M. 1970, *Progr. Cosmic Ray Phys.* (In press)

Oda, M., Bradt, H., Garmire, G., Spada, G., Sreekantan, B., Gursky, H., Giacconi, R., Gorenstein, P., Waters, J. 1967, *Ap. J.*, **148**, L5

Oke, J. 1969, *Ap. J.*, **156**, L49

Rao, U., Chitnis, E., Prakasarao, A., Jayanthi, U. 1969a, *Ap. J.*, **157**, L127

Rao, U., Prakasarao, A., Jayanthi, U. 1969b, *Nature*, **222**, 864

Riegler, G., Ramaty, R. 1969, *Ap. Lett.*, **4**, 27

Sandage, A., Osmer, P., Giacconi, R., Gorenstein, P., Gursky, H., Waters, J., Bradt, H., Garmire, G., Sreekantan, B., Oda, M., Osawa, K., Jugaku, J. 1966, *Ap. J.*, **146**, 316

Schnopper, H., Thompson, R., Watt, S. 1968, *Space Sci. Rev.*, **8**, 534

Sofia, S., Eichhorn, H., Gatewood, G. 1969, *Astron. J.*, **74**, 20

van Genderen, A. 1969, *Astron. Ap.*, **2**, 6

Wallerstein, G. 1967 *Ap. Lett.*, **1**, 31

Wallenstein, G. 1969, *Astron. J.*, **74**, 999

Wampler, E., Scargle, J., Miller, J. 1969. *Ap. J.*, **157**, L1

Warner, B., Nather, R., MacFarlane, M. 1969, *Nature*, **222**, 233

Westphal, J., Sandage, A., Krisitan, J. 1968, *Ap. J.*, **154**, 139

THE COSMIC ABUNDANCE OF HELIUM

I. J. Danziger

Harvard College Observatory
Cambridge, Massachusetts

Introduction

Up to the present time our attempts to demonstrate the existence of a cosmic abundance of helium have been unsuccessful. There is probably sufficient evidence to suggest some variation of the helium abundance among different types of celestial objects, but it is by no means clear what significance should be attached to this. Local fluctuations in a relatively high underlying abundance and variations in an initially low abundance seem to be the two possibilities that ought to be examined. Previous discussions of the helium question have been given by Tayler (1967) and Burbidge (1969). In what follows we will survey both the observational and theoretical results that may help to elucidate this question.

The possible cosmological significance of the helium abundance is demonstrated in the problem posed by Burbidge (1958). If we assume that the radiative luminosity of our Galaxy has been constant at its present level over its lifetime (10^{10} years), and that this radiative energy is provided by the thermonuclear conversion of hydrogen into helium in stars, this conversion should have resulted in a present abundance of helium about an order of magnitude less than that observed. One way to avoid this difficulty is to suppose that the initial abundance of helium in our Galaxy was reasonably high, an assumption that ultimately must be supported by cosmogonical or cosmological theory. A more detailed treatment of stellar helium production in our Galaxy by Truran et al (1965) makes the problem appear somewhat less severe without unambiguously resolving it. There are, of course, other possible explanations if we relax the assumptions used in the above estimates. For example, the bolometric luminosity may be far greater than previously assumed because of undetected excess infrared radiation.

The following discussion of the observations and theory has been subdivided according to the methods and particular nature of the determinations. We shall use X, Y, Z to represent the fractional mass of hydrogen, helium, and the heavier elements respectively. We note here, for convenience of reference, that a value of the helium-to-hydrogen ratio by number $N(\text{He})/N(\text{H}) = 0.10$ corresponds to a value of $Y = 0.28$ if we assume $Z \sim 0.02$.

Stellar Interiors

The theory of stellar interiors and observations provides us with the means of testing our knowledge of models of stars. Observations in the mass-luminosity plane and the luminosity-effective temperature plane are possible and useful in the determination of the helium abundance.

B stars in the mass-luminosity plane.—A number of workers have used masses and luminosities obtained for components of eclipsing binary systems to derive a mass-luminosity relation. Then theoretical models have been calculated with Y as the independent variable to fit the observations. In the cases discussed below the eclipsing binaries are those listed as reliable by Harris et al (1963) with unevolved Population I components on the upper main sequence.

Percy & Demarque (1967) obtained a mean value of $Y \sim 0.25 + 0.10$ for these systems. By assuming with Stothers (1965) that the occurrence of β Cephei instability corresponds to the region of hydrogen exhaustion in the core of their evolving models, they obtained $Y \sim 0.27$. Another study by Kelsall & Strömgren (1965) and Strömgren (1967) of eclipsing binaries from the list mentioned above gave an average value of $Y \sim 0.35$ for stars in the range of mass $\log M/M_\odot = 0.25$-0.73. Morton (1968) has determined luminosities for almost the same group of stars by using the determined radii and effective temperatures obtained from fitting model atmospheres to the observed continua. The average value of $Y \sim 0.24$ was obtained with the interior models of Kelsall & Strömgren. These results demonstrate without further discussion the uncertainties that might be expected from this method. Some of the differences appear to come from the interior models and some from the effective temperatures. Although the assumed abundance of heavy elements has some effect, it is fairly small. It will be seen in what follows that these results are in reasonable agreement with the high abundances of helium measured in the photospheres of Population I stars and in the interstellar medium.

Subdwarfs in the luminosity-effective temperature diagram.—The presence or absence of subdwarfs (spectral types F-M) in the L_{bol}-T_{eff} plane is closely related to the helium abundance of these stars. Consider the following relationships for luminosity and effective temperature obtained by homology arguments for solar-type stars on the main sequence, with energy supplied from the CNO cycle of hydrogen burning and bound-free opacities due to metals (see, for example, Strömgren 1952).

$$L \propto \mu^{7.22} Z^{-1.044} X^{-0.022} (1 + X)^{-1.022} M^{5.13}$$

$$T_{\text{eff}} \propto \mu^{1.528} Z^{-0.306} X^{-0.028} (1 + X)^{-0.278} M^{0.917}$$

where μ = mean molecular weight and is not sensitively dependent on Z; L = luminosity; T_{eff} = effective temperature; M = mass of star.

It is not strictly correct to use these equations quantitatively to study the effects of variation of abundances because as we reduce the metals, free-

free opacities become relatively more important and the same equations do not apply. However, qualitative effects can be understood. We see that reduction of Z increases both L and T_{eff} in such a way that a new sequence of stars with a given Z would be found to the left of the original main sequence. The equations also show that this effect of reducing Z can be compensated for by decreasing Y because a decrease in Y reduces L and T_{eff}. If we plot the positions of stars in this L-T_{eff} plane, and we know the values of Z spectroscopically, say, then we can determine Y. This type of work has been done by a number of authors and the results are not in agreement.

Sandage & Eggen (1959, 1962), using UBV photometry corrected for line blanketing, demonstrated that the 15 most extreme subdwarfs in the M_V, $B-V$ diagram moved onto the normal main sequence defined by the Hyades stars in the $M_v, [B-V]_0$ plane. Trigonometrical parallaxes greater than $0\farcs035$ were tolerated. Therefore, the subdwarf sequence seemed to be a result of differential line blanketing caused by differing metal abundances. An application of the interior models of Demarque, rather than the homology relations, gave $Y=0$ for the most extreme subdwarfs, when $Y=0.36$ for the Hyades was assumed.

A paper by Strom, Cohen & Strom (1967), who analyzed seven F and G type subdwarfs with large parallaxes, supported these previous conclusions. However, another paper by Strom & Strom (1967), using model atmospheres to obtain T_{eff} and trigonometric parallaxes $\geq 0\farcs010$ for 62 subdwarfs, seemed to rule out the possibility of $Y=0$ even for extreme subdwarfs. Thus it seemed that subdwarfs might in fact exist.

The most recent work by Cayrel (1968) is probably the most definitive in this area. He has obtained photometry of a group of later-type main-sequence stars, normal types and apparent subdwarfs. Three different photometric systems were used: GRI (Stebbins & Whitford), vby (Strömgren), VRIJK (Johnson). The indices were carefully chosen to give a measure of effective temperature. There are a number of advantages in the method. Evolutionary effects becoming important near type G0 are avoided. The photometric systems measure the radiation near the peak of the energy curves, which minimizes the bolometric correction. Line-blanketing effects at longer wavelengths are reduced. Cayrel finds that the most extreme subdwarfs now lie systematically below the normal main sequence by 0.7 ± 0.3 magnitudes. This result combined with theoretical models indicates that the helium abundance in these stars is not very different from that in the normal stars of similar mass, although a reduction by a factor ~ 2 cannot be excluded. The milder subdwarfs remain a puzzle, since if only the metals are reduced one would still expect them to be systematically and noticeably separated from the normal stars in the HR diagram. This does not appear to occur.

A more recent contribution by Dennis (1968) suggests a number of reasons for the different results so far reported. An underabundance of metals has other effects on model atmospheres of late-type stars that were not taken into account in the earlier work when model atmospheres were employed. (*a*) Convection is more efficient in subdwarfs because of their higher density. (*b*) In the subdwarfs the effect of photoionization of neutral hydrogen in the Balmer continuum is relatively more important than in other parts of the spectrum. This tends to offset the reduced metallic line blanketing in subdwarfs. (*c*) Rayleigh scattering may be more important in the subdwarfs.

The net result of accounting for these effects is that in the two-color ($U-B$, $B-V$) diagram, subdwarfs occupy a different sequence from the Hyades (metal-strong) standard sequence. This difference is such that the temperatures previously determined may have been too low, and hence the presence of true subdwarfs may have been obscured. Dennis shows that this subdwarf sequence is about 0.5 magnitudes below the Hyades sequence, which is now consistent with Cayrel's result for later-type stars and with the idea of a reasonably high helium abundance.

Even these last results are made to seem less definitive by a very recent result of Hegyi & Curott (1969), who have obtained a helium abundance for μ Cas A, an old Population II component of a binary system. By measuring the angular separation of the components of this system and combining it with the known distance and Kepler's third law for binary orbits, they estimated a lower limit to the mass of μ Cas A < 1.5 $M_\odot$. Masses for Population II stars have not been determined previously because few binary systems of Population II stars are known to exist, and those that do have not lent themselves to determinations of their masses. It can be readily seen from the first of the homology relations written above that if the luminosity, mass, and metal content Z are known, Y the helium content can be determined for a given star (essentially from its position in the mass-luminosity plane). Such an analysis employing the detailed models of Faulkner (1967) has allowed an upper limit for $Y \sim 0.0$ to be derived. Even allowing all uncertainties and errors to work in the appropriate direction keeps $Y \sim 0.10$. There is therefore a conflict with the most recent results for Population II stars placed in the luminosity-effective temperature plane. One might argue that the photospheric abundances of the heavy elements may not be applicable to the interiors of these stars, but this is little more than a speculation that will require greater insight to explain the different results for helium given by the different methods. This result also has important implications for the interpretation of globular cluster data, both main-sequence turnoff points and horizontal-branch stars, discussed in a later section. Clearly more observational and theoretical work is required to resolve this apparent contradiction.

Stars in moving groups.—In a series of papers Eggen (1962, 1963, 1965, 1969) has presented evidence that the mass-luminosity law for the Hyades-Pleiades moving group, and some other stars with no ultraviolet excess relative to the Hyades, is very different from that for the Sun-Sirius moving group. A similar idea was first put forward by Kuiper (1938). The masses in all cases have been obtained from the orbits of visual binary systems in the respective groups, and cover the range 0.5–2.0 solar masses.

The difference is such that the Hyades-Pleiades binaries are 1.65 magnitudes brighter than the Sun-Sirius stars for a given mass, or 0.6 times the Sun-Sirius masses at a given luminosity. Either homology relations or exact models can be used in the manner described previously. If $(X, Y, Z) = (0.72, 0.26, 0.02)$ for the Sun, the resulting value $(X, Y, Z) = (0.49, 0.49, 0.02)$ obtains for the Hyades-Pleiades stars where interior models computed by Iben (1963) have been used. This also means that (Z/X) Hyades ~ 1.5 (Z/X) Sun, a result consistent with spectroscopic determinations of these quantities by Wallerstein (1962).

A number of criticisms attempt to resolve this apparent problem but firm conclusions are not yet agreed upon. Hodge & Wallerstein (1966) proposed that the distance modulus of the Hyades has been underestimated by $+0.39$ mag. Demarque (1967) has shown that with either the Eggen or the Hodge-Wallerstein distance the interpretation of the mass-luminosity diagram according to conventional theory of stellar structure leads to an inconsistency in the luminosity-effective temperature plane. A compromise of an increased distance modulus of 0.20 mag for the Hyades is suggested by Wallerstein & Hodge (1967). Wilson (1967) has defended the older distance scale by reassessing the Wilson-Bappu K-line emission width and its correlation with absolute luminosity, as has Wayman (1967) for the convergent-point method of distance determinations. Eggen (1969) also supports his assumed distance by noting that in the HR diagram, with M plotted versus $(R-I)$ his red temperature index, the Hyades sequence coincides with that for the red field dwarfs.

Alexander (1968) suggested that a different choice of binary systems and possible orbital elements could eliminate the differences between the Hyades-Pleiades and Sun-Sirius groups. There has already been some variation in the choice of acceptable binary systems by Eggen (1969), and this seems to be a matter in which those working in the field have to agree on the merits and accuracies of orbital determinations. So far there has been no evidence that spectroscopic determinations of helium in the photospheres of hotter stars in these moving groups support any of the contentions mentioned above.

The Sun.—Our knowledge of the helium content of the Sun either now or at its time of formation is still uncertain. The neutrino flux expected from the solar interior has an important bearing on the helium abundance. With a correct solar model, and helium and metal abundance, one should be able in principle to calculate a self-consistent model of the present Sun (age 4.5×10^9 years) for which the neutrino flux from the thermonuclear conversion of hydrogen to helium can then be determined.

An upper limit to this flux of 3×10^{-36} neutrinos sec^{-1} per ^{37}Cl nucleus has been reported by Davis et al (1968). From the homology relations for stellar structure it can be seen that the central pressure in a star $P_c \propto M^2/R^4$ where $M=$ mass of star and $R=$ radius. The equation of state is

$$P_c = \frac{\rho_c}{\mu H} kT_c \propto \frac{M}{R^3 \mu} T_c$$

where $T_c=$ central temperature; $\mu=$ mean molecular weight; $\rho_c=$ density; $H=$ unit atomic mass. Therefore $T_c \propto (M/R)\,\mu$.

We see that the central temperature decreases with decreasing μ (or decreasing Y). Since T_c controls the thermonuclear reaction rates and hence the neutrino flux, we see how this neutrino flux depends on the helium content in the energy-producing regions of the Sun. (There is also a less important dependence of these quantities through the number density of interacting particles.)

In this way Bahcall et al (1968) obtained a current value of the helium abundance and thence through the evolutionary model an initial solar value of $Y=0.22 \pm 0.03$. Iben (1968a) has questioned this result, suggesting that an implied upper limit for $Y=0.16$–0.17 is consistent with the neutrino flux. He suggests that a specific choice of interior abundances of heavy elements at the present time may be inappropriate and that they can be considered a free parameter. Only a variation of nuclear cross sections in the appropriate direction beyond the limits suggested by the possible errors will allow the value of Y to reach 0.25. A result similar to that of Bahcall et al has been obtained by Torres-Peimbert et al (1969). If the Sun is totally mixed, a possibility suggested by Ezer & Cameron (1968), the current helium content at the center of the Sun would be lower than expected (dilution with a lower content in the envelope); hence, the temperature and neutrino flux being lower, a higher initial value $Y \sim 0.24$ of the helium abundance would be allowed. There may be uncertainties not only in the heavy metal content, but also in the numerical estimation of the opacity in the envelope that controls the central temperature necessary to maintain the observed luminosity of the Sun. Therefore the model of the Sun is not a self-consistent one if the current values of the photospheric abundances of the heavy elements are used. In fact a recent estimate of the iron abundance in the solar photosphere by Garz et al (1969), using revised oscillator strengths, raises the abundance and probably increases the degree of inconsistency with the neutrino flux discussed above.

These low values of the helium abundance tend to be supported (given the uncertainties) by results from a completely different approach. Lambert (1967) used the determination of $N(\text{He})/N(\text{CNO})$ in solar cosmic rays (all have the same charge-to-mass ratio) by Biswas & Fichtel (1964) and the values of $N(\text{H})/N(\text{CNO})$ determined for the solar photosphere. Assuming that the ratios are the same in the solar cosmic rays and the photosphere, he derived a value of $Y \sim 0.20$. A similar earlier discussion by Gaustad (1964) had given $Y=0.26$. The value of $Y \sim 0.20$ compares (or contrasts) with the value of $Y \sim 0.14$ from long-term measurements of the solar wind by Hundhausen et al (1967).

Although the reliability of the method is open to some question, Unsöld (1969) reported that an analysis of prominence spectra gave $Y \sim 0.38$, a value that would make the problem of the solar interior even more intractable.

Galactic clusters.—For younger galactic clusters the theory of stellar interiors does not allow a particularly sensitive test of the helium abundance. Fitting the observed upper main sequences of clusters with models is usually consistent with a value of $Y \sim 0.35$. A review is given by Iben (1967).

The older galactic clusters, M67 and NGC 188, present better possibilities. For example, for stars near one solar mass evolving from the main sequence, the occurrence of hydrogen exhaustion in the core causes a quickening in the evolution rate and a hook in the evolutionary track. This feature has been identified with a gap in the sequences of stars turning off the main sequence in M67 and NGC 188 by Demarque & Schlesinger (1969) and Aizenman et al (1969). It is shown to be sensitive to the helium abundance, and this effect gives $Y \sim 0.38 \pm 0.02$ for $Z = 0.03$ in M67 and a more tentative value of $Y \sim 0.30$ in NGC 188.

Globular clusters.—The calculation of models of horizontal-branch stars with double energy sources (helium-burning core with hydrogen-burning shell) by Faulkner (1966) has provided useful material for estimating helium content of stars in globular clusters. This work and that by Faulkner & Iben (1966) allowed the fitting of models on the horizontal branch of M92 with masses $\sim 0.7 M_\odot$ and helium abundance $Y \sim 0.35$ in the envelope. These parameters as well as a cluster age of 1.5×10^{10} years enabled a reasonable fit of the subgiant and giant regions, avoiding the necessity of invoking mass loss between the main-sequence and the horizontal-branch phase.

Iben & Faulkner (1968) and subsequently Rood & Iben (1968) have shown that with the above condition of no mass loss, the following quantities are sensitive to helium content and age: the ratio of the luminosities of the red end of the horizontal branch to the turnoff point from the main sequence, the corresponding effective temperatures, and the ratio of the luminosities of RR Lyrae stars to the turnoff point. In using these dependencies they obtain an average $Y \sim 0.30$ and an average age $\sim 8 \times 10^9$ years for globular clusters when neutrino losses are considered.

The number distribution of stars along the horizontal branch, which is dependent upon the helium content and the metal content, has been used by Iben (1968b) and Iben & Rood (1969), with better radiative opacities, to obtain $Y \sim 0.29$ for the globular cluster M15. The age is 9×10^9 years and both quantities depend a little on assumptions about the loss rate of neutrinos. The observational data for M15 were supplied by Sandage et al (1968).

However, Sandage & Wildey (1967) have brought attention to the globular cluster NGC 7006 where the metal abundance is low, but where the distribution of stars along the horizontal branch resembles that previously observed only in metal-rich clusters. These authors suggest that a different helium abundance may be responsible, thus pointing to variations of helium content among globular clusters. The possibility that it is only an effect of different ages has been suggested by Rood & Iben (1968).

Asano & Sugimoto (1968) also derive high abundances for helium in models of horizontal-branch stars.

Here we should note that reservations concerning the use of the techniques described previously have been expressed by Castellani & Renzini (1968), Castellani et al (1969), and Newell (1970). Further work is obviously required.

Support for the high abundance of helium in horizontal-branch stars comes from the work of Christy (1966a,b) on stellar pulsation. This work clearly demonstrates a dependence of the high-temperature boundary of the instability strip upon helium content such that 15 percent increase in the mass fraction of helium in the envelope of such stars makes the high-temperature boundary 500°K hotter. This theory has been so successful in explaining other phenomena related to stellar pulsation that there is a good case for applying it here. Indeed, Christy has applied it to the case of the globular cluster M3 to obtain $Y \sim 0.30$ as well as low masses ($\sim 0.5 M_\odot$) in agreement with the ideas presented above. Low masses of this order are also suggested by the more direct photometric and spectroscopic determinations of blue horizontal branch stars reported by Graham & Doremus (1968), Newell et al (1969), and Newell (1970).

Recently Sandage (1969) has applied these theoretical results to the new photometric data for horizontal-branch stars in the globular clusters M3, M15, M92. Assuming low masses (0.55 $M_\odot$), he derives $Y \sim 0.32 \pm 0.09$ with very little variation from cluster to cluster.

Any results from stellar interiors, both subdwarfs and horizontal-branch stars, must be considered tentative if not unreliable, at least until the problem of the neutrino flux from the Sun is solved.

Planetary Nebulae, Novae, and Supernovae

Abundance determinations of helium in planetary nebulae make use of the line-emission spectrum. A summary is given by Aller & Liller (1968). The average values of the helium abundance obtained by Mathis (1962) for two, and O'Dell (1963) for nine planetary nebulae are very similar, $N(\text{He})/N(\text{H}) \sim 0.14$. An interesting study is that of K648, a planetary nebula in the very metal-weak globular cluster M15. O'Dell et al (1964) obtained $N(\text{He})/N(\text{H}) \sim 0.18 \pm 0.03$, while the oxygen-to-hydrogen ratio was found to be reduced by a factor 61 compared to the Sun. This will be discussed further in another section. A similar trend was observed in the case of the field planetary nebula 49+88°1 in the halo for which Miller (1969) found $N(\text{He})/N(\text{H}) \sim 0.13$, while the heavier elements oxygen and neon were noticeably underabundant relative to the solar values.

Very few data on helium abundances in novae are available. Pottasch (1959) has determined the helium abundance from emission lines found in nova shells. The spectral data were old for the five novae—Aqr, Her, Lac, Per, Pic—and not particularly suitable for such an analysis. An average value for $N(\text{He})/N(\text{H}) \sim 0.15$ was obtained with individual values being as much as a factor of 2 greater or smaller than this. Although an interesting overbundance of helium is indicated, the fact that oxygen, nitrogen, sulfur, calcium, and neon abundances are greater than solar values by a factor of ~ 5 necessitates caution in interpreting such results literally.

For a supernova, the Crab Nebula, Woltjer (1958) has measured line-strengths on photographic spectra from which $N(\text{He}/N(\text{H}) \sim 0.45$ was obtained. The electron temperature was not well determined but a wide range of possible values centered on 10,000°K gave an overabundance. In view of the special nature of the Crab Nebula this abundance ought to be redetermined when the physical conditions existing there are better understood.

The correct interpretation of the results discussed in this section would be of great interest but our theoretical knowledge is limited. All three types of object may be responsible for ejecting material into the interstellar medium and enriching it with elements heavier than hydrogen. Because we do not know the initial composition of the progenitors of planetary nebulae and novae we cannot estimate their contribution to the helium enrichment of the interstellar medium. The limited observations and accuracy for the supernova also necessitate caution, but we do have some indication that such objects may be responsible for increasing the helium abundance in the Galaxy.

Interstellar Medium

Two different techniques are available for abundance determinations in the interstellar medium, optical and radio. The optical observations are made in H II regions and the results can be briefly summarized. All observers derive abundances $N(\text{He})/N(\text{H})$ in the range 0.10–0.14. Mathis (1957, 1962), Faulkner & Aller (1965), and Peimbert & Costero (1969) have contributed to this work. The last-named authors obtain a uniform value of $N(\text{He})/N(\text{H}) \sim 0.11$ for Orion, M8, and M17, in what can be taken to be the most recent and sophisticated analysis of H II regions.

High-order recombination lines of neutral hydrogen and helium are observed at radio frequencies. Lilley et al (1966), measuring the lines 156α, 158α, 159α of neutral helium, obtained $N(\text{He})/N(\text{H}) \sim 0.10 \pm 0.05$ in M17, assuming that most of the helium is singly ionized. Gordon & Meeks (1967) obtained $N(\text{He})/N(\text{H}) \sim 0.11$ in the Orion Nebula by measuring the lines 94α and 148δ of neutral hydrogen and 94α of neutral helium. The recent data by Palmer et al (1969) for the 109α lines of neutral hydrogen and helium give $N(\text{He})/N(\text{H}) = 0.084 \pm 0.003$ averaged over five H II regions. In NGC 2024, however, the He 109α line is not seen and the resulting abundance is very low. Other factors, such as lower excitation, may be causing the weakness of this line. Goss (1968), observing the He 158α line, has reported a low value of $N(\text{He})/N(\text{H}) \sim 0.063$ for W43, while Reifenstein et al (1969) have obtained results similar to those reported above by observing the H 109α and He 109α lines in M17 and Orion A. Even more interesting are the results of Churchwell & Mezger (1969) who not only obtained results similar to those above but showed that the $N(\text{He})/N(\text{H})$ ratio does not vary systematically with radial distance from the galactic center over the range 5 to 12 kpc. These last-mentioned abundances may be consistently lower than the photospheric abundances observed in Population I B stars but once again possible errors make a definitive conclusion difficult. In any case, these values might be lower limits if there is any neutral helium present in the H II regions. Radio observations so far cannot provide criteria for making some judgment on this question.

On the other hand, McGee et al (1969) have observed nine southern H II regions in the recombination lines H 158α and He 198β. Their results for Orion are significantly higher, $N(\text{He})/N(\text{H}) \sim 0.16$, than those reported above, and the average value 0.12 is higher as well as showing larger individual variations. If these results are taken literally, the differences between Population I B stars and optical and radio observations of H II regions are very small.

Interesting results have been obtained for the ³He isotope. The ground-state hyperfine splitting in singly ionized ³He gives a transition with a wavelength 3.46 cm. A preliminary theoretical discussion by Goldwire & Goss (1967) for eleven H II regions suggested that an upper limit of $N(^3\text{He})/N(\text{H})$ in the range $10^{-4} - 10^{-5}$ could be set, while Seling & Heiles (1969) have now set an upper limit for $N(^3\text{He})/N(\text{H}) \sim 4 \times 10^{-5}$ in the H II region M17.

Extragalactic Determinations

The helium-to-hydrogen ratio has been measured in a range of extragalactic objects, almost always from the emission-line spectra of H II regions. Aller & Faulkner (1962) obtained $N(\text{He})/N(\text{H}) \sim 0.11$ for NGC 346 in the Small Magellanic Cloud. Johnson (1959), Faulkner & Aller (1965), and Mathis (1965) obtained $N(\text{He})/N(\text{H}) \sim 0.13$, 0.08, 0.14, respectively, for the 30 Doradus Nebula in the Large Magellanic Cloud, where the last two results are the most reliable because the measurements were made photoelectrically. Mezger et al (1970) have recently observed the radio recombination lines H 109α and He 109α from 30 Doradus. They obtain a higher electron temperature than any reported for other H II regions with the same technique, and in contrast with the above results, a very high value of the helium abundance, $N(\text{He})/N(\text{H}) \sim 0.17$. The reason for these differences is not yet understood. Schmidt (1962a) obtained $N(\text{He})/N(\text{H}) \sim 0.10$ for NGC 6822, an irregular galaxy. Other irregular galaxies, NGC 4214 and IC 1569, have been observed by Mathis (1965) to have $N(\text{He})/N(\text{H})$ 0.1–0.2. Uncertain space reddening made these determinations less secure. Mathis (1962) obtained $N(\text{He})/N(\text{H}) \sim 0.11$ for NGC 604 in the spiral galaxy M33. Aller et al (1968) obtained 0.12, while Peimbert & Spinrad (1970a) obtained 0.13 for this object. In the exploding galaxy M82 the last-named authors (1970b) obtained $N(\text{He})/N(\text{H}) \sim 0.13$ and for NGC 4449, NGC 5461, NGC 5471, NGC 7679 a uniform value close to 0.10. These last-mentioned galaxies are not members of the local group.

Osterbrock & Parker (1965) have estimated $N(\text{He})/N(\text{H}) \sim 0.14$ in the Seyfert galaxy NGC 1068. From the above results it appears that helium does not vary enormously from galaxy to galaxy, though whether the fluctuations reported here are real or due to observational errors is not easy to ascertain.

Although only limited data are available, some evidence indicates a systematic decrease of the helium-hydrogen ratio in H II regions in M31 correlating with distance from the center. Schmidt (1962b) and Rubin & Ford (unpublished) have observed this effect, which, if it is a real effect, probably has implications for element production at the centers of galaxies and subsequent diffusion throughout the disk. We noted previously that such effects were not observed in our Galaxy with radio-recombination lines.

One outstanding characteristic of quasistellar sources is the apparent weakness or absence of helium lines. Osterbrock & Parker (1966) drew attention to this phenomenon and suggested a low abundance of helium. Since then Peimbert & Spinrad (1970a), using the published data of Wampler (1967, 1968), have derived some upper limits for the helium abundance in 3C 249.1, PKS 2251+11, and Ton 1542 that are well below the Population I values. Bahcall & Kozlovsky (1969a,b), using models of 3C 273 and 3C 48 in which the continuum comes from a condensed central source and the lines from an extended envelope, obtained $N(\text{He})/N(\text{H}) \sim 0.005$ for 3C 273 and approximately one-half the solar value, or less, for 3C 48. It has been pointed out by some of these authors that the hydrogen Balmer decrement in the spectra of many quasistellar objects and Seyfert galaxies is too steep to be consistent with the population of the upper levels of hydrogen by a recombination mechanism. It remains to be noted, as others have, that these low values should not be taken literally until the physical processes and conditions in quasistellar sources are better understood. For example, Burbidge et al (1966) have proposed multizone models of quasars that may avoid the conclusion that helium is underabundant.

Stellar Photospheres

Population I stars.—We mention here in detail only the *recent* work to determine helium abundances in hot Population I stars. The uncertainties in the line-broadening theory for helium are such that comparisons of small differences are probably meaningless. A range in $N(\text{He})/N(\text{H}) \sim 0.15\text{--}0.20$ was found by Mihalas (1964) for six O stars and one B star. Other earlier work is reviewed by Underhill (1966) where it is clear that the derived helium abundances tend to be as high as the values given above. However, an underestimate of the turbulent velocities in these stars, which is possible, results in a determination that is too high. Shipman & Strom (1969) have reported a uniform abundance $N(\text{He})/N(\text{H}) \sim 0.10$ for a selection of B stars of Population I later than B3, using both weak and strong lines. This value is somewhat lower than that quoted above for the O stars studied by Mihalas. Whether this difference is real or whether it is a result of different physical conditions not specifically allowed for in present models (non-LTE, theory of line broadening) is not clear and will probably require more work.

Recently Norris (1970) has determined helium abundances in 14 Population I stars in the range of spectral type B0—B8. He obtains a mean value of $N(\text{He})/N(\text{H}) = 0.09 \pm 0.015$ while demonstrating that the apparent singlet-triplet anomaly can be satisfactorily accounted for within the framework of local thermodynamic equilibrium and line-saturation effects.

Sargent & Strittmatter (1966) have discussed a class of sharp-lined B stars in Orion that have weak helium lines for their color. The authors suggested that these stars are slowly rotating and are an extension of the Ap stars, whose characteristics, particularly the underabundances of helium, have been documented by Searle & Sargent (1964). Norris (1970) has demonstrated that the Orion B stars are not in fact underabundant in helium. However, his work shows that there is a bounded region in the gravity-effective temperature plane where weak helium lines (and apparently low abundances of helium) can exist. There are B stars of Population I that lie in this region. Therefore, as in the case of the Ap stars confined to a similar region, as elucidated by Sargent & Searle (1967a), there is some reason to suspect the abundances are not real (homogenous over the photosphere) or may be confined to the photospheric layers. The most mysterious example of a star in this class is 3 Cen A whose total helium content is low and almost entirely in the form of the light isotope ^{3}He. Norris also extends these conclusions to Population II stars, which will be discussed later.

Another example of a possibly anomalous abundance of helium and peculiar spectroscopic behavior has been reported by Norris (1968) for HD 125823 in which the helium lines periodically vary in strength. Other examples exist.

There are well-known examples (the helium stars) of O and B stars very overabundant in helium for which the spectral characteristics are well documented by Underhill (1966), but whose evolutionary history is not understood. Apparently helium-rich (or hydrogen-poor) stars extend across the HR diagram, through the A supergiants (v Sgr and HD 30353) and R Corona Borealis stars to the cooler carbon stars. The evolutionary history of these stars is not understood either. We will not comment further on this type of object.

Hot Population II stars.—The most suitable objects to observe in order to gain some insight into the abundance of helium present early in the lifetime of the Galaxy are hot Population II stars where helium lines may be visible.

Sargent & Searle (1966) investigated three halo B stars in the field,

Feige 11, 36, 65, and found them to have very weak helium lines signifying an underabundance of helium by a factor 100 compared to normal B stars. These stars also have very high gravities (log $g \sim 5.5$) compared to normal main-sequence stars (log $g \sim 4.0$). Greenstein & Münch (1966) came to essentially similar conclusions concerning a mixture of seven halo and globular cluster B stars.

Later, Sargent & Searle (1968) in a larger survey of 30 halo stars subdivided the group into four identifiable classes: (a) 14 normal A and B stars which require no further comment in this context; (b) 4 subdwarf O stars ($T_{eff} \sim 42,000°K$) with normal helium lines and high gravities (log $g \sim 5.4$); (c) 5 subdwarf B stars ($T_{eff} \sim 25,000$–$30,000°K$) in which are included the Feige stars discussed above—all have the high gravities mentioned above; (d) 7 B stars with weak helium lines for their colors and roughly normal gravities. One of these stars, Feige 86, was examined spectroscopically at higher resolution by Sargent & Searle (1967b) who showed it to be peculiar, possibly in the same way that 3 Cen A is peculiar, i.e. weak helium lines and lines of ionized phosphorus present. They concluded that the surface material was not representative of the primeval material from which it formed.

Baschek & Norris (1970) have subsequently analyzed in more detail stars from group (c) and a newly discovered one of the same type, HD 205805. They show the helium abundance to be about one order of magnitude lower than in the Population I star γ Peg. Feige 36 has the same abundance of helium as HD 205805 while in Feige 11 and 65 the abundance is another factor 4 lower. In addition, the anomalous singlet-triplet ratio among the helium lines is accounted for by the high gravity (log $g \sim 5.0$) and low abundance of helium. Carbon, nitrogen, and silicon appear to have a normal abundance while magnesium is underabundant. The relationship of these stars to those in globular clusters is not yet understood.

Although the earlier investigations of the helium abundance in hot horizontal-branch stars in NGC 6397 by Searle & Rodgers (1966) proved unjustified (Newell et al 1969), some useful data are available. Sargent (1967) published details of the spectra of 11 horizontal-branch stars in the clusters M13, M15, and M92. Some of these stars are sufficiently hot that underabundances of helium by factors of 5–10 were noted by that author. In a recent analysis of horizontal-branch stars in M15, M92, M13, and M3, Newell (1970) consolidates the previous conclusions. For example, a more detailed investigation of the horizontal-branch star S-18 in M15 shows it to be underabundant in helium by at least a factor of 10 ($T_{eff} \sim 16,200$, log $g \sim 3.8$). Furthermore, there appears to be a real difficulty in explaining the presence in the HR diagram of the hottest horizontal-branch stars in a number of clusters with the models of Iben & Faulkner (1968). These models have low ratios of core mass to total mass, and high (normal) abundances of helium, whereas models with large core masses and low helium abundances in the envelope seem called for. We have already mentioned the high abundance of helium observed in the planetary nebula in M15. This result and those above are not easy to reconcile without at least greater knowledge of the evolution of planetary nebulae.

We note again here the suggestion by Norris (1970) that the horizontal-branch stars also lie in a restricted region of the gravity-temperature plane where anomalous helium strengths occur even in Population I stars. Hence, the information they contain for the primeval abundance of helium may be misleading. Thus without explaining these anomalies physically, but assuming empirically that this argument has meaning, we are left with the subdwarf B stars of Population II, which have low abundances of helium. These do not lie in the same region of the gravity-temperature plane, and represent, even by the empirical standards applied in other cases, an enigma.

There are also other classes of hot stars of Population II where the helium abundance does not appear low. Barnard 29 in M13 is a hot star ($T_{eff}\sim$ 19,700°K) analyzed by Stoeckly & Greenstein (1968) and Traving (1962). It has a low gravity (log $g\sim$2.6) and normal helium abundance. Traving obtains a similar result for the hot field star BD+33°2642. HD 137569 is a halo field star analyzed by Danziger & Jura (1970). It appears to be a lower-temperature ($T_{eff}\sim$12,000°K, log $g\sim$2.3) analog of Barnard 29, also having a normal helium abundance. Przybylski (1969) derived a normal helium abundance and low metal abundance for the Population II A0 Ib star HD 214539. These types of stars are less well understood than the ones we have discussed above because a well-defined sequence has not yet been identified. Yet another puzzling example is the O star in the globular cluster M3, analyzed by Strom & Strom (1970). The helium, nitrogen, and oxygen abundances all appear similar to those observed in Population I stars, even though M3 is a noticeably metal-poor cluster. The composition of the photosphere may not be the primordial one.

To summarize the important information on photospheric abundances in Population II stars we note that there are now two classes of old stars in which the helium abundance appears to be considerably lower than normal. One group comprises the weak-line field stars (type Bw) and possibly their analogs on the horizontal branch of globular clusters. Both have normal gravities and low abundances of helium. The other group includes the subdwarf B stars with high gravities and a low helium abundance, and so far is found only amongst the field stars.

THEORETICAL CONSIDERATIONS AND CONCLUSION

We have now noted that the problem of the primeval helium abundance remains to be clarified observationally. A certain ambiguity also exists in the results predicted theoretically for the production of helium. It was pointed out at the beginning that the work of Truran et al (1965) did not rule out the possibility that most of the helium in our Galaxy may have been produced in the still-uncertain lifetime of the Galaxy in stars of conventional mass.

The discovery of the 3°K microwave background radiation by Penzias & Wilson (1965) has led to increased interest in the possibility of a primeval fireball in which helium could be produced. Here we do not give a complete summary of the relevant theoretical work but mention some of the main points. The early work of Alpher et al (1953) and Hayashi (1950) has been improved by more detailed calculations of Hoyle & Tayler (1964), Smirnov (1965), Peebles (1966a,b), and Wagoner et al (1967). In all these calculations a high abundance of helium ($Y \sim 0.30$) appeared to be one product of the primeval explosion. The encouraging aspect of these calculations was that this abundance appeared to be close to a limiting one, which was approached asymptotically and did not depend on highly special conditions.

The existence of objects with a low abundance of helium caused some workers to investigate conditions under which a primeval fireball might produce little helium. Hawking & Tayler (1966) and Thorne (1967) demonstrated that a large anisotropy in the initial expansion could limit the amount of helium produced. Misner (1968) suggests that such an anisotropy would be damped out quickly by the neutrino viscosity, and so far no well-established anisotropy has been observed in the microwave background radiation (Wilkinson & Partridge 1967). This question of the theoretical effects of anisotropies is still being investigated, for example by Stewart (1968).

The possibility of spatial fluctuations in the primeval helium abundance correlating with density fluctuations and the sites of formation of galaxies has been discussed by Harrison (1968), who suggested that it may be supported by an observed variability of helium content among globular clusters. As we have seen, the interpretation of data for globular clusters is by no means unambiguous, and extragalactic variations remain to be well established. Wagoner et al (1967) have shown that neutrino degeneracy giving a large excess of neutrons over protons or vice versa can result in only a small production of helium. Dicke (1968) has calculated that the scalar-tensor theory of gravitation will also result in very little helium being produced in an initial fireball.

Another site of helium production has been investigated by Wagoner et al (1967, 1968), who showed that supermassive objects ($> 10^4 M\odot$) undergoing explosions at high temperatures can produce even higher abundances of helium than the universal fireball. For example, a value for $Y \sim 0.40$ seems possible as does the general heavy-element abundance seen in Population II stars if we accept that Y is large.

The present theoretical results have some ambiguity concerning the expected abundance of helium. The observational results also do not provide clear answers. One important question is whether many of the variations reported previously would be reduced by the strict application of the same observing techniques, reduction of data and atomic data. Fluctuations in results for the same object indicate that this may be a very relevant consideration. An outstanding problem is the helium abundance in the Sun, because the neutrino experiment and the other determinations give an abundance that is not only lower than expected on more general grounds, but too low to be consistent even with our understanding of the physics of the solar interior. To explain the difference between this low abundance in the Sun and Sun-Sirius group, and the measured values for very young Population I objects and also the Hyades-Pleiades stars (if correct), seems to be an even greater problem than to explain the general increase of helium over the lifetime of the Galaxy because the time scales in the former case may be shorter ($\sim 10^8$ years).

Although there is on balance at the present time perhaps a little stronger support for a high primeval helium abundance, it should be carefully noted that the strongest observational test, the helium abundance in Population II B stars, indicates the opposite. Reservations about the interpretation of these results have already been expressed. As well, there has been an attempt to explain the phenomenon as gravitational diffusion of helium below the stellar photospheres, as proposed by Greenstein, Truran & Cameron (1967). Apart from more detailed considerations it is clear that a condition of no convective or rotational mixing is required and this may be reasonable for the stars under consideration. However, there is a real difficulty in explaining why gravitational diffusion is not even more effective for the heavier elements.

The production of substantial amounts of helium in supermassive objects (e.g. galaxies or the centers of galaxies) provides some attractive possibilities. This theory can, with our present understanding, accommodate the apparent variation of helium abundance within our Galaxy. The helium abundance in the high-energy cosmic rays [$N(\text{He})/N(\text{H}) \sim 0.11$; see Meyer (1969)], and the energy density of cosmic rays may be consistent with their production in explosions of supermassive objects, if as Burbidge (1969) speculates, the microwave background is also generated in such events. Without a satisfactory theory of formation of galaxies and their evolution, it is not yet obvious why galaxies of different types (spirals and irregulars discussed above) should have helium abundances that are very similar.

To conclude, we emphasize that there is support for and objection to most of the ideas for production of helium. It may emerge that all three main processes discussed above contribute somewhat to its production. A unique cosmic abundance of helium is not yet identifiable although the more widespread abundance seems to be a high one. Certainly the past changes to produce this remain undetermined. Therefore, if the term "cosmic" is meaningful in this context, the appropriate value may be that observed in the interstellar medium with the presently unavoidable errors indicated previously.

I am grateful to Professor Fred Hoyle for the hospitality of the Institute of Theoretical Astronomy, Cambridge, England, where most of this paper was prepared during the summer of 1969.

LITERATURE CITED

Aizenman, M. L., Demarque, P., Miller, R. H. 1969, *Ap. J.*, **155**, 973
Aller, L. H., Czyzak, S. J., Walker, M. F. 1968, *Ap. J.*, **151**, 491
Aller, L. H., Faulkner, D. J. 1962, *Publ. Astron. Soc. Pacific*, **74**, 219
Aller, L. H., Liller, W. 1968, *Nebulae and Interstellar Matter*, 483 (Middlehurst, B. M., Aller, L. H., Eds., Univ. Chicago Press)
Alexander, J. B. 1968, *Quart. J. Roy. Astron. Soc.*, **9**, 136
Alpher, R. A., Follen, J. W., Herman, R. C. 1953, *Phys. Rev.*, **92**, 1347
Asano, N., Sugimoto, D. 1968, *Ap. J.*, **154**, 1127
Bahcall, J. N., Bahcall, N. A., Shaviv, G. 1968, *Phys. Rev. Lett.*, **20**, 1209
Bahcall, J. N., Kozlovsky, B. 1969a, *Ap. J.*, **155**, 1077
Bahcall, J. N., Kozlosvky, B. 1969b, *Ap. J.*, **158**, 529
Baschek, B., Norris, J. 1970, *Ap. J. Suppl. No. 176*
Biswas, S., Fichtel, C. E. 1964, *Ap. J.*, **139**, 941
Burbidge, G. R. 1958, *Publ. Astron. Soc. Pacific*, **70**, 83
Burbidge, G. R. 1969, *Comments Ap. Space Phys.*, **1**, 101
Burbidge, G. R., Burbidge, E. M., Hoyle, F., Lynds, C. R. 1966, *Nature*, **210**, 774
Castellani, V., Giannone, P., Renzini, A. 1969, *Nature*, **222**, 153
Castellani, V., Renzini, A. 1968, *Ap. Space Sci.*, **2**, 310
Cayrel, R. 1968, *Ap. J.*, **151**, 997
Christy, R. F. 1966a, *Ann. Rev. Astron. Ap.*, **4**, 353
Christy, R. F. 1966b, *Ap. J.*, **144**, 108
Churchwell, E., Mezger, P. 1969, *131st Meet. AAS, New York*
Danziger, I. J., Jura, M. A. 1969, *Ap. J.* (In press)
Davis, R. Jr., Harmer, D. S., Hoffman, K. C. 1968, *Phys. Rev. Lett.*, **20**, 1205
Demarque, P. 1967, *Ap. J.*, **150**, 943
Demarque, P., Schlesinger, B. 1969, *Ap. J.*, **155**, 965
Dennis, T. R. 1968, *Ap. J.*, **151**, L47
Dicke, R. H. 1968, *Ap. J.*, **152**, 1
Eggen, O. J. 1962, *Quart. J. Roy Astron. Soc.*, **3**, 259
Eggen, O. J. 1963, *Ap. J. Suppl. No. 76*, **8**, 125
Eggen, O. J. 1965, *Ann. Rev. Astron. Ap.*, **3**, 235
Eggen, O. J. 1969, *Ap. J.*, **156**, 241
Eggen, O. J., Sandage, A. R. 1962, *Ap. J.*, **136**, 735
Ezer, D., Cameron, A. G. W. 1968, *Ap. Lett.*, **1**, 177
Faulkner, D. J., Aller, L. H. 1965, *MNRAS*, **130**, 393
Faulkner, J. 1966, *Ap. J.*, **144**, 978
Faulkner, J. 1967, *Ap. J.*, **147**, 617
Faulkner, J., Iben, I., Jr. 1966, *Ap. J.*, **144**, 995
Garz, T., Holweger, H., Kock, M., Richter, J. 1969, *Astron. Ap.*, **2**, 446
Gaustad, J. E. 1964, *Ap. J.*, **139**, 406
Goldwire, H. C., Goss, W. M. 1967, *Ap. J.*, **149**, 15
Gordon, M. A., Meeks, M. L. 1967, *Ap. J.*, **149**, L21
Goss, W. M. 1968, *Ap. J. Suppl.*, **15**, 131
Graham, J. A., Doremus, C. 1968, *Astron. J.*, **73**, 226
Greenstein, J. L., Münch, G. 1966, *Ap. J.*, **146**, 618
Greenstein, G. S., Truran, J. W., Cameron, A. G. W. 1967, *Nature*, **213**, 871
Harris, D. L., Strand, K. Aa., Worley, C. E. 1963, *Basic Astronomical Data*, Chap. 15 (Strand, K. Aa., Ed., Univ. Chicago Press)
Harrison, E. R. 1968, *Astron. J.*, **73**, 533
Hawking, S. W., Tayler, R. J. 1966, *Nature*, **209**, 1278
Hayashi, C. 1950, *Progr. Theor. Phys.*, **5**, 224
Hegyi, D., Curott, D. 1969 (Preprint)
Hodge, P. W., Wallerstein, G. 1966, *Publ. Astron. Soc. Pacific*, **78**, 411
Hoyle, F., Tayler, R. J. 1964, *Nature*, **203**, 1108
Hundhausen, A. J., Asbridge, J. R., Bame, S. J., Gilbert, H. E., Strong, I. B. 1967, *J. Geophys. Res.*, **72**, 87
Iben, I., Jr. 1963, *Ap. J.*, **138**, 452
Iben, I., Jr. 1967, *Ann. Rev. Astron. Ap.*, **5**, 571
Iben, I., Jr. 1968a, *Phys. Rev. Lett.*, **21**, 1208
Iben, I., Jr. 1968b, *Nature*, **220**, 143
Iben, I., Jr., Faulkner, J. 1968, *Ap. J.*, **153**, 101
Iben, I., Jr., Rood, R. 1969 (Preprint)
Johnson, H. M. 1959, *Publ. Astron. Soc. Pacific*, **71**, 245
Kelsall, T., Strömgren, B. 1965, *Vistas Astron.*, **8**, 159
Kuiper, G. P. 1938, *Ap. J.*, **88**, 429
Lambert, D. L. 1967, *Observatory*, **87**, 199
Lilley, A. E. Palmer, P., Penfield, H., Zuckerman, B, 1966, *Nature*, **211**, 174
Mathis, J. S. 1957, *Ap. J.*, **125**, 328
Mathis, J. S. 1962, *Ap. J.*, **136**, 374
Mathis, J. S. 1965, *Publ. Astron. Soc. Pacific*, **77**, 90

McGee, R. X., Batchelor, R. A., Brooks, J. W., Sinclair, M. W. 1969, *Australian J. Phys.*, **22**, 631
Meyer, P. 1969, *Ann. Rev. Astron. Ap.*, **7**, 1
Mezger, P. G., Wilson, T. L., Gardner, F. F., Milne, D. K. 1970, *Ap. Lett.*, **5**, 117
Mihalas, D. 1964, *Ap. J.*, **140**, 885
Miller, J. S. 1969, *Ap. J.*, **157**, 1215
Misner, C. W. 1968, *Ap. J.*, **151**, 431
Morton, D. C. 1968, *Ap. J.*, **151**, 285
Newell, E. B. 1970, *Ap. J.*, **159**, 443
Newell, E. B., Rodgers, A. W., Searle, L. 1969, *Ap. J.*, **156**, 597
Norris, J. 1968, *Nature*, **219**, 1342
Norris, J. 1970, *Ap. J., Suppl. No. 176*
O'Dell, C. R. 1963, *Ap. J.*, **138**, 1018
O'Dell, C. R., Peimbert, M., Kinman, T. D. 1964, *Ap. J.*, **140**, 119
Osterbrock, D. E., Parker, R. A. 1965, *Ap. J.*, **141**, 892
Osterbrock, D. E., Parker, R. A. 1966, *Ap. J.*, **143**, 268
Palmer, P., Zuckerman, B., Penfield, H., Lilley, A. E. 1969, *Ap. J.*, **156**, 887
Peebles, P. J. E. 1966a, *Phys. Rev. Lett.*, **16**, 410
Peebles, P. J. E. 1966b, *Ap. J.*, **146**, 542
Peimbert, M., Costero, R. 1969, *Bol. Obs. Tonanzintla Tacubaya*, **5**, 3
Peimbert, M., Spinrad, H. 1970a, *Ap. J.*, **159**, 809
Peimbert, M., Spinrad, H. 1970b, *Ap. J.* (In press)
Penzias, A. A., Wilson, R. W. 1965, *Ap. J.*, **142**, 419
Percy, J. R., Demarque, P. 1967, *Ap. J.*, **147**, 1200
Pottasch, S. 1959, *Ann. Ap.*, **22**, 412
Przybylski, A. 1969, *MNRAS*, **146**, 71
Reifenstein, E., Wilson, T., Burke, B., Altenhoff, W., Mezger, P. 1969 (In press)
Rood, R., Iben, I., Jr. 1968, *Ap. J.*, **154**, 215
Sandage, A. 1969, *Ap. J.*, **15**, 515
Sandage, A., Eggen, O. J. 1959, *MNRAS*, **119**, 278
Sandage, A., Katem, B., Kristian, J. 1968, *Ap. J.*, **153**, L129
Sandage, A. R., Wildey, R. 1967, *Ap. J.*, **150**, 469
Sargent, W. L. W. 1967, *Ap. J.*, **148**, 147
Sargent, W. L. W., Searle, L. 1966, *Ap. J.*, **145**, 652
Sargent, W. L. W., Searle, L. 1967a, *The Magnetic and Related Stars*, 209 (Cameron, R. C., Ed., Baltimore: Mona Book Corp.)
Sargent, W. L. W., Searle, L. 1967b, *Ap. J.*, **150**, L33
Sargent, W. L. W., Searle, L. 1968, *Ap. J.*, **150**, 443
Sargent, W. L. W., Strittmatter, P. A. 1966, *Ap. J.*, **145**, 938
Schmidt, M. 1962a, *Symposium on Stellar Evolution, La Plata Observatory*, 61
Schmidt, M. 1962b, *Annual Report of the Director, Mt. Wilson and Palomar Observatories, 1961-62*, 27
Searle, L., Sargent, W. L. W. 1964, *Ap. J.*, **139**, 793
Searle, L., Rodgers, A. W. 1966, *Ap. J.*, **143**, 809
Seling, T. V., Heiles, C., 1969 *Ap. J.*, **155**, L163
Shipman, H. S., Strom, S. E., 1969, *Ap. J.*, **159**, 183
Smirnov, Yu. N., 1965, *Sov. Astron.*, **8**, 864
Stewart, J. M., 1968, *Ap. Lett.*, **2**, 133
Stoeckly, R., Greenstein, J. L., 1968, *Ap. J.*, **154**, 909
Stothers, R. 1965, *Ap. J.*, **141**, 671
Strom, S. E., Strom, K. M. 1967, *Ap. J.*, **150**, 501
Strom, S. E., Strom, K. M. 1970, *Ap. J.*, **159**, 195
Strom, S. E., Cohen, J. G., Strom, K. M., 1967, *Ap. J.*, **147**, 1038
Strömgren, B. 1952, *Astron. J.*, **57**, 65
Strömgren, B. 1967, *Modern Astrophysics* (Paris: Gauthier Villars)
Tayler, R. J. 1967, *Quart. J. Roy. Astron. Soc.*, **8**, 313
Thorne, K. S. 1967, *Ap. J.*, **148**, 51
Torres-Peimbert, S., Simpson, E., Ulrich, R. K. 1969, *Ap. J.*, **155**, 957
Traving, G. 1962, *Ap. J.*, **135**, 439
Truran, J. W., Hansen, C. J., Cameron, A. G. W. 1965, *Can. J. Phys.*, **43**, 1616
Underhill, A. B. 1966, *The Early Type Stars* (Dordrecht, Holland: Reidel)
Unsöld, A. O. J. 1969, *Science*, **163**, 1015
Wagoner, R. V. 1968, *Ap. J.*, **151**, L103
Wagoner, R. V., Fowler, W. A., Hoyle, F. 1967, *Ap. J.*, **148**, 3
Wallerstein, G. 1962, *Ap. J. Suppl.*, **6**, 407
Wallerstein, G., Hodge, P. W. 1967, *Ap. J.*, **150**, 951
Wampler, E. J. 1967, *Publ. Astron. Soc. Pacific*, **79**, 210
Wampler, E. J. 1968, *Ap. J.*, **153**, 19
Wayman, P. 1967, *Publ. Astron. Soc. Pacific*, **79**, 156
Wilkinson, D. T., Partridge, R. B. 1967, *Nature*, **215**, 719
Wilson, O. C. 1967, *Publ. Astron. Soc. Pacific*, **79**, 46
Woltjer, L. 1958, *Bull. Astron. Neth.*, **14**, 39

Copyright 1970. All rights reserved

NEUTRON STARS 2007

A. G. W. CAMERON

Belfer Graduate School of Science
Yeshiva University
New York, N.Y.
and
Institute for Space Studies
Goddard Space Flight Center, NASA
New York, N.Y.

INTRODUCTION

In the 1930s neutron stars were conceived as theoretically possible stable structures in astrophysics (Landau 1932, Oppenheimer & Volkoff 1939). It was suggested that neutron stars might be formed in supernova explosions (supernovae were discovered at about the same time), and it was also suggested that neutron stars could form the cores of more ordinary types of stars. Thereafter, for a number of years, little interest was displayed in neutron stars, since there seemed to be no way in which they could be directly observed, and astrophysical theory was insufficiently developed to predict unambiguously the path by which they might be formed. Nevertheless, neutron stars hold the distinction of being probably the only major astrophysical object whose properties and existence were predicted long before the discovery of the object.

Slightly more than 10 years ago a new interest in neutron-star models arose, in part because of the work by Wheeler and his collaborators (Harrison, Wakano & Wheeler 1958) in the issue of the final state reached by a collapsing massive astrophysical body, and in part because of the realization that neutron stars might very well be the endpoints of supernova explosions and hence might play a role also in nucleosynthesis (Cameron 1959, Ambartsumyan & Saakyan 1960). The original model calculations of Oppenheimer & Volkoff used a Fermi gas of noninteracting neutrons. The new work in the late 1950s involved an improved treatment of this equation of state, together with an attempt to determine more precisely the composition of the neutron star, including not only such essential particles as protons and electrons, but also mesons and hyperons.

Research on neutron stars received an impetus in the early 1960s when celestial X-ray sources were discovered. It was at first thought that the X rays might be thermally emitted from the surfaces of neutron stars, and hence there was a period in which the atmospheric structure and rate of cooling of neutron stars were intensively investigated (Morton 1964, Tsuruta & Cameron 1966a, Chiu 1964). It was soon shown that the celestial X-ray sources could not in general result from thermal emission from a neutron-star surface, both because in at least one instance (the Crab Nebula) the main source of X rays was emitted from too large an angular dimension, and because the theoretical calculations of the neutron-star cooling indicated that the temperatures would be too low to be of interest at reasonable ages for astronomical objects. Nevertheless, research on neutron stars continued, for a variety of reasons. The writer (Cameron 1965a) considered it possible that the X-ray emission arises by indirect nonthermal mechanisms due to the vibrations of a neutron star, and he initiated a number of calculations relating to this possibility. Some general relativistic aspects of neutron stars, both stable and unstable models, also became of some interest, and calculations of the properties of both the unstable and stable configurations continued under the direction of J. A. Wheeler and his colleagues at Princeton; these authors considered possible emission of gravitational radiation from spinning and vibrating neutron stars. The situation as of the end of 1965 is described by Wheeler in an article in a previous volume of these reviews (Wheeler 1966), and in the book by Harrison, Thorne, Wakano & Wheeler (1965). See also the excellent review by Thorne (1967).

The most important event in the history of neutron-star studies was the discovery of the pulsars. By a process of elimination, it was shown that only rotating neutron stars possessed the necessary characteristics to account for the type of celestial clock observed in the cases of pulsars (for a survey of these arguments see Maran & Cameron 1968, 1969).

During 1969 there was a large increase in the rate of publication of papers on the properties of neutron stars, most of them concerned with some aspects of neutron stars related to the pulsar problem, but nevertheless the issues under discussion are very fundamental. It is to be expected that a rapid rate of publication on this subject will continue for some time, since the properties of neutron stars represent extreme extrapolations from the experience obtained by the physicist in his laboratory, and hence many of the phenomena presently under discussion have only been investigated in a very preliminary way as they apply to neutron-star interiors.

In this review the principal concern is with papers published since the earlier review by Wheeler (1966). No attempt is made to discuss the properties of pulsars as such, but reference is made to certain pulsar properties in the course of this review, both because their discovery has motivated certain investigations of neutron-star properties and because the subjects of neutron stars and of pulsars will be inextricably linked from this point forward.

Equations of State

Most of the equations of state that have been used in neutron-star investigations have been calculated at zero temperature. Some of the equations of state have been schematic, rather than physical, since they were designed to test the properties of neutron-star models corresponding to the mathematical form of the equation of state that was used. Many of these investigations have been described in the reviews by Wheeler and his colleagues. See in particular Barker, Bhatia & Szamosi (1966); Gratton & Szamosi (1964); Pacini (1965); Cazzola, Lucaroni & Scarini (1968); Gerlach (1968); and Durgopal & Gehlot (1968).

In a physical equation of state, it is necessary to minimize the internal energy of a specified number of baryons subject to the restraint of charge neutrality. Complications arise in this procedure because it is necessary to make some assumption about the potential energy of interactions between the neutrons and other baryons that are present, and these potential energies must be included in the internal energy under consideration. Also, they are very poorly known in general. However, it appears that calculations carried out with any reasonable nuclear potential, which is attractive at large interbaryonic separations and repulsive at small interbaryonic separations, should certainly give more believable results than calculations done with no nuclear potential at all.

Let us first consider the qualitative features of the composition of matter as an increasing function of density that will govern the properties of the equation of state. At the lowest densities the material will consist of ^{56}Fe together with electrons, forming a degenerate electron gas. As the density is increased, the electron Fermi level will rise, and electron capture becomes possible on nuclei that are stable under normal laboratory conditions. After such electron capture has taken place, the nuclei that will exist in equilibrium in the high-density matter will be unstable under ordinary laboratory conditions, and will be neutron rich, but will be stable in the presence of the high electron Fermi level. Increase in the electron Fermi level leads to further electron capture, hence to a further increase in the neutron-proton ratio, and to an increase in the mass number of a nucleus that will have the greatest binding energy per nucleon under these conditions. Figure 1 shows the charge and mass numbers of nuclei which are stable at various densities. These results are due to Langer, Rosen, Cohen & Cameron (1969) and are based upon a simple nuclear-mass equation that does not contain nuclear-shell effects, due to A. E. S. Green (1954). An earlier investigation, due to Tsuruta & Cameron (1965), indicates that the charge and mass numbers of the nuclei present at higher densities will not increase in this smooth way, but that nuclei in the vicinity of the closed neutron shell $N=50$ and with mass numbers in the vicinity of 80 will predominate at densities in the range 10^{10} to 10^{11} g/cm^3, and that nuclei near the closed neutron shell $N=82$ and with mass numbers in the vicinity of 120 will predominate at densities in excess of 2×10^{11} g/cm^3.

When the electron Fermi level becomes sufficiently large, about 25 MeV, the nuclei that are present under equilibrium conditions reach the neutron drip line, at which neutron binding energies become zero. At still

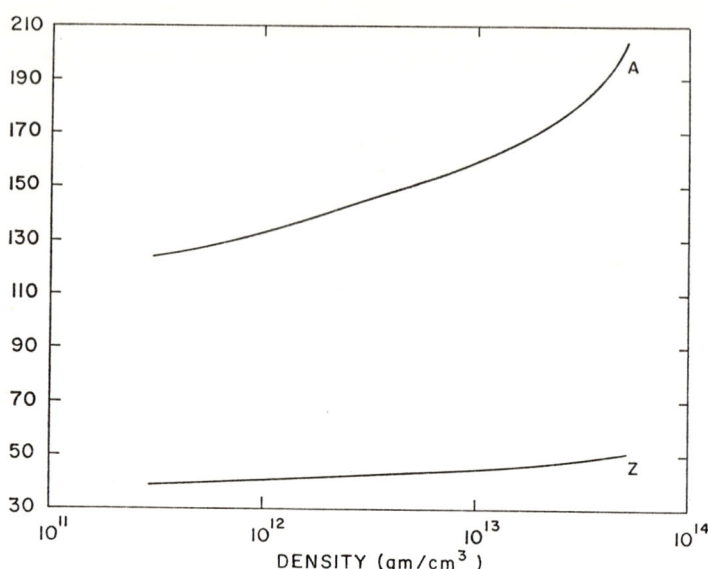

FIGURE 1. The charge and mass numbers of nuclei that are most stable as a function of density, based on a simple nuclear-mass formula without shell corrections.

higher Fermi levels, it can be expected that electron capture will produce nuclei in which neutrons are unbound, and these neutrons would then be emitted from the nucleus. However, after a sufficient number of neutrons has been emitted, the neutrons will form a degenerate neutron gas, and this will stabilize nuclei with apparently unbound neutrons. For example, if the neutron Fermi level becomes 5 MeV, then nuclei in which the last neutrons are unbound by energies < 5 MeV will not be able to emit these neutrons, owing to a lack of available phase space. The calculations of Langer, Rosen, Cohen, and Cameron indicate that the threshold for production of free neutrons lies in the vicinity of 2.85×10^{11} g/cm³.

As the neutron Fermi level continues to increase, the number density of neutrons becomes many orders of magnitude greater than the number density of nuclei under equilibrium conditions. The number densities are shown in Figure 2, also taken from the calculations of Langer et al (1969). At all times it is necessary to have enough electrons present, so that their Fermi level is sufficiently high to prevent the neutrons from decaying into protons. Initially this is no problem, since the number of electrons present is just that needed to provide charge equality for the numbers of protons present in the nuclei, and this is more than sufficient to prevent the neutrons from decaying into protons. However, when the neutron Fermi energy approaches the electron Fermi energy, this is no longer true. Also, under such conditions, the matter density becomes several percent of normal nuclear density, and

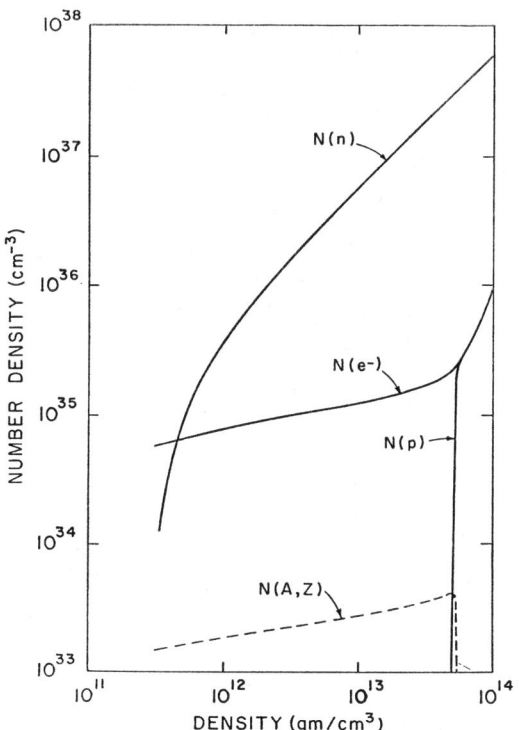

FIGURE 2. Number densities of neutrons, electrons, protons, and nuclei under equilibrium conditions as a function of density.

a new situation arises with regard to the distribution of protons in the matter. Protons are present in the nuclei in individual nuclear potential wells in which they occupy a variety of orbitals with a distribution of energies. Eventually, it becomes energetically more favorable to extract protons from the uppermost orbitals in nuclei and to include them in the internuclear region where they are strongly bound to the neutron gas. Langer et al found the threshold density at which this should occur to be about 4×10^{13} g/cm³. As the density is increased still further, the total binding energy of the protons to the internuclear gas increases rapidly, while that of the protons in the nuclei remains roughly constant. Consequently, there is a rapid shift in the proton distribution from being all in the nuclei at the proton threshold to being all in the internuclear region at the density at which the nuclei disappear. The number densities of the nuclei diminish rapidly in the intervening region, as may be seen in Figure 2. Langer et al find that the nuclei disappear at 6×10^{13} g/cm³. With a more realistic mass formula, these densities for the proton threshold and nuclear disappearance would probably be increased.

After the nuclei have disappeared, the material has a relatively simple composition of neutrons, protons, and electrons. The calculations shown in Figure 2 used a particular nuclear potential, V_α, due to Levinger & Simmons (1961). Because of the attractiveness of nuclear forces at large internucleonic separations, the number density of protons present in the medium at a given density is higher than that which would normally exist if there were no nuclear interactions between the nucleons. In the vicinity of ordinary nuclear density, about 4×10^{14} g/cm^3, the calculations done without nuclear potentials indicate that the protons and electrons will have number densities 1% of that of the neutrons. However, the attractiveness of the nuclear forces raises these number densities to about 3% of the neutron number densities. These results have been found in a variety of ways, in addition to the calculations of the V_α potential shown in Figure 2; similar results have been obtained by Wolf (1966) and by Nemeth & Sprung (1968). See also Bahcall & Wolf (1965c).

When the Fermi level of the electrons exceeds the rest mass of a muon, then it becomes energetically more favorable to introduce a negative muon with zero kinetic energy than to introduce another electron having a kinetic energy in excess of the muon rest mass. Thus one expects that negative muons will become present in matter at a sufficiently high density. Langer et al find the negative muon threshold to be at 2.2×10^{14} g/cm^3. This is lower than in a gas without nuclear interactions, owing to the fact that the electron number densities are higher for a given matter density than in a noninteracting case, which in turn results from the increased number density of protons and from the requirement for charge neutrality.

As the Fermi levels of the particles continue to rise, the time comes when the neutron Fermi level exceeds mass differences between the neutron and certain hyperons. At this point, it becomes energetically more economical to introduce new baryons into the system, which initially can have zero kinetic energy, than to continue to introduce neutrons with still higher kinetic energy. The first hyperon likely to appear in this fashion is the Σ^- hyperon, which appears before the somewhat lower-mass Λ^0 hyperon because the negative charge of the Σ^- allows the elimination of an electron or muon simultaneously with the appearance of the Σ^- hyperon. The composition that corresponds to the Levinger & Simmons V_α potential is shown in Figure 3. It may be seen in this figure that the rapid rise in the number density of the Σ^- hyperon is accompanied by a decline in the number densities of both electrons and negative muons. Our qualitative arguments would not have predicted this decline in the number densities of the leptons; this is a particular result of the nuclear potentials used in the calculation.

At still higher densities, as the neutron and proton Fermi energies continue to increase, a number of new hyperons will put in an appearance. The calculations of Langer & Rosen (1970), the results of which are shown in Figure 3, indicate that the Λ^0, Δ^-, Σ^0, and Δ^0 hyperons should all put in an appearance by the time the density has increased slightly above 10^{15} g/cm^3,

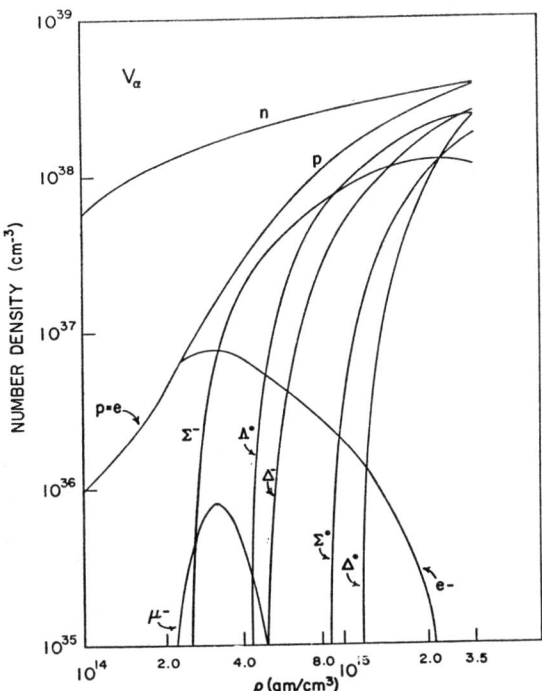

FIGURE 3. Number densities of leptons and baryons as a function of density, calculated with the V_α potential. Similar results are obtained with the V_γ potential.

or at only three times normal nuclear density. Once again, the effect of a nuclear potential interaction is to lower the thresholds at which these hyperons will appear. This is because the neutrons have now been squeezed to a sufficiently high density so that the potential interaction between them is repulsive and gives a less negative contribution to the energy per particle, whereas the potential interaction of a neutron with a newly appearing hyperon is expected to be strongly attractive and hence to give a large negative contribution to the energy per particle. Table 1 shows the results of Langer & Rosen for the threshold densities at which the muons and various hyperons appear for systems without nuclear interactions and for systems with the V_α and V_γ potentials due to Levinger & Simmons (1961).

The interested reader may refer to the papers of Langer et al (1969), Langer & Rosen (1970), and Cohen, Langer, Rosen & Cameron (1970). Generally speaking, the equation of state is very soft at densities below 3×10^{13} g/cm³, but becomes very stiff at higher densities. The adiabatic exponent,

TABLE 1. Thresholds for the appearance of various kinds of particles, calculated for various interaction potentials

Particle	Mass MeV	Threshold density (units of 10^{14} g/cm³)		
		noninteracting	V_α	V_γ
n	939.6		0.003	0.003
p	938.3		0.4	0.4
μ^-	105.7	7.75	2.30	2.35
Σ^-	1197	11.2	2.60	2.83
Λ^0	1115	19.1	4.02	4.17
Δ^-	1236	19.1	4.45	4.49
Σ^0	1197	70.6	9.28	8.93
Δ^0	1236	112	12.1	11.6

$$\Gamma = \frac{P + \rho}{P} \frac{\partial P}{\partial \rho}$$

typically has values less than unity for densities of less than 2×10^{13} g/cm³, but at densities in the vicinity of 10^{14} and 10^{15} g/cm³, the exponent typically has values fluctuating above and below about 3. It is the stiffness of the equation of state, characterized by these high values of Γ, that is responsible for the relatively small variation of density throughout much of the interior of a typical neutron-star model.

It must be emphasized that a good equation of state is necessary for the calculation of believable neutron-star models. The calculations described above are still very crude, and they have not been carried out by the most fundamental possible methods. Many physicists are now interested in calculating equations of state applicable to neutron-star interiors by the more fundamental many-body methods pioneered by Brueckner and by Bethe. Such methods at present usually involve errors $\sim 10\%$ in the estimate of the total potential energy; this is equivalent to nearly a factor of two error in determining the binding energy per particle, since the latter is the small difference between the two large potential and kinetic energies. Thus these more fundamental methods cannot be trusted to give reliable results for neutron stars unless the parameters have been adjusted to give good values for the binding energy per particle, density, symmetry energy coefficient, and compressibility of ordinary nuclei. This test is met for the modified V_α and V_γ potentials used in the calculations described above (see Weiss & Cameron 1969a,b).

Similar remarks are appropriate relating to the use of nuclear potentials that fit high-energy scattering cross sections to calculate a neutron-star equation of state (Libby & Thomas 1969a,b). At nuclear densities and above, many of the interactions involve lower energies, and three-body and many-

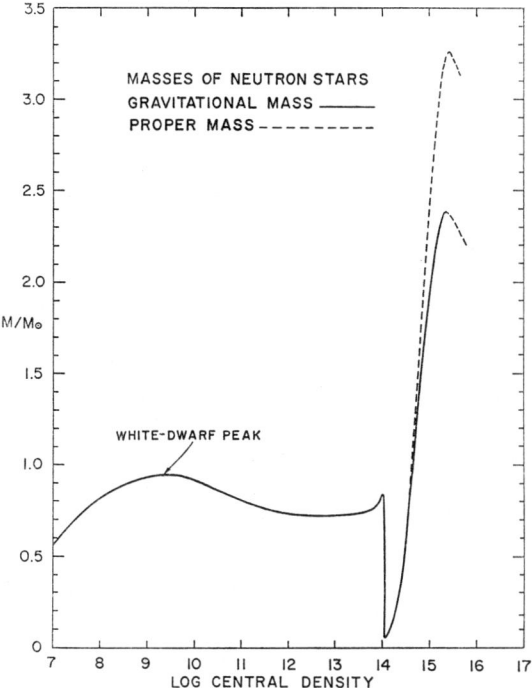

FIGURE 4. The gravitational mass of cold degenerate stars in the range of central density including white dwarfs and neutron stars, calculated with the V_α equation of state. The proper mass for the neutron stars is shown by the dashed line.

body forces become important; these cannot be included from the use of two-body scattering data. Hence the resulting calculated neutron-star models cannot be considered reliable unless realistic finite nuclei can be constructed from the same potentials used in the equation of state.

Neutron-Star Models

Neutron-star models must be computed using a general relativistic equation of hydrostatic equilibrium. Discussions of the form of these equations appear in many places; see for example the review by Harrison et al (1965). Figure 4 shows the stellar mass as a function of the central density for the composite equation of state of Langer et al, based upon the Levinger & Simmons V_γ potential. These results have been presented by Cohen & Cameron (1970). The gravitational mass is that which would be measured by means of the orbital motion of a test particle about such a mass, while the proper mass is the number of baryons contained in the star if dispersed to infinity and multiplied by the mass of the hydrogen atom.

The difference between these two masses is the gravitational binding energy of the neutron star.

On the left-hand side of Figure 4, the rising curve corresponds to the stable white-dwarf stars, with electron degeneracy. The equation of state used to obtain these is that with nuclei catalyzed to the endpoint of nuclear statistical equilibrium, and the resulting maximum mass derived for the white-dwarf star with the corresponding equation of state gives lower masses than would be obtained with physically more realistic compositions, such as carbon, oxygen, and helium mixtures. Beyond the peak of the white-dwarf curve, in the vicinity of a central density of 10^9 g/cm^3, the models are unstable, characterized by an imaginary value for their fundamental periods of vibration. It may be noted that there is a minimum in the curve in the vicinity of 10^{12} to 10^{13} g/cm^3, followed by a slight rise in the mass with increasing density. It might be expected that this rising portion of the curve would also correspond to a region of stability of stable masses. However, on this rising portion of the curve, not only is the fundamental vibrational period imaginary, but also the first overtone. Beyond the second peak just below 10^{14} g/cm^3, the second-overtone vibrational period also becomes imaginary. See Harrison et al (1965) for discussion of the meaning of the instabilities in the overtones of stellar models as one progresses through successive rises and falls in the mass as a function of central density. The discussion of Harrison et al refers explicitly to the behavior of the curve beyond the principal neutron-star peak, but it appears that similar phenomena are occurring beyond the peak of the white-dwarf stability curve.

The sharp drop in the mass curve in the vicinity of 10^{14} g/cm^3 central density corresponds to a very rapid shrinkage in the radius of the configuration calculated from the equations of hydrostatic equilibrium. Following this rapid shrinkage, the curve turns up and gives the region of stable neutron stars. This region extends to a central density of about $10^{15.3}$ g/cm^3. It may be seen that in this relatively narrow range of the central density the mass of the neutron star rises to about 2.3 solar masses. At the peak of the stability curve the radius of the model is only about 1.6 times the Schwarzschild radius, so that general relativistic effects have become very large in the model. Beyond that density general relativistic effects crush the stability of the models, and produce the downturn in the curves sketched schematically in Figure 4. In this review paper the large number of fascinating phenomena associated with the configurations beyond the maxima in the stable neutron-star curves will not be discussed. The reader is referred to the review by Harrison et al (1965).

The original calculation of Oppenheimer & Volkoff (1939), based upon a noninteracting Fermi gas of neutrons, gave a maximum stable mass for a neutron star of 0.7 solar masses. The much higher value for the maximum obtained in the calculations shown in Figure 4 is a direct result of the inclusion of nuclear forces in the equation of state. These nuclear forces stiffen the equation of state very much compared to that of a noninteracting

Fermi gas, and therefore it must be expected that the maximum stable mass of a neutron star will certainly be in excess of the amount calculated originally by Oppenheimer & Volkoff, although the precise value shown in Figure 4 is clearly subject to considerable uncertainty as yet. The density at which the maximum occurs is an order of magnitude greater than the normal density that a purely neutron gas would have if it were to be bound in the same manner as ordinary nuclear matter. Such ordinary nuclear matter, containing comparable numbers of neutrons and protons, has a density of some 4×10^{14} g/cm^3, just five times less than the maximum in the neutron-star curve. Thus the equation of state used in the calculation of the models involves a physical extrapolation by a significant factor beyond the kind of phenomena studied in nuclear physics at normal nuclear densities. Furthermore, because of the stiffness of the equation of state derived from the assumed nuclear potentials, the pressure of the equation of state at the center of the star when the maximum of the stability curve is reached is very slightly in excess of one third of the proper energy density. There is some question whether the pressure should become quite so high in a fluid with normal isotropic properties. Furthermore, the material at the densities involved appears to contain at least five and possibly more hyperon species, and the nuclear potentials among the hyperons and between the hyperons and nucleons are scarcely known with any accuracy; this increased the uncertainty in the precise results presented. At the same time, it presents a challenge to the physicist to improve the knowledge of the equation of state in the presence of the various hyperons involved. It should be remarked that the equation of state is changed by at most a few tens of percent in the pressure by the inclusion of these hyperons, as compared to the equation of state in which their appearance is ignored. Thus it appears that the basic properties of the nuclear potential at greater than normal nuclear density are more important than the precise composition of the material in determining the pressure at the relevant densities.

In any event, it appears that the maximum mass of a neutron star probably is considerably in excess of the maximum mass of a white-dwarf star. This may be of some comfort to those astrophysicists who are concerned with what should happen if a stellar body having a mass just slightly in excess of the maximum mass of a white dwarf should collapse to higher densities than those characteristic of a white dwarf. Such objects can almost certainly form stable neutron stars. The principal question that might arise in such a case would be whether a significant amount of mass can be ejected as a result of energy release in such a collapse, leading to masses of neutron stars significantly below the white-dwarf upper limit.

Figure 5 shows the radius of the neutron-star configurations plotted in Figure 4. Note that there is remarkably little variation in radius over most of the range of stable neutron-star masses.

The interior distribution of density in these neutron stars is very flat, as may be judged from the density distributions for several representative

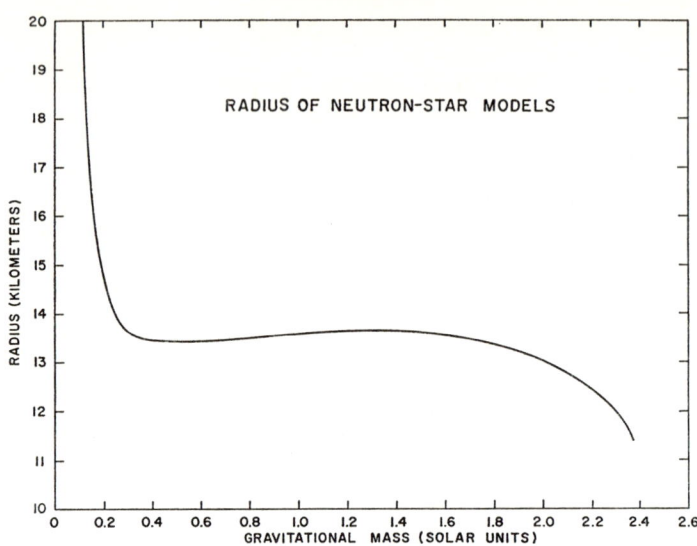

FIGURE 5. Radius of the neutron-star models as a function of gravitational mass. The radius is the circumference at the equator divided by 2π.

neutron-star models, calculated with the V_γ nuclear potential, shown in Figures 6, 7, and 8. The radial-distance scale is linear in these figures, while the density scale is logarithmic. It is worth noting that the surface layer of a neutron star, except for the most massive models, has a thickness a significant fraction of the total radius, in which the density is markedly less than that of the deeper interior. This is the region in which the outer crust of the neutron star contains ordinary nuclei.

The thickness of the outer crust is shown in more detail as a function of the mass of a neutron star in Figure 9. If we define the crust as that region which contains nuclei mixed with neutrons and possibly protons, which may be too broad a definition, then it may be seen that the thickness of the crust region varies from ∼200 m near the upper limit to the neutron-star stable mass to several kilometers at lower neutron-star masses. The thickness of the crust containing only nuclei and electrons, but no free neutrons, constitutes in general more than half of the total thickness of the crust, and can typically amount to more than 1 km. The corresponding masses of the material in the various layers of the crust are shown in Figure 10.

With the equation of state of Langer et al, the Σ^- hyperon appears in the interior for those models having a mass of 0.3 solar masses and greater. This is of some interest in connection with the discussion of vibrational damping given later in this review. It seems likely that any neutron star formed in nature will probably contain some content of hyperons.

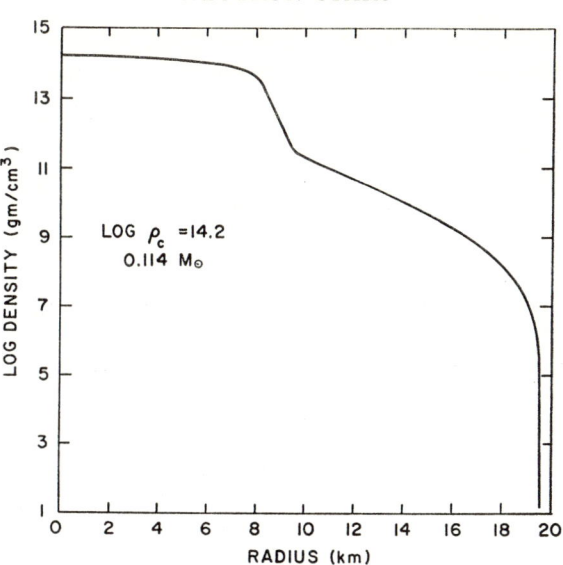

FIGURE 6. Density distribution in the interior of a neutron-star model of 0.114 solar masses.

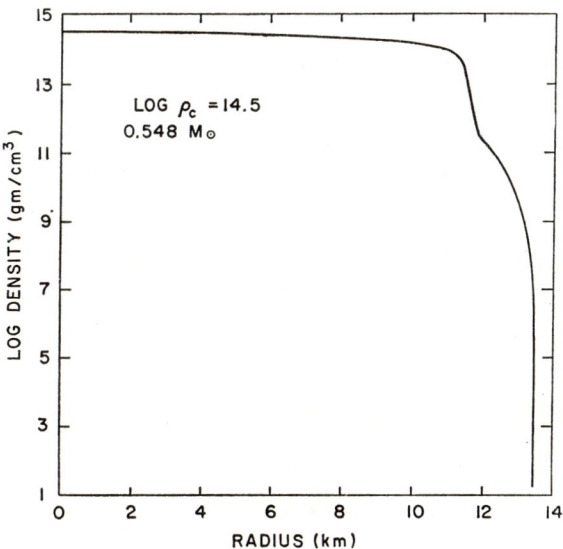

FIGURE 7. Density distribution in the interior of a neutron-star model of 0.548 solar masses.

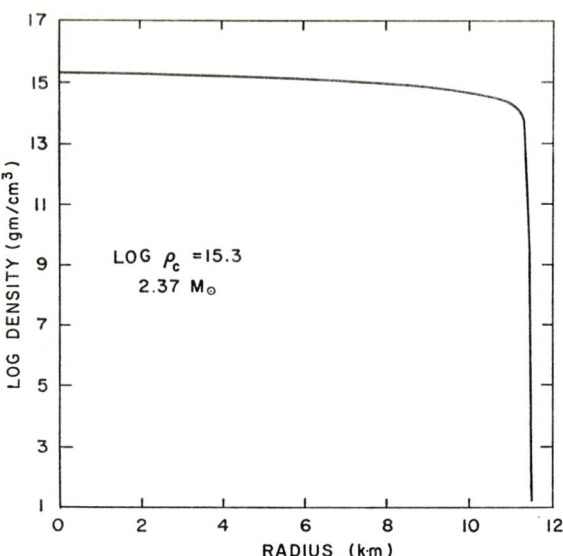

FIGURE 8. Density distribution in the interior of a neutron-star model of 2.37 solar masses.

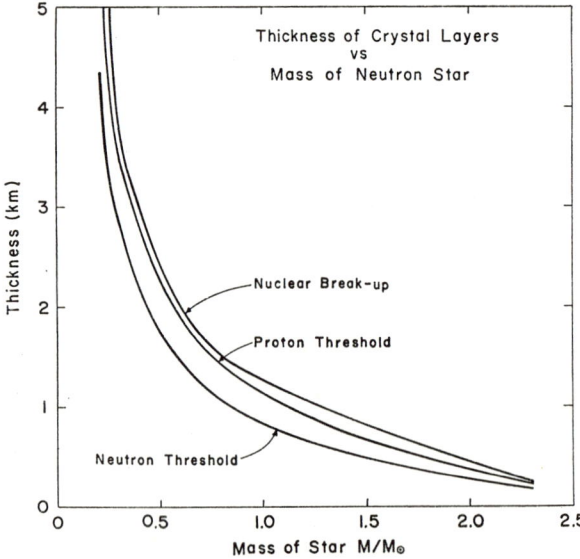

FIGURE 9. Thickness of the surface layer of a neutron star containing nuclei (line showing nuclear breakup). The depths at which neutrons and protons appear are also shown.

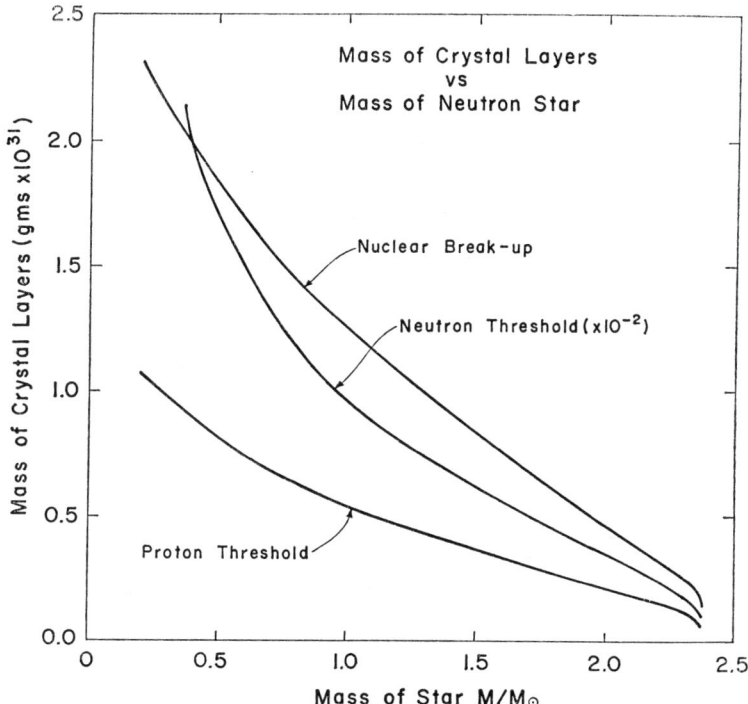

FIGURE 10. The masses of material contained down to the depths in the crust corresponding to the lines in Figure 9.

Some general characteristics of the neutron-star models characterized by the V_γ equation of state are listed in Table 2.

All the results presented in this review of neutron-star models are based on the assumed validity of the equation of hydrostatic equilibrium based on general relativity. Salmona (1967) has calculated neutron-star models with an equation of hydrostatic equilibrium based on the Brans-Dicke scalar-tensor theory. He finds small differences in the neutron-star structure that are due to the differences in the theories.

Bhatia, Bonazzola & Szamosi (1969) have shown that the simultaneous hydrostatic and chemical equilibrium of a neutron star requires the presence of a small radial electric field, $\sim 10^2$ V per centimeter. This results from a small outward displacement of the electrons in the star.

The Atmosphere

Unless the neutron star is completely at absolute zero temperature, it will have a nondegenerate atmosphere. In order to calculate the structure of this atmosphere, it is necessary to assume, in addition to the equations of

TABLE 2. Characteristics of neutron-star models. Masses are in units of the solar mass. The proper mass is the sum of the gravitational mass and the mass binding energy. The surface redshift is the fractional wavelength shift $\Delta\lambda/\lambda$

Log_{10} central density (g/cm^3)	Gravitational mass	Binding energy	Radius (km)	Redshift
14.1	0.0632	0.0004	99.64	0.0009
14.2	0.114	0.0014	19.56	0.0087
14.3	0.206	0.0041	14.54	0.022
14.4	0.356	0.0113	13.49	0.042
14.5	0.551	0.0258	13.42	0.067
14.6	0.828	0.0577	13.50	0.11
14.7	1.129	0.111	13.64	0.15
14.8	1.415	0.183	13.64	0.20
14.9	1.697	0.284	13.47	0.26
15.0	1.97	0.414	13.10	0.34
15.1	2.13	0.556	12.62	0.43
15.2	2.30	0.692	12.08	0.55
15.3	2.37	0.807	11.48	0.62

hydrostatic equilibrium, that the rate of energy flow through the atmosphere is constant. If the energy transfer is by radiation, then the temperature gradient in the atmosphere will depend upon the opacity of the material. If the opacity were to be high enough, then convection could occur instead of radiative transfer. However, Tsuruta & Cameron (1966a) found that convection is very unlikely in neutron-star atmospheres with the range of opacities that seems likely to exist there.

At lower temperatures, bound-free transitions will give a major contribution to the opacity. At higher temperatures, Thomson or Compton scattering will be the principal contributor to the opacity, and the resulting opacity is considerably less than in the lower-temperature region. Tsuruta & Cameron (1966a) found that it made very little difference to the resulting calculated opacities whether the surface was assumed to be composed of magnesium or of iron and hence there was little effect on the structure of the atmosphere.

At the base of the atmosphere electron degeneracy will set in, and the very high thermal conductivity characteristic of electron degeneracy will cause very small temperature gradients to exist. In some early work on the structure of neutron-star atmospheres, Morton (1964) assumed that the temperature gradient could be neglected once electron degeneracy set in. However, Tsuruta & Cameron (1966a) found that a significant total temperature drop still occurred between the center of a neutron star and the point of onset of electron degeneracy at the base of the atmosphere. They found that for a typical neutron star (Tsuruta and Cameron 1966c) it could be expected that the surface temperature would be $\sim 1\%$ of the central temperature at lower temperatures where the atmospheric opacity is higher, and that the surface temperature might be $\sim 10\%$ of the central temperature at higher temperatures where Thomson scattering predominates.

These results may be strongly modified by the presence of a magnetic field. A very strong magnetic field will constrain the motion of the electrons in such a way that the motion is still free along the direction of the magnetic field, but it is quantized into discrete energy levels for perpendicular motions about the magnetic lines of force. If the magnetic field has values as high as 10^{12} G, a value commonly suggested for neutron stars, then the energy levels corresponding to the perpendicular motions of the electrons may have spacing ~ 10 keV, which would be equivalent to a temperature of 10^7 degrees—higher than the temperatures generally considered probable at the surface of a neutron star in the more interesting later stages of its cooling. Therefore, the Thomson or Compton scattering processes must be modified to include the restriction that the interaction of photons with the electrons can lead predominantly only to a recoil of the electron along the magnetic lines of force. Such calculations have yet to be carried out in detail.

The strong magnetic field can also influence the motion of electrons about atoms or ions forming the surface layers of the neutron star. Extremely strong perturbations of the electron motions about the central nucleus can take place as a result of the magnetic interactions, and this will certainly cause strong modifications to the bound-free source of opacity.

It has been suggested (Cameron 1965b, Gold 1969) that much of the cosmic-ray flux accelerated within our Galaxy may come from neutron-star surface plasma accelerated by the magnetic fields associated with neutron-star surfaces and ejected from the neutron star by the basic pulsar mechanism, whatever it is. It is therefore of some interest to determine what is the likely composition of such neutron-star atmospheres.

A first look at this problem was taken by Chiu & Salpeter (1964) who estimated that any hydrogen or helium in the surface layers of neutron stars would quickly diffuse into the interior, where it would be rapidly destroyed by thermonuclear reactions in the hot interior on a time scale short compared to the cooling time of the star. Calculations of this sort were done in considerably more detail by Rosen (1968, 1969), who considered the diffusion of a wide variety of nuclear species from an assumed surface abundance into the interior, with the diffusion equations coupled to the thermonuclear rate equations that change one nuclear species into another. He confirmed the results of Chiu & Salpeter on the rapid destruction of hydrogen and helium, and showed that under virtually all probable conditions of neutron-star cooling, the diffusion of any species into the neutron-star interior would be sufficiently rapid to change the great bulk of the material of that species into nuclei in the iron statistical equilibrium peak. Thus it seems very probable that the overwhelmingly abundant constituent of neutron-star atmospheres will be ^{56}Fe nuclei. Much of the iron in the galactic cosmic rays might be accelerated by the pulsar mechanism from neutron-star surfaces, but it seems unlikely in view of these diffusion calculations that any significant part of the remainder of the observed abundance spectrum of the cosmic rays could arise in this manner.

The possibility that the emitted X-ray spectrum of a neutron star could contain absorption edges or lines has been discussed by Orszag (1965), among others. Onyejuba & Gaustad (1967) have shown that such features would be broadened beyond recognition by the Stark effect. Strong magnetic fields would also obliterate recognizable spectral features.

The Crust

It is expected that an outer layer of a neutron star of substantial thickness, lying below the atmosphere, will form a crystalline solid. This is the part of the neutron star having densities in the range characteristic of the interiors of white-dwarf stars and somewhat higher. It is for this reason that the solid crystalline properties expected for this region have first been discussed in connection with white-dwarf stars by Salpeter (1961) and Van Horn (1968).

In a crystalline solid of the type existing on the Earth at low pressure and density, the nuclei are almost completely screened by the surrounding clouds of electrons, and the interaction between the atoms that gives rise to the crystal depends entirely on the way in which two such adjacent electron clouds can interact. The residual energies associated with such interactions amount only to a few electron volts. In contrast to this, in a white-dwarf star or the crust of a neutron star the electrons are detached from the nuclei and form a degenerate Fermi sea in which most of the electrons will be relativistic. The nuclei exert only very small perturbations upon the motions of the electrons in the Fermi sea, so that the electrons are not effective in screening the nuclei from the Coulomb fields of each other until distances of several internuclear spacings are reached. This means that the nuclei will repel one another because of their strong Coulomb fields, and the energies of interaction at average internuclear spacings may easily amount to a value $\sim$1 MeV. The medium can thus form a state of lowest energy by arranging the nuclei in a regular lattice array so that the nuclei stay at all times as far away as possible from one another.

This is the basis for the expectation that the nuclei in white-dwarf interiors and neutron-star crusts will form a body-centered cubic lattice when temperatures are sufficiently small so that the energy of interaction between the nuclei exceeds the thermal energies per particle. Actually, solids melt when their thermal energy is $\sim$1% of the lattice interaction energy, since the resulting fluid can allow a general mobility of the ions without the individual ions coming too close to one another. The melting temperature to be expected in neutron-star crusts is clearly a function of the density and composition, but values in the vicinity of $10^9 °K$ are in general to be expected (Ruderman 1969a).

As we shall see, a neutron star can be expected to cool below $10^9 °K$ very soon after formation. At such a time the star may be rotating quite rapidly. Ruderman (1969b) has pointed out that a Coulomb crystal formed in this way may be significantly nonspherical because of the rotation of the neutron star and also that it will be very rigid. It should have a very high shear modulus. Therefore, with the slowing down of the neutron star, presumably associated with the pulsar mechanism, high internal strains will arise in the crystalline crust. Ruderman (1969b) has suggested that these stresses will eventually cause a fracture of the crust, resulting in a "starquake" that will readjust the figure of the star so that once again surfaces of equal density will also be equipotential surfaces. In this way, one might expect a series of such starquakes to occur during the history of the neutron star as it slows down and becomes more nearly spherical. Ruderman (1969b) has interpreted the sudden decrease of period of the Vela X pulsar as resulting from such a starquake.

Such a suggestion is not without its theoretical difficulties. If there is no very strong magnetic field associated with a neutron star, then the surface of the star, which is always gaseous, can flow from the equator in the direction of the pole, as the star slows down, in order to maintain an equipotential surface in the atmosphere. This would vaporize crystalline material at the equator and form new crystalline material at the poles. Ruderman has pointed out that a strong surface magnetic field would retard this redistribution of matter for sufficiently long periods of time that the conditions needed for starquakes might still be attained. This would apply a stress to the deeper layers containing the currents which maintain the magnetic field.

However, it is also necessary to consider the strong density variation in the crust of a neutron star and the ability of the star to change its composition as the pressure at a point in the interior changes with the slowing down of the star. Thus, at the base of the crystalline crust near the equator, the pressure will decrease as slowing down occurs, and new nuclei can be expected to be formed out of the nuclear fluid at that position, which will add to the crystal structure at the base. Similarly, the pressure at the base of the crust near the poles will increase, leading to the dissolution of nuclei into the surrounding nuclear medium. There will still be strong stresses in the crystal, but it is now not clear that they must be relieved by starquakes, and a more detailed analysis seems required. It seems likely that there will be dislocations and vacancies in the crystal, and that creep phenomena may be important. The crust of a neutron star presents a problem in astrophysics that will require considerable attention on the part of solid-state physicists to elucidate.

We have called the crystalline layer of a neutron star the crust because it resembles a terrestrial feature much more than a stellar one. The terrestrial analogy has been extended by Dyson (1969a,b) who has suggested that seismic and volcanic activity may occur in neutron-star crusts. Dyson has suggested the possibility that if a channel is opened through a neutron-star crust, the material at low depths rises through the channel, gaining enough energy to maintain the motion both because it may have a high initial temperature and because beta decays occurring in the nuclei that compose the material can release enough energy to maintain the flow through the channel. He assumes that this flow may cause material to spout through the channel, forming an analogy to the terrestrial volcano, and after it has spread out on the surface, the additional loading of the surface may cause a great deal of seismic activity and starquakes. The idea is interesting, but it must be justified by further analysis. In particular, it is necessary to show that a channel through a crystalline crust can stay open for the flow of material through it in the manner envisaged by Dyson, and that the beta decays occurring in the rising material can release energy fast enough to maintain the volcano in operation. Basically, such a process merely provides an accelerated means for cooling a neutron star, since the volcano must operate on the basis that the internal temperature is high enough to provide enough energy to maintain the flow of material through a channel in the crust, and thus to bring very hot material to the surface of the star.

The Superfluid Interior

Below the neutron-star crust the material consists of a neutron-proton-electron fluid up to densities slightly more than 2×10^{14} g/cm^3, at which time additional constituents become present in the fluid.

In the range below normal nuclear densities, the neutrons lying at the top of the Fermi sea have an attractive potential for one another. Thus pairs of neutrons can be expected to act like bosons, just as pairs of electrons can act like bosons to form a superconductor under normal laboratory conditions. It is thus expected that the neutron and proton fluids in the outer part of the neutron star will probably form superfluids because of the mutual attraction between pairs of particles (Migdal 1959, Ginzburg & Kirzhnits 1965, Canuto & Marx 1965, Wolf 1966, Clark & Chao 1969, Ruderman 1969a, Itoh 1970).

In a neutron superfluid a gap forms in the spectrum of single-particle energy states immediately above the Fermi surface. This gap may amount to about 1 MeV. Certain states of collective motion may, however, exist in the gap region. Nevertheless, the presence of the energy gap can be expected to decrease the heat capacity of the neutron-star interior by a large factor, thus possibly allowing the star to cool very quickly once the temperature is low enough for the superfluid to form.

The gap energy and the resulting transition temperature to the superfluid state can be expected to vary widely as a function of density in the neutron-star interior. Calculations quoted by Ruderman (1969a) show that the transition temperature may lie as high as $2\times 10^{10}\,°K$ at certain densities. The transition temperature and the gap energy decrease as normal nuclear density is approached.

The superfluidity just discussed arises from the $^1s_0$ interaction between neutrons, which changes its phase shift from attractive to repulsive at normal nuclear densities, so that at higher than normal nuclear densities the superfluidity can be expected to vanish. However, another form of superfluidity may arise from the p-state interactions. The $^3p_1$ phase shift appears to be repulsive at all densities; the $^3p_0$ phase shift changes from attractive to repulsive at about the same density as the $^1s_0$, but the $^3p_2$ phase shift appears to be attractive even at higher than normal nuclear densities. Hence it is possible that superfluidity can be maintained at higher than normal nuclear densities by the $^3p_2$ interaction. However, this would imply a superfluid state of a type not hitherto investigated in the laboratory, having anisotropic properties. Ruderman (1969a) suggests that in a nonrotating neutron star the lowest energy would be achieved with the anisotropic-gap directions arranged so that the compressibility of the matter would be greatest in the radial direction. This would correlate the directions of the anisotropic superfluid gap throughout the interior of the star, which would have some important effects on the heat capacity of the deeper interior where the anisotropic superfluidity might occur.

An alternative possibility in the interior of a neutron star is ferromagnetism. This might occur in a region where there is a repulsion between singlet neutron pairs, which favors the alignment of their spins. At lower densities, where the singlet phase shifts are known, it appears that ferromagnetism certainly cannot arise. However, at higher densities, the singlet phase shifts are not known, and this gives rise to the possibility that in the deep interior of a neutron star, at higher than normal nuclear densities, ferromagnetism is possible (Brownell & Callaway 1969, Silverstein 1969, Rice 1969, Clark 1969). If this is the case, it would probably interfere with the anisotropic superfluidity. Magnetic fields $>10^{15}$ G could be maintained by such ferromagnetism. However, Pearson & Saunier (1970) have shown that such ferromagnetism is very unlikely.

If the protons form a superfluid, then one can also expect that they will be a superconductor. A superconductor can be expected to exclude a magnetic field from its interior, a phenomenon known as the Meissner effect. However, Baym, Pethick & Pines (1969a,b) have argued that the superfluid protons in a neutron star cannot exclude the magnetic field from the interior. This is because the electrons in the interior are expected to have so high a conductivity that they will maintain the field against the operation of the Meissner effect. The conductivity of the electrons is so high in the interior because the electrons can scatter only from protons in the top of the Fermi sea, in the portion of the Fermi surface that has been rounded by thermal effects. Therefore electrons have a large mean free path in the neutron-star interior, and the corresponding conductivity is very high. For typical conditions, Baym,

Pethick & Pines have estimated that the decay time of the magnetic field may be some 10^3 times the age of the universe.

If the interior of a neutron star does exhibit superfluid properties, then the rotation of the neutron star imposes some interesting conditions on the superfluids. The rotation of the superfluid is manifested by the establishment of quantized vortex lines throughout the interior. The superfluid itself moves irrotationally; the curl of the fluid velocity is zero except within the core of a vortex line. Such vortex cores are composed of normal (nonsuperfluid) material. In a rotating neutron-star interior it is to be expected that the radius of such a core will be $\sim 10^{-12}$ cm and the separation between quantized vortex lines $\sim 10^{-2}$ cm (Baym, Pethick & Pines 1969b). The mixture of superfluid and quantized vortex lines can mimic rigid-body rotation since the curl of the fluid velocity averaged over many vortex lines is equal to twice the fluid angular velocity, the usual relation. Conditions of energy minimization require that the vortices be arranged in a triangular lattice array in the interior of the fluid in the corotating frame of the rigid-body rotation. Vortex lines in such arrays may allow certain types of oscillations in the neutron-star interior, which will be discussed below.

Cooling of Neutron Stars

Neutron stars are expected to be formed at a very high temperature as a result of a supernova implosion process. The initial cooling should proceed predominantly by the emission of neutrino-antineutrino pairs. Three processes that can give rise to these have been discussed by Tsuruta & Cameron (1966a). These include neutrino-pair emission from the plasma process, the URCA process, and the bremsstrahlung process. In the first of these a plasmon is converted into a neutrino-antineutrino pair. In the second the neutrons decay into protons, electrons, and antineutrinos, and in turn the protons and electrons recombine to form neutrons and neutrinos. In the third process an electron passes close to a charged baryon, yielding a neutrino-antineutrino pair in place of the more usual photon.

The cooling of the neutron-star models thus depends on the combined effects of neutrino-antineutrino emission from the interior and electromagnetic radiation from the surface. Neutrino emission dominates at higher temperatures, and quickly brings the star down to relatively low internal temperatures. Surface photon emission dominates in the later stages of the cooling history of a neutron star. The results of the calculations of Tsuruta & Cameron (1966a) are shown in Figure 11. These are for a variety of models calculated by Tsuruta & Cameron (1966c), based upon the Levinger & Simmons V_γ potential, but without the elaborate treatment reported earlier in this review. Neutrino emission dominates the cooling for the first 10^4 to 10^5 years, and photon cooling dominates after that. Temperatures fall very low by ages $\sim 10^8$ years. Additional discussions of neutron-star cooling have been given by Morton (1964), Finzi (1965a), Bahcall & Wolf (1965a,b,d), Ellis (1965), and Wolf (1966).

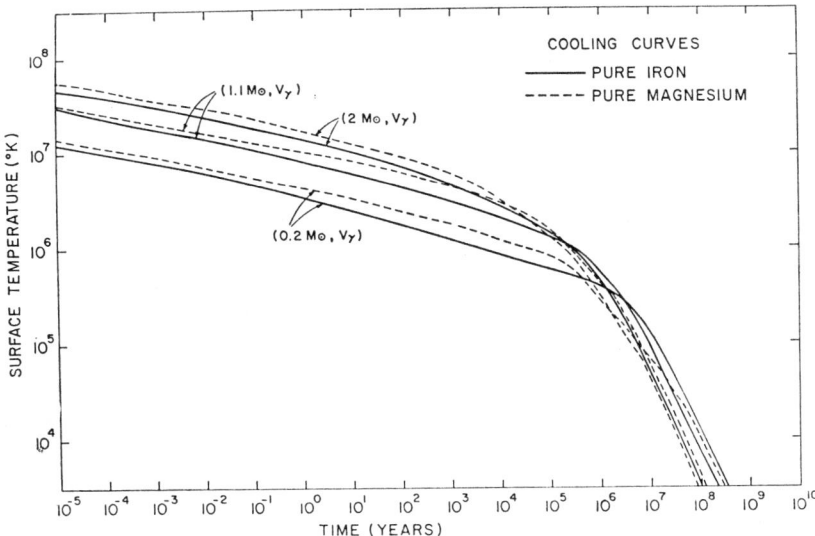

FIGURE 11. Surface temperatures as a function of time of a variety of neutron-star models due to the combined effects of neutrino-antineutrino emission and of electromagnetic radiation from the surface, as calculated by Tsuruta & Cameron.

It is necessary to give a word of caution concerning these calculations. A strong magnetic field in the interior may substantially alter the opacities in the atmospheric layers, thus changing the surface cooling rates significantly. If there are superfluid energy gaps in the interior, the heat capacity may be very much less than that assumed in the calculations of Tsuruta & Cameron, and cooling may proceed very much faster than shown in Figure 11. However, if the calculations are reasonably representative, then neutron stars having the ages commonly attributed to pulsars can be expected to have surface temperatures $\sim 10^6$ °K, or a little less.

If there is a strong magnetic field in the interior of a neutron star, then there can be a contribution to the cooling by synchrotron radiation of neutrinos and antineutrinos (Landstreet 1967), but this is probably not important compared to other processes.

VIBRATION OF NEUTRON STARS

Since a neutron star must be formed in a hydrodynamically violent event, it is conceivable that it is left with some vibrational energy following the supernova implosion. The writer (Cameron 1965a) suggested that such neutron-star vibrations might excite motions in an associated magnetosphere that could accelerate charged particles and perhaps be responsible for celestial X-ray sources. This led to some investigations of the vibrational properties of neutron stars, particularly their damping times.

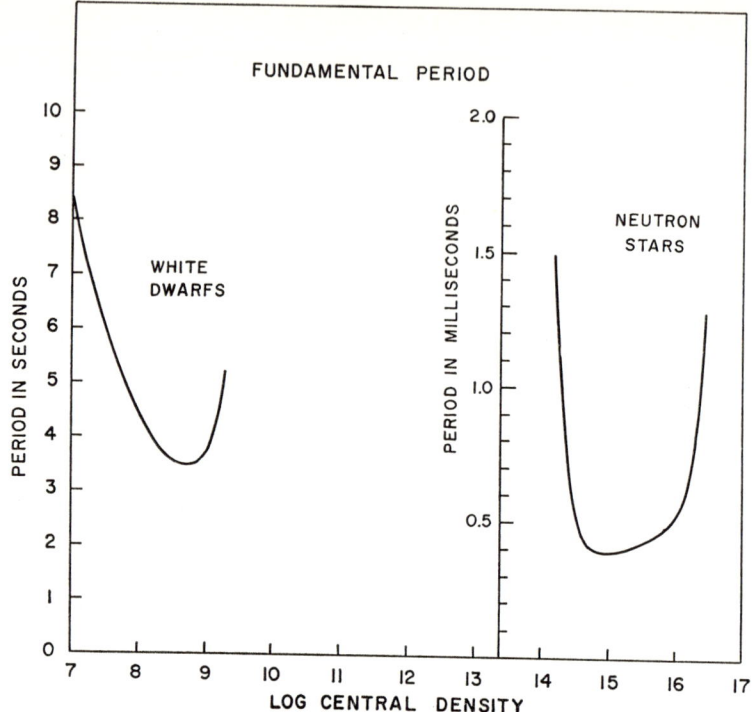

FIGURE 12. Fundamental vibrational period of the white-dwarf and neutron-star models shown in Figure 4. These are based on the customary equation of state, but in an actual case weak interactions in the neutron-star interiors would be slower than the oscillations, leading to a stiffer equation of state and slightly shorter periods.

The fundamental periods of radial oscillation of the neutron-star models shown above in Figure 4 are shown in Figure 12. On the left-hand part of the figure are the fundamental vibrational periods of the white dwarfs that correspond to the left portion of Figure 4; in the intervening range of central densities the fundamental vibrational periods of the models are imaginary, as has previously been stated, and in the neutron-star range the majority of the vibrational periods lie in the vicinity of 0.4 msec.

Nonradial vibrations have been studied by Thorne (1969a). He studied the first few quadrupole pulsation periods and calculated the damping of these modes due to the emission of gravitational radiation (see also Thorne 1969b). These calculations were carried out for several different types of neutron-star model. Generally speaking, the first few quadrupole oscillations have periods that are also a fraction of a millisecond, and these periods are damped by gravitational radiation with damping times $\sim$1 sec. These results are in general agreement with earlier calculations by Zee & Wheeler (Wheeler

1966), Chau (1967), and Occhionero (1968). Zee & Wheeler had pointed out that if a vibrating neutron star is also rotating, then rotation can provide a coupling between the radial vibration mode and the nonradial vibration modes, thus leading to rapid loss of radial vibrational energy through gravitational radiation from the nonradial modes.

The radial vibrational modes can also be damped by other mechanisms. The damping by the vibrational URCA process has been studied by Finzi (1965b), Hansen (1966), Hansen & Tsuruta (1967), and Finzi & Wolf (1968). In this process the Fermi levels of the neutrons, protons, and electrons oscillate about their equilibrium values as the density changes, thus opening up phase space first for the decay of the neutron into protons, electrons, and antineutrinos, and then for the capture of electrons on protons to form neutrons and neutrinos. Hansen (1966) found that there was a half-life of approximately 100 years for the emission of vibrational energy from a neutron-star interior by these processes. A slightly more elaborate treatment of this problem was carried out by Hansen & Tsuruta (1967) and by Finzi & Wolf (1968) in which they also took account of the part of the vibrational energy that went into kinetic energies of the emitted particles and hence into internal heat.

Numerical calculations dealing with neutron-star vibrations were carried out with a general relativistic hydrodynamic code by Mock (1967). Studying the steepening of the vibrational motion into a shock in the atmosphere of a neutron star, he found that quite large amounts of atmospheric heating were possible in this way, which for larger vibration amplitudes could dissipate the vibrational energy with a half-life of about 30 years. However, even for large vibrational amplitudes, the atmospheric heating was too small to make such vibrating neutron stars of interest as celestial X-ray sources.

An additional process applicable to that part of a neutron-star interior having hyperons was studied by Langer & Cameron (1969). The process by which two neutrons are converted into a proton and a Σ^- hyperon is a weak interaction, since the strangeness quantum number is not conserved in the transformation. However, it appears to be a weak interaction that can take place without the emission of neutrinos or antineutrinos. The transformation rates in the weak interaction are comparable to the vibration rates of neutron stars. Hence the transformations among the hyperons get out of phase with the vibration of the star. This leads to a dissipation of vibrational energy into internal heat. Langer & Cameron found that the half-life for conversion of vibrational energy into internal heat was ~ 1 sec. Similar results were independently obtained by Jones (1970).

Ruderman (1968, 1969a, 1970) has discussed various additional possible oscillations of the neutron star. One mode that results from the probable crystalline character of the crust depends upon the high shear modulus of the crust; Ruderman has estimated that torsional oscillations of the crust should have a period $\sim 10^{-1}$ sec. One can also expect that there will be a magnetic coupling between the conducting crust and the electrons and protons in the deeper interior if magnetic fields are present. Baym, Pethick, Pines & Ruderman (1969) have estimated that the magnetic coupling can give oscillations with a period $\sim 10^{14}/B$ sec, where B is the magnetic-field strength in G; if the magnetic field is $\sim 10^{12}$ G, then the resulting period is $\sim 10^2$ sec.

If the interior has superfluid properties, then two additional types of oscillation are possible. If the fluid is uniformly rotating, then a classical torsional oscillation, corresponding to a bending of the vortex cores, has a period approximately the inverse of the angular velocity. This would give oscillations ~ 1 sec, or somewhat less, for most of the observed pulsars. A second mode of vibration consists of purely axial motion of the lines in the vortex lattice. Each vortex line would remain parallel to the axis of rotation but there would be a redistribution of the density of the vortices and of the corresponding angular momenta within the superfluid. Ruderman (1970) has estimated that there is an oscillation period associated with this phenomenon $\sim 10^8/\Omega^{1/2}$, where Ω is the angular velocity in radians/sec. For the pulsar in the Crab Nebula, this would predict an oscillation of a point on the surface with respect to uniform angular rotation having a period of a few months. Ruderman has suggested that this "wobble" is responsible for the "sine-wave" behavior of the residuals in the phases of the pulsar signals from the Crab Nebula.

Magnetic Fields on Neutron Stars

There has been a great deal of recent discussion in the literature about the external configurations of magnetic fields to be expected in pulsars. This lies outside the scope of the present review, which will be concerned only with the character of magnetic fields in the interior of the neutron star.

When a star having a reasonable internal magnetic field collapses to form a neutron star, flux conservation may produce initial magnetic fields in the interior of the neutron star $\sim 10^{12}$ G. This is significantly less than a critical value of the magnetic field, near 10^{14} G, at which the field energy density becomes great enough so that it is equivalent to two electron rest masses per cubic electron Compton wavelength. For fields near the critical value and above, a fascinating variety of physical phenomena are possible. These have been explored in a series of papers by Chiu and the Canutos (Canuto & Chiu 1968a,b,c; Chiu & Canuto 1968; Chiu, Canuto & Fassio-Canuto 1968).

It was pointed out above that the electrical conductivity in a neutron-star interior is expected to be so high that there will be no significant ohmic dissipation of the magnetic field during any possible present age of a neutron star. However, two mechanisms have been suggested whereby magnetic fields may be intrinsically maintained in the interior of a neutron star. One of these is the possible neutron ferromagnetism described above. The other is Landau orbital ferromagnetism of the electrons, suggested by Lee, Canuto, Chiu & Chiuderi (1969), and named by them "lofer." In this mechanism account is taken of the fact that in a high magnetic-field strength the electrons have quantized perpendicular motions. The magnetic moments associated with the perpendicular motions of the electrons can give rise to the field that maintains the motion. Magnetic fields $\sim 10^{12}$ G could be maintained in the interior if the Landau orbital ferromagnetism of the electrons operates, whereas if there is a neutron ferromagnetic interior, fields $\sim 10^{15}$ G or greater may be generated. Most suggestions for pulsar slowing-down mechanisms prefer the lower value for a surface magnetic field assumed to be associated with a neutron star.

If a general magnetic field extends throughout the neutron-star interior, then the electrons that are tied to the field throughout the interior can be expected to rotate uniformly. The protons must also rotate with the electrons in order to prevent the establishment of strong currents in the interior, which would require a rapid buildup of a new component of the magnetic field, which is prevented since the dissipation time for a field is so long. However, the neutrons in the interior are not coupled to the motion of the electrons and protons except through viscous effects. Since the neutrons constitute the greater part of the mass in a star, it is of some interest to learn how rapidly the motions of the neutrons will be coupled to those of the electrons and protons.

Rotation of the Neutron Star

Some of the rotational properties of a neutron star were considered by Tsuruta & Cameron (1966b) before the discovery of pulsars. They considered a very rapidly rotating neutron star, and showed that the interior density distribution was likely to be sufficiently flat so that a neutron star would deform into a Jacobi ellipsoid if it were rotating fast enough. Such an ellipsoid has a time-varying quadrupole moment and therefore it radiates gravitational radiation. Tsuruta & Cameron therefore argued that any very fast rotating neutron star would be slowed down by gravitational radiation until the Jacobi ellipsoid relaxed to a MacLaurin ellipsoid having equatorial flattening but no time-varying quadrupole moment. They suggested that a further rapid slowing down of the spin would occur if the rotational energy thus lost could be utilized in the ejection of mass. However, the discovery of relatively fast rotating pulsars indicates that such mass ejection did not occur with any significant efficiency.

The actual further slowing down of neutron stars occurs by the unknown pulsar emission mechanism. All pulsars that have been observed with sufficient accuracy are slowing down. However, two pulsars have been observed to have had discontinuous decreases of period, which presumably give information about the properties of the interior provided the observations can be properly interpreted. These are the Crab and Vela X pulsars. In both of the observed events the rate of slowing down was higher after the discontinuity than before, although it gradually returned to the former value.

In the model of Ruderman (1969b), and of Baym, Pethick, Pines & Ruderman (1969), the period discontinuities result from starquakes. There is a small decrease in the equatorial radius, which results in a decreased moment of intertia and hence a decreased period following the starquake. The amount of decrease of the equatorial radius would be about 1 cm for the Vela X pulsar and about 10 μ for the Crab pulsar. This would obviously require that the crystalline shell in the Crab pulsar be very much weaker than that in the Vela X pulsar. In these models the increased rate of slowing down following the starquake arises from the fact that the superfluid neutrons in the interior would lag behind the motion of the electrons and protons following the discontinuity. If both the protons and neutrons formed superfluids, the frictional effects between them could only occur through the normal cores of the vortices. Baym et al suggest the recovery period of apparently a few years for the Vela X pulsar to return to its initial slowing-down rate, and of a few months for the Crab Nebula pulsar, to be the time required for frictional effects to make the neutron and proton superfluids corotate. This requires that the internal temperature of both the Crab and the Vela X pulsars lie in the vicinity of 10^8°K. The friction is proportional to the square of this temperature. Sutherland, Baym, Pethick & Pines (1970) have suggested a method for determining the neutron-star mass based on the above model.

An alternate suggestion, due to Greenstein & Cameron (1969), is that the friction throughout the deep interior of the neutron star is sufficiently low so that the core of the neutron star can be left spinning much faster than the surface after the surface has been slowed down for a long time. If a finite negative gradient of angular momentum per unit mass is established away from the axis of rotation, then fluid instabilities may transport angular momentum from the core to the surface in an abrupt event, thus leading to a discontinuity in the period. In this model the outer part of the star, with most of the moment of inertia, would have to be relatively viscous, and the viscosity near the center would have to be very small. This would probably require that the protons not have superfluid properties in the outer part of the neutron star owing to interference with the proton-pairing interaction by the very much larger number of neutrons present, and also that the interior temperature be very much less than 10^8°K, so that ordinary frictional effects would not slow down the center of the star too rapidly.

A neutron star has a radius only a few times greater than its Schwarzschild radius. Therefore general relativistic effects can be quite significant in the properties of neutron stars. Thus (see Table 2) gravitational redshifts at the surface may become quite large, and the model binding energies can amount to as much as 25% of the proper mass of the star. Also, the star can drag its inertial frames quite rapidly as it rotates. The inertial frames may be dragged with the motion of the star a few tens of percent as rapidly as the rotation itself at the surface, and even more rapidly on the axis of rotation. These latter properties have been discussed by Cohen & Cameron (1969). They showed that the period and rate of change of period of the Crab pulsar, together with the minimum energy emission rate of 10^{38} ergs/sec from the Crab Nebula, required rotational energy loss from a neutron-star model having at least 0.4 solar masses.

This work has been supported in part by grants from the United States Atomic Energy Commission, the National Science Foundation, and the National Aeronautics and Space Administration.

LITERATURE CITED

Ambartsumyan, V. A., Saakyan, G. S. 1960, Sov. Astron. AJ, 4, 187
Bahcall, J. N., Wolf, R. A. 1965a, Ap. J., 142, 1254
Bahcall, J. N., Wolf, R. A. 1965b, Phys. Rev. Lett., 14, 343
Bahcall, J. N., Wolf, R. A. 1965c, Phys. Rev., 140, B1445
Bahcall, J. N., Wolf, R. A. 1965d, Phys. Rev., 140, B1452
Barker, B. M., Bhatia, M. S., Szamosi, G. 1966, Nuovo Cimento, 44, 109
Baym, G., Pethick, C., Pines, D. 1969a, Nature, 224, 673
Baym, G., Pethick, C., Pines, D. 1969b, Nature, 224, 674
Baym, G., Pethick, C., Pines, D., Ruderman, M. 1969, Nature, 224, 872
Bhatia, M. S., Bonazzola, S., Szamosi, G. 1969, Astron. Ap., 3, 206
Brownell, D. H., Callaway, J. 1969, Nuovo Cimento, 60B, 169
Cameron, A. G. W. 1959, Ap. J., 130, 884
Cameron, A. G. W. 1965a, Nature, 205, 787
Cameron, A. G. W. 1965b, Nature, 206, 1342
Canuto, V., Chiu, H. Y. 1968a, Phys. Rev., 173, 1210
Canuto, V., Chiu, H. Y. 1968b, Phys. Rev., 173, 1220
Canuto, V., Chiu, H. Y. 1968c, Phys. Rev., 173, 1229
Canuto, V., Marx, G. 1965 (Preprint)
Cazzola, P., Lucaroni, L., Scarini, C. 1968, Nuovo Cimento, 52B, 411
Chau, W. Y. 1967, Ap. J., 147, 664
Chiu, H. Y. 1964, Ann. Phys. (N. Y.), 26, 364
Chiu, H. Y., Canuto, V. 1968, Phys. Rev. Lett., 21, 110
Chiu, H. Y., Canuto, V., Fassio-Canuto, L. 1968, Phys. Rev., 176, 1438
Chiu, H. Y., Salpeter, E. E. 1964, Phys. Rev. Lett., 12, 413
Clark, J. W. 1969, Phys. Rev. Lett., 23, 1463
Clark, J. W., Chao, N. C. 1969, Nuovo Cimento, 2, 185
Cohen, J. M., Cameron, A. G. W. 1969, Nature, 224, 566
Cohen, J. M., Cameron, A. G. W. 1970 (To be published)
Cohen, J. M., Langer, W. D., Rosen, L. C.,
Cameron, A. G. W. 1970, Ap. Space Sci., 6, 228
Durgopal, M. C., Gehlot, G. L. 1968, Phys. Rev., 172, 1308
Dyson, F. J. 1969a, Nature, 223, 486
Dyson, F. J. 1969b, Comm. Ap. Space Phys., 1, 198
Ellis, D. G. 1965, Phys. Rev., 139, B754
Finzi, A. 1965a, Phys. Rev., 137, B472
Finzi, A. 1965b, Phys. Rev. Lett., 15, 599
Finzi, A., Wolf, R. A. 1968, Ap. J., 153, 835
Gerlach, U. H. 1968, Phys. Rev., 172, 1325
Ginzburg, V. L., Kirzhnits, D. A. 1965, Sov. Phys. JETP, 20, 1346
Gold, T. 1969, Nature, 221, 25
Gratton, L., Szamosi, G. 1964, Nuovo Cimento, 33, 1056
Green, A. E. S. 1954, Phys. Rev. 95, 1006
Greenstein, G. S., Cameron, A. G. W. 1969, Nature, 222, 862
Hansen, C. J. 1966, Nature, 211, 1069
Hansen, C. J., Tsuruta, S. 1967, Can. J. Phys., 45, 2823
Harrison, B. K., Wakano, M., Wheeler, J. A. 1958, in La Structure et l'evolution de l'universe (Brussels: Stoops)
Harrison, B. K., Thorne, K. S., Wakano, M., Wheeler, J. A. 1965, Gravitation Theory and Gravitational Collapse (Univ. Chicago Press)
Itoh, N. 1970 (Preprint)
Jones, P. B. 1970, Ap. Lett., 5, 33
Landau, L. 1932, Phys. Z. Sowjetunion, 1, 285
Landstreet, J. D. 1967, Phys. Rev., 153, 1372
Langer, W. D., Cameron, A. G. W. 1969, Ap. Space Sci., 5, 213
Langer, W. D., Rosen L. C., Cohen, J. M., Cameron, A. G. W. 1969, Ap. Space Sci., 5, 259
Langer, W. D., Rosen, L. C. 1970, Ap. Space Sci., 6, 217
Lee, H. J., Canuto, V., Chiu, H. Y., Chiuderi, C. 1969, Phys. Rev. Lett., 23, 390
Levinger, J. S., Simmons, L. M. 1961, Phys. Rev., 124, 916
Libby, L. M., Thomas, F. J. 1969a, Phys. Lett., 30B, 88
Libby, L. M., Thomas, F. J. 1969b, Phys. Lett., 30B, 400

Maran, S. P., Cameron, A. G. W. 1968, *Phys. Today*, **21**, No. 8, 41
Maran, S. P., Cameron, A. G. W. 1969, *Earth Extrater. Sci.*, **1**, 3
Migdal, A. B. 1959, *Nucl. Phys.*, **13**, 655
Mock, M. 1967 (Thesis, Columbia Univ.)
Morton, D. C. 1964, *Ap. J.*, **140**, 460
Nemeth, J., Sprung, D. W. L. 1968, *Phys. Rev.*, **176**, 1496
Occhionero, F. 1968, *Mem. Soc. Astron. Ital.*, **39**, 351
Onyejuba, P. E., Gaustad, J. E. 1967, *Ap. J.*, **147**, 806
Oppenheimer, J. R., Volkoff, G. M. 1939, *Phys. Rev.*, **55**, 374
Orszag, S. A. 1965, *Ap. J.*, **142**, 473
Pacini, F. 1965, *Nuovo Cimento*, **37**, 767
Pearson, J. M., Saunier, G. 1970, *Phys. Rev. Lett.*, **24**, 325
Rice, M. J. 1969, *Phys. Lett.*, **29A**, 637
Rosen, L. C. 1968, *Ap. Space Sci.*, **1**, 372
Rosen, L. C. 1969, *Ap. Space Sci.*, **5**, 150
Ruderman, M. 1968, *Nature*, **218**, 1128
Ruderman, M. 1969a, *N. Y. Univ. Phys. Dept. Tech. Rept. No. 6/69*
Ruderman, M. 1969b, *Nature*, **223**, 597
Ruderman, M. 1970, *Nature*, **225**, 619
Salmona, A. 1967, *Phys. Rev.*, **154**, 1218
Salpeter, E. E. 1961, *Ap. J.*, **134**, 669
Silverstein, S. D. 1969, *Phys. Rev. Lett.*, **23**, 139
Sutherland, P., Baym, G., Pethick, C., Pines, D. 1970, *Nature*, **225**, 353
Thorne, K. S. 1967, in *High Energy Astrophysics*, **3** (Dewitt, C., Schatzman, E., Veron, P., Eds., New York: Gordon & Breach)
Thorne, K. S. 1969a, *Ap. J.*, **158**, 1
Thorne, K. S. 1969b, *Ap. J.*, **158**, 997
Tsuruta, S., Cameron, A. G. W. 1965, *Can. J. Phys.*, **43**, 2056
Tsuruta, S., Cameron, A. G. W. 1966a, *Can. J. Phys.*, **44**, 1863
Tsuruta, S., Cameron, A. G. W. 1966b, *Nature*, **211**, 356
Tsuruta, S., Cameron, A. G. W. 1966c, *Can. J. Phys.*, **44**, 1895
Van Horn, H. M. 1968, *Ap. J.*, **151**, 227
Weiss, R. A., Cameron, A. G. W. 1969a, *Can. J. Phys.*, **47**, 2171
Weiss, R. A., Cameron, A. G. W. 1969b, *Can. J. Phys.*, **47**, 2211
Wheeler, J. A. 1966, *Ann. Rev. Astron. Ap.*, **4**, 393
Wolf, R. A. 1966, *Ap. J.*, **145**, 834

Copyright 1970. All rights reserved

ASTRONOMICAL FOURIER SPECTROSCOPY

PIERRE CONNES

Laboratoire Aimé Cotton, CNRS, Campus d'Orsay, 91, France

INTRODUCTION

The literature on astronomical Fourier spectroscopy is undergoing the exponential increase characteristic of fashionable subjects. Of even greater comfort to the happy few who believed in the technique from its very infancy is the publication of papers dealing purely with scientific results extracted from Fourier spectra. Thus the time for reviewing the subject seems ripe; hopefully a few years from now it may be impossible to cover entirely.

We shall not attempt to give here a general treatment of Fourier spectroscopy since the interested reader will find the subject treated by other papers. Jacquinot (5) mainly points out the relationship between Fourier interferometric spectroscopy and other interferometric methods. Sakai & Vanasse (8) give a fuller theoretical treatment, covering such points as apodization, sampling, phase corrections, signal-to-noise ratio and computation. Loewenstein (7) sketches the history of the technique. Mertz in his book *Transformation in Optics* (16) devotes many pages to Fourier spectroscopy, in particular from the astronomer's point of view. Fellgett (2) also gives a history of early development and reviews the work presented at the 1966 Instrumental Spectroscopy Conference (13), some of it of astronomical interest. He most clearly points out the different factors that have made what should be properly termed Fourier interferometric multiplex spectroscopy a success.

Two other papers dealing with related subjects can also be consulted. Johnson (9) described the 1966 status of infrared astronomical measurements; however, spectrometry was not discussed and it is indeed worth noticing that in those not so distant pre-Fourier days most of the results were the fruits of broadband photometry. Vaughan (10) describes Fabry-Perot spectroscopy as used on astronomical sources. Most of the applications are in the visible and ultraviolet ranges.

The viewpoint adopted in this paper is one of a physicist who applied to astronomical problems a technique developed first for laboratory use. The aim is to answer the questions Why, When, and How use Fourier spectroscopy? The emphasis will be put on the techniques or experiments that have produced the greatest improvements compared to classical methods; already collected or potential scientific results will not be discussed (the author being incompetent to do so) but the references will be given.

ELEMENTS OF FOURIER SPECTROSCOPY

Let us remind the reader that to produce a Fourier spectrum one goes through two distinct steps:

1. An interferogram is recorded by feeding the light through a two-beam interferometer onto a suitable receiver and varying the path difference Δ; if the spectrum is described by $B(\sigma)$, where $\sigma = 1/\lambda$ designates wavenumbers, the interferogram is

$$I(\Delta) = \int_0^\infty B(\sigma) \cos 2\pi\sigma\Delta d\sigma$$

2. The spectrum is then computed by performing the operation

$$B(\sigma) \propto \int_0^\infty I(\Delta) \cos 2\pi\sigma\Delta d\Delta$$

which we shall write in abridged form as

$$B(\sigma) = T_{\cos}[I(\Delta)]$$

By contrast with the original fringe visibility method used by Michelson, *any* kind of spectrum (i.e. narrow or wideband, emission or absorption) can be reconstructed in that way.

In practice Δ is always varied up to some maximum value $\Delta_{\max}$ and the actually reconstructed spectrum is

$$B'(\sigma) = \int_0^\infty R(\Delta) \cdot I(\Delta) \cos 2\pi\sigma\Delta d\Delta$$

where $R(\Delta)$ is a rectangular function of width $\Delta_{\max}$. By virtue of the convolution theorem

$$B'(\sigma) = T_{\cos}[R(\Delta) \cdot I(\Delta)] = T_{\cos}[R(\Delta)] * T_{\cos}[I(\Delta)] = F_R(\sigma) * B(\sigma)$$

where $F_R(\sigma)$, the Fourier transform of $R(\Delta)$, is the usual sinc function.

If the spectrum $B(\sigma)$ contains just one monochromatic line of wavenumber σ_0 and negligible width, the computed spectrum is given by sinc $\pi(\sigma-\sigma_0)\Delta_{\max}$; this is the instrumental lineshape (ILS). Since the intense secondary maxima are very inconvenient, an apodization has to be performed during the computation. This can be done in two mathematically equivalent ways (6): Before performing the transform one multiples the interferogram by a weighting function $P(\Delta)$; the result is

$$B''(\sigma) = T_{\cos}[P(\Delta)] * T_{\cos}[I(\Delta)] = F_P(\sigma) * B(\sigma)$$

Alternately, one computes the spectrum first and then performs a numerical convolution with $F_p(\sigma)$ which is the apodized ILS; this has the advantage that several different ILS can be used doing just one transform (44).

In this manner the secondary maxima can be reduced to negligible importance. Different $F_p(\sigma)$ functions can be used; in all practical cases their width $\delta\sigma$ at half-intensity of the central peak differs little from the reciprocal of the maximum path difference; the fundamental result is thus:

$$\delta\sigma \simeq 1/\Delta_{\max}$$

So far the spectral range appears unlimited, except by technical considerations such as beamsplitter transmission or receiver characteristics. However one should also at this stage consider the method of computation. In the great majority of cases the transformation is performed on a digital computer. The interferogram is sampled at N equally spaced points; the sampling interval $d = \Delta_{\max}/N$ is a function of the limits σ_1 and σ_2 of the spectral range under study. The exact rules for selecting the value of d need not be repeated here (43); however, a simplified one can be given. The minimum possible number N is equal to the number of spectral elements M. The width of a spectral element is defined as $\delta_{\sigma'} = 1/2\Delta_{\max}$ which means $\delta_\sigma \simeq 2\delta_{\sigma'}$ with the usual apodizations; thus $N = M = \Delta_\sigma/\delta_{\sigma'}$, where $\Delta_\sigma = \sigma_2 - \sigma_1$ is the spectral range. According to the sampling theorem the continuous spectral curve can be fully reconstructed from just M samples. In practice some oversampling is always necessary and one has $N \geq M$.

The two fundamental limitations are thus: 1. the maximum interferometer carriage displacement which determines $\Delta_{\max}$ and the instrumental width $\delta\sigma$; 2. the maximum number N of points that can be recorded and transformed; it fixes the number of spectral elements that can be studied at the same time.

PRINCIPAL ADVANTAGES OF FOURIER SPECTROSCOPY

We shall presently list the principal reasons that have made Fourier spectroscopy a success. Some are fundamental, others more technical—which does not mean less useful in practice. The importance of these various factors for astronomical purposes will be discussed separately.

Multiplexing Ability

When recording an interferogram no dispersion of the spectrum occurs and no loss comparable to the one due to the *exit* slit of a spectrometer takes place. The entire spectral range is recorded in coded form within the interferogram; performing the Fourier transform can be looked at as a kind of decoding. The system is equivalent to a spectrometer with a multidetector array; however, only one is actually used, and it can be said to be multiplexed. The multiplexing factor is simply $M = \Delta\sigma/\delta\sigma'$.

Fellgett, who originally pointed out the importance of multiplexing in the infrared (1), has also shown that systems that are neither interferometric nor based on Fourier transformations can indeed multiplex a spectrum (2) (inversely it is entirely possible to use a Fourier interferometer without taking advantage of the multiplexing ability). None of these systems has actually been used in astronomy and none appears to hold promise comparable to the Fourier interferometer to which we will restrict this discussion.

The signal-to-noise ratio gain arising from the multiplexing ability can vary widely:

1. In the visible and very near infrared, or more precisely where photoemitters can be used (roughly $\lambda < 1.1\ \mu$), little or no advantage can be expected. In nearly all practical cases the photocurrent will be large compared to the dark current of a properly cooled photomultiplier (a few electrons/sec) and the resultant increase in noise due to the photocurrent itself cancels the multiplexing advantage. This case has been treated by Kahn (32); he finds that if we compare a multiplexing and a scanning system all factors (in particular input energy) being equal, the multiplexer will be better only for those parts of the spectrum where the intensity is more than twice the average. For continuous spectra with weak absorptions—by far the most common case in astronomy—the multiplexer is slightly inferior to the scanner.[1]

However, this discussion does *not* prove that Fourier spectroscopy is useless in the visible region because of the other advantages to be considered later.

2. In the infrared, or to be more exact where photoconductors and thermal receivers are the only choice, the situation is very different. The lowest NEP actually achieved are $\sim 10^{-14}$ W for PbS cells (1 to 3.5 μ range). This means the photon flux producing a *signal* equal to the NEP would be 10^5 photons/sec at 2 μ, and the flux producing a *noise* equal to the NEP, about 10^{10} photons/sec.[2] Thus noise will be increased—and the multiplexing advantage reduced—only by very intense signals.

The infrared multiplex advantage, while at present the most important single one of Fourier spectroscopy, may of course not last forever. We can dream of the day when a near 0°K spectrometer equipped with an ideal infrared photon counter will be flown outside the atmosphere.[3] Then obviously no basic difference will remain between the visible and infrared ranges. It is a safe prediction however that because of the other factors to be discussed presently the Fourier interferometer will remain the best tool.

Luminosity, or "Throughput" Advantage

This is a purely geometrical factor, characteristic of interferometers that

[1] Furthermore we should not forget that precisely within that same spectral range photographic plates and image tubes are available; these can be looked at as very simple multidetector arrays.

[2] This is about the order of magnitude of the photon flux available from α Orionis (the brightest near-infrared object apart from the Sun, the Moon, and Venus) through a 1-m² telescope within the 2 to 2.5 μ window.

[3] For a beginning a small cooled telescope (without a spectrometer) has actually been flown in a rocket (86).

make use of beam division by semitransparent surfaces; it has been indicated by Jacquinot (3,4). This factor is independent of spectral range; it is the same for the Fabry-Perot etalon, and for the classical form of Michelson interferometer. In either case the solid angle of the light beam which can be sent through the interferometer is linked to the resolving power R by the simple relation $\Omega = 2\pi/R$.

By contrast, with a grating spectrometer (Littrow mounting, incidence $i \simeq 60°$) this same solid angle is $\Omega' = \beta/2R$ where β is the angular length of the slit. One can then define a solid-angle gain $G = \pi/\beta$ of the interferometer versus the slit spectrometer. If the source is sufficiently extended in one direction at least, β is limited only by aberrations within the spectrometer; it can hardly exceed 10^{-1} rad, and the corresponding gain is $G \simeq 30$. However (for high-resolution instruments anyway), β is usually smaller; also if the source is more or less circular it is considerably easier to match to the interferometer than to the grating spectrometer (which requires image slicers), and the gain can be larger. Furthermore field-compensated interferometers of various types (12) have been described for which Ω is greatly increased. They have not so far been used in astronomy but hold definite promise for extended objects[4] (85).

Resolving Power Advantage

In a grating spectrometer the wavenumber resolution is simply the reciprocal of the maximum path difference between interfering rays; in a Littrow mounting, under grazing incidence $\delta\sigma = 1/2L$ where L is the ruled length of the grating and $R = 2L/\lambda$. The present maximum is $L = 25$ cm, and it is hoped that gratings with $L = 60$ cm will be available, but this is obviously a major undertaking. With Fabry-Perot or Michelson-type interferometers the resolving power is not limited by the size of any optical component and Δ_{max} can be much greater.

In a Fabry-Perot etalon this is achieved by increasing the plate separation ℓ; the wavenumber resolution can be written $\delta\sigma = 1/2N\ell$ where N is the "finesse" coefficient and $R = 2N\ell/\lambda$. However when ℓ becomes very large the free spectral range $\Delta\sigma = 1/2\ell$ becomes too small for useful purposes. The real limitation lies in the factor N, which depends not only on plate reflectivity, as required by elementary theory, but also on surface imperfections. In practice, in the visible or near-infrared range N_{max} is ~ 50, and no technical breakthrough promising order-of-magnitude improvement is in view. Since $N = \Delta\sigma/\delta\sigma$ the finesse is simply the number of spectral elements that can be studied from a single Fabry-Perot record; and if the spectrum is complex, an auxiliary system is required for isolating the spectral range $\Delta\sigma$. This can be a grating monochromator or a combination of Fabry-Perot etalons.

[4] On the other hand one should not forget the existence of the grid and SISAM spectrometers (13). Both are scanning instruments that make use of gratings, and their light-gathering power is comparable to the one of the interferometer. Of course they do not have the multiplex advantage.

In a Fourier interferometer the spectral range is not limited by the accuracy of optical surfaces. The number of interferogram samples N plays exactly the same role as the finesse coefficient of the Fabry-Perot. The largest number of samples that have actually been transformed up to now is 10^6 (51) and this is by no means a limit. Thus it is feasible in Fourier spectroscopy to increase the path difference without unduly restricting the spectral range. The greatest Δ_{max} reached so far is 170 cm and the corresponding instrumental width of 6.10^{-3} cm^{-1} has been demonstrated in emission and absorption laboratory spectra (48); again there is no reason why these figures should not be exceeded in the future.

Accuracy of Wavenumber Measurements

The regular practice for accurate wavenumber measurements with a scanning spectrometer is to interpolate between secondary standards. Precision depends on the linearity of the scanning device and on the accuracy of the standard lines themselves.

In Fourier spectroscopy the spectrum is not scanned and all lines in the selected spectral range are simultaneously recorded within the interferogram. Then the Fourier inversion—if performed on a digital computer anyway—does not contribute any further errors. Only one standard line is needed (to control the interferometer motion) and the accuracy with which that line is known is directly transferred to the entire Fourier-computed spectrum. Thus it is not surprising that the best published results by Fourier spectroscopy ($1.2 \; 10^{-4}$ cm^{-1} rms error for a near-infrared N_2O band) greatly exceed the best corresponding grating-spectrometer performance (2.10^{-3} cm^{-1}) (48).

Size and Weight Reductions

While this advantage is not fully independent of luminosity and resolving power, it is convenient to discuss it under a separate heading. Obviously no exact rules can be given concerning the relative size and weight of a Fourier interferometer or scanning spectrometer suitable to tackle a given problem. However, it will be found in practice that the interferometer is far smaller and lighter and this can be easily explained:

Diameter of optical elements.—In many instances the luminosity advantage cannot be fully used (when the $S \times \Omega$ product of the light beam is limited by detector, source, or auxiliary optics considerations). But if this is the case the diameter of the interferometer elements (mirrors and beamsplitter) needed to accommodate that beam will be much smaller than the diameter of the grating-spectrometer elements.

Focal lengths.—The focal lengths of the lenses or mirrors associated with a given grating are normally many times as long as the grating. In a Fourier interferometer, since no sharp slits are needed, aberrations outside the interferometer matter considerably less. The focal length of the collimator for instance can be very short; sometimes it could be dispensed with (11).

Consequences for Astronomy

The various advantages just discussed are not of equal importance for astronomical use or for different astronomical problems. Taking them in reverse order: the size and weight reductions are not essential for ground-based observations—unless one has only the poor choice of working from a Cassegrain focus. However, aircraft, balloon, and satellite astronomy are fast-growing fields and a large fraction of the work we shall discuss later falls in that category.

Extreme resolving power and very high accuracy of line positions are needed only for sharp lines; and in any case only the Sun, the major planets, and a moderate number of bright stars can provide sufficient energy. While these advantages have only been used so far in the infrared (52 to 63), one should not forget they are also potentially important in the visible or ultraviolet ranges. Even the largest Coudé spectrographs cannot provide the quality and quantity of data which are available in the latest laboratory Fourier spectra (51). We can also expect solar atlases recorded by Fourier spectroscopy.

The *luminosity* advantage is more amenable to quantitative discussion. Figure 1 (from 49) shows one way of presenting the result; supposing a given telescope diameter, one plots as a function of angular source diameter θ the resolving power R at which the spectrometer throughput just matches the beam. For all points falling between the two lines the interferometer accepts the total available flux while the grating spectrometer does not.

The practical gain to be realized from the multiplex advantage can vary widely even in the infrared. The main consideration is of course the number of spectral elements M one wants to study. If it is very small, little can be gained. This may happen irrespective of resolving power. For instance, broadband photometry can be looked at as ultralow-resolution spectrometry. The use of a multiplexing device would provide only a small improvement in S/N over a set of filters used in succession (moreover it would be difficult to cover an extremely wide range with just one beamsplitter and receiver).

Another extreme case is the study of the profile of just *one* line (absorption or emission). This is a high-resolution but low-M factor problem. Very little improvement can be expected compared to a scanning Fabry-Perot interferometer—a much simpler system.

When a source is just detectable, multiplexing is useless. To be more precise let us suppose an object gives an S/N ratio of unity when looked at with a given telescope, the best infrared detector in the range of interest, and the largest practical integration time—say a few hours. Clearly we cannot hope by any technique to divide the spectral range in separate elements, and still get a workable S/N ratio. From this admittedly extreme

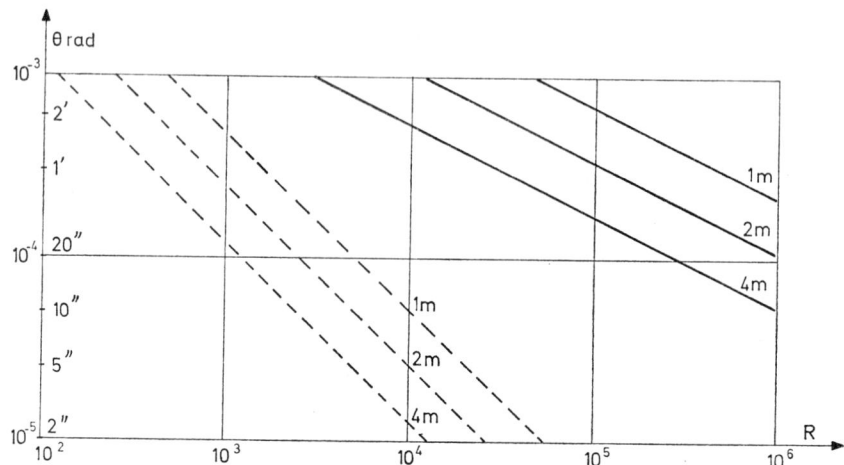

FIGURE 1. (Adapted from 49.) Angular diameter θ of source filling the aperture of a spectrometer through a telescope of given diameter (indicated in meters) as a function of resolving power. Broken lines: grating spectrometer, width 25 cm, Littrow mount, $i=64°$. Solid lines: Interferometer, beam diameter 80 mm (as described in 51).

case one understands that the multiplexing gain will be large only for relatively bright sources. It can be expressed in a simple manner. Let us suppose that in a given recording time T a scanning spectrometer able to accommodate the entire beam available from the telescope has covered the spectral range $\Delta\sigma$ with resolution $\delta\sigma$ and a given S/N ratio. Then the multiplexing system will produce the *same* result in time T/M, where again $M=\Delta\sigma/\delta\sigma$. Alternately the advantage can be used to improve the S/N ratio by a factor $\sqrt{M}$ (all other factors being kept constant) or to improve resolution, or to increase spectral range.[4]

To conclude, we shall examine the one published case in which Fourier spectroscopy has provided the greatest gain over classical techniques, and show how much of it can be attributed to multiplexing and how much to increased throughput.

The best near-infrared spectrum of Venus given by a grating spectrometer has been recorded by Kuiper (75) (Figure 3) and we shall compare it with the Fourier spectrum of Connes et al (46,52). The spectral range is 1.1 to 2.5 μ in both cases. Energy available at the spectrometer input was slightly greater (by a factor of 1.5) in the grating case, because of larger telescope size (210 versus 190 cm) and more favorable observing period (visual magnitude -3.8 versus -3.4). However, recording time was longer in the second case (9 hr[5] versus approximately 3 hr). The best available CO_2-cooled PbS receivers (from the same manufacturer) were used in both cases. Resolution varies from 6 to 8 cm^{-1} for the grating spectrometer; the interferometer gave a constant 0.08 cm^{-1}—about 100 times less. Signal-to-noise ratio is somewhat difficult to compare when the resolution is so different; however it appears to be ~100/1 in both cases.

We can then ask the following question: what increase in recording time could have produced an improvement of 10^2 in resolving power with the scanning spectrometer? A resolution of 0.08 cm^{-1} can easily be achieved with a medium-size grating: it is simply a matter of closing down the slits. However since the entrance slit width had been chosen to accommodate the entire planetary image, reducing it by 100 would have reduced by the same factor the energy entering the spectrometer. A similar reduction of the output slit would have been needed, giving an overall reduction of 10^4 in the energy falling on the detector. Since the detector was already the smallest available no reduction in size was possible and, noise being constant, S/N would have fallen by the same factor 10^4. To bring it back to the previous level an increase in recording time of 10^8 per spectral element would have been required. And since the same spectral range now contains 10^2 times more spectral elements the increase in overall scanning time would have been 10^{10}.

An extended throughput scanning instrument (SISAM, grid spectrometer or PEPSIOS-type combination of Fabry-Perot etalons), if applied to the same problem, would eliminate the loss due to the entrance slit. If we suppose the overall transmission to be the same as for the slit spectrometer (which is clearly optimistic in the Fabry-Perot case at least), an improvement of 10^2 in energy would be realized. Still the increase in recording time over the low-resolution case would be a not so modest 10^6.

Thus we conclude that—when applied to suitable problems—the combined luminosity and multiplexing advantages are truly enormous and open new fields to infrared spectroscopy.

PRINCIPAL DIFFICULTIES OF FOURIER SPECTROSCOPY

We cannot of course get all these improvements for nothing; many technical refinements were needed before the simple principles described above actually led to the expected gains. The basic accuracy requirements will be discussed first and the specifically astronomical problems later.

Basic Accuracy Requirements

When recording the interferogram $I(\Delta)$ one can introduce errors in intensity and errors in path difference.

Path difference errors.—These will produce distortions of the instru-

---

[5] This recording time of 9 hr arises from incomplete use of the multiplex principle. Because of limitations in recording speed and computing facilities, three separate recordings were needed to study the three windows in the range. With the latest version of the interferometer (51) only one recording would be necessary and the same S/N ratio could be obtained in 3 hr.

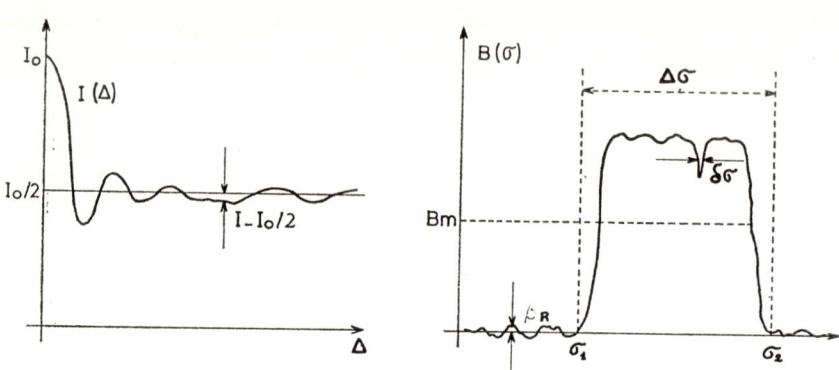

FIGURE 2. Broadband spectrum $B(\sigma)$ and corresponding interferogram $I(\Delta)$. B_m is the mean spectral intensity across range $\Delta\sigma$ and β_R the noise level.

mental lineshape (43), and will be acceptable only if their amplitude is a very small fraction of the smallest wavelength. Since the difficulty in measuring Δ with a given accuracy increases with Δ_{max}, it is greatest when a high resolving power $R = \Delta_{max}/\lambda$ is sought.

This problem is now solved. With an interferometrically controlled stepping drive (44), extreme accuracy in path difference measurement—and thus in instrumental lineshape definition—has been demonstrated.[6] However, this is needed only for large resolving powers, which means mostly in the near infrared. At longer wavelengths and/or moderate R a continuous interferometer carriage motion appears satisfactory, provided the sampling is triggered by a reference fringe system.

Intensity errors.—An interferogram from a broadband spectrum (Figure 2) generally contains a few large fringes near $\Delta = 0$ and then many small ripples of decreasing amplitude. The maximum intensity I_0 is proportional to the total energy $\int_0^\infty B(\sigma)d\sigma$ but all the information about $B(\sigma)$ is contained in the *difference* $I - I_0/2$, which is for the greater part of the interferogram a small fraction of the *mean* intensity $I_0/2$. This is a fundamental difficulty in Fourier spectroscopy and one easy to visualize. An interferogram is the sum of M elementary sine waves, each contributed by *one* spectral element; they all add in phase for $\Delta = 0$, then get out of phase and add randomly when Δ increases. Their amplitude relative to the main peak is inversely proportional to M. Thus the accuracy required is greatest for a

[6] The other advantages of stepping are often misunderstood. Stepping provides optimum signal-to-noise ratio with the *minimum* number of samples N; however, continuous motion can give the same s/n provided oversampling is used (43) and this is easy if N is not too large. Stepping permits stopping at will, varying the integration time (44), and producing computer-compatible magnetic tape without buffers even for very large N (51).

large M; it is precisely when the multiplexing property promises a large gain that much care is needed to get it in practice—a rather natural conclusion.

Systematic intensity errors, coming for instance from nonlinearities in the recording system, introduce zero-level distortions in the spectrum, together with harmonics and cross-modulation terms (i.e. false lines) in emission spectra. Elementary precautions, such as the use of gain changes (46), give enough dynamic range to accommodate even the highest M cases (51).

More or less random intensity fluctuations, due to source intensity, receiver sensitivity, or recording system gain variations, *add* noise to the interferogram and to the spectrum. For instance let us suppose the total intensity $I_0 = \int_0^\infty B(\sigma)d\sigma$ fluctuates according to $I_0 = \bar{I}_0[1+\epsilon(t)]$ where $\epsilon(t)$ is a random function of time, with $\bar{\epsilon} = 0$. Let $\epsilon_R \ll 1$ be the rms value of $\epsilon(t)$. If the interferogram is scanned according to $\Delta = Vt$ the perturbation can be considered as a function of Δ. For the greater part of the interferogram, where $I \simeq I_0/2$, to a good approximation, a supplementary noise $\alpha = I_0/2 \cdot \epsilon(\Delta/\gamma)$ has simply been added to the interferogram with $\alpha_R = \epsilon_R I_0/2$.

The effect on the spectrum can then be understood by using results derived for receiver noise (43). A supplementary "source noise" $\beta(\sigma)$ will be added to the computed spectrum, and the rms value β_R will be proportional to α_R and to I_0. This noise will be best measured—like receiver noise—outside the occupied spectral range $\Delta\sigma = \sigma_2 - \sigma_1$. However if $\epsilon(t)$ is not white the source noise will not be spread uniformly along the spectrum.

It will be convenient to define a quality factor Q for a Fourier-produced spectrum. This will be simply the ratio between the total energy I_0 falling on the receiver for $\Delta = 0$, and the energy of a single just detectable line, i.e. one whose intensity is equal to the rms noise β_R. If B_M is the mean spectral intensity across $\Delta\sigma$ we have:

$$Q = M \frac{B_M}{\beta_R}$$

If source noise is nonexistent, β_R is due to receiver noise alone and Q reaches the maximum possible value. The factor Q indicates the perfection with which multiplexing has been accomplished. It depends not on the source alone but very much on the recording technique and on the degree of care used in avoiding or compensating for intensity fluctuations. Actual figures for Q will be quoted later.

DIFFICULTIES ASSOCIATED WITH ATMOSPHERIC TRANSMISSION

Nearly achromatic fluctuations, such as those caused by haze or thin cirrus clouds, are easily canceled either by a ratio recording system (67) or by varying the integration time (44,46). A more delicate problem is the one of chromatic variations, for instance coming from fluctuations in the amount

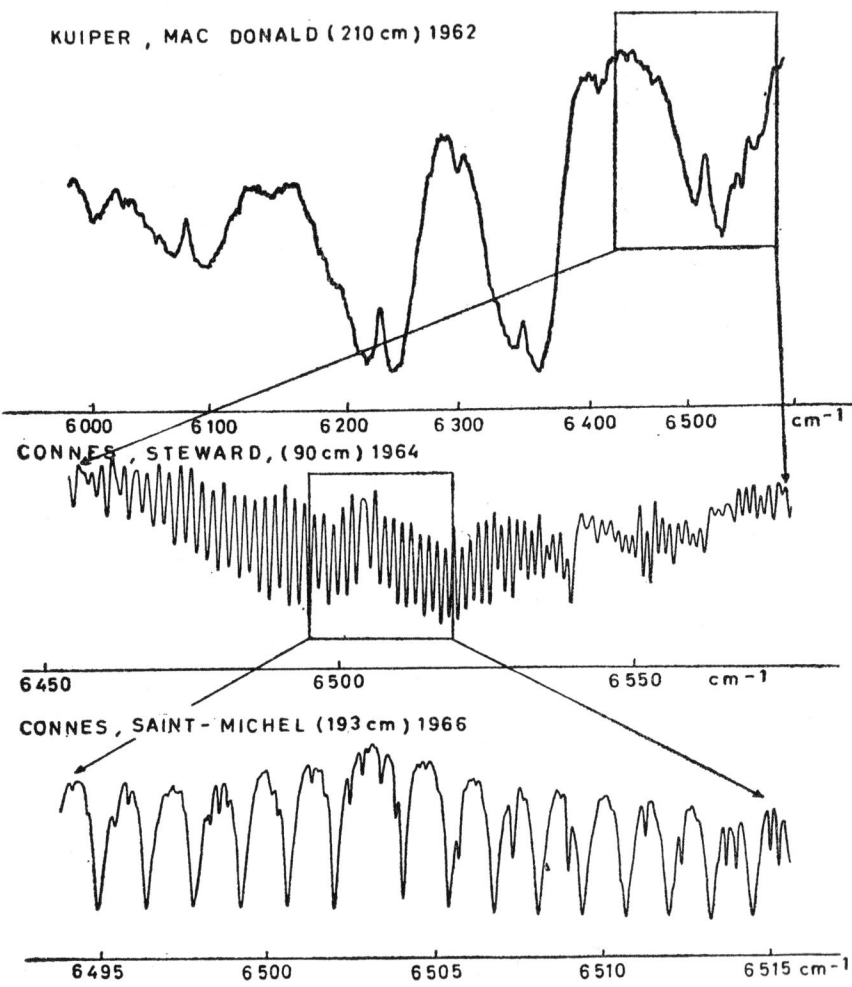

FIGURE 3. (From 50.) *Upper curve:* Best spectrum of Venus recorded with a grating spectrometer (75, Figure 11b) in the 1.6-μ window. Four strong CO_2 bands are visible. *Center curve:* First results by Fourier spectroscopy (44) (without internal modulation; $\delta\sigma = 0.7$ cm^{-1}). *Lower curve:* Latest Fourier results (46,52) (with internal modulation; $\delta\sigma = 0.08$ cm^{-1}). The smallest visible features are real and have been interpreted as CO_2 lines. The complete spectrum is given by (52).

of precipitable H_2O in the path. A simplified theoretical treatment of this type of perturbation has been given (49) but we have something better—experimental proof. The planetary atlas of Connes et al (52) systematically presents two superimposed Venus spectra; one is recorded near the meridian through a minimal and nearly constant air mass and the second at low elevation through a much larger and rapidly varying one. The results show *telluric* line profiles to be distorted; *planetary* lines are not affected. Admittedly if the resolving power were much lower it would be difficult to separate the two; but the culprit then is low resolution, not Fourier spectroscopy.

DIFFICULTIES ASSOCIATED WITH ATMOSPHERIC TURBULENCE

Two different cases should be considered: near and intermediate infrared.

1. *In the near infrared* (up to 2.5 μ), thermal sky emission is negligible but turbulence modulates the optical beam, thus producing an additional noise of the type discussed above. A few years ago this appeared to be of considerable importance (39–41) and the somewhat hasty conclusion was sometimes reached that multiplexing was nearly useless even in the infrared because this noise would cancel out the advantage just as photon noise does in the visible; this turned out to be wrong on two counts. First, the magnitude of the phenomenon cannot be predicted from any simple turbulence model, it has to be found from experiment. Second, not enough attention had been paid to proper recording techniques.

The first technique is the use of two detectors that receive the two interferometer outputs, as done by Fellgett (1)—now common practice. The difference between the two signals is taken; in principle the mean interferogram level should be zero instead of $I_0/2$ and modulation noise should be no longer $\alpha_R = \epsilon_R I_0/2$ but $\alpha_R' = \epsilon_R(I - I_0/2)$—that is to say much smaller.

The limitations have been discussed elsewhere (44); perfect cancellation is impossible because the problem is not one of pure intensity modulation. Turbulence modulates the beamshape and position as well and it is very difficult to get well-correlated outputs from two separate receivers. Once optimum balancing has been achieved, a residual noise from the uncorrelated fraction of the two outputs is observed. Contrary to simple expectations, this noise does not diminish markedly with respect to the signal when using a larger telescope, and it is not less for a planet than for a star.

For the same reason ratio recording does not give the expected improvement either; the total intensity I_0, whether measured from the sum of the two signals or from an independent receiver, will again exhibit fluctuations poorly correlated with those of the interferogram.

However, the difficulty should not be exaggerated. The first spectrum of Mars published by Connes (44) (now obsolete—see 52) was recorded with two balanced receivers, an external chopper and a rather low scanning speed of 1.5 sample/sec; it gave $M = 10^3$ from $N = 12,000$ $B_M/\beta_R = 300$ and $Q = 3.10^5$. The reproducibility between two separate spectra is possibly as good as anything achieved by a scanning spectrometer on any source in any spectral range. Clearly, for a problem where a lesser Q is expected because of receiver noise, a more complex recording technique is not needed.

Two different methods, both originated by Mertz, give further improvements. The first is fast scanning (14 to 21). No chopper is used, and path difference is varied according to $\Delta = Vt$. The detector output contains all frequencies between $F_1 = V\sigma_1$ and $F_2 = V\sigma_2$. It is amplified by a broadband ac amplifier and recorded. The speed V is selected to get frequencies as high as the infrared detector will allow in order to separate them from the $1/F$ frequency spectrum of turbulence. Since V is thus determined one is no longer free to choose the total recording time, which is $T = \Delta_{\max}/V = R_2/F_2$. If $R_2 = 10^3$ and $F_2 = 100$ Hz one gets $T = 10$ sec. In all practical cases T is a small fraction of the time (a few hours) normally available for ground-based observations, which has to be used in full to get a workable S/N ratio. One must then add a large number K of identical fast scanned interferograms; the final S/N is the same as if *one* interferogram of duration KT had been recorded [coherent adding (17,21)]. Of course high instrumental stability is needed for long-duration adding.

Up to now the method has been used only for low M and R problems with moderate integration times. The highest reported figures are $KT = 30$ min and $R = 10^3$, corresponding to $\delta\sigma = 8$ cm^{-1} at 8000 cm^{-1}, in the stellar and planetary spectra given by the Lunar and Planetary Laboratory group (77, 79); from $\sigma_1 = 3700$ cm^{-1} and $\sigma_2 = 8500$ cm^{-1} one finds $M = 600$. A crude estimate for the highest Q gives 5.10^4 (for a ground-based Venus spectrum (77, Figure 14). For similar problems the use of fast scanning is not justified from the sole turbulence point of view. However, the technique has the appreciable advantage of maximum optical simplicity because the chopper is eliminated. Signal averagers (with $N \leq 2048$) are common and not too expensive.

Fast scanning could in principle be extended to high R with heterodyning (16), and to high M with disk stores plus buffer memories. But the second technique, internal modulation (14,46), is then clearly preferable. During scanning, the path difference is oscillated at a suitable frequency F with amplitude $a = \lambda'/4$, where $\lambda' = 1/\sigma'$ and $\sigma' = (\sigma_2 - \sigma_1)/2$. The amplitude of the first harmonic (of frequency F) in the interferometer output is recorded. To a first approximation the interferogram is the derivative $dI/d\Delta$ of the normal interferogram $I(\Delta)$. In principle the spectral range is limited. However one can still get better than 70% efficiency in a 3 to 1 spectral range ($\sigma_2 = 3\sigma_1$); this is more than receivers or beamsplitters will allow in many cases. The spectrum is recovered by a sine transform.

No restriction exists as to recording time and the method is compatible with the stepping drive. It has been used on the brightest near-infrared object (Venus, 46) under the worst possible turbulence conditions (daytime and low elevation); this is a maximum-difficulty case. Results can be summarized by $M = 19,000$, $B_M/\beta_R = 160$, and $Q = 3.10^6$. It has also been used in the laboratory (since astronomers have no monopoly on troublesome sources) for the highest performances so far [$R = 1.110^6$ (48), $M = 6.10^5$, $N = 10^6$, $T = 10$ hr, $Q = 2.510^7$ (51)].

2. *In the intermediate infrared*, i.e. in the 8 to 12 μ and longer wavelengths windows, and to a lesser extent in the 3 to 5 μ ones, fluctuating thermal emission from the atmosphere adds to the interferogram a noiselike signal independent of the source under study.

While this sky noise is being investigated by several groups, no data have been published on which to base a discussion since it was first mentioned (33). However the best technique for cancellation is already clear: two interferometer inputs are always available and they should look at two sky patches as small and as close together as possible; one of them of course will contain the object (44). The additional noise on the interferogram will depend not on the absolute level of the fluctuating power, but on the *difference* between the two signals. Thus the main question is: what is the angular

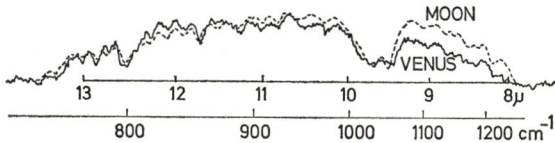

FIGURE 4. Venus and lunar spectra from a grating spectrometer (Sinton & Strong 34, Figure 15). The linear λ scale given by the authors has been inverted and a σ scale added for ease of comparison with Figure 5.

correlation of sky noise? Before it is answered no quantitative prediction of the effect on a Fourier spectrum can be made. Macroscopic fluctuations in thermal emission from the telescope should cancel in the same way. However, the actual degree of cancellation will again only be determined from experiment.

Meanwhile we can take some comfort from published results. The fluctuating nature of the signal did not prevent Hanel et al (67) from producing a Fourier sky emission spectrum in the 8 to 12 μ window with $\delta\sigma = 0.6$ cm^{-1}, $M = 750$, and $Q \simeq 5000$. Emission lines are clearly shown, with a flat background beginning to appear between them. The Venus spectra recorded by the same authors (66) with a 150-cm telescope show 0.6-cm^{-1} resolution (Figure 5)—a great improvement over the previous best spectrum by grating spectroscopy [Sinton & Strong, 500-cm telescope (34)] (Figure 4). However, the gain can be credited only in part to Fourier spectroscopy since the grating spectra were recorded with a thermopile and the others with a 4°K Ge-Hg detector. While a clear case cannot be presented, the situation is reversed, compared to the early days: The burden of proof now rests with the grating-spectrometer champions.

Nevertheless a basic point should—again—not be forgotten. Within the

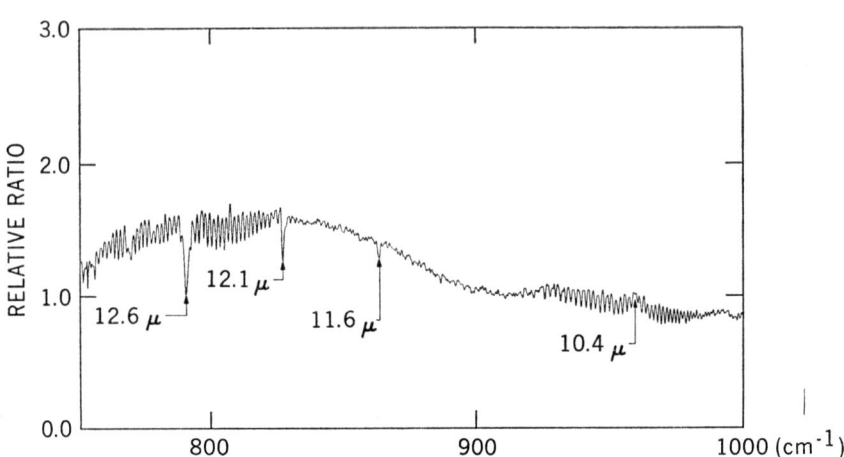

FIGURE 5. Ratio of the Venus to the lunar spectrum by Fourier spectroscopy (Hanel et al 92). Resolution is 0.67 cm^{-1} and spectral range extends from 400 to 1300 cm^{-1}, with usable transmission between 450 and 600 cm^{-1}, and between 700 and 1200 cm^{-1}. The portion reproduced here shows four identified CO_2 Venusian bands and very good cancellation of telluric lines.

mean-infrared windows we cannot expect from astronomical sources the large energy we get in the *near*-infrared windows. According to pp. 215–16, this means that—even if the sky-noise problem is licked—we cannot hope that the multiplexing advantage will provide the very large gains realized in the near infrared.

DIFFICULTIES ASSOCIATED WITH THE COMPUTATION

The problem of computing the Fourier transform is outside the scope of this paper, but something must be said about the astronomer's need for immediate computation. Since the untransformed interferogram is almost meaningless, it is easy to waste valuable observing time, unknowingly. The technique of materially carrying the paper or magnetic tapes to the nearest available large general-purpose computer must be improved upon. The optimum is real-time computation, by which the experimenter sees on an oscilloscope the spectral resolution improve while Δ increases. However, a computation performed immediately after Δ_{max} is reached is also very valuable. In both cases it is sufficient to check a small well-chosen slice of the spectrum, but the complete interferogram must be used.

In one case at least (46), the data were sent through a telephone line to a large computer and results sent back to the experimenter about an hour after the end of the recording. While largely responsible for the success of these observations, the method is somewhat cumbersome, prone to errors, and expensive for routine observing.

Several very fast—but not real-time—special-purpose digital computers in which the fast-Fourier transform algorithm is wired in are now available.[7] These highly expensive devices are limited to 1024 input samples. They might be used to transform the output of a fast scanning interferometer.

Several special-purpose real-time or nonreal-time (13,65) computers have also been built. When they contain analog elements their accuracy is not comparable to that of the digital computer. They are so far limited to low N and low R ($\sim 10^3$). An interesting solution lies in the use of a small general-purpose computer. Using one of the cheapest machines available[8] it has been found possible to present $M = 1\,900$ spectral points in real time while accepting interferogram samples at a rate of one every 2 sec. There is no limitation on N and thus on R, but if the input rate is increased, M and the spectral width must be reduced. While the problem is not completely solved at present—especially in the high-N and high-input rate case—we can expect it will soon be, by considering the curve of growth in the small computer field.

REVIEW OF EXPERIMENTAL RESULTS
Ground-Based Work

The very first attempts by Fellgett (1), Mertz (14), and Gebbie (22,23) are now of only historical interest but their instruments and techniques included many essential features of later work. Development has followed several lines. Polarization interferometers have been built by Mertz (14), Sinton (34 to 37), and Moroz (38). They are attractive because of simplicity and stability but limited to very low Δ_{max}; thus they will show improvement over classical techniques only for very faint sources (which of course means the largest number of them and quite possibly the problems with the greatest scientific impact).

All the other interferometers derive from the Michelson type. Gebbie, Delbouille & Roland (24 to 27) gave the first spectra of Venus and Jupiter and pointed out many of the basic difficulties when recording astronomical spectra.

The small fast scanning interferometer of Mertz is clearly best used off the ground, but can also be used at a Cassegrain focus; it has produced very low-resolution spectra of such faint objects as Saturn's ring, the Orion Nebula, and χ Cygni (17 to 21). The instrument is commercially available; with some improvements and using the coadding technique Johnson et al (79 to 81) have recorded higher-resolution spectra (8 cm^{-1}) of many stars with a 150-cm telescope.

The increasingly complex interferometers of Connes et al (44 to 51) were primarily developed for laboratory work and need a coudé focus for

---

[7] For instance Hewlett-Packard 5450 A; Time Data 100.

[8] A Varian 620 I. MM. Le Toullec, Bras, and Levy, Laboratoire de Physique des Solides, Faculté des Sciences de Paris, R. Milward (Soc. CODERG) (private communication).

operation. Most of the observing time has been devoted to planets and the spectra published in atlas form (52). Resolution is 2 cm^{-1} for Saturn, 0.3 cm^{-1} for Jupiter, 0.08 cm^{-1} for Mars and Venus. In this last case the line position rms error has been found equal to 1.7 10^{-3} cm^{-1} at 4300 cm^{-1}, corresponding to a 0.12 km/sec radial velocity error. The spectra and implications concerning the planetary atmospheres have been discussed in several papers (53 to 60). The Venus spectra have led to improved understanding of the CO_2 molecule (90). The spectra of a few bright stars were also recorded rather on a spare-time basis (61 to 64)[9]; resolution went down to 0.1 cm^{-1} for α Orionis. The first results have been reviewed by Spinrad (91); prospects are discussed in (93).

The already discussed Venus spectra by Hanel et al (66) in the 8 to 12 μ window were recorded by fast scanning, with later averaging of the spectra—not of the interferograms—in the computer. In the far infrared the Sun is the only source up to now, and ground-based spectra yield information almost entirely about the Earth's atmosphere. The latest Gebbie solar spectra (28) show 0.5-cm^{-1} resolution from 10 to 35 cm^{-1}.

FOURIER SPECTROSCOPY FROM AIRCRAFT, BALLOONS, OR SATELLITES

Low-resolution infrared spectra recorded from the ground are distinctly difficult to interpret because of the small number of independent pieces of information one gets within narrow and rather ill-defined windows. Mistakes have been made when trying to identify absorption bands under these conditions, and the situation improves only when the bands are resolved into lines, which means $\delta\sigma \simeq 1$ cm^{-1} or better.

Above most or all of the atmosphere the situation is very different and even the simple, moderate-resolution interferometers so far built for aircraft, balloon, or satellite operation give highly interesting results and striking improvements when compared to classical techniques.

The first balloon-borne Michelson interferometer was built by Gush (42) for recording near-infrared night-sky emission. So far the Sun is the only astronomical source of which Fourier spectra have been recorded from a balloon; this is of course only because of the small collecting optics needed and relative ease of guiding. The first far-infrared solar spectrum was obtained by Beer (72) with only 10-cm^{-1} resolution—because of guiding troubles. Gay et al (74) recorded two with 0.25-cm^{-1} resolution; their equipment had first been used on the ground (73). Stettler et al (89) prefer a lamellar grating interferometer. No far-infrared spectrum appears to have been recorded with a slit spectrometer from a balloon, thus direct comparison is not possible.

[9] Of special interest to this paper is the radial-velocity measurement of Arcturus made by Edmonds & Bopp (63) from one of these spectra; the accuracy is comparable to the best classical results. One should realize that (a) before Fourier spectroscopy no infrared *line* had been even detected from any source except the Sun beyond the photographic range and (b) signal-to-noise ratio considerations would be in favor of an infrared radial-velocity measurement only for a much cooler star.

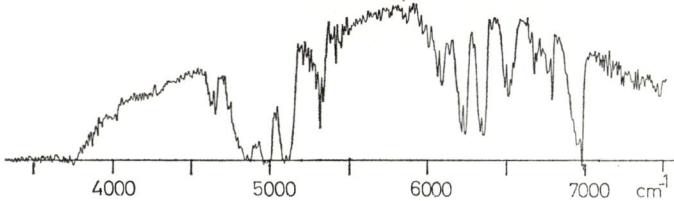

FIGURE 6. Near-infrared Venus spectrum recorded from an aircraft with a Fourier spectrometer by Kuiper et al (77, Figure 3). Resolution $\delta\sigma = 8$ cm^{-1}, comparable to Figure 3, upper curve. Authors also give a lunar reference spectrum (showing telluric absorption to be very small) and a ratio Venus/Moon.

Several groups have successfully solved the difficult problem of operating a Michelson interferometer aboard an aircraft. The National Physical Laboratory group records solar and sky emission far-infrared spectra (29 to 31). The Boulder group has obtained a 0.25-cm^{-1} resolution solar spectrum (82); the results concern both the Sun and the Earth's atmosphere (83, 84).

At the Lunar and Planetary Laboratory, Kuiper et al (76, 77) recorded near-infrared Venus spectra, with a Mertz-type interferometer, Figure 6; resolution was 8 cm^{-1}. While this would seem to be no improvement over the already mentioned ground-based grating recorded spectra (75), the advantages of Fourier spectroscopy have nevertheless been fully used: the airborne telescope diameter was only 30 cm versus 210 cm, the recording time somewhat shorter, and the environment far more severe. The PbS receivers were not cooled. Of course the amount of information extracted from the telluric absorption-free spectra is greater (77, 78). The spectrum of α Orionis was also obtained by Johnson (79); a comparison with the one

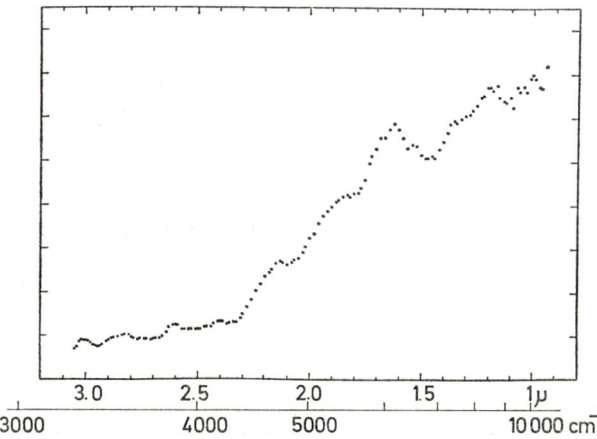

FIGURE 7. Spectrum of α Orionis (prism spectrometer, 90-cm telescope, cooled InAs receivers) from the Stratoscope balloon, given by Woolf et al (70, Figure 5). Scales as for Figure 4.

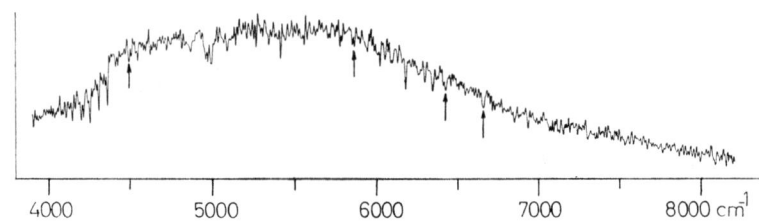

FIGURE 8. Spectrum of α Orionis by Fourier spectroscopy from an aircraft, by Johnson et al (79, Figure 8). Noise level cannot be estimated from this trace alone; however, comparison with another ground-recorded spectrum showing better S/N ratio proves features as small as the marked ones to be real. 30-cm telescope, uncooled PbS receivers.

recorded from the Stratoscope balloon (70) with a 90-cm telescope is given by Figures 7 and 8.

The near-infrared spectrum of the solar corona has been recorded during an eclipse from an aircraft by Mangus et al (71); it shows emission lines above a continuum. Clearly this could not even be attempted with a grating.

A small fast scanning interferometer was placed aboard a Gemini spacecraft but no results have been made available. However on April 14, 1969 the Nimbus 3 satellite was launched carrying a Fourier spectrometer; the Goddard Space Center group of Hanel is responsible for the experiment (68) (Figure 9). The spectral range is 400 to 2000 cm^{-1} (5 to 25 μ) and resolution 5 cm^{-1}; the instrument incorporates a monochromatic source for accurate path difference calibration. Earth emission spectra from a patch 150 km in diameter are being recorded at a rate of one every 16 sec—an enormous amount of information. Thus Fourier spectroscopy is now encircling the planet and all of us are continually being Fourier analyzed whether we like it or not.

This achievement is obviously a milestone in the history of a still not very old technique; and no better way of concluding this review can be found than to mention the similar experiment scheduled for the Mars 1971 Orbiter (69).

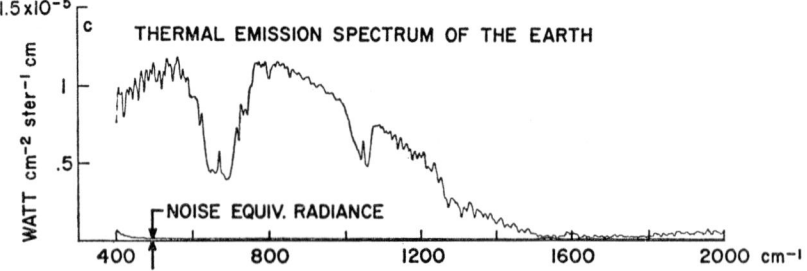

FIGURE 9. Earth emission spectrum recorded by a Fourier interferometer carried by the Nimbus 3 satellite, and published by Hanel et al (68).

CONCLUSION

The amount of astronomical and laboratory information collected from Fourier spectra already exceeds what the most enthusiastic Fourierists hoped for 10 years ago. The rate of increase is now chiefly a function of telescope time (and size) available. An aircraft-borne 90-cm telescope is being built by NASA (87) and large ground-based light collectors have been proposed (88). In view of this considerable effort we can well ask the question: Is Fourier spectroscopy here to stay? According to a commonly heard argument, since the multiplex advantage is transient, we shall some day revert to the good old tools we were born and raised with; consequently we only have to sit and wait until better detectors become available. We hope to have shown that this might involve indefinite sitting.

LITERATURE CITED

1. Fellgett, P. B. 1958, *J. Phys.*, **19**, 187, 237
2. Fellgett, P. B. 1967, *J. Phys.*, **28 C_2**, 165
3. Jacquinot, P. 1954, *J. Opt. Soc. Am.*, **44**, 761
4. Jacquinot, P. 1958, *J. Phys.*, **19**, 223
5. Jacquinot, P. 1960, *Rep. Progr. Phys.*, **23**, 267
6. Jacquinot, P., Roizen-Dossier, B. 1964, *Progr. Opt.*, **3**, 31
7. Loewenstein, E. 1966, *Appl. Opt.*, **5**, 845
8. Vanasse, G. A., Sakai, H. 1967, *Progr. Opt.*, **6**, 259
9. Johnson, H. L. 1966, *Ann. Rev. Astron. Ap.*, **4**, 193
10. Vaughan, A. H., Jr. 1967, *Ann. Rev. Astron. Ap.*, **5**, 139
11. Steel, W. H. 1964, *J. Opt. Soc. Am.*, **54**, 151
17. Steel, W. H. 1967, *Interferometry*, 84 (Cambridge)
13. Methodes Nouvelles de Spectroscopie Instrumentale (Orsay 1966). 1967, *J. Phys.*, **28 C_2**
14. Mertz, L. 1958, *J. Phys.*, **19**, 233
15. Mertz, L. 1964, *Mem. Soc. Roy. Sci. Liège*, **9**, 120
16. Mertz, L. 1965, *Transformations in Optics* (New York: Wiley)
17. Mertz, L. 1965, *Astron. J.*, **70**, 548
18. Mertz, L., Coleman, I. 1966, *Astron. J.*, **71**, 747
19. Mertz, L., Coleman, I. 1966, *Ap. J.*, **143**, 1000
20. Mertz, L. 1967, *Infrared Phys.*, **7**, 17
21. Mertz, L. 1967, *J. Phys.*, **28 C_2**, 87
22. Gebbie, H. A. 1957, *Phys. Rev.*, **97**, 1174
23. Gebbie, H. A. 1958, *J. Phys.*, **19**, 230
24. Gebbie, H. A., Delbouille, L., Roland, G. 1962, *MNRAS*, **123**, 497
25. Gebbie, H. A., Delbouille, L., Roland, G. 1964, *MNRAS*, **9**, 125
26. Delbouille, L., Roland, G., Gebbie, H. A. 1964, *Astron. J.*, **69**, 334
27. Roland, G. 1967, *J. Phys.*, **82 C_2**, 120
28. Gebbie, H. A., Chamberlain, J., Burroughs, W. J. 1968, *Nature*, **220**, 893
29. Gebbie, H. A., Burroughs, W. J., Harris, J. E., Cameron, R. M. 1968, *Ap. J.*, **154**, 405
30. Bader, M., Cameron, R. M., Burroughs, W. J., Gebbie, H. A. 1967, *Nature*, **214**, 377
31. Gebbie, H. A., Burroughs, W. J., Chamberlain, J., Harris, J. E., Jones, R. G. 1969, *Nature*, **221**, 143
32. Kahn, F. D. 1959, *Ap. J.*, **129**, 518
33. Westphal, J. A., Murray, B. C., Martz, D. E. 1963, *Appl. Opt.*, **2**, 749
34. Sinton, W. M., Strong, J. 1960, *Ap. J.*, **131**, 470
35. Sinton, W. M. 1963, *J. Quant. Spectrosc. Radiat. Transfer*, **3**, 551
36. Sinton, W. M., Boyce, P. B. 1964, *Astron. J.*, **69**, 558
37. Sinton, W. M. 1968, *Infrared Astronomy*, 55 (New York: Gordon & Breach)
38. Moroz, W. I. 1964, *Arch. Astron. Ac. Sci.*, **302**
39. Bowers, H. C. 1964, *Appl. Opt.*, **3**, 627
40. Woolf, N. J. 1964, *Appl. Opt.*, **3**, 1195
41. James, J. F. 1964, *J. Quant. Spectrosc. Radiat. Transfer*, **4**, 793
42. Gush, H. P., Buijs, H. L. 1964, *Can. J. Phys.*, **42**, 1037
43. Connes, J. 1961, *Rev. Opt.*, **40**, 45, 116, 171, 231

44. Connes, J., Connes, P. 1966, *J. Opt. Soc. Am.*, **56**, 896
45. Connes, J., Connes, P. 1968. *Infrared Astronomy*, 193 (New York: Gordon & Breach)
46. Connes, J., Connes, P., Maillard, J. P. 1967, *J. Phys.*, **28 C_2**, 120
47. Pinard, J. 1967, *J. Phys.*, **28 C_2**, 136
48. Pinard, J. 1970, *Ann. Phys.*, **2**
49. Maillard, J. P. 1967 (Thèse, Univ. Paris)
50. Connes, P. *Opt. Acta* (In press)
51. Connes, J., Delouis, H., Connes, P., Guelachvili, G., Maillard, J. P., Michel, G. 1970, *Nouv. Rev. Opt. Appl.*, **1**, 1
52. Connes, J., Connes, P., Maillard, J. P. 1969, *Atlas des spectres planétaires infrarouges* (Paris: CNRS)
53. Benedict, W., Connes, J., Connes, P., Kaplan, L. D. 1967, *Ap. J.*, **147**, 1230
54. Connes, J., Connes, P., Kaplan, L. D., Benedict, W. S. 1968. *Ap. J.*, **152**, 731
55. Kaplan, L. D., Connes, J., Connes, P. 1969, *Ap. J.*, **157**, L187
56. Young, L. D. J. 1969, *Icarus*, **2**, 66
57. Young, L. D. J. *Icarus* (In press)
58. Welch, W. J., Rea, D. G. 1967, *Ap. J.*, **148**, L 151
59. Swings, P. 1969, *Proc. Am. Phil. Soc.*, **113**, 229
60. Lewis, J. S. 1968, *Ap. J.*, **152**, L79
61. Connes, P., Connes, J., Bouigue, R., Querci, M, Chauville, J. 1968, *Ann. Ap.*, **31**, 485
62. Montgomery, E. F., Connes, P., Connes, J., Edmonds, F. N. 1969, *Bull. AAS*, **1**, 201 and *Ap. J. Suppl.*, **19**, 167, 1
63. Bopp, B. W., Edmonds, F. N. 1970, *Publ. Astron. Soc. Pac.*, **82**
64. Greene, Th. F. (In press)
65. Yoshinaga, H., Fujita, S., Minami, S., Suemoto, Y., Inoue, M., Shiba, K., Nakano, K., Yoshida, S., Sugimori, H. 1966, *Appl. Opt.*, **5**, 1159
66. Hanel, R., Forman, M., Meilleur, T., Stambach, G. 1968, *Atmos. Sci.*, **25**, 586
67. Hanel, R., Forman, M., Meilleur, T., Westcott, R., Pritchard, J. 1969, *Appl. Opt.*, **8**, 2059
68. Hanel, R., Conrath, B. 1969, *Science*, **165**, 1258
69. Hanel, R., Conrath, B., Hovis, W. A., Kunde, V., Lowman, P. D., Prabhakara, C., Schlachman, B., Levin, G. V. *Goddard Space Flight Cent. Preprint x-620-69-35*
70. Woolf, N. J., Schwarzschild, M., Rose, W. K. 1964, *Ap. J.*, **140**, 833
71. Mangus, J., Strockhausen, R. 1-965, *Proc. Solar Eclipse Symp.*, *NASA/Ames*, 329
72. Beer, R. 1967, *J. Phys.*, **28 C_2**, 113; 1967, *Appl. Opt.*, **6**, 209
73. Biraud, Y., Gay, J., Verdet, J. P., Zeau, Y. 1969, *Astron. Ap.*, **2**, 413
74. Gay, J., Lequeux, J., Verdet, J. P., Turon-Lacarrieu, P., Bardet, M., Roucher, J., Zeau, Y. 1968, *Ap. Lett.*, **2**, 169
75. Kuiper, G. P. 1962, *Comm. Lun. Plan. Lab. Univ. Ariz.*, **1**, 83
76. Kuiper, G. P., Forbes, F. F., Johnson, H. L. 1967, *Comm. Lun. Plan. Lab. Univ. Ariz.*, **6**, 155
77. Kuiper, G. P., Forbes, F. F., Steinmetz, D. L., Mitchell, R. I. 1968, *Comm. Lun. Plan. Lab. Univ. Ariz.*, **6**, 209
78. Kuiper, G. P. 1968, *Comm. Lun. Plan. Lab. Univ. Ariz.*, **6**, 229
79. Johnson, H. L., Coleman, I., Mitchell, R. I., Steinmetz, D. L. 1968, *Comm. Lun. Plan. Lab. Univ. Ariz.*, **7**, 83
80. Johnson, H. L. 1968, *Ap. J.*, **154**, L 125
81. Thompson, R. I., Schnopper, H. W., Mitchell, R. I., Johnson, H. L. 1969, *Ap. J.*, **158**, L55
82. Eddy, J. A., Lee, R. H., Lena, P. J., MacQueen, R. M. *Appl. Opt.* (In press)
83. Eddy, J. A., Lena, P. J., Macqueen, R. M. *Space Sci. Rev.* (In press)
84. Eddy, J. A., Lena, P. J., Macqueen, R. M. *Solar Phys.* (In press)
85. Schofield, J. W., Ring, J. 1969, *ICO 8 Meet.*, *Reading*
86. McNutt, D. P., Shivanandan, K., Feldman, P. D. 1969, *Appl. Opt.*, **8**, 2205, 2199
87. 1969, *Report on Planetary Exploration Nat. Acad. Sci.*
88. 1969, *Report on Planetary Astronomy Nat. Acad. Sci.*
89. Stettler, P., Kneubuhl, F., Muller, E. A. 1969, *Helvet. Phys. Acta*, **42**
90. Cihla, Z., Chedin, A. *J. Mol. Spectrosc.* (To be published)
91. Spinrad, H., Wing, R. F. *Ann. Rev. Astron. Ap.*, **7**, 249
92. Hanel, R. A., Kunde, V. G., Meilleur, T., Stambach, G. 1969, Comm. to the *Planetary atm. symp.*, *Marfa*
93. Connes, P. 1969, *Theory and Observation of Normal Stellar Atmospheres*, 323 (Cambridge: MIT Press)

Copyright 1970, All rights reserved

RADIOFREQUENCY RECOMBINATION LINES 2009

A. K. DUPREE AND LEO GOLDBERG

Harvard College Observatory, Cambridge, Massachusetts

INTRODUCTION

This review deals with spectral lines in the radiofrequency domain emitted as a result of transitions between atomic levels with high principal quantum numbers. They are called recombination lines because the energy levels are populated primarily by electronic recombination, although a given atom may suffer one or more electron collisions after it recombines and before it emits a photon. In 1959, Kardashev predicted that radio recombination lines of hydrogen and helium should be emitted by H II regions with sufficient intensity to be observable above the thermal continuous background. Confirming observations were made a few years later, first at the Pulkovo Observatory in the USSR (Dravskikh & Dravskikh 1964) and shortly thereafter at the National Radio Astronomy Observatory in the USA (Höglund & Mezger 1965). Until Kardashev's paper was published, it seemed not to have been appreciated that transitions between levels of very high quantum number could be observed without overlap because (1) the widths of the energy levels are determined by their radiative lifetimes and (2) the lines are broadened principally by the Doppler effect and therefore the linewidth is proportional to the frequency.

The observation and interpretation of radio recombination lines now constitutes an important branch of radio astronomy. Transitions in hydrogen have been observed for many different values of the principal quantum number in the range from $n=56$ to $n=253$ corresponding to the frequency range from 36.5 GHz to 400 MHz (Sorochenko et al 1969, Penfield et al 1967). In addition, numerous transitions in He I and a so-called "anomalous" line provisionally ascribed to carbon have been detected in a considerable number of galactic H II regions (Dieter 1967, Reifenstein et al 1970, Wilson et al 1970). A line has also been found in an extragalactic object (Mezger et al 1970b), and possibly a planetary nebula (Rubin & Turner 1969) although in NGC 7027 its intensity appears well below the predicted value (Terzian & Balick 1969). Figure 1 shows the radiofrequency spectrum of Orion near 5000 MHz which contains recombination lines from hydrogen, helium, and the anomalous line.

Recombination lines can provide important information on the behavior of atoms in highly excited states, on the physical properties of the interstellar

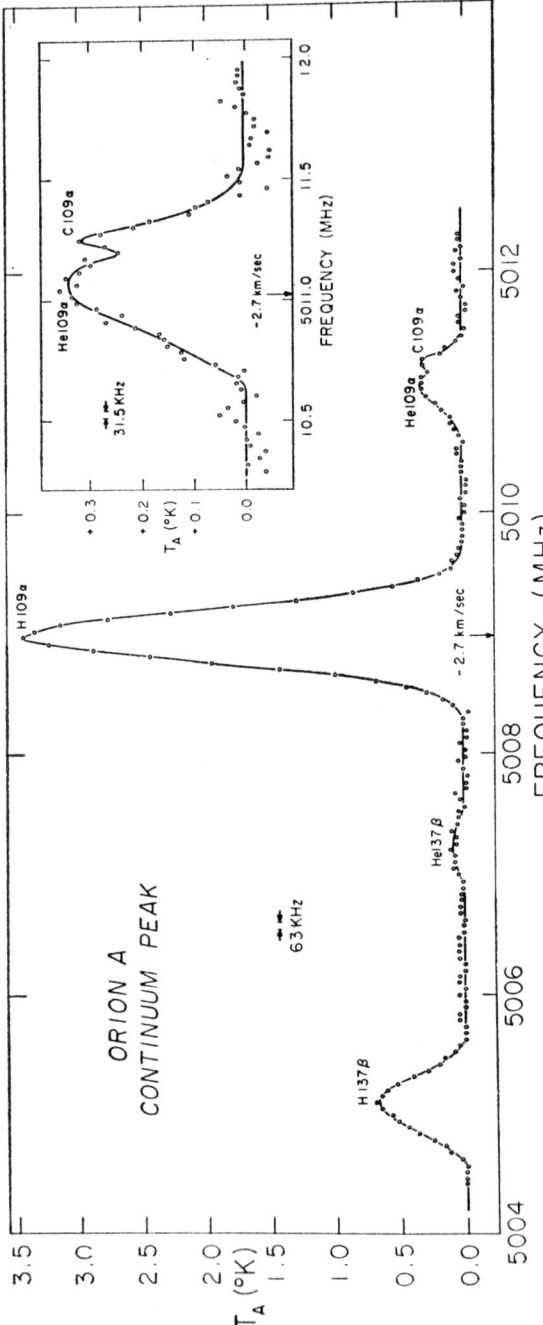

FIGURE 1. A broadband spectrogram showing recombination lines at the center of Orion A (M42). These observations were made with the 400-channel autocorrelator at the 140′ telescope of the NRAO in Green Bank, West Virginia by Churchwell & Mezger (1970). Five recombination lines are indicated: H 137β, He 137β, H 109α, He 109α, and the narrow anomalous line labeled C 109α. The bandwidth corresponds to 3.8 km sec^{-1} in the large figure, and 1.9 km sec^{-1} in the insert.

medium, and on the large-scale structure of the Galaxy but only if a reasonably accurate theory of line intensities is available. Hence this review will concentrate most strongly on recent developments connected with the theoretical interpretation of observations and on those observations that seem most relevant to the evaluation of theories. The most important elements in the theory are the mechanisms of line formation and the calculation of level populations. A crucial test of the theory is whether it can account for the observed intensities of the hydrogen lines over a wide range of frequency including the transitions for which the quantum number changes by two or more. We may then infer with confidence the other parameters that enter into the line formation such as the electron temperature, the electron density, and the emission measure. Another important issue is the identification of the anomalous recombination line and an answer to the question of whether it originates in H II or H I regions. We shall also comment briefly on the abundance of helium and the use of recombination lines to infer the internal structure and distances of H II regions.

THEORY OF LINE INTENSITIES

We consider a transition from an upper level of principal quantum number m to a lower level n, the frequency being given by the usual formula:

$$\nu = cRZ^2(n^{-2} - m^{-2}) \qquad 1.$$

where the Rydberg constant R is taken for an excited atom of charge Z. When $m = n+1$, the line is denoted an $n\alpha$ transition; likewise when $m = n+2$, $n+3 \cdots$, the transition is known as $n\beta$, $n\gamma \cdots$. It is accurate enough for most purposes to approximate Equation 1 by

$$\nu \simeq 2cRZ^2 n^{-3}(m - n)$$

The emission per unit volume at the center of the line is made up of contributions from both line and continuum. Thus,

$$j_\nu = j_C + j_L = B_\nu(T_e)(k_C + b_m k_L^*) \qquad 2.$$

where B_ν is the Planck function, k_C (cm^{-1}) is the continuous absorption coefficient, k_L^* (cm^{-1}) is the central line-absorption coefficient for conditions of thermodynamic equilibrium (LTE), and b_m is the ratio of the actual population N_m in level m, to that given by the Saha-Boltzmann equation. Similarly, the absorption coefficient is given by

$$k_\nu = k_C + k_L = k_C + b_n \beta_{nm} k_L^* \qquad 3.$$

where β_{nm} is introduced to correct the LTE line-absorption coefficient for stimulated emission under non-LTE conditions (Goldberg 1966), viz:

$$\beta_{nm} = \frac{1 - (b_m/b_n) \exp(-h\nu/kT_e)}{1 - \exp(-h\nu/kT_e)} \qquad 4.$$

The solution of the equation of transfer for a central ray in a gaseous nebula is

$$I_\nu = I_L + I_C = \int_0^s j_\nu \exp(-\tau_\nu) ds' \qquad 5.$$

where I_ν is the specific intensity, the sum of contributions from the line I_L and the neighboring continuum I_C. The continuum intensity is

$$I_C = \int_0^s j_C \exp(-\tau_C) ds'$$

and the optical depth τ is defined by

$$\tau(s') = \int_0^{s'} k\, ds'' \qquad 7.$$

It is usually assumed that the emitting volume is isothermal and of uniform density, so that the previous integrals can be evaluated. The ratio of the emergent intensity at line center to that in the continuum is given by

$$r = (I_L/I_C) = \eta_\nu \frac{(1-e^{-\tau_\nu})}{(1-e^{-\tau_C})} - 1 \qquad 8.$$

where $\eta_\nu = (j_\nu/k_\nu)/B_\nu(T_e)$ and is evaluated from Equations 2 and 3:

$$\eta_\nu = \frac{1 + b_m(k_L^*/k_C)}{1 + b_n \beta_{nm}(k_L^*/k_C)} \qquad 9.$$

The quantity r is frequently observed and is equivalent to (T_L/T_C) when the Rayleigh-Jeans approximation to the Planck function is substituted for the specific intensity of line and continuum. Even when T_e is as small as 20°K, the condition $h\nu/kT_e \ll 1$ is satisfied for all frequencies in the radio-frequency domain.

Because $h\nu/kT_e \ll 1$, β_{nm} may be written exactly as:

$$\beta_{nm} = \frac{b_m}{b_n}\left[1 - \left(\frac{kT_e}{h\nu}\right)\left(\frac{b_m - b_n}{b_m}\right)\right] \qquad 10.$$

which, for small $b_m - b_n$ is equivalent to

$$\beta_{nm} \simeq \frac{b_m}{b_n}\left[1 - \left(\frac{kT_e}{h\nu}\right)\left(\frac{d\ln b_n}{dn}\right)\Delta n\right] \qquad 11.$$

where $\Delta n = m - n$. When atoms recombine by radiative recombination, b_n is less than one but approaches unity as $n \to \infty$. Hence $b_m - b_n$ is always positive. Furthermore, since b_n is a very slowly changing function of n, $b_m - b_n$ is proportional to Δn, as is the frequency ν, and therefore β_{nm} is independent of

Δn as long as Δn is not too large. For densities and temperatures characteristic of H II regions, it is often found that $\Delta b/b_m > h\nu/kT_e$ in which case $\beta < 0$ and the line-absorption coefficient is negative.

The value of β passes through a minimum and then begins to increase with increasing n, so that at some value of n, which is a function of both T_e and N_e, $\beta = 0$ and the line-absorption coefficient vanishes. At still higher values of n, β is positive and approaches unity as $n \to \infty$.

We may derive simple expressions for r in three limiting cases in order to illustrate the behavior of the line-to-continuum ratio in and out of equi-

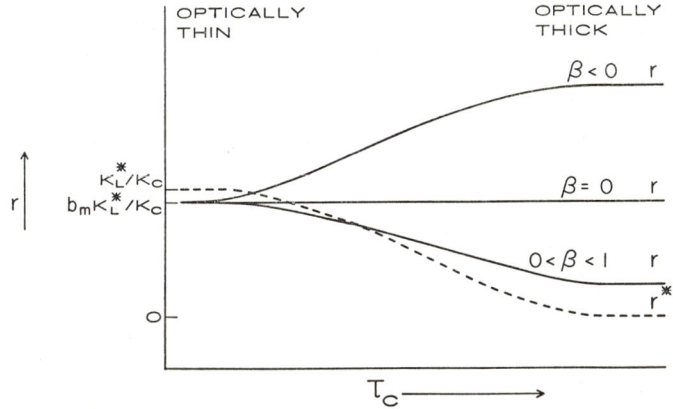

FIGURE 2. Schematic behavior of r as a function of optical depth. The optically thin and optically thick approximations are given by Equations 12 and 13 respectively. The broken line indicates the LTE value r^*.

librium (see Figure 2). First, for an optically thin gas, τ_ν and $\tau_C \ll 1$ and

$$\frac{r}{r^*} = b_m - \frac{1}{2} b_n \beta_{nm} (\tau_C + b_m \tau_L^*) \qquad 12.$$

where Equation 8 has been divided by its LTE value, $r^* = k_L^*/k_C$ in the optically thin approximation. In LTE the b_n factors equal one, and $r/r^* = 1$. When the gas departs from LTE, the line intensity is modified in two ways. The transition rate is directly proportional to b_m and is also dependent on the ratio of the populations of the upper and lower levels, which determines the influence of stimulated emission in strengthening or weakening the line. Note that if τ_C is sufficiently large, the intensity may be greater or less than its LTE value, depending on whether β is negative or positive. However, since β can take on only relatively small positive values, and $\tau_C \ll 1$, line weakening is not significant in the thin-layer approximation.

Next, we approximate r for an optically thick gas, τ_ν and $\tau_C \gg 1$, by substituting for η_ν from Equation 9, viz:

$$r = \frac{k_L^*}{k_C} \frac{b_m - b_n\beta_{nm}}{1 + b_n\beta_{nm}(k_L^*/k_C)} \qquad 13.$$

The intensity ratio is independent of τ_C and in LTE the line vanishes because $b_m = b_n\beta = 1$. Furthermore, if $k_L^*/k_C \ll 1$ and $b_n\beta < 0$, the line intensity is enhanced with respect to its thin-layer LTE value. In other words, non-LTE amplification by stimulated emission more than compensates for absorption by the optically thick free-free continuum.

Of particular interest is the case when $\beta = 0$ and the intensity ratio reduces to $r = b_m(k_L^*/k_C)$. For the conditions in H II regions, β usually passes through zero at a sufficiently high value of n so that $b \simeq 1$, and the intensity ratio has the same value it would have in the optically thin LTE approximation. At this value of n, which is determined by the temperature and density, the intensity ratios of at least the first few members of the series of lines terminating in level n are independent of the optical depth in the continuum.

The optical depth in continuum radiation is provided by free-free transitions in hydrogen and helium. The continuous absorption coefficient is given numerically (Oster 1961) by:

$$k_C = \frac{3.01 \times 10^{-2}}{T_e^{3/2}\nu^2} \sum_j Z_j^2 N_j N_e \ln\left(\frac{4.95 \times 10^7 T_e^{3/2}}{\nu Z_j}\right) (\text{pc}^{-1}) \qquad 14.$$

where ν(GHz) is the frequency and the summation is taken over all ions j. Numerical approximations valid for the temperatures and ionization of H II regions have been made for the logarithmic term (Mezger & Henderson 1967); the optical depth can be expressed for a homogeneous isothermal region of size L (pc) as:

$$\tau_C = 8.235 \times 10^{-2} a(\nu, T_e) T_e^{-1.35} \nu^{-2.1} E_C \qquad 15.$$

where $a(\nu, T_e) = 1$ approximately. The continuum emission measure E_C (pc cm^{-6}) is given by

$$E_C = \left(N_e \sum_j Z_j^2 N_j\right) L \qquad 16.$$

and the summation is taken over all ions j.

The optical depth in the line center is given, in LTE, by

$$\tau_L^* = 1.01 \times 10^4 Z^2 \Delta n (f_{nm}/n) T_e^{-5/2} \exp(X_n) E_L/\Delta\nu_L \qquad 17.$$

where $X_n = 157800\, Z^2/n^2 T_e$, $\Delta\nu_L$ (kHz) is the full linewidth at half power, and E_L (pc cm^{-6}) is the line emission measure of the recombining ion. At temperatures of H II regions, the exponential factor is close to unity and can be neglected. Hydrogenic oscillator strengths for transitions between adjacent levels have been calculated (Goldwire 1968, Palmer 1968, Menzel 1969);

for the transition 109→110, $(f_{nm}/n) = 0.193$. The two emission measures are related:

$$\frac{E_L}{E_C} = \frac{N_i/N_{H^+}}{1 + (N_{He^+} + 4N_{He^{++}})/N_{H^+}} \qquad 18.$$

where N_i is the number density of the recombining ions.

In most H II regions the abundance of doubly ionized helium relative to hydrogen is less than 3×10^{-3} (Palmer et al 1969) but in some high-excitation planetary nebulae it may be as high as 10% and thus may contribute nearly as much to the emission measure as does hydrogen. Observed values of the ratio N_{He^+}/N_{H^+} may be inferred from the observed relative intensities of hydrogen and helium recombination lines and are found to average about 0.1 so that $E_C = 1.10\, E_L$. In NGC 7027, the abundance of doubly ionized and singly ionized helium relative to hydrogen is about 0.06 and 0.12 respectively and accordingly $E_C = 1.36\, E_L$.

When the levels are populated in LTE, insertion of numerical values leads to an expression for r^*,

$$\Delta\nu_L r^* \simeq 2.33 \times 10^4 \nu^{2.1} T_e^{-1.15} (E_L/E_C) \qquad 19.$$

which applies to $n\alpha$ transitions from neutral atoms in the optically thin approximation.

Equation 19 was first given by Kardashev (1959) in a slightly different form in which the linewidth was defined at the $1/e$ level, the oscillator strength was taken as $n/6$, and the Gaunt factor for free-free emission was set equal to unity.

Emission and absorption by H I clouds.—The ionized constituents of H I regions can be expected to produce recombination lines as well (Zuckerman & Palmer 1968, Dupree & Goldberg 1969, Greenberg 1969). In such clouds, atoms with ionization potentials less than that of hydrogen (13.6 eV) can be ionized by the ambient radiation field. In order of abundance, atoms of this type include carbon, silicon, magnesium, and sulfur. In addition, a small fraction of hydrogen and helium may be ionized by cosmic rays (Spitzer & Tomasko 1968) or by soft X rays (Silk & Werner 1969). Hence H I regions may absorb and emit recombination radiation in lines and continuum.

We first consider the emission of radiation from an isolated H I cloud and denote the LTE value of the central optical depth by t_L^*. Then the temperature at line center for an optically thin gas is similar to Equation 12, viz:

$$T_L^I = T_e^I t_L^* b_m \left[1 - \frac{1}{2} \frac{b_n}{b_m} \beta_{nm}(t_C + b_m t_L^*) \right] \qquad 20.$$

Here, however, the departure coefficients and other parameters are those appropriate to an H I region.

If an H I region is observed against a strong source of continuum radiation, Equation 20 must be modified to account for the background continuum. For small optical depths, the line temperature is given by:

$$T_L{}^I = t_L{}^*(b_m T_e{}^I - T_C b_n \beta_{nm}) \qquad 21.$$

where T_C is the brightness temperature of the continuum source. In effect, the term $T_C b_n \beta$ replaces the second term on the right-hand side of Equation 20 because the line emission is now stimulated by the background continuum rather than by emission in the local continuum or in the line itself. In comparing the relative importance of the two modes of line amplification, we note that usually $T_e{}^I \ll T_C$. For example, in Orion A, T_C is considerably larger than $T_e{}^I$ at 5 GHz and increases rapidly with decreasing ν. Thus a background continuum will affect the line intensity substantially more than the local radiation field in the H I region. In addition, β varies with frequency and can be positive, negative, or zero. Hence the background continuum may produce emission or absorption lines, depending upon the relative sizes of the two terms in Equation 21.

Under LTE conditions in the H I region the line temperature reduces to the well-known simple form:

$$T_L{}^I = t_L{}^*(T_e{}^I - T_C) \qquad 22.$$

and production of an emission or absorption line depends on the difference between the local kinetic temperature and the brightness temperature of the background continuum.

It is more likely, however, that levels producing recombination lines in H I regions are not populated in LTE. There is considerable evidence that the so-called anomalous recombination line is emitted in H I regions (Dupree & Goldberg 1969) and that enhancement by stimulated emission under non-LTE conditions plays an important role. Recombination lines formed in H II regions may contain a contribution arising from foreground H I regions. A line of sufficient intensity formed in the H I region should appear as a relatively sharp bump on the line profile from the H II region, and should be displaced from the center of the line profile by the amount of the difference in radial velocity between the two regions. Such components of the hydrogen recombination lines have not yet been detected.

DERIVATION OF LEVEL POPULATIONS
Hydrogen and Helium

High temperatures.—Interpretation of the observed intensities of recombination lines requires accurate values of the departure coefficients b_n and of the differences Δb. The first calculations of b_n for very high values of n were carried out by Seaton (1964), primarily to illustrate the dependence of the departure coefficients on the electron density. The requirement of statistical equilibrium, that atoms enter and leave a given level at the same rate, leads to an infinite set of simultaneous equations for the populations of the levels. Various analytic and numerical procedures have been developed to obtain a solution to these equations (Baker & Menzel 1938, Seaton 1964, Burgess & Summers 1969, Dupree 1969, Sejnowski & Hjellming 1969). When the rate coefficients for population and depopulation are known, the departure coefficients b_n can be determined as a function of N_e and T_e.

The following processes regulate the populations of levels with $n \gtrsim 30$: (1) electronic recombination and subsequent cascade; (2) spontaneous transitions; (3) collisional ionization and three-body recombination by electron impact; (4) collisional transitions between levels of different principal quantum number, also by electron impact. Under the conditions found in H II regions, photoexcitation may be neglected.

In the early b_n calculations, Seaton (1964) included only collisional transitions for which $\Delta n = 1$ and, in order to calculate the cascade rate, assumed that $b_n = 1$ for all higher levels. Both procedures can introduce substantial errors. Finally Seaton approximated the resulting equations of statistical equilibrium by a second-order differential equation that could be solved to obtain departure coefficients b_n.

More recently the b_n problem has been investigated in great detail by Hjellming and his collaborators (Sejnowski & Hjellming 1969). An iterative method is used to solve a truncated set of simultaneous equations, making it unnecessary to invoke the $b_n = 1$ approximation in the cascade term. Collisions between adjacent levels are included for $\Delta n = 1$ to 20. Solutions have been obtained for Case B (in which the medium is assumed to be completely opaque to Lyman radiation) with T_e in the range 5000–12,500°K and $N_e = 10 - 3 \times 10^5$ cm^{-3}. Figure 3 shows the results of the b_n calculations for $T_e = 10,000$°K and values of $N_e = 10, 10^2, 10^3,$ and 10^4 cm^{-3}. In most respects, the solutions are of rather high accuracy but uncertainties in the electron-hydrogen cross sections for excitation still leave room for considerable error. The solution is very sensitive to the choice of T_e and N_e. Independent calculations of b_n values have been completed by Brocklehurst (1970). He used collision cross sections based on different approximations depending on Δn of the transition and a matrix condensation technique to solve the simultaneous equations of statistical equilibrium. The solutions show agreement in $d \ln b/dn$ to within a factor of 2 as compared to the results of Sejnowski & Hjellming (1969).

The b_n values have been used for extensive calculations of intensities of alpha, beta, gamma, delta, and epsilon lines in the interval $40 \leq n \leq 225$ (Andrews & Hjellming 1969; Hjellming, Andrews & Sejnowski 1969). The intensity ratio T_L/T_C depends on T_e, N_e, and E, the emission measure. Figure 4 shows a portion of the results for alpha and delta lines with $T_e = 10,000$°K and $N_e = 10^2$ and 10^5. The dashed lines represent the LTE solution for an optically thin gas. The results are given for different values of E_C from 10^4 to 10^8 pc cm^{-6}. When n is small, the optical depth is too small to produce enhancement and the intensity is actually reduced by the factor b_n. When

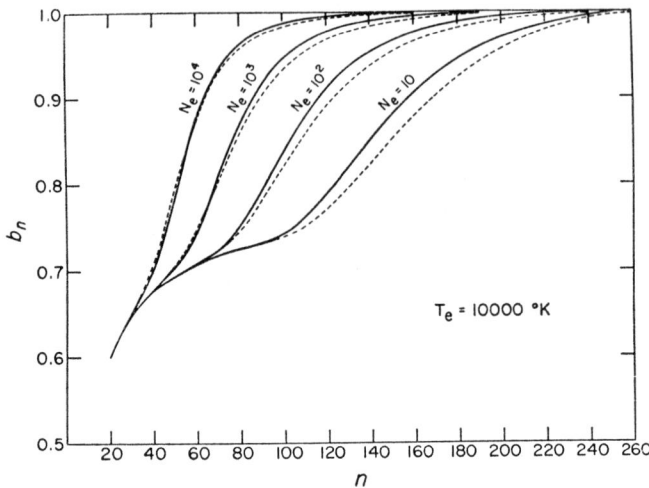

FIGURE 3. The b_n values for hydrogen. The calculations were carried out by Sejnowski & Hjellming (1969) for two different sets of cross sections based on (1) an impact-parameter approximation (solid lines) and (2) a dipole approximation (broken lines).

n is large the enhancement depends on E but the amount of enhancement is sensitive to both N_e and T_e, being large when N_e is small. The dependence on T_e is not as great, but large amplifications are associated with low electron temperatures and vice versa. At very low frequencies (high n) the gas becomes optically thick. And, when the electron density is high, the lines become weaker with increasing emission measure. At low N_e, departures from LTE produce enhancements that compensate for the attenuation due to optical thickness, and the intensity increases with increasing E.

Low temperatures.—Interpretation of the anomalous recombination line appears to require a knowledge of b_n values under low-temperature conditions. Unfortunately, at present the rate coefficients for collisional processes between high levels are not known for temperatures of 10–200°K. Furthermore, the radiation incident upon an H I region from a background source can substantially modify the populations. Before proceeding with detailed calculations it may be necessary to have a model of the geometric configuration of H I clouds with respect to local H II regions. Solutions for b_n have been obtained for the zero-density case (Dupree 1970); however, none are as yet available for the conditions expected in H I regions.

Complex Atoms

Discovery of the anomalous recombination line (Palmer et al 1967) motivated theoretical investigation of emission arising from complex atoms in

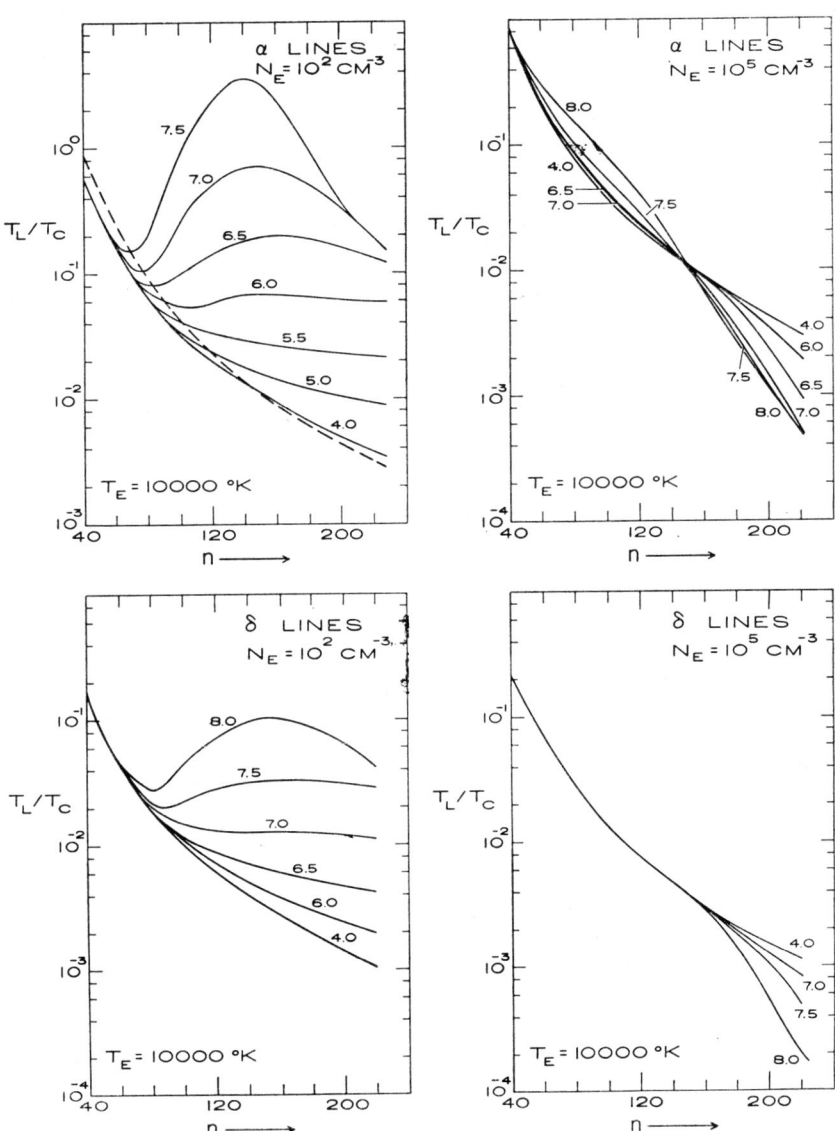

FIGURE 4. T_L/T_C for α and δ transitions of hydrogen. Note the change of scale for α lines at $N_e = 10^2$ cm^{-3}. The broken curve indicates the LTE solution.

H II regions. Departures from LTE in such atoms can be quite different from hydrogen, principally because an alternative mode of recombination is available. Dielectronic recombination, so termed because it proceeds through doubly excited atomic levels, can populate high levels of complex atoms at a greater rate than radiative recombination. Large overpopulations of high levels can result under appropriate conditions (Goldberg & Dupree 1967).

One of the most likely elements to produce recombination lines is neutral carbon, and b_n values have been obtained (Dupree 1969, Gayet et al 1969) for this atom. The processes of population and depopulation of excited levels are the same as for hydrogen, with the addition of the dielectronic recombi-

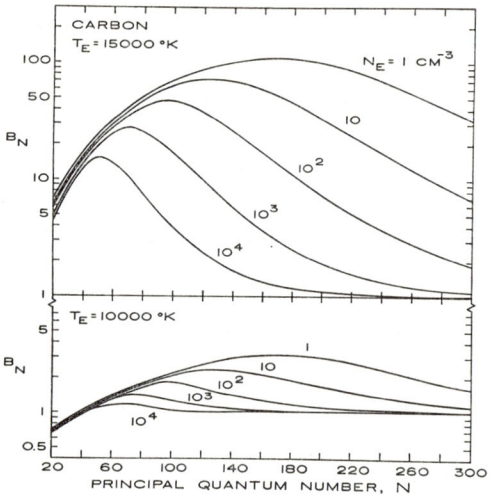

FIGURE 5. The b_n values for neutral carbon at $T_e = 10,000$ and $15,000°K$ and various values of the electron density. (Dupree 1969)

nation process. One method of solution incorporates a direct matrix inversion of the equations of statistical equilibrium, which is then combined with iteration on the b_n values (Dupree 1969). In this way, the cascade term can be evaluated exactly. Some results for $T_e = 10,000$ and $15,000°K$ are shown in Figure 5. The departure coefficients are extremely sensitive to temperature and density.

The temperature dependence occurs as a result of the dielectronic recombination process. Recombining electrons must have energies commensurate to those of the doubly excited levels through which they recombine. In carbon, these levels occur at 10 eV or higher; hence, the Boltzmann factor is rapidly varying at temperatures near 1 eV. With increasing temperature, dielectronic recombination proceeds at a faster rate resulting in larger overpopulations. With increasing electron density, collisional interchange be-

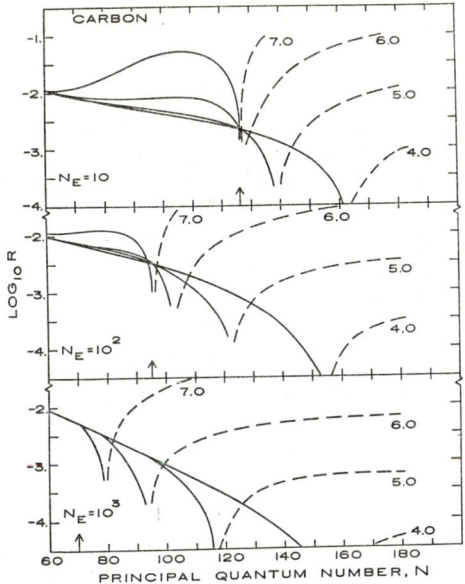

FIGURE 6. The line-to-continuum ratio at line center, $r = T_L/T_C$ for $n\alpha$ transitions in neutral carbon. The calculations are for $T_e = 15{,}000°$K and for electron densities $N_e = 10$, 100, and 1000 cm^{-3}. Four values of the emission measure are considered: $\log E_C = 4., 5., 6., 7.$ pc cm^{-6}. *Solid line*, emission lines; *broken line*, absorption lines; *arrow*, maximum of the corresponding b_n curve.

tween adjacent levels becomes more effective in reducing the population to its equilibrium value. The b_n's become smaller and quickly approach 1 with increasing n.

In contrast to the results for hydrogen, the b_n values for carbon increase and decrease with increasing n. When the b_n factors are decreasing, negative values of $(b_m - b_n)/b_m$ occur that can lead to positive values of β_{nm}, as seen from Equation 10. If β_{nm} is sufficiently large and positive, the line may appear in absorption. Figure 6 demonstrates the behavior of r (Equation 8) for $T_e = 15{,}000°$K. With increasing density r decreases for all n because the b_n values become smaller and stimulated emission decreases. When β is negative and approaches zero, the emission line rapidly weakens. With increasing n, β becomes large and positive, and the line turns into absorption.

Such overpopulations can occur in any atom or ion with two or more electrons. Optimum conditions for dielectronic recombination are determined by the placement of the resonance transition in the next stage of ionization. When the electrons in the thermal plasma have energies comparable to this transition, recombination through doubly excited levels can occur at a rate greater than direct radiative recombination. At the relatively low densities found in H II regions, enhancements of high-level populations can be expected.

INTERPRETATION OF OBSERVATIONS

Recent and current observations of recombination lines are principally designed to answer the following questions: (1) what are the physical properties of H II regions? (2) what is the abundance ratio He:H in the interstellar medium? (3) what is the identity of the so-called anomalous line and by what mechanism is it formed? (4) what can be learned about the internal structure of H II regions and the large-scale structure of the Galaxy from recombination-line data?

HYDROGEN LINES

If the levels of hydrogen in H II regions are populated as in LTE the intensities of $n\alpha$ lines are given by Equation 19, and accordingly electron temperatures may be derived from observed values of $\Delta\nu_L r^*$. Such temperatures, which we may designate by T_e^*, have been obtained from α-line measurements on a large number of thermal galactic sources (e.g. Dieter 1967, Mezger & Höglund 1967, McGee & Gardner 1968, Reifenstein et al 1970). Values of T_e^* derived in this way are in the range 3000–10,000°K and are often a few thousand degrees smaller than the temperature of the H II region derived from ratios of forbidden line intensities in the optical spectrum (Peimbert 1967). The differences between radio and optical temperatures have been attributed to one cause or a combination of various causes, including (1) temperature fluctuations in H II regions (Peimbert 1967), (2) differences between the solid angles subtended by radio and optical telescopes, (3) possible errors in the cross sections for the collisional excitation of forbidden lines (Flower & Seaton 1969), and (4) deviations from thermodynamic equilibrium in the populations of high levels (Goldberg 1966).

The first three suggestions, however, now appear not to be the principal explanation of the temperature discrepancy. For one of the most well-studied objects, the Orion Nebula (Orion A), electron temperatures determined from the optical spectrum are found not to vary by more than 15% from point to point in the nebula (Peimbert 1967, Münch 1968). Recent determinations of cross sections for the observed O II and III forbidden lines do not significantly differ from earlier values (Eissner et al 1969). The different spatial resolution achieved by radio and optical techniques may not be a great obstacle to interpretation of line intensities. In Orion, the profile of the H 109α line agrees in width with an Hβ profile constructed by a superposition of optical profiles over the angle subtended by the antenna used for radio observations (Weedman 1966). Even when $n\alpha$ and higher-order transitions occurring at nearby frequencies are observed with the same beamwidth, different temperatures are still derived from the individual line intensities (Zuckerman et al 1967).

To investigate departures from LTE several empirical tests have been devised. Equation 12 suggests that values of T_e^* derived from observations of a given $n\alpha$ line in a sizable number of H II regions might decrease with increasing emission measure (Mezger & Höglund 1967, Dieter 1967). The absence of such a correlation in the NRAO-MIT survey data led Mezger (1968) to conclude that line enhancement does not occur. But it can also be argued that, since high emission measure is correlated with high density, the amount of line enhancement should decrease and T_e^* should increase with increasing emission measure.

TABLE 1. Observations of higher-order recombination lines

Source	Transitions (hydrogen)	ρ_{Obs}	ρ_{TE}	Ref.
Orion	$148\delta/94\alpha$	0.062 ± 0.009	0.0734	a
	$137\beta/109\alpha$	0.19 ± 0.01	0.279	b
	$158\beta/126\alpha$	0.13 ± 0.03	0.224	c
	$197\beta/156\alpha$	0.11 ± 0.02	0.271	d
	$225\gamma/156\alpha$	0.092 ± 0.015	0.126	d
M17	$137\beta/109\alpha$	0.24 ± 0.03	0.279	b
	$158\beta/126\alpha$	0.22 ± 0.04	0.224	c
	$197\beta/156\alpha$	0.137 ± 0.017	0.271	d
	$225\gamma/156\alpha$	0.068 ± 0.010	0.126	d
NGC 2024	$137\beta/109\alpha$	0.22 ± 0.02	0.279	b

a Gordon & Meeks (1967).
b Zuckerman et al (1967).
c Gardner & McGee (1967).
d Williams (1967).

A second test for non-LTE is the determination of T_e and $b_n\beta_{nm}$ by comparison of the intensities of α and β lines originating from or terminating in the same level (Dyson 1967, Gardner & McGee 1967, Palmer 1967). Equation 12, with $b_m\tau_L^* = 0$ and $b_m = 1$ is written for each of the two lines and it is assumed that r, $\Delta\nu_L$, and E_C are known from observation. Hence $b_n\beta_{nm}$, which is the same for both lines, and T_e are unknowns to be obtained from the solution of two simultaneous equations. The method gives an unambiguous determination of both T_e and $d \ln b_n/dn$ but only for a source of uniform temperature, density, and emission measure. These conditions are not likely to be met.

The most decisive evidence for departures from LTE has been given by intensity ratios of $n\alpha$ and higher-order lines observed at approximately the same frequency. The antenna has the same beamwidth for both transitions, which eliminates any complications due to an inhomogeneous nebula. Let n be the lower level of an $n\alpha$ line and m the lower level of a higher-order line at nearly the same frequency. Then the ratio of central line intensities in thermodynamic equilibrium is

$$\rho_{TE} = \frac{m^2 f_{m,m+\Delta m}}{n^2 f_{n,n+1}} \qquad 23.$$

where f is the absorption oscillator strength. To a good approximation the oscillator strengths may be written in the form $f_{n,n+\Delta n} = n\ M(\Delta n)$, where $M(\Delta n)$ is independent of n, and has been tabulated by Menzel (1969). Hence

$$\rho_{TE} = \Delta m\ \frac{M(\Delta m)}{M(1)} \qquad 24.$$

since $m^3/n^3 \sim \Delta m$.

Measured values of ρ are usually smaller than ρ_{TE} as first shown by the relative intensities of the 137β and 109α lines in Orion A, M17, and NGC 2024 (Zuckerman et al 1967). Some representative values of ρ are compiled in Table 1 for a number of lines at different frequencies in the three sources. Values of T_e^* derived from the measurements are presented in Table 2.

According to Zuckerman et al (1967) the higher values of T_e^* given by 137β lines suggest that departures from LTE are smaller for this line than for the 109α line and the derived temperatures are therefore lower limits to the true temperature. Nevertheless Mezger & Ellis (1968) presented plausible arguments for the view that the α lines in Orion were formed in LTE and that the higher-order lines are weakened by some unexplained process. Table 2 shows, for example, that values of T_e^* derived from the α lines show relatively little change with frequency whereas those derived from higher-order lines usually show more scatter than can be attributed to observational error. Menon & Payne (1969) find the constancy in T_e^* to hold for all α lines observed in Orion A over the frequency range 8000–1500 MHz. Furthermore, only very slight enhancements are predicted (Mezger & Ellis 1968) from the calculated b_n values and a theoretical model of Orion A.

In our view there is no possibility of explaining the relative weakness of the higher-order lines without invoking departures from LTE in the level populations. As shown by Hjellming and his collaborators (Hjellming, Andrews & Sejnowski 1969; Hjellming & Churchwell 1969; Hjellming & Davies 1970), most of the observed line intensities including values of Δn from 1 to 4 can be predicted with reasonable accuracy by non-LTE theory with suitably chosen values of the three parameters T_e, N_e, and E_C. The results for the Orion Nebula and for several other H II regions are given in Table 3. Since both the electron density and the emission measure are considerably higher than the average values derived by other optical and radio methods, it must be supposed that the electron density occurs in clumps. Hjellming & Churchwell (1969) estimate that a typical clump size in Orion is $0.04/N$ pc where N is the number of clumps in the line of sight. If $N \sim 5$, then the clump size is ~ 0.008 pc. Knots of optical emission have been observed

TABLE 2. Some electron temperatures determined under LTE assumption

Source	Transition	T_e^* (°K)	Ref.
Orion	H 94α	6610 ± 400	a
	H 109α	6730 ± 610	b
	H 126α	7000	c
	H 158α	6460 ± 500	d
	H 137β	9250 ± 1390	b
	H 197β	21400 ± 3600	e
	H 225γ	11700 ± 1700	e
M17	H 56α	9250 ± 1100	f
	H 109α	5750 ± 1350	b
	H 126α	6600	c
	H 158α	3310 ± 460	d
	H 137β	6190 ± 1900	b
	H 197β	10900 ± 1500	e
	H 225γ	6600 ± 1000	e
NGC 2024	H 94α	7124 ± 207	g
	H 109α	7090 ± 1060	b
	H 126α	9600	c
	H 158α	6310 ± 1070	d
	H 137β	8660 ± 1730	b

a Gordon & Meeks (1968).
b Zuckerman et al (1967).
c McGee & Gardner (1968).
d Dieter (1967).
e Williams (1967).
f Sorochenko et al (1969).
g Gordon (1969).

TABLE 3. Parameters of H II regions[a]

Object	$\langle T_e \rangle$ (°K)	$\langle N_e \rangle$ (cm^{-3})	$\langle E_C \rangle$ (pc cm^{-6})
Orion	10000.	2.5×10^4	1.3×10^7
M17	8000.	3.8×10^4	1.0×10^7
W51	9000.	1.0×10^5	3.2×10^7
W3	11000.	7.7×10^4	4.0×10^7

a Hjellming & Davies (1970).

with sizes less than 0.01 pc (Münch & Wilson 1962). Radio interferometric observations of Orion indicate that the central component has a diameter of ~0.3 pc. Several elements have been found that are as small as 0.03 pc, but these components contribute less than 5 % of the total flux from the nebula (Webster 1970).

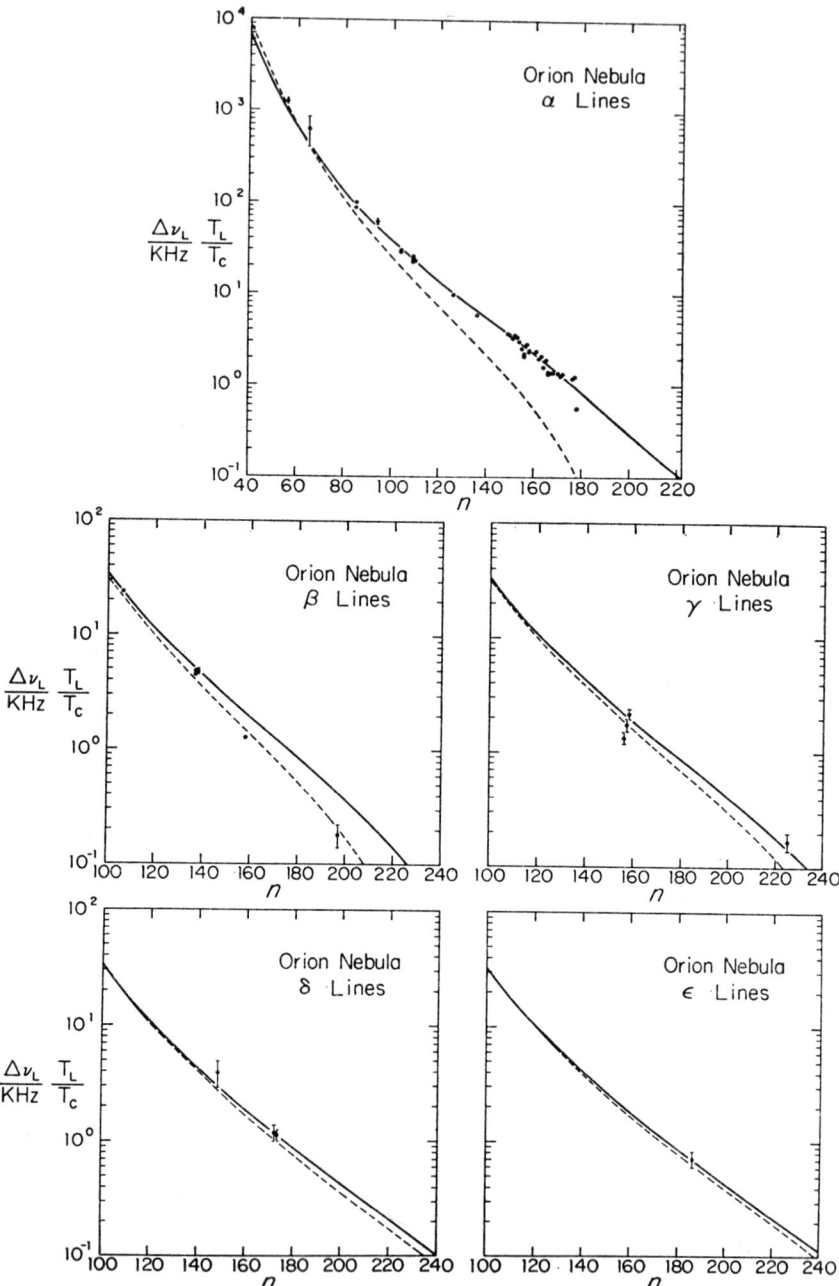

FIGURE 7. Comparison between the non-LTE model of Orion (solid line) and observations. The solid lines were computed for $T_e = 10{,}000°K$, $N_e = 2.5 \times 10^4$ cm^{-3}, and $E_C = 1.3 \times 10^7$ pc cm^{-6}. The broken line indicates the LTE solution. (Hjellming & Churchwell 1969, Hjellming & Davies 1970)

The comparison between theory and observation is shown for the Orion Nebula in Figure 7. The agreement is generally within the observational error, but the β lines are notably discrepant. The number of higher-order lines that are measured must be substantially increased before the adopted model can be considered to be at all definitive.

In the method of analysis employed by Hjellming and collaborators, all three parameters, T_e, N_e, and E_C, are derived simultaneously in the process of finding the best fit between theory and observation. It is suggested (Hjellming & Davies 1970), however, that if a series of higher-order lines can be observed at a frequency such that $\tau_C \ll 1$, values of T_e^* inferred from these lines will converge upon the correct value of the temperature.

An alternative method of data analysis, which involves T_e^* obtained directly from the observations, has recently been proposed (Goldberg & Cesarsky 1970). Values of T_e^* are calculated from each line by using Equation 19 and the observed values $\Delta \nu_L T_L / T_C$. The derived quantities T_e^* are then fitted to theoretical curves of T_e^* versus n, which are computed from non-LTE theory. In this way, both the temperature and the density may be determined independently of the emission measure. Figure 8 shows two sets of theoretical curves giving T_e^* as a function of n for α, β, γ, δ, and ϵ lines. For comparison, the broken curves give the variation of T_e^* that would be expected for an α line if the gas were radiating as in LTE. In this case, as n increases, T_e^* will equal T_e as long as $\tau_C \ll 1$ but eventually becomes infinite as the gas becomes optically thick in continuum radiation and the line intensity vanishes.

The curves in Figure 8 show a number of noteworthy features.

1. At very small values of n, when τ_C is too small to cause appreciable line enhancement, the line-to-continuum ratio is given by $r = b_m k_L^*/k_C$ and is therefore *smaller* than the LTE value. Hence $T_e^* > T_e$ for all lines.

2. As n increases, τ_C becomes large enough to cause line enhancement and the intensity ratio is always greater than the LTE value.

3. For given N_e and T_e, the curves for the different series members all have a common intersection at the same value of n that we call n_0. At this value of n, β is equal to zero, and line enhancement by stimulated emission balances attenuation by absorption in the continuum. Hence, $r = b_m k_L^*/k_C$ and since b_m can usually be taken as unity, T_e^* is the true temperature T_e. By Equation 11,

$$\left(\frac{d \ln b_n}{dn}\right)_{n=n_0} = \frac{1}{\Delta n} \frac{h\nu}{kT_e} \qquad 25.$$

and therefore the electron density may be deduced from comparison with theoretical b_n values. The value of n_0 is much more sensitive to N_e than to T_e.

4. When $\beta < 0$, the level populations are inverted but the amount of line enhancement decreases with increasing Δn. Hence if a series of α and higher-order lines is observed at the same frequency and all lines have negative

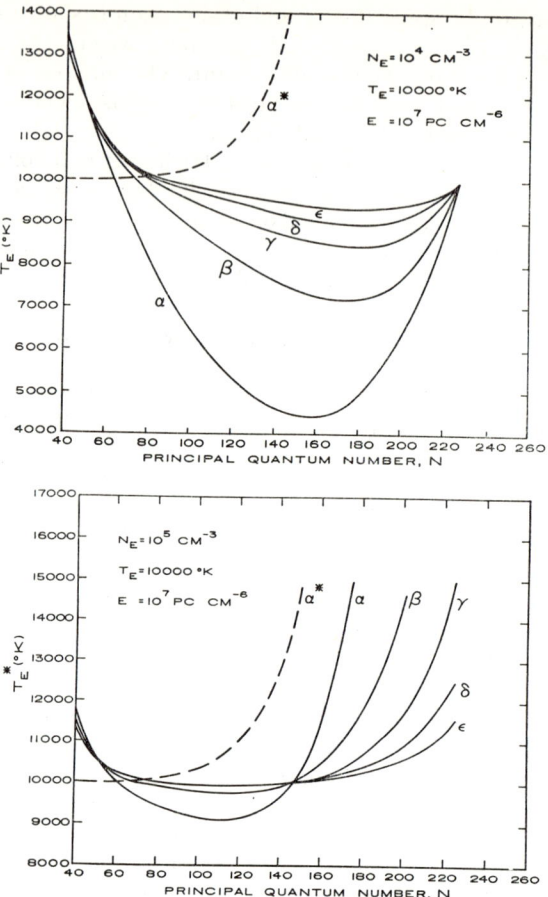

FIGURE 8. Behavior of T_e^* as a function of n for $n\alpha$-$n\epsilon$ transitions. The solid lines are calculated by evaluating $\Delta\nu_L T_L/T_C$ from non-LTE solutions for b_n (Sejnowski & Hjellming 1969) and deriving an LTE temperature T_e^* from Equation 19. The broken line α^* indicates the temperatures that result for an α transition in the LTE case.

values of β, T_e^* will increase with increasing Δn and will converge to the true temperature.

5. When $0 < \beta < 1$ the line-absorption coefficient is still smaller than the LTE value. In the limit of large n, $b_n \to 1$, $\beta \to 1$, and by Equation 13, $r \to 0$. The rate at which the line intensity approaches zero depends primarily upon the emission measure and secondarily upon the density and temperature. Since for a given value of n the frequency is proportional to Δn, the rate of attenuation is highest for α lines and decreases with increasing Δn.

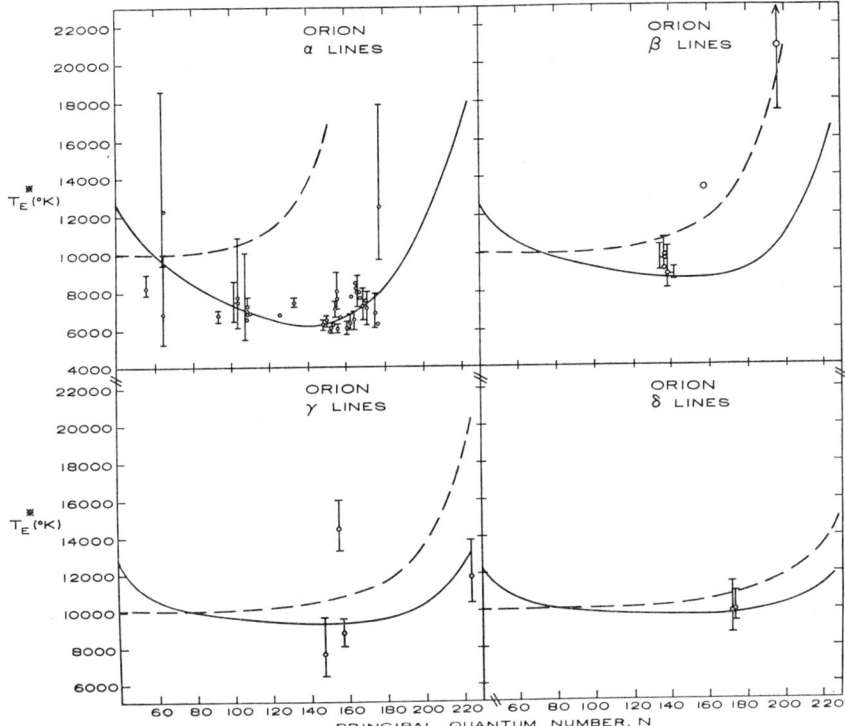

FIGURE 9. T_e^* as determined from the observations in Orion. The solid curve is a non-LTE solution, calculated by using the best-fit parameters of Hjellming & Davies (1970). The broken curve is the LTE solution.

6. Figure 8 shows that in comparing values of T_e^* derived from observations of α and higher-order lines at the same frequency, the values of ρ are likely to deviate most from ρ_{TE} when the value of β is negative or zero for the α line and positive for the higher-order line. This will be particularly true when τ_C at the given frequency is of order unity or greater. In this case, absorption by the continuum may be balanced or exceeded by stimulated emission, whereas when β is positive, the amount of stimulated emission is relatively small. The value of n_0 at which β changes sign decreases with increasing N_e and the observed large scatter in determinations of T_e^* from higher-order lines may imply that electron densities and emission measures of H II regions are generally larger than has been supposed.

In Figure 9, we have plotted values of T_e^* computed from values of $\Delta\nu_L T_L/T_C$ observed in the spectrum of the Orion Nebula and compiled by Hjellming & Churchwell (1969). The curves drawn through the points have been calculated from non-LTE theory with the parameters derived by the "best-fit" method (Hjellming & Davies 1970). The error bars are considerably larger in some cases than would be required for a rigorous test of the theory. Nevertheless the discrepancies between theory and observation are more clearly apparent than in Figure 7, in which the data were plotted on a logarithmic scale. In particular, the observed intensity of the 56α line is substantially less than the predicted value and the agreement for the β and γ lines is not at all satisfactory. More observations are needed, but the Orion Nebula may not provide the best testing ground for the theory because the profiles of the hydrogen recombination lines at low frequencies are known to

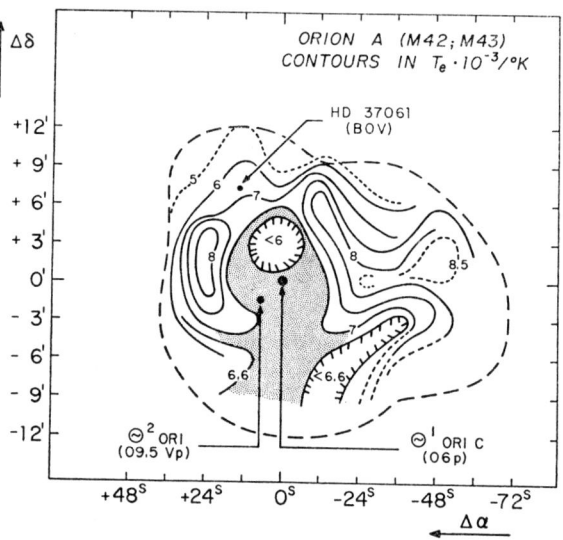

FIGURE 10. T_e^* across Orion as determined from the H 109α transition (Mezger & Ellis 1968). In the shaded area $T_e^* = 6800 \pm 200°K$. The angular resolution is 6 arc min. The broken line corresponds to a contour of $T_A = 1°K$ in continuum emission.

be non-Gaussian, probably because of the superposition of two or more components in the antenna beam (McGee & Gardner 1968, Menon & Payne 1969).

The Orion Nebula has been mapped in the H 109α transition, and the observed line intensities used to derive T_e^* over the nebula (Figure 10). Surrounding the central star (θ^1 Ori C) and the region with relatively constant T_e^* is a shell of gas, open at the north and south, of higher T_e^*. The results of the above discussion indicate that an increase of T_e^* in an isothermal H II region could be caused by a decrease in emission measure and/or an increase in electron density from the value at the center of the nebula. However, optical observations show that the density decreases with distance from the center (Osterbrock & Flather 1956, Davies et al 1964). Such structure would increase the line enhancement for a given emission measure. The observed behavior of T_e^* is the result of these two effects influencing the intensity in opposite ways. When additional observations at offset points are available, it should be possible to determine the physical conditions and structure of the emitting regions.

Helium Lines

Kardashev's (1959) original prediction of recombination lines noted that in addition to hydrogen transitions, helium lines might also be detected from H II regions. They would be weaker than the hydrogen lines because the abundance of helium is less. Transitions in helium were first detected in M17 by Lilley et al (1966), and have now been observed in numerous H II regions (Palmer et al 1969, Churchwell & Mezger 1970, Reifenstein et al 1970). The principal aim of interpretation of the helium line intensities is to derive the helium-to-hydrogen abundance ratio. Abundances determined from radiofrequency lines have distinct advantages over those derived from the optical spectrum. Radiofrequencies do not suffer from interstellar reddening and no correction to the line intensities is necessary. In addition, relative abundances can be obtained from the radio spectrum without the knowledge of b_n values, which, however, are required for interpretation of the optical lines.

Certain simplifying assumptions must be made in order to derive the He:H ratio from radio lines. Observations yield the total energy in the line:

$$E = \int_0^\infty T_L(\nu)d\nu \qquad 26.$$

which is obtained by direct integration of the antenna temperature of the line over the line profile. This antenna temperature is proportional to the line brightness temperature integrated over the source, which itself has been weighted previously by the emission measure of the ion along the line of sight. It is usual to compare the quantity E for the same transition arising from hydrogen and helium. The ratio E_{He}/E_H then represents a weighted average of the helium and hydrogen level populations and respective ion densities over the emitting regions. It is only with three assumptions that a He:H ratio can be obtained, viz:

1. The relative abundance of H and He is constant through the nebula. There is no reason to suspect the presence of abundance variations through the nebula.

2. The helium is all singly ionized. A search for the 173α recombination line from He^+ in several sources has set upper limits of 3×10^{-3} for the ratio $N(He^{++})/N(H^+)$ (Palmer et al 1969). Estimates of the amount of neutral helium present rely on theoretical calculations of the ionization structure in H II regions (Rubin 1968). For stars of sufficiently early spectral type, it appears that all of the helium will be singly ionized, and the Strömgren spheres for hydrogen and helium will coincide. For exciting stars of later type, the volume of ionized helium may be more than two orders of magnitude smaller than the corresponding hydrogen volume. The nebula NGC

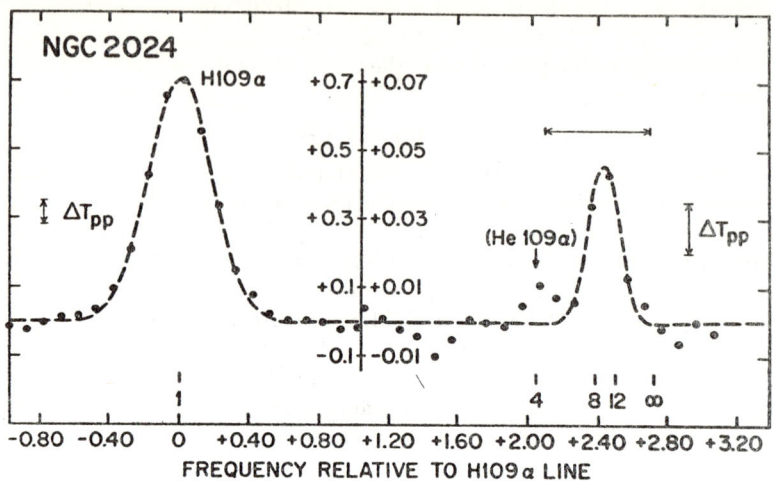

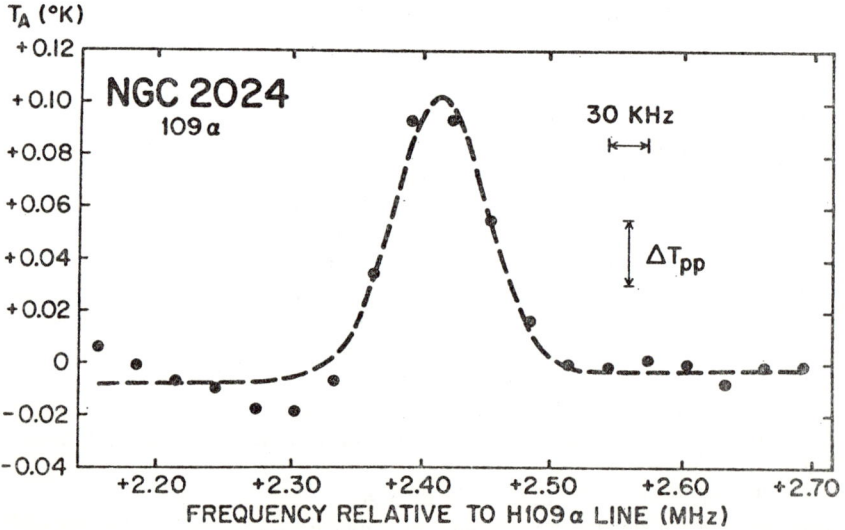

FIGURE 11. *Upper*: H 109α and the anomalous recombination line in NGC 2024 observed by Palmer et al (1969). Numbers below the spectrum indicate the positions of recombination lines of emitters of mass 4, 8, and 12 m_H and ∞, if their radial velocity is the same as hydrogen. Note the weakness of the helium line, which is not significantly larger than the noise. *Arrow*, frequency range covered in lower figure.

Lower: The anomalous recombination line.

2024 is undoubtedly an example of this, because the exciting star is of later type than Orion, and the helium line is exceptionally weak (see Figure 11 in comparison to Figure 1). The ratio E_{He}/E_H does not vary with position in Orion and M17 (Churchwell & Mezger 1970), which indicates that the relative ion densities are constant over the source.

3. The departures from LTE are identical for hydrogen and helium. There are no theoretical reasons to doubt this assumption. The high levels of helium involved in these transitions are sufficiently hydrogenic so that hydrogenic rate coefficients are satisfactory. Temperatures in H II regions are ~ 1 eV so that dielectronic recombination processes do not affect the helium populations. Observational evidence to support this assumption is the agreement to within experimental error, of the $109\alpha/137\beta$ ratio for helium

TABLE 4. Determinations of the helium abundance in Orion from radiofrequency observations

Transition	$N(\text{He})/N(\text{H})$	Remarks
85α	0.089 ± 0.004	a
109α	0.081 ± 0.02	a
109α	0.083 ± 0.004	b
109α	$0.078^{+0.016}_{-0.008}$	c
134α	0.087 ± 0.01	d

a Churchwell & Mezger (1970).
b Palmer et al (1969).
c Reifenstein et al (1970).
d Quoted by Churchwell & Mezger (1970) from Zuckerman & Palmer (1970).

with the measured hydrogen value (Palmer & Zuckerman 1968). Additionally, the quantity E_{He}/E_H from 85α to 134α is effectively constant in several sources (see Table 4 and Churchwell & Mezger 1970).

With these assumptions, the ratio of energy in the lines becomes simply

$$\frac{E_{He}}{E_H} = \frac{N(\text{He}^+)}{N(\text{H}^+)} = \frac{N(\text{He})}{N(\text{H})} \qquad 27.$$

and the relative abundances are found directly from the observed quantities. In some nebulae an anomalous recombination line is blended with the helium line (see Figure 1). However, with sufficient frequency resolution, the lines can be separated to obtain a helium profile.

Some values of $N(\text{He})/N(\text{H})$ as determined for Orion are given in Table 4. Within the experimental errors, there is good agreement among different authors measuring the same transition, and as noted earlier, the derived abundance is constant for all observed transitions. When the helium line is not exceptionally weak as in NGC 2024, there is also agreement in the He:H ratio among different H II regions. Radio observations of five H II regions (Palmer et al 1969) yielded an average value $N(\text{He})/N(\text{H}) = 0.084 \pm 0.003$. Optical values of the helium abundance show little change from one region to another as well (Mathis 1962). However, optically determined abundances show larger scatter of values among various authors (Aller & Liller 1959, Mathis 1962, O'Dell et al 1964), in part because different reddening corrections and b_n calculations have been applied. This scatter was largely eliminated by Palmer et al (1969) who used uniform parameters in reducing three sets of published observations of Orion and derived $N(\text{He})/N(\text{H}) = 0.119 \pm 0.014$. Peimbert & Costero (1969) have incorporated a correction factor for neutral helium into their results to find $N(\text{He})/N(\text{H}) = 0.103 - 0.105$ at three different points in Orion. A review of the helium abundances obtained for other galactic and extragalactic sources is given by Danziger (1970) in this volume. The optical values are in reasonable agreement with the radio determinations, especially since unresolved problems exist (Kaler 1966) with interpretation of hydrogen and helium series in the optical spectrum.

The Anomalous Line

While studying helium recombination lines in H II regions, Palmer et al (1967) discovered a new emission line, occurring at a higher frequency than the hydrogen and helium 109α transition, and with an intensity of 1–3% of the neighboring hydrogen lines. That this new line is indeed a recombination line was verified by detection of the 110α transition. Subsequent observations covering 85α to 166α have detected the emission line in seven H II regions. All objects show the anomalous line to be narrower by a factor of 3 to 7 than corresponding transitions in hydrogen or helium.

The line cannot be identified from its observed frequency alone, because of the uncertainty in the Doppler shift of the emitter. If the transition is attributed to hydrogen, the observed displacement from the strong adjacent hydrogen transition indicates its Doppler shift. However, in a number of sources, this supposition requires the line to be formed in an H II cloud moving with very high velocities (80–120 km sec^{-1}) relative to the principal H II region (Palmer et al 1967). There is contradictory evidence on whether velocities this large could be associated with H II regions (Wilson et al 1959, Sheglov 1968). If the lines were due to helium, the required Doppler velocities would be less; however, the absence of a similar component on the hydrogen profile is puzzling. The closest higher-order line from hydrogen or helium is expected to be much weaker than the observed line.

The appearance of this anomalous line at frequencies higher than the corresponding hydrogen or helium transition suggested that a heavy element might be responsible for the emission. For the line to arise from another element in an H II region, some sort of nonequilibrium process would be required. The most abundant heavy elements (carbon, nitrogen, and oxygen) are approximately $10^{-3} - 10^{-4}$ less abundant than hydrogen. To produce a line from a heavy element with an intensity 1–3% of the hydrogen line requires an overpopulation of the levels by a factor of 10–300.

High levels of complex atoms can be overpopulated to this extent, as a result of dielectronic recombination (Goldberg & Dupree 1967). Considerations of abundance, ionization potential, and the recombination rate showed the atom of neutral carbon to be the most likely candidate. Detailed calculations indicated that overpopulation of carbon levels would indeed occur, but the amount of the overpopulation at temperatures of $10^{4°}K$ is not sufficient to explain the observed intensities under reasonable conditions of emission measure and electron density (Dupree 1969). In Orion, for instance, a combination of high emission measure ($\sim 4 \times 10^6$ pc cm^{-6}) and low density ($N_e \sim 1$ cm^{-3}) is necessary to produce the 109α and 137β lines in emission with the observed line temperature. Furthermore, the theoretical results predict that with increasing n, the emission line should become weaker, and finally turn into an absorption line. Subsequent observations do not confirm this prediction. Hence it is extremely difficult to explain both the intensity and behavior of the anomalous line if it arises in an H II region.

It has been noted that recombination lines can arise from the ionized components in H I regions. Inability to understand the presence of the anomalous line in an H II region led to proposals that it is formed in an H I region, either in a dense cloud under LTE conditions (Zuckerman & Palmer 1968) or under non-LTE conditions, and enhanced by a background continuum (Dupree & Goldberg 1969). In either case, the most likely candidate to produce the line again appears to be carbon. It is the most abundant element with an ionization potential less than hydrogen and is expected to be singly ionized in H I regions by the local radiation field. Ultraviolet lines of C II have been observed in absorption in stellar spectra (Stone & Morton 1967) supporting the presence of singly ionized carbon in the interstellar medium. Carbon may be depleted by grain formation (Field et al 1969), but this appears speculative at present. In the following, we shall refer to the anomalous line as a carbon line.

Accumulating observations seem to indicate that the line is formed in an H I region (Dupree & Goldberg 1969). The linewidth in all cases is a factor of 3 to 7 times less than the width of the adjacent hydrogen transition. In Orion, the width of the C 109α transition is 4.4 ± 1.6 km sec^{-1} (Churchwell & Mezger 1969), in good agreement with the 21-cm absorption profile on the source (Clark 1965), and a factor of 6.8 smaller than the corresponding hydrogen transition. Two sources, Orion A and NGC 2024, have been mapped at several points in both the hydrogen and carbon line. The radial velocity of the hydrogen line varies with position on the source, whereas the velocity of the anomalous line is constant to within experimental error (Zuckerman & Palmer 1968, Churchwell & Mezger 1969). Moreover in every source measured to date, the radial velocity of the anomalous line shows reasonable agreement with the velocity of at least one of the 21-cm absorption profiles on the source. These observations strongly suggest that the anomalous line is formed in a region different from that of the hydrogen and that this region is an H I cloud.

Two sources, Orion A and NGC 2024, have been observed at several points in the nebula in the C 109α transition. In both sources, the energy ($T_L \Delta \nu_L$) of the C 109α line can change by a factor of 3 relative to H 109α in NGC 2024 or to He 109α in Orion. This variation in intensity, coupled with a different radial velocity from the H and He lines, has been attributed to emission arising from a source that is more localized in comparison to the source of the hydrogen or helium emission (Palmer 1968, Zuckerman & Palmer 1968, Churchwell & Mezger 1969).

To date, the carbon line has been found only in emission against the background continuum of an H II region. If we assume that the carbon line arises in a foreground H I region, the line temperature in LTE, in the optically thin approximation, is given by:

$$T_L^* = t_L^*(T_e - T_C) \qquad 28.$$

where t_L^* is the LTE optical depth and T_e is the electron temperature in the H I region, and T_C is the brightness temperature of the continuum directly behind the H I region. Observations and theoretical models of H I clouds indicate that electron temperatures may be 10 to 100°K. The brightness temperatures of H II regions are in excess of these values, especially at low frequencies. Therefore, if the H I region is optically thin, the observation of an emission line argues strongly for emission under non-LTE conditions.

Since the line and continuum emission would appear to originate in different regions, and to have different spatial distributions, it is difficult to choose an appropriate observational parameter to characterize the line. The majority of observations have been made on the 140' telescope of NRAO at Green Bank, West Virginia, and therefore it is convenient to compare the line intensities by using the quantity $T_L \Delta \nu_L$. Here $T_L(°K)$ is the antenna temperature of the line and $\Delta \nu_L(kHz)$ is the full linewidth at half intensity. In the following discussion, we ignore the antenna and beam efficiencies that introduce a slight frequency dependence into the antenna temperature. These quantities are required to convert antenna temperatures to brightness temperatures. Some observations of the carbon line at the continuum maximum in three sources are given in Table 5. Of particular interest is the decrease of $T_L \Delta \nu_L$ between 10.5 and about 5 GHz, and then its relative constancy to 3 GHz. At frequencies less than 3 GHz, its behavior is obscure, for there are insufficient and conflicting observations.

If the carbon emitter is an extended source radiating in LTE, $T_L \Delta \nu_L$ is expected to decrease by an exponential factor, exp X_n (see Equations 20 and 17), which is important at low temperatures ($T \sim 10°K$). On the other hand, if the emission arises from a point source again in LTE, there would be an additional decrease in $T_L \Delta \nu_L$ as ν^2 due to the antenna beamwidth. Yet two arguments weigh against an LTE interpretation. In NGC 2024, the line at 134α appears as strong as at 109α (Zuckerman & Palmer 1970). And, as noted

TABLE 5. Observations of the carbon recombination line

Transition	ν(GHz)	$T_L \Delta \nu_L$ (°K kHz) Orion	NGC 2024	Remarks
C 85α	10.5	30.5 ±3	—	a
C 109α	5.01	13. ±3	7.04	a, b
C 134α	2.70	7.23±1.4	8.68±1.3	c
C 157α	1.68	18.7 ±6.3	—	d
C 166α	1.42	5.6 ±0.5	3.6 ±0.2	e

[a] Mezger & Churchwell (1970).
[b] Palmer et al (1969).
[c] Zuckerman & Palmer (1970).
[d] Churchwell & Edrich (1970).
[e] Zuckerman & Ball (1970).

above, because of the close relation of the emitting cloud and H II regions, it appears unlikely that a line would be observed in emission from an H I region that is in LTE.

A quantitative non-LTE interpretation is difficult partly because information is lacking on the spatial distribution of the carbon emitter. Also, b_n values for H I regions have not been calculated. However, the observations of the carbon-line strength are compatible with behavior possible under non-LTE conditions. The intensity of the line is given by Equation 21 in which there are two terms, perhaps of opposite sign. The first term in parentheses is expected to increase with decreasing frequency of the α transition as the populations approach their LTE values at high principal quantum number. The second term contains the factor T_C which also increases at low frequencies, due to the optical depth of the background H II region. The behavior of the factor $b_n \beta_{nm}$ is difficult to predict because it is strongly dependent on temperature and density. It may take on a positive, a negative, or a zero value, and change sign and value rapidly with n. Hence the line temperature may be decreased, increased, or unchanged by the $b_n \beta T_C$ term. When the two terms are of comparable size, T_L will be extremely sensitive to the value and sign of $b_n \beta_{nm}$. However, upper limits on the values of $b_n \beta$ required by the observed line intensities show reasonable agreement with the values that may be expected from detailed calculations (Dupree & Goldberg 1969). There may also be an additional dependence on frequency if the carbon source does not fill the antenna beam. In the extreme case of a point source, a ν^2 frequency dependence is introduced. At low temperatures, the Boltzmann factor may be important in adding a frequency dependence to t_L^*. It appears possible then to explain qualitatively the observed intensities under non-LTE conditions.

Clearly, more observations are required to determine the source of the carbon line. The spatial distribution of the emission and especially the behavior of the line at high n values are necessary for theoretical interpretation of the emission. If the line originates under conditions of stimulated emission from a background continuum, emission lines would be expected to occur from all ionized components in an H I region. Detection of such lines would give a direct measurement of the degree of ionization and abundances in H I regions. Such results have important implications for the heating mechanisms present in the interstellar medium.

KINEMATICS OF H II REGIONS

Spectral lines, because they can be shifted by the Doppler effect, offer information on the kinematical structure of H II regions. Moreover radiofrequencies, unlike optical frequencies, do not suffer attenuation by the interstellar medium. Therefore radio recombination lines are particularly valuable for investigating large-scale galactic structure.

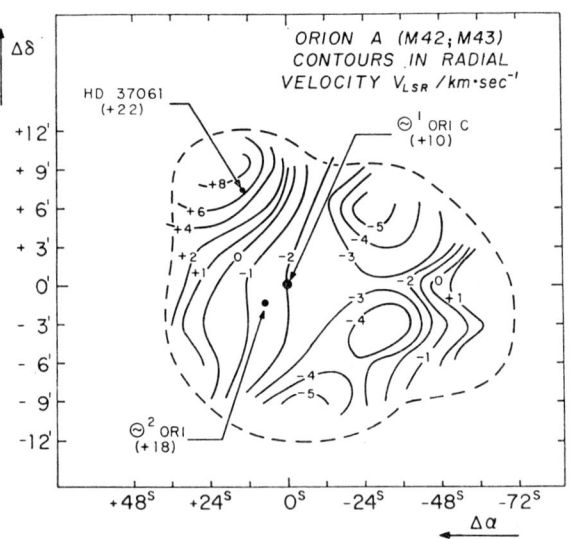

FIGURE 12. Contours of constant radial velocity in Orion as determined by the Doppler shift of the H 109α line (Mezger & Ellis 1968). The outer broken line corresponds to a contour of $T_A = 1°K$ in continuum emission.

Structure of H II regions.—Several extended H II regions have been mapped in a recombination line to determine their internal velocity structure (Gordon & Meeks 1968, Mezger & Ellis 1968, McGee & Gardner 1968, Gordon 1969, Wilson 1969). The results of such a mapping of Orion in the H 109α line (Figure 12) with a 6-arc min beam show that the radial velocity of the emitting region varies by less than 15 km sec^{-1} across the nebula (Mezger

& Ellis 1968). The distribution of velocities within about 9 arc min from the center can be interpreted as a rotation of the nebula around an axis of symmetry tilted slightly to the northwest. The rotation period is 10^7 years or less depending on the inclination of the axis. The westernmost part of the nebula apparently does not participate in the rotation suggested by the observations near the center. It is unclear however just how meaningful this rotation rate is because the rotational period is greater than the probable age of Orion (Mezger & Ellis 1968). In addition, the nebula is expected to have a radial expansion velocity $\sim$10 km sec^{-1}.

The linewidths appear to be determined by Doppler broadening that results from thermal and turbulent motions in the nebula. Studies of line profiles of high principal quantum number, the H 253α line in a low-density object, NGC 7000 (Penfield et al 1967), and the H 164α transition in the dense H II region Orion (Menon & Payne 1969) show that the profiles may be non-Gaussian but Stark broadening is not present in the amount predicted by current theories (Griem 1967). From a compilation of Orion observations, Churchwell & Edrich (1970) find that low-frequency lines appear broader than would be predicted from an extrapolation of high-frequency observations. However, this may be caused not by pressure broadening, but by the combined effects of the increased optical depth of the source at low frequencies and the presence of components with different motions that are included in the beamwidth. If Stark broadening is negligible and the nebula is assumed to be isothermal, the linewidth gives some indication of the expansion and turbulent velocities. Five-point observations in Orion with a 6-arc min beam at H 109α suggest that these velocities are constant over the central portion of the nebula (Churchwell & Mezger 1969), yet measurements of the H 94α transition with a 4-arc min beam indicate an increase in the velocity dispersion with distance from the central portions (Gordon & Meeks 1968).

A symmetrical distribution of radial velocities has been found in NGC 2024 that also implies rotational motion (Gordon 1969). The H 94α line profiles mapped at 2-arc min intervals across the source indicate a decrease in linewidth away from the source center. If the nebula is isothermal with $T_e = 10^{4\circ}$K, the dispersion in radial velocities obtained from the linewidth is about 10 km sec^{-1} or less, which can be interpreted as an expansion or contraction of the nebula at 5 km sec^{-1} or less.

Galactic structure.—Because H II regions occur near hot early-type stars, they may be expected to be tracers of the spiral-arm patterns in the Galaxy. Two surveys of the H 109α line in both northern and southern skies have provided valuable data on the large-scale distribution of H II regions (Reifenstein et al 1970, Wilson et al 1970). The measured radial velocity of the line can be used in conjunction with an adopted model of galactic rotation to derive a kinematic distance to the H II region. Within the solar velocity circle, there is an ambiguity between two distances although it can be re-

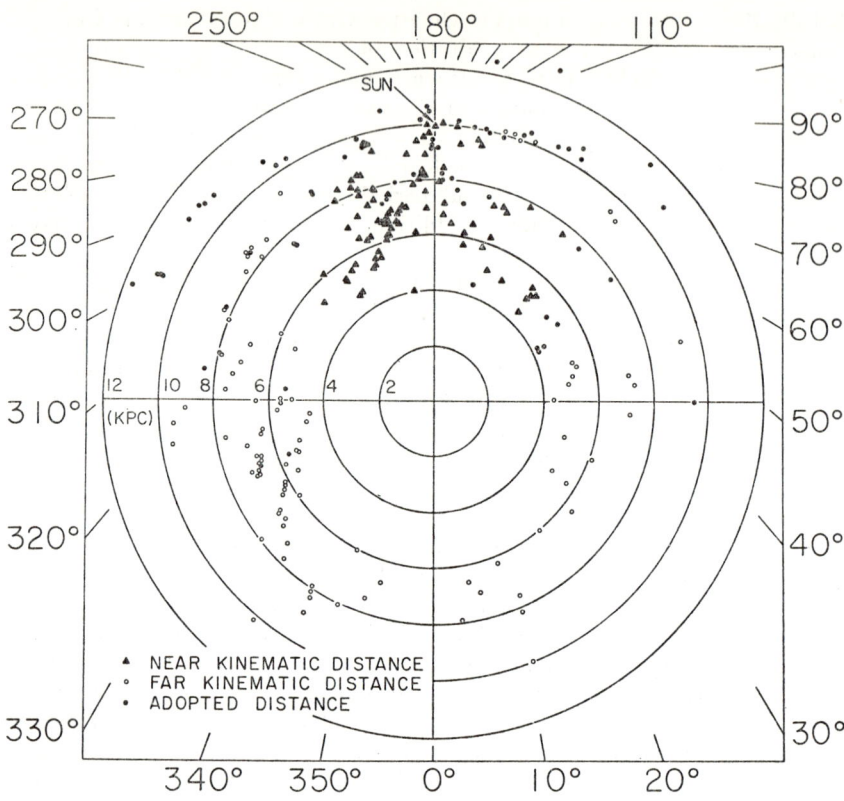

FIGURE 13. Galactic distribution of H II regions as determined from H 109α emission. Results from the northern sky survey (NRAO-MIT) (Reifenstein et al 1970) and the southern survey (NRAO-MIT-CSIRO) have been combined in this figure (Wilson et al 1970). Seven sources that lie within the 4-kpc arm are not shown. The Schmidt model of galactic rotation was assumed. "Adopted" distances were obtained by resolution of the distance ambiguity.

solved in some cases by consideration of the galactic latitude or by a comparison with optical observations or with 21-cm line data. In Figure 13, results of surveys of the northern and southern skies (Reifenstein et al 1970, Wilson et al 1970) show the distribution of H II regions determined from kinematic distances and projected on the galactic plane. Most of the H II regions lie between 4 and 12 kpc from the galactic center. Seven sources (not shown in Figure 13) inside the 4-kpc expanding arm appear to be associated with the galactic center. At distances greater than 12 kpc hardly any H II regions are found. If only the intrinsically brightest regions in the northern survey are considered, there is a tendency for them to be located between 4–7 kpc (Reifenstein et al 1970). It is not possible to draw a unique spiral structure from these data on H II regions because of the distance ambiguity that is present (Wilson et al 1970). However, comparisons may be made between H II regions and other galactic components.

The radial velocities of H II regions as determined by the southern H 109α survey are in fairly good agreement with available optically measured velocities (Wilson et al 1970). The distribution of H II regions has been compared to the H I distribution as determined by 21-cm surveys (Kerr et al 1968, Mezger et al 1970a, Reifenstein et al 1970). There is good correspondence between the kinematics and large-scale distribution of neutral and ionized hydrogen. H II regions generally appear close to areas of high H I surface brightness and hence to the local density maximum of neutral hydrogen. In our Galaxy this peak occurs between 7 and 11 kpc. However, in the northern survey most of the brightest H II regions are within 7 kpc of the center and do not coincide with the maximum H I density.

ACKNOWLEDGMENTS

We are grateful to E. Churchwell, R. Hjellming, P. Mezger, T. Wilson, and B. Zuckerman for providing us with many of the figures and their results prior to publication. This article was written in part with the support of a United States National Aeronautics and Space Administration grant NGL-22-007-006 and a United States National Science Foundation grant GP 6750.

LITERATURE CITED

Aller, L. H., Liller, W. 1959, *Ap. J.*, **130**, 45
Andrews, M. H., Hjellming, R. M. 1969, *Ap. Lett.*, **4**, 159
Baker, J. G., Menzel, D. H. 1938, *Ap. J.*, **88**, 52
Brocklehurst, M. 1970, *MNRAS* (Submitted)
Burgess, A., Summers, H. P. 1969, *Ap. J.*, **157**, 1007
Churchwell, E., Edrich, J. 1970, *Astron. Ap.* (In press)
Churchwell, E., Mezger, P. G., 1969, *Bull. AAS*, **1**, 337
Churchwell, E., Mezger, P. G. 1970, *Ap. Lett.* (In press)
Clark, B. G. 1965, *Ap. J.*, **142**, 1398
Danziger, I. J. 1970, *Ann. Rev. Astron. Ap.*, **8**, 161
Davies, L. G., Ring, J., Selby, M. J. 1964, *MNRAS*, **128**, 339
Dieter, N. H. 1967, *Ap. J.*, **150**, 435
Dravskikh, Z. V., Dravskikh, A. F. 1964, *Astron. Tsirk.*, **282**, 2
Dupree, A. K. 1969, *Ap. J.*, **158**, 491
Dupree, A. K. 1970 (Unpublished)
Dupree, A. K., Goldberg, L. 1969, *Ap. J. (Lett.)*, **158**, L49
Dyson, J. E. 1967, *Ap. J. (Lett.)*, **150**, L45
Eissner, W., Martins, P. de A. P., Nussbaumer, H., Saraph, H. E., Seaton, M. J. 1969, *MNRAS*, **146**, 63
Field, G. B., Goldsmith, D. W., Habing, H. J. 1969, *Ap. J. (Lett.)*, **155**, L149
Flower, D. R., Seaton, M. J. 1969, *Mem. Soc. Roy. Sci. Liège*, **17**, 251
Gardner, F. F., McGee, R. X. 1967, *Nature*, **213**, 480
Gayet, R., Hoang Binh, D., Joly, F., McCarroll, R. 1969, *Astron. Ap.*, **1**, 365
Goldberg, L. 1966, *Ap. J.*, **144**, 1225
Goldberg, L., Dupree, A. K. 1967, *Nature*, **215**, 41
Goldberg, L., Cesarsky, D. 1970 (In preparation)
Goldwire, H. C., Jr. 1968, *Ap. J. Suppl.*, **17**, 1171
Gordon, M. A. 1969, *Ap. J.*, **158**, 479
Gordon, M. A., Meeks, M. L. 1967, *Ap. J. (Lett.)*, **149**, L21

Gordon, M. A., Meeks, M. L. 1968, *Ap. J.*, **152**, 417
Greenberg, D. W. 1969, *Ap. J. (Lett.)*, **155**, L51
Griem, H. R. 1967, *Ap. J.*, **148**, 547
Hjellming, R. M., Andrews, M. H., Sejnowski, T. J. 1969, *Ap. Lett.*, **3**, 111
Hjellming, R. M., Churchwell, E. 1969, *Ap. Lett.*, **4**, 165
Hjellming, R. M., Davies, R. D. 1970, *Astron. Ap.* (In press)
Höglund, B., Mezger, P. G. 1965, *Science*, **150**, 339
Kaler, J. B. 1966, *Ap. J.*, **143**, 722
Kardashev, N. S. 1959, *Astron. Zh.*, **36** 838 (Engl. Transl. *Soviet Astron. AJ*, **3**, 813)
Kerr, F., Burke, B., Reifenstein, E., Wilson, T., Mezger, P. 1968, *Nature*, **220**, 1210
Lilley, A. E., Palmer, P., Penfield, H., Zuckerman, B. 1966, *Nature*, **211**, 174
Mathis, J. S. 1962, *Ap. J.*, **136**, 374
McGee, R. X., Gardner, F. F., 1968, *Austr. J. Phys.*, **21**, 149
Menon, T. K., Payne, J. 1969, *Ap. Lett.*, **3**, 25
Menzel, D. H. 1969, *Ap. J. Suppl.*, **18**, 221
Mezger, P. G. 1968, *Interstellar Ionized Hydrogen*, 477 (Terzian, Y., Ed., New York: Benjamin)
Mezger, P. G., Churchwell, E. 1970 (Preprint)
Mezger, P. G., Ellis, S. A. 1968, *Ap. Lett.*, **1**, 159
Mezger, P. G., Henderson, A. P. 1967, *Ap. J.*, **147**, 471
Mezger, P. G., Höglund, B. 1967, *Ap. J.*, **147**, 490
Mezger, P. G., Wilson, T. L., Gardner, F. F., Milne, D. K. 1970a, *Astron. Ap.*, **4**, 96
Mezger, P. G., Wilson, T. L., Gardner, F. F., Milne, D. K. 1970b, *Ap. Lett.*, **5**, 117
Münch, G. 1968, *Interstellar Ionized Hydrogen*, 507 (Terzian, Y., Ed., New York: Benjamin)
Münch, G., Wilson, O. C. 1962, *Z. Ap.*, **56**, 127
O'Dell, C. R., Peimbert, M., Kinman, T. D. 1964, *Ap. J.*, **140**, 119
Oster, L. 1961, *Ap. J.*, **134**, 1010
Osterbrock, D., Flather, E. 1956, *Ap. J.*, **129**, 26
Palmer, P. 1967, *Ap. J.*, **149**, 715
Palmer, P. 1968 (Ph.D. thesis, Harvard Univ.)
Palmer, P., Zuckerman, B. 1968, *Astron. J.*, **73**, S196
Palmer, P., Zuckerman, B., Penfield, H., Lilley, A. E., Mezger, P. G. 1967, *Nature*, **215**, 40
Palmer, P., Zuckerman, B., Penfield, H., Lilley, A. E., Mezger, P. G. 1969, *Ap. J.*, **156**, 887
Peimbert, M. 1967, *Ap. J.*, **150**, 825
Peimbert, M., Costero, R. 1969, *Bol. Obs. Ton. Tac.*, **5**, 3
Penfield, H., Palmer, P., Zuckerman, B. 1967, *Ap. J. (Lett.)*, **148**, L25
Reifenstein, E. C. III, Wilson, T. L., Burke, B. F., Mezger, P. G., Altenhoff, W. 1970, *Astron. Ap.*, **4**, 357
Rubin, R. H. 1968, *Ap. J.*, **153**, 761
Rubin, R. H., Turner, B. E. 1969, *Ap. J. (Lett.)*, **157**, L41
Seaton, M. J. 1964, *MNRAS*, **127**, 177
Sejnowski, T. J., Hjellming, R. M. 1969, *Ap. J.*, **156**, 915
Sheglov, P. V. 1968, *Ap. Lett.*, **1**, 145
Silk, J., Werner, M. W. 1969, *Ap. J.*, **158**, 185
Sorochenko, R. L., Puzanov, V. A., Salomonovich, A. E., Shteinshleger, V. B. 1969, *Ap. Lett.*, **3**, 7
Spitzer, L., Jr., Tomasko, M. G. 1968, *Ap. J.*, **152**, 971
Stone, M. E., Morton, D. C. 1967, *Ap. J.*, **149**, 29
Terzian, Y., Balick, B. 1969, *Ap. Lett.*, **4**, 195
Webster, W. J., Jr. 1970 (PhD thesis, Case Western Reserve Univ.)
Weedman, D. W. 1966, *Ap. J.* **145**, 965
Williams, D. R. W. 1967, *Ap. Lett.*, **1**, 59
Wilson, O. C., Münch, G., Flather, E. M., Coffeen, M. F. 1959, *Ap. J. Suppl.*, **4**, 199
Wilson, T. 1969 (PhD thesis, Massachussetts Inst. Technol.)
Wilson, T. L., Mezger, P. G., Gardner, F. F., Milne, D. K. 1970, *Astron Ap.* (In press)
Zuckerman, B., Ball, J. A. 1970 (In preparation)
Zuckerman, B., Palmer, P. 1968, *Ap. J. (Lett.)*, **153**, L145
Zuckerman, B., Palmer, P. 1970, *Astron. Ap.*, **4**, 244
Zuckerman, B., Palmer, P., Penfield, H., Lilley, A. E. 1967, *Ap. J. (Lett.)*, **149**, L61

PULSARS

A. HEWISH

Cavendish Laboratory, University of Cambridge

INTRODUCTION

Pulsars are a new and remarkable class of galactic objects discovered towards the end of 1967. Undoubtedly their most intriguing feature is that they radiate short pulses of energy at exceedingly well-maintained intervals, typically in the neighborhood of 1 second. But this regular behavior is associated with great variability in almost every other respect, so pulsars present an exciting challenge to observer and theoretician alike.

The first pulsars (1) were revealed at Cambridge during a systematic sky survey undertaken to gain information about the angular structure of compact radio sources through the method of interplanetary scintillation. A large phased array was constructed for this survey and the data analysis took note of all sources that exhibited intensity variations on a short time scale. Pulses from CP 1919 were first recorded on November 28, 1967. Variable radio emission, not then recognized as pulses, had been noted on several occasions during preceding months, but the sporadic nature of the radiation argued against a celestial orgiin. Indeed, the detection of individual pulses during the more detailed observations carried out later did little to engender confidence in the source, and its reality was accepted only when the difficulty of a terrestrial origin appeared greater than that of a celestial one.

The following year witnessed a phenomenal output of observational and theoretical work on pulsars, as well as unbounded speculation concerning their nature. As 1968 drew to a close, the discovery of the rapid pulsars in Vela and the Crab Nebula (2–4) caused an overwhelming swing of opinion in favor of the spinning neutron-star model. Further confirmation was added early in 1969 when NP 0532, the pulsar in the Crab Nebula, was optically identified with the star from which the supernova is believed to have originated (5, 6).

Toward the end of 1969 the number of pulsars reported had risen to 50 and it does not seem likely that the number will increase significantly until new radio telescopes of greater collecting area are available. Initial hopes that computer integration techniques might afford substantial increases of sensitivity have largely faded on account of the great variability of source intensity, and present instruments appear to have nearly exhausted their capacity. This should not be important in determining the nature of pulsars, however, because the sample currently available shows that the parameters are falling into a meaningful pattern.

The main purpose of this review is to present what is known about pulsars from observation. The theoretical situation will then be surveyed briefly and it will be seen that while no model can yet be regarded as satisfactory the field is narrowing. The reviewers' task in this rapidly expanding subject may be likened to swimming up a waterfall. An attempt has been made to present a broad picture based on a literature survey conducted in November 1969, with occasional updating where possible. The references cited are intended to be representative rather than exhaustive, and indulgence is craved for serious omissions or oversimplifications. Thanks are due to the many authors who have made available material prior to publication.

In addition to providing a new class of stars, whose study holds great promise for furthering our understanding of stellar evolution and the properties of highly condensed matter, pulsars will undoubtedly be important as a tool for investigating the Galaxy. Measurements of Faraday rotation, combined with those of dispersion, are being used in new determinations of the galactic magnetic field, and both dispersion and interstellar scintillation give information about the interstellar plasma and its fine structure. Pulsars also provide a new type of astronomical clock which may be of sufficient accuracy for tests of general relativity. These further topics will not be discussed in this review.

OBSERVATIONS

Search techniques and limitations.—The swept-frequency nature of pulsar radiation, caused by dispersion in the interstellar plasma, is one of the dominant features of the received signals. Accurate measurements at different frequencies (7) have shown that the rate of change of frequency ($\dot{\nu}$) obeys the relation predicted for a wideband pulse (all frequencies emitted simultaneously at the source) which has been dispersed by traveling through an ionized gas having a plasma frequency much lower than the radio frequency. Thus

$$\dot{\nu} = \frac{1.23 \times 10^{-4} \nu^3}{\int n_e dl} \text{ MHz sec}^{-1}$$

The integrated electron content ($\int n_e dl$) along the line of sight will be referred to as the dispersion measure (DM) and has units electron centimeter^{-3} parsec. The dynamic spectrum of the received pulse is therefore as shown in Figure 1, for a pulse of duration is W_p. At some frequency the signal has an instantaneous bandwidth $\delta\nu \sim W_p \dot{\nu} \propto \nu^3$ and this bandwidth is an important parameter in the detection and study of pulsars.

A receiver having a bandwidth B cannot resolve fine structure on a time scale shorter than $B/\dot{\nu}$. This means that very small bandwidths are required to investigate highly dispersed pulsars at low frequencies, which entails a corresponding loss of sensitivity. These difficulties increase in direct proportion to DM and hence to the distance of the source.

Unless special steps are taken to eliminate pulse smearing, the maximum usable bandwidth is set by the condition that individual pulses must be resolved, which implies $B < P\dot{\nu}$, where P is the period of the source. It may also be shown that there is little gain in sensitivity when B exceeds $\delta\nu = W_p\dot{\nu}$. Such considerations determine the optimum bandwidth, and the dependence of B upon $\dot{\nu}$ (which varies as ν^3) means that considerably larger bandwidths are possible at higher frequencies. When allowance is also made for the sky background temperature, which decreases roughly as $\nu^{-2.7}$, it follows that if a fixed collecting area is used, detection sensitivity increases rapidly with

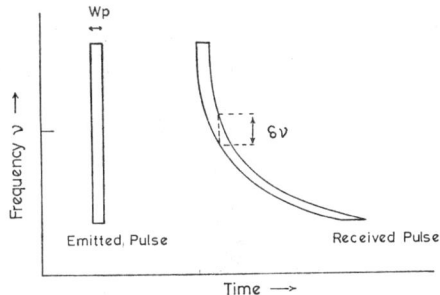

FIGURE 1. Schematic diagram of the dynamic spectrum of a single pulse.

increasing frequency until about 500 MHz, where the pulsar energy spectrum begins to fall off steeply. It is possible to increase the bandwidth beyond the limit $B \sim P\dot{\nu}$ by multichannel techniques. This method has been used at Molonglo to achieve maximum sensitivity at DM $\sim$ 400.

Pulsar searches have been carried out at several observatories. The 50 known pulsars are listed in Table 1. The success of the Molonglo search may be ascribed to the use of a large collecting area at a near-optimum frequency of 408 MHz. Digital techniques designed to hunt for pulse trains, and methods utilizing pulse dispersion to discriminate against impulsive interference, have been employed at Greenbank, Arecibo, and Jodrell Bank. While it is difficult to compare different techniques that have not been used with the same radio telescope, there is no evidence that computer searches have been more successful than scanning records for individual large pulses.

In the absence of a detailed list of search areas and observational limits, the situation may be summarized by stating that a large portion of the sky in both hemispheres has been surveyed without serious observational bias within the period range $0.1 < P < 10$ sec for DM < 100. To investigate the possibility that sources of higher dispersion were being excluded, an additional search was made (8) of the galactic plane with the sensitivity optimized at DM $\sim$ 400. The greatest dispersion measure so far obtained is DM $\sim$

270, so it is clear that the number of sources excluded on account of high dispersion is exceedingly small.

Galactic distribution of pulsars.—The positions of the known pulsars in galactic coordinates are shown in Figure 2. This distribution may be slightly influenced by observational selection effects, but there is clearly a significant clustering of pulsars towards the galactic plane. The concentration is most pronounced just below the plane. It has been suggested (9) that this peculiarity is due to a thin layer of material causing enhanced dispersion along the plane itself. A search (8) sensitive to highly dispersed sources has not revealed the hidden pulsars, however, and this explanation does not account for

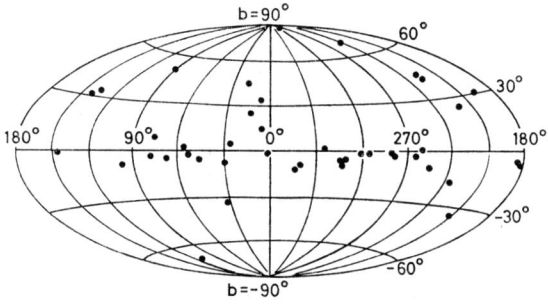

GALACTIC DISTRIBUTION OF PULSARS

FIGURE 2. Pulsar positions in galactic coordinates.

the absence of a similar concentration of sources just above the plane. It has also been suggested that the effect may be due to the position of the Sun relative to the plane.

While the details of the distribution are still clouded by statistical uncertainty, it is clearly beginning to reveal galactic structure. This indicates, in the absence of other data, that pulsars are galactic sources situated at distances probably within 1 or 2 kpc. A histogram of the distribution of pulsars in galactic latitude is plotted in Figure 3. This again reveals the marked concentration towards low latitude. It is notable that the distribution cannot be explained by a simple disk population of sources of comparable intrinsic luminosity. It appears necessary to assume a considerable spread of luminosity; this implies that the high-latitude pulsars are relatively weak and numerous as compared with the rarer, more intense, and more distant pulsars seen in the direction of the plane. Alternatively, the distribution may be influenced by the spiral structure of the Galaxy (13).

Pulsar periods.—The histogram of pulsar periods given in Figure 4 shows a marked concentration of values in the range 0.5 to 1.0 sec. The shape of this distribution is not seriously modified by selection effects, at least within

PULSARS 269

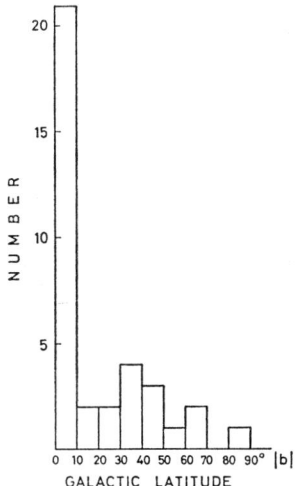

FIGURE 3. The distribution of pulsars in galactic latitude.

the range 0.1–10 sec. The abrupt cutoff above 3 sec and the more gradual decrease towards the short periods therefore represent genuine population effects. As will be discussed in a later section, the distribution might be explained by an evolutionary sequence in which pulsars are initially created

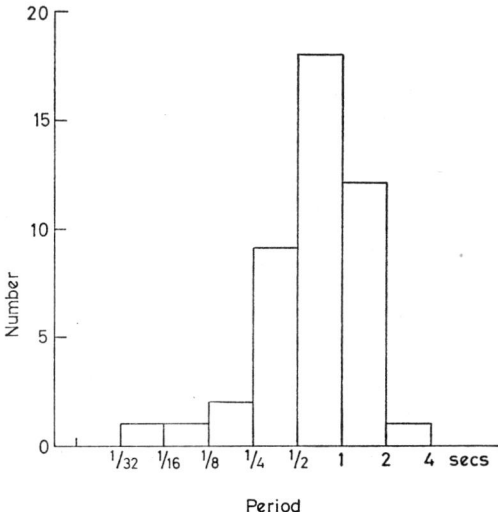

FIGURE 4. Histogram of pulsar periods.

TABLE 1. Pulsar catalogue

	Right Ascension (1950.0)			Declination			Period (P_{sec})	$\frac{dP}{dt}_{(sec/sec)} \times 10^{15}$	T (year)	DM (cm^{-3} pc)
	h	m	s	°	′	″				
MP	00	31	37	−07	37	—	0.940	—	—	12
MP	02	54	24	−54	—	—	0.448	—	—	10
CP	03	29	11.17	−54	24	37	0.714 518 625	2.03	1.1×10⁷	26.75
MP	04	50	22	−18	00	—	0.54978	—	—	25
NP	05	25	45	−21	58	—	3.745 491 780	—	—	50.2
NP	05	31	31.46	−21	58	54.8	0.033 099 324 (28–6–69)	422.66	2.5×10³	56.88
PSR	06	28	53	−28	33	—	1.244 436	—	—	34.4
MP	07	36	51	−40	35	—	0.375	—	—	100
CP	08	08	58.00	−74	38	10	1.292 241 315	<0.07	>6×10⁸	5.77
MP	08	18	6	−15	—	—	1.237	—	—	25
AP	08	23	52	−26	48	00.0	0.530 659 625	124.26	23×10⁴	19.4
PSR	08	33	39	−45	00	05	0.089 209 298 (24–3–69)	6.81	5.9×10⁶	63
CP	08	34	26.3	−06	20	47.0	1.273 763 256	—	—	12.8
MP	08	35	34	−40	—	—	0.765	—	—	120
HP	09	04	—	−77	40	—	1.579 05	—	—	—
MP	09	40	40	−56	—	—	0.662	—	—	145
PP	09	43	19.6	−10	05	33	1.097 707	—	—	15.35
CP	09	50	30.85	−08	09	49.8	0.253 065 046	0.238	3.4×10⁷	2.98
MP	09	59	51	−54	37	—	1.436 551	—	—	90
CP	11	33	27.39	−16	07	30.4	1.187 911 129	3.75	1.0×10⁷	4.87
MP	11	54	45	−62	—	—	0.400	—	—	270
AP	12	37	17	−25	09	30	1.382 451	—	—	8.5
MP	12	40	21	−63	36	—	0.388	—	—	220
MP	12	59	43	−50	—	—	0.690	—	—	20
MP	13	26	35	−66	30	—	0.788	—	—	60
MP	14	49	22	−65	—	—	0.180	—	—	90
PSR	14	51	29	−68	32	—	0.264	—	—	8.6

PULSARS

							Period (s)			
HP	15	08	03.27	55	42	50	0.739 677 806	5.12	4.6×10⁶	19.6
MP	15	30	23	−53	30	—	1.368 852	—	—	20
AP	15	41	10	09	38	—	0.748 45	—	—	35
MP	16	04	37	−03	—	—	0.421	—	—	10
MP	16	42	—	−03	12	—	0.387 688	—	—	35
MP	17	06	35	−16	37	—	0.654	—	—	10
MP	17	27	50	−47	40	—	0.825 683	—	—	140
MP	17	47	56	−46	56	—	0.742 349	—	—	40
PSR	17	49	49	−28	05	57	0.562 553 758	71.8	2.5×10⁵	50.88
MP	18	18	14	−04	25	—	0.598 08 361	—	—	70
MP	18	57	44	−25	—	—	0.611 8	—	—	35
MP	19	11	15	−04	47	—	0.825	—	—	75
CP	19	19	36.1	21	47	12.0	1.337 301 134	1.35	3.2×10⁷	12.43
PSR	19	29	52	10	52	49	0.226 516 75	—	—	8.0
JP	19	33	31.9	16	09	58.8	0.358 735 236	6.1	1.9×10⁶	143
MP	19	44	38	17	—	—	0.440	—	—	35
JP	19	46	—	35	30	—	0.717 35	—	—	100
AP	20	16	00.07	28	30	31	0.557 953 403	0.127	1.4×10⁸	14.2
JP	20	22	20	51	44.5	—	0.529 1	—	—	10
PSR	20	45	47.6	−16	27	50	1.961 566 604	11.7	5.3×10⁶	11.4
JP	21	13	30	46	36	—	1.014 78	—	—	100
HP	22	18	20	47	40	—	0.538 461	—	—	43.8
AP	23	03	30	30	45	—	1.575 869	—	—	46

Notes:
CP   Cambridge
MP } Molonglo
PSR }
JP   Jodrell Bank
HP   Harvard
NP   NRAO, Greenbank
AP   Arecibo
PP   Pushchino

Data for dP/dt refer to epoch July 1, 1969 as given by Hunt (14).

More accurate values for NP 0532 are given in text.

with short periods that subsequently increase until some further process "turns off" the radiation.

The astonishing regularity of individual pulsar periods was recognized at the outset as one of the most interesting features of these sources. Extended measurements of pulse arrival time over many months are necessary to detect systematic variations in period for typical pulsars (10, 11), although PSR 0833 and NP 0532 exhibit larger variations that are more readily detectable (12, 30).

Two factors usually limit the accuracy of the observations. One is the Doppler shift of period arising from the orbital motion of the Earth. A correction for this factor demands either a highly accurate position for the source or extended measurements that enable periodic annual variations to be distinguished from genuine source variations. The second, and more fundamental, limitation is the presence of rapid and irregular changes of pulse shape that may often be important.

Two quantities are useful for specifying systematic variations of period to first order. These are (dP/dt), the mean rate of change of period P averaged over some suitable interval, and the characteristic time $T = P(dP/dt)^{-1}$. Some recent values given by Hunt (14) are included in Table 1. The periods of all pulsars so far studied are systematically increasing, and the two sources of shortest period have characteristic times roughly in proportion to their periods. The relation between T and P is plotted in Figure 5. There is little relation between these parameters for the bulk of the sources, which exhibit a wide scatter of characteristic times within the range $10^6 - 10^8$ years.

The greatest timing accuracy has been obtained for the source NP 0532, which has a sharp feature on both the radio and optical pulses. Following some initial uncertainty due to a small discontinuous change in period, the optical (15) and radio (16) measurements give results as follows:

DATA

	Arecibo (radio)	Princeton (optical)
$f = P^{-1}$	30.2137051 Hz (JD 2440352.253)	30.2155298589 Hz (JD 2440297.50904)
$\dfrac{df}{dt}$	-0.38594×10^{-9} Hz sec^{-1}	$-0.3859294 \times 10^{-9}$ Hz sec^{-1}
$\dfrac{d^2f}{dt^2}$	$1.1 \pm 0.11 \times 10^{-20}$ Hz sec^{-2}	1.3×10^{-20} Hz sec^{-2}

Excluding the rapid pulsars NP 0532 and PSR 0833, there is no evidence that the systematic variations of period are not completely stable. Apparent variations may, of course, arise from relativistic effects when pulsars are viewed close to the Sun (20), or from changes of integrated electron content along the line of sight (21).

Evidence for sudden changes of period was first obtained on the source

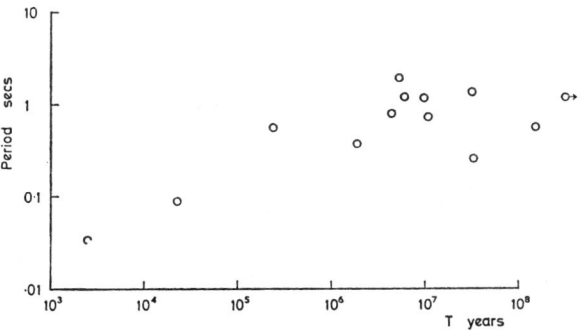

FIGURE 5. The relation between

$$T = \left(\frac{1}{P}\frac{dP}{dt}\right)^{-1}$$

and period P.

PSR 0833 (22, 23). Precisely how the change took place is not known, but the period decreased by $\sim$200 msec (about one part in 5×10^5) between February 24 and March 3, 1969. It is significant that (dP/dt) increased by about 1% following this event and estimates suggest that the period will revert to its original value with a relaxation time of a few years. More recently a similar discontinuity has been observed for NP 0532 (15, 16). During this event, which took place on September 28 ± 1, 1969, P^{-1} initially increased by 2.6×10^{-7} Hz and then reverted to a new value, 4.3×10^{-8} Hz higher in frequency than the original value, with a relaxation time of 4.3 days. It is clearly important to watch all pulsars closely for similar events.

There is also evidence (17) for a quasisinusoidal modulation of the period of NP 0532. This component has an amplitude of 300 ± 50 μsec and a period of 72 ± 3 days. Such an effect might be caused by the orbital motion of a satellite associated with the pulsar.

Mean pulse shape.—Individual pulses from a typical pulsar are extremely variable, both in shape and intensity, and the variations exist on a time scale ranging from fractions of a millisecond to several weeks or longer. As discussed later, there is good evidence that some of this variability may be ascribed to the interstellar plasma, but on the shortest and longest time scales it will be shown that changes must be intrinsic to the source. When pulses are averaged for a few minutes, however, the mean pulse envelope obtained is stable and exhibits characteristic differences between one pulsar and another (7). Some typical pulse envelopes are displayed in Figures 6a and 6b. The radio data in Figure 6b are from Arecibo (17).

One obvious feature of the pulse envelope is that it has a duration that increases approximately in direct proportion to the pulse period. This rela-

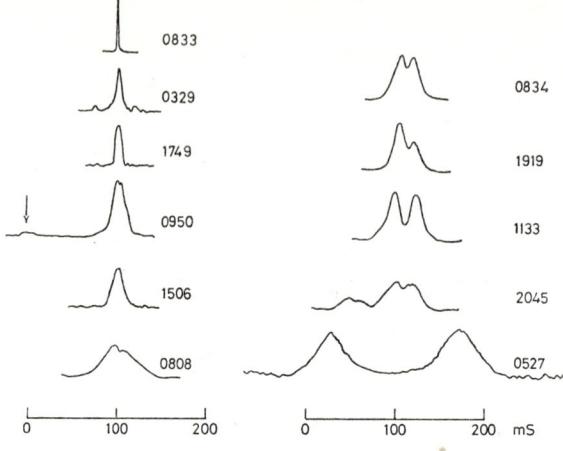

FIGURE 6a. Some typical mean pulse envelopes.

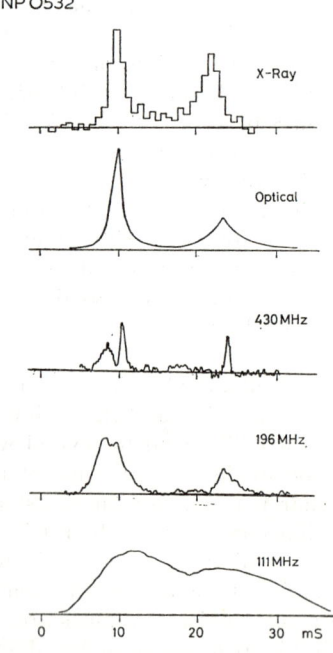

FIGURE 6b. Mean pulse envelope of NP 0532 at radio, optical, and X-ray wavelengths.

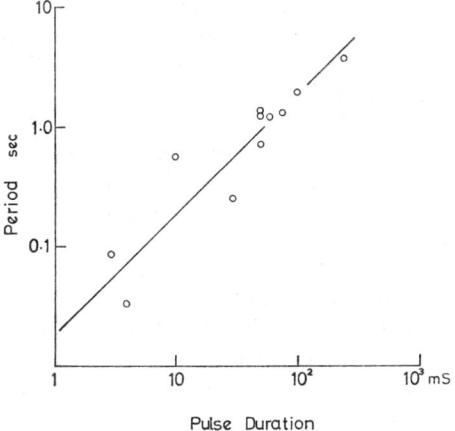

FIGURE 7. Pulse duration as a function of period P. The duration is defined as the interval outside which the envelope is always less than 10% of the peak value.

tionship is shown in Figure 7. The points are considerably scattered but a close correspondence is not expected since the envelopes have such different shapes.

Few generalizations are yet possible with regard to the mean pulse envelopes. A double-peaked curve is a typical feature of pulsars having periods of about 1 sec or greater. Even for simple envelopes, such as that of PSR 0833, there is considerable asymmetry, It is notable that the leading edge of the sharp peak in the envelope of NP 0532 is unresolved with a time constant of 100 μsec.

CP 0950 was the first pulsar for which a secondary pulse, or interpulse, was detected in the mean envelope at a point roughly midway between the main pulses (24). In this instance the interpulse is weak, having an intensity of only 1.5% of the main pulse. Another source currently known with certainty to have an interpulse is NP 0532 and here its strength is comparable to that of the main pulse. Radio observations (17) show that it is the sharp feature of the main pulse which is repeated in the interpulse. The source HP 0904 may also have an interpulse. Since interpulses have been detected in only 2 or 3 of the known pulsars this property is not typical. It is interesting that the interpulse is displaced by about one tenth of the period from the symmetrical midway position for CP 0950 and NP 0532.

Dependence of pulse shape on wavelength.—Observations outside the radio spectrum have been possible only for NP 0532. The mean pulse envelope for this source as obtained at X-ray, optical, and radio wavelengths is shown in Figure 6b. The X-ray and optical envelopes appear to be substantially the same, but the envelope changes markedly at radio wavelengths. At a frequency of 430 MHz the shape has a complexity absent in the optical and X-ray pulse. It is also interesting that below 430 MHz the pulsewidth rapidly broadens until at 111 MHz the pulsed nature of the radiation is almost fully smeared out. This rapid broadening can be satisfactorily explained in terms of time delays resulting from interstellar scintillation (17). Bradt et al (25) showed that the X-ray and optical pulses are simultaneous to an accuracy of 1 msec; at radio wavelengths some initial uncertainty in the correction arising from dispersion has now been removed (17) and again the optical and radio pulses are simultaneous (21) to within 0.3 msec.

The increase of pulse width at frequencies below about 400 MHz has been found for some other pulsars, although to a lesser degree than for NP 0532. Observations made at Arecibo (17, 26) show that several sources have a mean pulse envelope which increases in duration approximately as (frequency)$^{-1/4}$. This law is not always obeyed, however, since CP 0950 and CP 1919 show no significant increase in width at lower frequencies (27). The second peak of CP 1919, for example, has a radio spectrum which appears to differ from that of the first peak, but the width stays constant. Again, observations of PSR 0833 at 4800 MHz show that the pulse is wider than at 1720 MHz and it has been suggested that this may be due to a double-peaked pulse at the higher frequency (31).

Polarization.—Soon after the discovery of pulsars it was found that individual pulses were often highly polarized and both linear and elliptical components were reported (28, 29). But when many pulses were averaged the net polarization usually vanished or fell to a low value. The picture was undoubtedly confused by the varying polarization across the pulse, and by the rapidly changing fine structure to be described later.

The first results indicating a systematic and stable polarization were obtained by Radhakrishnan et al (30) on PSR 0833. This source is distinguished by a simple mean pulse envelope and is unique in producing approximately 100% linear polarization. The polarization vector swings through nearly 90° between the leading and trailing edges of the pulse as shown in Figure 8a. When allowance is made for Faraday rotation, the angle is independent of wavelength. This regular behavior appears to be a constant feature of this source and it was not affected by the sudden change in period already discussed.

A strikingly similar polarization behavior has also been found (32) in the optical radiation from NP 0532. The degree of polarization is substantially less than for 0833, varying from about 5–10% near the peaks of each pulse to about 20–30% in the wings. Its variation with time is shown in Figure 8b which indicates how the polarization vector swings through about 130°, in the same sense, during both the main and secondary pulses. The radio polarization of NP 0532 is somewhat more complex. Data at 430 MHz show (19) that the initial broad feature (precursor) of the main pulse is ∼100% linearly polarized while the sharp feature is ∼18% linearly polarized in the

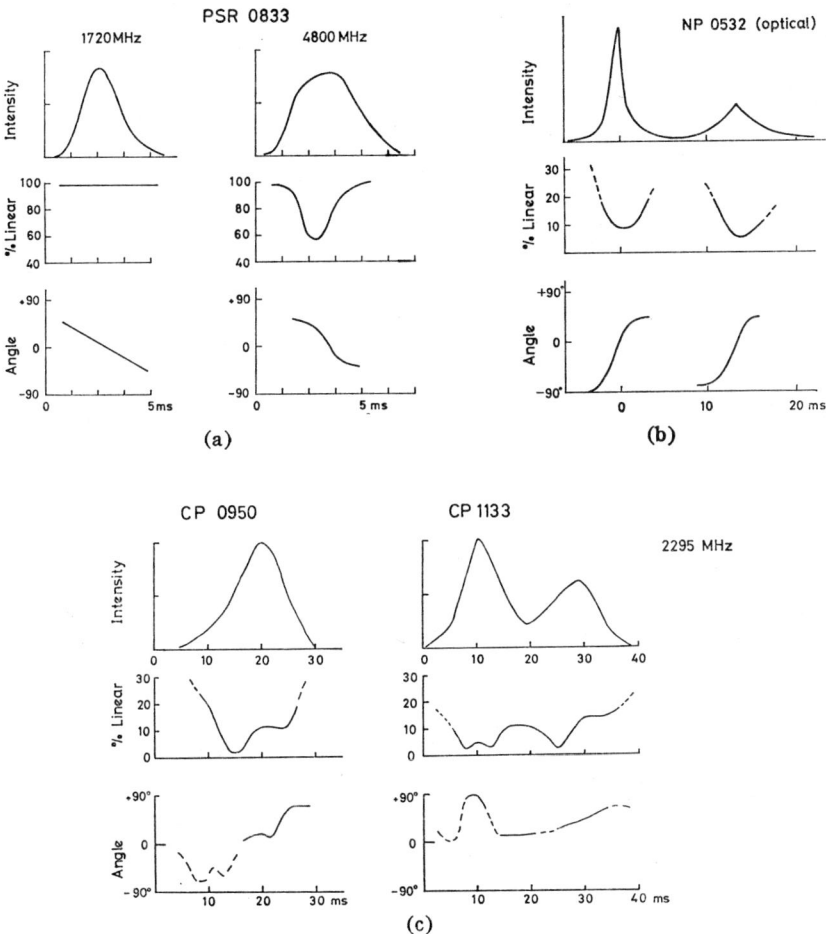

FIGURE 8. Mean polarization. (a) PSR 0833. (b) NP 0532. (c) CP 0950 and CP 1133.

same direction. The interpulse, on the other hand, is more weakly polarized and contains no linear or circular component greater than 11%.

Polarization studies of this kind undoubtedly provide one of the strongest clues concerning the radiation mechanism, but unfortunately no unique picture emerges. Ekers & Moffet (33) have determined the time-averaged polarization in a number of other cases. As shown in Figure 8c both CP 0950 and CP 1133 possess linearly polarized components, whose angle rotates in a manner reminiscent of PSR 0833, but the percentage polarization is small and variable, and the rotation is more irregular. PSR 1749 is the only pulsar that exhibits a circularly polarized component in the time average. In the

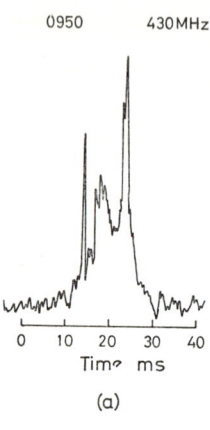
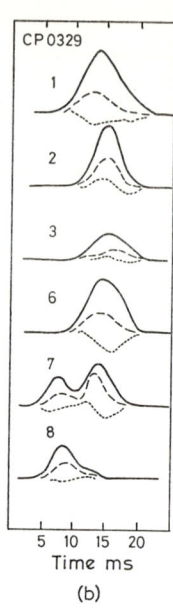
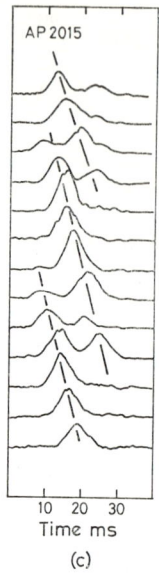

FIGURE 9. (a) Fine structure in an individual pulse. (b) Pulse-to-pulse variation of polarization; (——total intensity - - - - linear component · · · · circular component). (c) Progressive drift of fine structure in successive pulses.

case of CP 1133 the angle of polarization appears to be related to the double-peaked structure of the envelope and there may be a regular swing within the second peak. It is clear that a simple behavior cannot be expected of pulsars having a complex pulse envelope. No evidence has yet been obtained for the time-averaged behavior of the interpulse of CP 0950.

Individual pulse fine structure.—Averaging many pulses to obtain the mean envelope smooths out a wealth of fine detail sometimes described as the subpulse structure. This behavior is certainly typical and has been observed over a frequency range 100–2300 MHz (29, 34, 35), although PSR 0833 is exceptional in showing little or no fine structure. The time scale of these variations differs from source to source with no obvious relation to the pulse period, and fluctuations having a duration of a few milliseconds are typical. In CP 0950 structure is present down to about 0.1 msec, and even this may be an instrumental limit. No evidence has yet been obtained on the correlation of the subpulse structure at different frequencies, although the width of the subpulses appears to be independent of frequency (34).

The polarization of the fine structure varies on a similar time scale. Instantaneous linear and circular polarization approaching 100% has been observed. A series of consecutive pulses from CP 0329 obtained by Clark &

Smith (34) which illustrate some of these phenomena is shown in Figure 9b, along with an example (Figure 9a) of the extremely fine detail exhibited by CP 0950.

Recognizable features appear to persist for several pulse periods and in several cases there is a progressive time shift in relation to the main pulse (36). This drift can take place in either direction, but maintains its sense for a given source. The latter effect, shown in Figure 9c for two pulsars, led Drake & Craft to suggest that in these cases the fine structure was an additional periodic oscillation not commensurate with the main period. The phenomenon was called class II pulsation and periods ~10–15 msec were deduced. A subsequent and more extensive study (20) showed that these time shifts are more correctly interpreted as repetitive phenomena having a basic

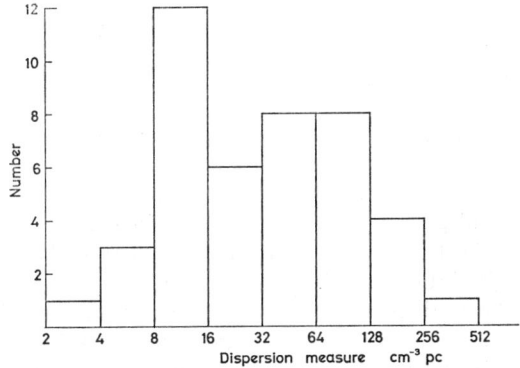

FIGURE 10. Histogram of dispersion measure (DM) for 45 pulsars.

period several times longer than the pulsar period. Further evidence on this point will be mentioned later.

Pulse dispersion.—A histogram of dispersion measure (DM) is given in Figure 10. The histogram shows a broad maximum near DM~30, although considerably higher values have been recorded. The effect of dispersion upon pulsar detection has already been discussed; the histogram is believed not to be seriously influenced by selection, except possibly for pulsars having periods <0.1 sec.

If the interstellar plasma were of constant density and if pulsars were distributed evenly throughout the Galaxy, the number of pulsars in any dispersion range would increase steadily with increasing DM. What is actually observed is plotted in Figure 11. If most pulsars were at a distance small compared to the thickness of the galactic disk, a slope of +3 for the log N-log DM relation would be expected, while a slope of +2 would be consistent with a disk population of pulsars in which sources were detected at distances greater than the disk thickness. The results given in Figure 11 show that a

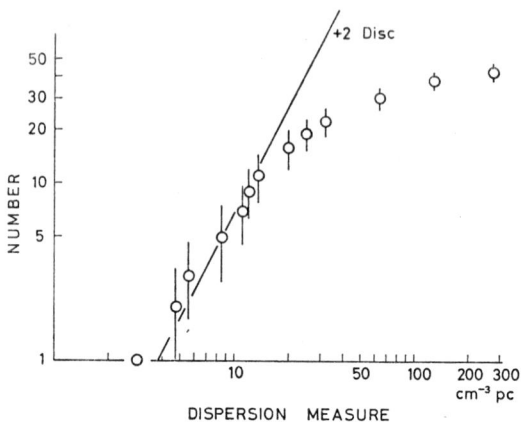

FIGURE 11. The log N−log DM relation.

slope of 2, or possibly 3, is consistent at the lowest values of DM but beyond DM∼20 a much smaller slope is obtained.

Further information obtained by investigating the relation between dispersion measure and galactic latitude is shown in Figure 12. The tendency for large values of DM to be observed near the galactic plane was noticed in the early Molonglo data (2) and this trend is obvious from the figure. The possibility that large dispersion measures could be caused by the line of sight passing through individual H II regions has been discussed by Davidson & Terzian (38) and by Prentice & ter Haar (39). The latter have considered the available optical evidence on nearby H II regions, and DM values for pulsars believed to lie behind H II regions are shown separately in Figure 12. This plot gives strong evidence that beyond DM∼20 there is a high likelihood of the line of sight intersecting an H II region. It follows that the use of pulse dispersion to estimate pulsar distances is difficult, except perhaps for the nearest sources.

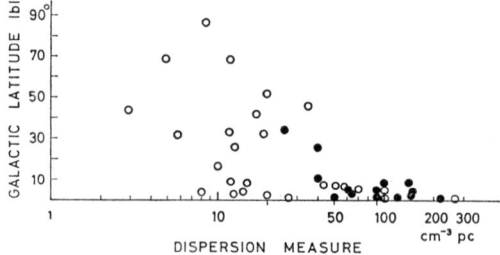

FIGURE 12. The relation between dispersion measure and galactic latitude $|b|$. Filled points indicate sources believed to lie behind H II regions (39).

Distance estimates.—The distribution of pulsars in galactic latitude shows clearly that they must be a galactic phenomenon. If it is assumed that they are distributed like Population I objects, then they must be at distances comparable to, but somewhat larger than, the width of the disk.

The first attempt (1) to use pulse dispersion as a quantitative distance indicator assumed a mean interstellar electron content of 0.2 cm^{-3}; later a value of 0.1 cm^{-3} was generally adopted. On this basis, if pulsars believed to lie beyond H II regions are ignored, the bulk of the sources are in the range 100–500 pc. Even without the difficulty of H II regions, however, the correct value for the mean density is subject to considerable uncertainty. The basic problem is to determine what fraction of the plasma is contained in low-temperature clouds that are largely neutral and whose distribution may be studied by 21-cm observations, and what fraction is contained in hot, highly ionized, intercloud regions. The difficulties were first discussed by Habing & Pottasch (40). Further work, assuming a variety of models, has been described by Shuter et al (41), Grewing et al (42), and Prentice & ter Haar (39). The results still display considerable discrepancies, and it is clear that dispersion is a far from reliable distance measure; but for pulsars that do not lie behind H II regions the assumption of a mean density of about 0.03 appears to yield distances correct within a factor of two. This implies that more than 50% of the known pulsars are within a distance of 2 kpc.

A further method of estimating distance relies upon the detection of absorption by neutral hydrogen in spiral arms at a wavelength of 21 cm. This technique has been applied to only a few intense pulsars lying in the galactic plane, but the results are important in view of uncertainties in the dispersion method. The first observations were carried out by de Jager et al (43) on CP 0329. Absorption was reported in both the local arm and the Perseus arm, which implies that this source is at a distance exceeding 4 kpc. The measurements were later repeated by Guélin et al (44) who found absorption only in the local arm placing CP 0329 at a distance > 800 pc, but certainly <4 kpc. More recent observations by Gordon et al (45) support the results of Guélin et al; the most distant feature in the absorption profile is at 2 kpc and the reality of even this feature is questionable, while absorption in the local arm is strongly evident. Gordon et al also found significant absorption in the case of PSR 1749–28 corresponding to a sharp feature at 1 kpc. The lack of detectable absorption in several other cases may be used to set upper limits if a neutral hydrogen density is assumed. This again leads to values ~1 kpc.

In the case of NP 0532 the distance is known to be about 2 kpc from optical observations of the Crab Nebula. The dispersion measure of 56 thus leads to a mean density $n_e \sim 0.03$ e cm^{-3}. On the other hand PSR 0833, which is associated with the Vela supernova remnant, is believed to be at a distance of 500 pc and the dispersion measure of 69 gives $n_e \sim 0.14$ e cm^{-3}. This again illustrates the difficulty of using dispersion measure as a distance indicator. It is, however, reassuring that all methods of distance measurement are in

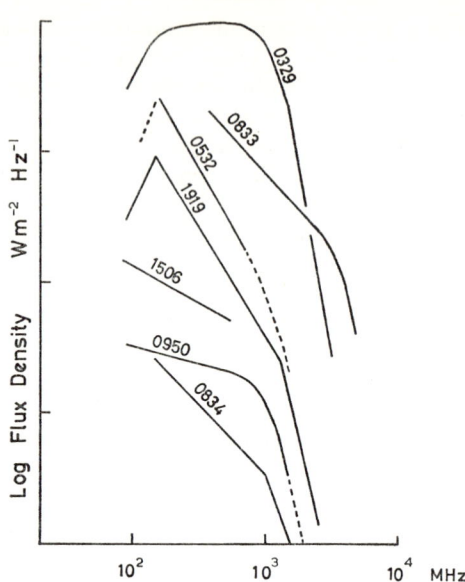

FIGURE 13. Radio spectra of pulsars. The absolute intensity is arbitrary.

rough agreement, and the accuracy is at least sufficient for reasonable estimates of the energy radiated by the sources.

Spectra.—Pulsars radiate such a variable intensity that simultaneous observations over a wide range of frequencies are desirable to ascertain their spectra. Few such observations have been made. The extreme frequencies at which pulsars have been detected vary from 40 to 5000 MHz (27, 46). A compilation of spectral data currently available leads to the results shown in Figure 13. There are some general features which typify pulsar spectra, such as rapid cutoffs above about 1000 MHz and below about 100 MHz. Both CP 0950 and CP 0329 are unusual in having remarkably flat spectra in the neighborhood of 500 MHz; the latter source has a most pronounced low-frequency cutoff. If we put $S = \nu^{-\alpha}$ where S is the average flux density of the source near some frequency ν, then α varies from about 0.6 to 2 in the range 100–1000 MHz. For frequencies exceeding 1000 MHz, α increases typically to 3 or 4.

For NP 0532 the radio data may be combined with optical and infrared results (47, 48) and with X-ray measurements (18, 49) to give the composite spectrum shown in Figure 14. The spectrum is obviously complex and different mechanisms appear to be at work in the radio and optical regions, but the optical and X-ray spectra may be related. The featureless optical and infrared spectrum follows a curve similar to that of blackbody radiation at 10 000°K, but the intensity must be far greater in view of the small surface

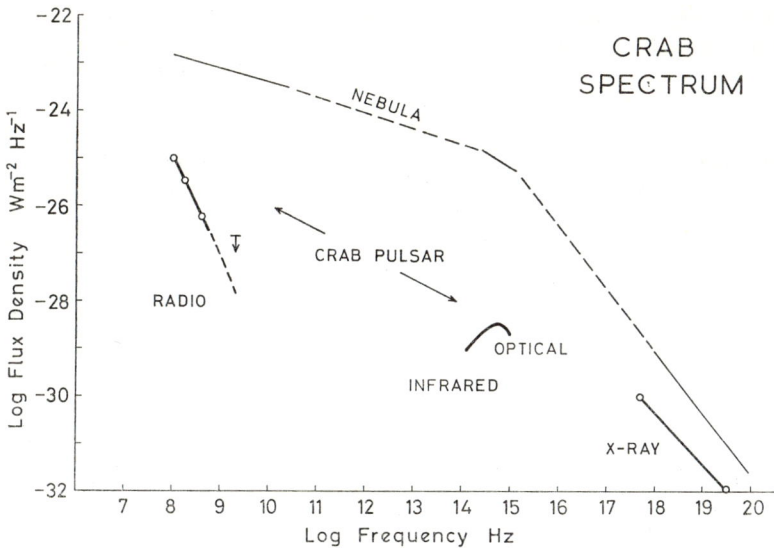

FIGURE 14. The spectrum of NP 0532 compared with that of the whole of the Crab Nebula.

area of the source. It has been suggested that the cutoff in the radio spectrum below 100 MHz is a spurious phenomenon caused by the increasing pulsewidth (19), but this cannot account for the cutoff in other sources.

For CP 0950 and NP 0532, the two pulsars with interpulses, it is interesting to know whether the spectrum of the interpulse is the same as that of the main pulse. Moffet & Ekers (35) obtain the same relative intensities for the main pulse and the interpulse of CP 0950 at 2295 MHz as was obtained by Rickett & Lyne (24) at 408 MHz which indicates that the spectra are, in fact, the same. On the other hand Comella et al (4) report a variable pulse-to-interpulse ratio at different frequencies for NP 0532. In this case the situation is complicated by the different structure of the main pulse and the interpulse, and it is possible that the sharp features in the main pulse and in the interpulse have identical spectra while the precursor has a different spectrum (17).

Irregular intensity variations.—Pulsars exhibit intensity variations on time scales ranging from a fraction of a second to several weeks or longer. Much of this variation must be intrinsic to the source, but there is strong evidence that some of it can be ascribed to interstellar scintillation. It is helpful in understanding these phenomena to consider the dynamic spectrum of the variations.

The dynamic spectrum of the radiation from a typical pulsar is shown schematically in Figure 15. First there is the systematic sweep-frequency effect caused by dispersion. Then there is the rapid pulse-to-pulse intensity variation that is highly correlated over the whole frequency range (50) and must therefore be intrinsic to the source. Pulsars differ considerably with respect to this kind of variation; CP 0808 and CP 1919 emit pulse trains which usually contain five or more pulses of comparable intensity, while CP 0329 and NP 0532 often emit single pulses of large intensity spaced by several pulse periods, and rarely emit trains containing more than two or three intense pulses. This behavior is well illustrated in results obtained by Huguenin & Taylor (51) Autocorrelation functions of pulse trains have been studied by Taylor et al (52).

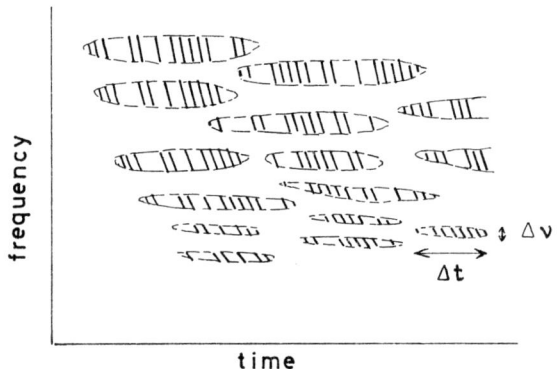

FIGURE 15. Schematic diagram of the variation of pulse intensity with frequency and time.

Superimposed upon the wideband pulse-to-pulse variations is a relatively narrowband modulation which correlates over a range $\Delta\nu \sim 0.1$ to 10 MHz and has a time scale $\Delta t \sim$ minutes to hours. As shown in Figure 15 both $\Delta\nu$ and Δt increase as the frequency is increased. An important result obtained by Rickett (53) is that $\Delta\nu$ is related to the dispersion measure, and this has been confirmed by Huguenin & Taylor (54). These authors adopt slightly different definitions of the bandwidth $\Delta\nu$. Their combined results are given in Figure 17. The relation $\Delta\nu \propto (DM)^{-2}$ gives a reasonable fit to the points, although there is appreciable scatter. Huguenin et al (54) have also shown that $\Delta\nu$ is strongly dependent upon frequency, and obtain $\Delta\nu \propto \nu^\alpha$ where $\alpha \sim 3$ although the index varies from source to source within a range 2.6–3.5. Values for Δt are less well determined but data obtained by Rickett (55) are roughly consistent with $\Delta t \propto \nu$ for individual sources and $\Delta t \propto DM^{-2}$. Typical magnitudes are such that $\Delta t \sim 30$ min for $DM \sim 10$ at 408 MHz. It is possible that these variations diminish at frequencies above 1000 MHz

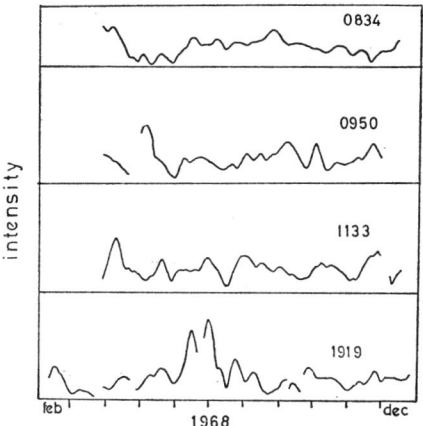

FIGURE 16. Long-term variation of source intensity. Daily values of intensity have been smoothed by an 8-day running mean to remove variations caused by interstellar scintillation.

where further data are badly needed, but PSR 0833 gives evidence for the presence of intensity fluctuations up to 4800 MHz (31).

To complete the description of irregular intensity variations it is also necessary to include long-term effects not shown in Figure 15. It is well known to those involved in pulsar searches that some sources often show enhanced intensity for 1 or 2 days and then remain undetectable for perhaps a week or more, while others are more dependable. Some typical results for a few pulsars at 81.5 MHz are shown in Figure 16. It is not known with certainty whether these long-term variations, which may have time scales > 1 week, are correlated over the whole of the radio spectrum. A small amount of evidence suggests that there may be some correlation. Not all sources show the same degree of variation. PSR 0833, for example, was noteworthy as a remarkably constant source at 1720 MHz, although more recent observations at 1720 and 4800 MHz exhibit variations (31, 37). CP 0950 has exhibited very large variations at 81.5 MHz and on one occasion during November 1967, the flux density increased by a factor of 200 above the mean flux density.

Periodic intensity variations.—It was mentioned earlier that the observation of systematic time displacements of subsidiary peaks in successive individual pulses led to the idea of a class II pulsation considerably more rapid than the pulse period (36). A related phenomenon, the occurrence of repetitive effects in the pulse-to-pulse intensity variations, is shown both in the autocorrelation of pulse trains from CP 1919 and in the corresponding power spectra (56, 57) which indicate a modulation at roughly 4.4 pulse periods. Further observations of power spectra by Taylor et al (52) have shown similar behavior for CP 0834 (2.16 pulse periods) and CP 0823 (5.5 pulse periods) and possibly in three other cases as well. Thus the effects are by no means rare and may be typical, although the evidence on CP 0834 suggests that the modulation is much weaker at frequencies above 200 MHz (52).

Further studies of the position of subpulses in successive pulses led Drake to conclude that his results are better explained by a period several times longer than the pulse period, rather than by a more rapid oscillation. This would be in good agreement with the data provided by power spectra. The reason for these effects is not yet known. Vila (99) has suggested that rotation combined with precession might account for a double periodicity, but the relatively small difference between the periods would then imply a markedly nonspherical rotator.

Interstellar scintillation.—The possibility that intensity variations might be caused by irregular diffraction in the interstellar medium was first put forward by Lyne & Rickett (7), and preliminary theoretical studies have been made by Scheuer (58) and Salpeter (59). Further work based on the observed values of $\Delta \nu$ and Δt has been carried out by Rickett (55).

The strongest evidence that intensity variations on a time scale of several minutes to a few hours are caused by interstellar scintillation is the dependence of $\Delta \nu$ upon DM shown in Figure 17. To account for this we suppose that the plasma irregularities are situated roughly midway along the line of sight. If the radiation is scattered through an angle θ we then have $\Delta \nu \propto (\theta^2 z)^{-1}$ where z is the distance of the pulsar (Little 60). Now θ^2 is proportional to z if the plasma irregularities are scattered randomly and if we assume $z \propto$ DM we obtain $\Delta \nu \propto (DM)^{-2}$ as observed. Since the interstellar plasma irregularities may well be a variable fraction of the mean density in different regions of the Galaxy it is not surprising that the points show some scatter.

The same model may also be used to predict the variation of $\Delta \nu$ and Δt with frequency. For some source at a fixed distance we have $\Delta \nu \propto \theta^{-2} \propto \nu^{-4}$. The agreement with observation is not so good in this case as the results discussed earlier give $\Delta \nu \propto \nu^{-3}$ approximately, but the observations are not very precise. Finally we have $\Delta t \sim (V\theta\nu)^{-1}$, where V is some velocity which defines the relative motion (systematic or random) of the plasma with respect to the line of sight. Since $\theta \propto \nu^{-2}$ we obtain $\Delta t \propto \nu$ for a given source, which is again consistent with observation.

It is interesting that there are severe restrictions on the sizes of the plasma irregularities which can give rise to scintillation (58). These are best understood with reference to the scale l of the spatial intensity variations. Quite generally we have $l \sim \lambda/\theta$ where λ is the wavelength. Now θ clearly cannot exceed the angular diameters of the smallest extragalactic radio sources when these are observed at low frequencies with a bandwidth $> \Delta \nu$. On the

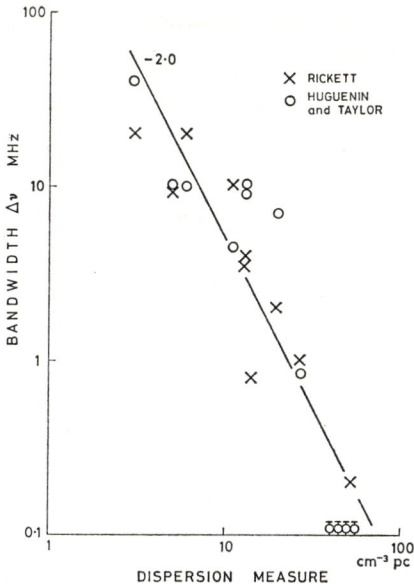

FIGURE 17. The relation between the scale ($\Delta\nu$) of fine structure in the dynamic spectrum and dispersion measure (DM) for a number of sources.

other hand θ cannot be too small, since phase-changing irregularities do not produce large intensity variations within a distance $z \sim \lambda \theta^{-2}$, and z must not exceed a few kiloparsecs. These limits allow little freedom and require l to be roughly in the range 10^4 to 10^6 km. The corresponding plasma irregularities may, of course, be larger than this by a factor of the order of the phase modulation (in radians). Since bulk velocities in the interstellar medium will not exceed about 100 km sec^{-1} and the Earth itself has a velocity of 30 km sec^{-1}, it follows that the time scale of the fluctuations must be in the range of about 2 min to 10 hr. This agrees well with the observations. It is also interesting that small time delays in the arrival of pulses caused by scintillation, which vary as ν^{-4} (59), can account for the rapid increase of pulsewidth observed in the case of NP 0532 at low frequencies (17).

Considering all the evidence, it is difficult to avoid the conclusion that scintillation affords the simplest explanation of intensity variations on an intermediate time scale, but clearly it is not possible to account for variations on a time scale of seconds or days owing to the restrictions upon l and V. Rapid variations might arise if significant scintillation were produced by high-velocity regions close to the source, but they could not be correlated over a wide frequency range as is observed.

The estimates of l and V discussed above suggest that these quantities might be measured by comparing intensity variations at widely separated points on the Earth. Some initial measurements have been made (57, 62) but the necessity for using a receiver having a sufficiently narrow bandwidth was not then appreciated and the fluctuations recorded could not have been scintillation. Successful observations between Jodrell Bank and Arecibo have been reported by Lang & Rickett (101), who obtain $2.7 \times 10^4 > l > 5.10^3$ km and $92 > V > 22$ km sec^{-1}.

THEORY

The physical nature of the source.—An upper limit to the dimensions of the source, or at least to that part which regulates the period, is given by the velocity of light multiplied by the period. For NP 0532 this limit is about 10^4 km. The emitting region may be considerably smaller if the pulse duration, or its fine structure, defines the relevant time interval, but since the radiation must result from some coherent mechanism it might be argued that size limits obtained in this way are invalid (35).

An estimate of the space density of the sources is given by assuming $n_e = 0.03$ e cm^{-3} and using the observed dispersion data shown in Figures 10 and 11. Thus eleven pulsars are scattered nearly isotropically within 400 pc giving a density $\sim 5 \times 10^{-8}$ pc^{-3}. This value is comparable to that of O stars. If the radiation is beamed in two directions the density must be increased, but the factor cannot reasonably be greater than about ten.

Some idea of the energy requirements of pulsars may be gained from NP 0532 and CP 0329 since their distances are known with fair accuracy. These are somewhat extreme cases involving the most rapid pulsar and that which gives the greatest energy per pulse. The mean received energy per pulse is approximately 1.0×10^{-26} J m^{-2} Hz^{-1} at 500 MHz for CP 0329 which corresponds, at a distance of 1 kpc, to a continuous energy loss from the source of about 10^{30} ergs sec^{-1} integrated over the radio spectrum. Similarly for NP 0532 the received energy per pulse is $\sim 0.2 \times 10^{-26}$ J m^{-2} Hz at 200 MHz which leads to an energy loss of about 6×10^{30} ergs sec^{-1}. For NP 0532, however, the energy loss is greatest at X-ray wavelengths and the spectrum shown in Figure 14 indicates a radiated energy of $\sim 10^6$ ergs sec^{-1}. Beaming in two coordinates will again reduce the foregoing estimates by a factor of less than ten.

Little evidence is available concerning the lifetime of the sources, but it seems reasonable to adopt the characteristic time $T = P(dP/dt)^{-1}$, in view of the good agreement, in the case of NP 0532, with the known age of the Crab supernova event. If CP 0329 is taken as a typical pulsar the energy requirement for a lifetime of 10^7 years is therefore $\sim 10^{44}$ ergs for the radio emission alone, but the total energy loss would be several orders of magnitude larger than this if young pulsars are similar to NP 0532.

Combining the observed radiation density with the limiting dimension of 10^4 km leads to a brightness temperature of 10^{23}°K for NP 0532 at radio wavelengths. This is probably a conservative estimate since 10^4 km is larger than the radius of the light cylinder $= cP/2\pi$ which features in rotating models. It is therefore clear that coherent radiation mechanisms must be involved.

The possibility that pulsars might be members of binary systems is readily tested because orbital motion would introduce a Doppler shift of the observed period. Extended timing measurements have so far revealed no such shifts, except for the slight modulation of period of NP 0532 mentioned earlier.

The clock mechanism.—Some of the ideas which have been suggested to account for pulsars are listed in Table 2. From the start it has been generally agreed that only massive condensed objects such as white dwarfs or neutron stars must be involved and the clock mechanism has been ascribed to rotation, vibration, or binary orbital motion. With the discovery of PSR 0833 and NP 0532, both of which have periods beyond the capabilities of white dwarfs (82–84) and are associated with supernova remnants, opinion has hardened in favor of neutron stars. The remaining possibilities for white dwarfs appear to be an atmospheric oscillation, or oscillation in some high-order mode, neither of which accounts for the extreme stability of the periods, nor for their systematic increase with time. Further evidence against white dwarfs is the failure to detect them optically despite careful searches in pulsar positions (Kristian 85).

Neutron stars are generally considered to have masses $\sim 1 M_\odot$ and radii $\sim$ 10 km; see, however, Wang et al (102), who obtain masses in the range 0.13 to $0.26 M_\odot$. The material density is then so great that radial vibration in the fundamental mode has a period of the order of milliseconds (86). It has been suggested (80) that a sufficiently light neutron star ($M \sim 0.1 M_\odot$ in the case of NP 0532) which undergoes a steady mass loss could give a vibration period of the right order and also explain the increase of period with time, but radiation induced by shock disturbances resulting from pulsation cannot easily account for the rotating polarization found in PSR 0833. Binary orbital motion also appears to be an unreasonable clock mechanism for neutron stars since the energy lost in gravitational radiation would result in a steady decrease of period with time which contradicts the observations.

The rotation of neutron stars, as originally suggested by Gold (67), remains as the simplest and most flexible method of achieving periods in the observed range. Moreover, it accounts for the systematic increase of period predicted before the measurements were made. Further evidence for rotation is the swing of the polarization vector found in both NP 0532 and PSR 0833.

Energy considerations.—The next stage is to investigate rotation as a source of energy and to seek some plausible mechanism whereby such kinetic energy may be converted into pulsed radiation. Given the dimensions of a neutron star and its period, and assuming solid-body rotation, it is a simple matter to estimate the energy store. The rate of change of period then

TABLE 2. Pulsar theories

References	Theories
Hewish et al (1)	Radial pulsation of neutron star or white dwarf. Shock-wave excitation
Saslaw et al (63)	Gravitational focusing of radiation from neutron-star binary
Ostriker (64)	Active spot on rotating white dwarf
Hoyle & Narlikar (65)	Reversible collapse of supernova
Burbidge & Strittmatter (66)	Neutron star with satellite, analogous to Jupiter-Io effect
Gold (67)	Synchrotron-type radiation from density fluctuations in corotating neutron-star magnetosphere
Israel (68)	Repetitive mass loss from neutron star on verge of gravitational instability
Pacini (69)	Neutron star or white dwarf, oblique magnetic rotator
Layzer (70)	New type of condensed star, gravity balanced by magneto-turbulence
McIlraith (71)	Plasma interaction between binary neutron stars
Eastlund (72)	Synchrotron radiation from corotating neutron-star magnetosphere
Black (73)	White-dwarf atmospheric pulsation
Gunn & Ostriker (74)	Particle acceleration in magnetic dipole radiation from spinning neutron star
Chiu et al (75)	Coherent emission from population inversion in quantum magnetic states of rotating and vibrating neutron star
Bertotti et al (76)	Neutron-star oblique magnetic rotator, radiation pressure induced shockwaves
Ginzburg & Zaitsev (77)	Radio emission by induced scattering from plasma waves near neutron-star magnetic poles
Piddington (78)	Interaction between boundary of neutron-star magnetosphere and expanding magnetic shell
Michel & Tucker (79)	Radiation from tangential field discontinuities in plasma outflow from spinning neutron star
Stothers (80)	Radial pulsation of neutron star undergoing mass loss
Dyson (91)	Radiation from neutron-star volcano

specifies the power available from this source alone (74, 87). Putting $M \sim M_\odot$ and $r \sim 10$ km leads to an upper limit of $\sim 10^{52}$ ergs for the initial rotation energy (87). The same model gives an energy of $\sim 10^{49}$ ergs for NP 0532 and the increase of period corresponds to a steady energy loss of $\sim 10^{38}$ ergs sec^{-1}. This power is greater than the mean energy radiated as pulses but falls close to the value required to sustain all the radiation from the Crab Nebula (88). The significance of this conclusion has been emphasized by Gold (87) and others.

In a more typical case, for example CP 0329, the observed decrease of period corresponds to an energy loss of about 2×10^{31} ergs sec^{-1} on the basis of the same model. This value is uncomfortably close to the estimated isotropic radiation loss of 10^{30} ergs sec^{-1}, and the same is true for CP 0808 which has the most constant period of any pulsar yet studied. The energy is, however, sufficient, provided that a conversion efficiency of a few percent is attainable.

Radiation mechanisms.—Neutron stars are expected to possess surface magnetic-field strengths $\sim 10^{12}$ G resulting from magnetic flux conservation during collapse. The resultant magnetic moment, if it is not aligned with the rotation axis, will generate low-frequency radiation as a classical dipole radiator. It has been shown that the power radiated in vacuo is sufficient to account for pulsars and that the decay time of the magnetic field is sufficiently long (69, 74, 89). Furthermore, as discussed by Ostriker (89), the eventual decay of the magnetic field provides a simple explanation for the absence of long-period pulsars. Recent estimates of the conductivity of neutron-star material suggest, however, that the decay time of the field is about 10^{13} years (100). If this is so, another means must be sought for turning-off pulsars. On the oblique rotator model it can also be shown that the age of the source is just one-half the characteristic time T. This is in good agreement with the known age of the supernova in the case of NP 0532.

The electrodynamics of a spinning neutron star is obviously a subtle matter. An initial attack on the problem has been made by Goldreich & Julian (90) who consider only the axially symmetric case. It is shown that electric fields induced in the surface of the neutron star produce forces on charged particles which far exceed gravity. This causes an outflow of plasma which gives rise to a corotating magnetosphere within the light cylinder. Beyond this boundary a radial stellar wind sets in and convects outwards a substantially toroidal magnetic field until the motion is eventually halted by the interstellar gas. The theory predicts considerable acceleration of charged particles, especially near the extremity of the stellar wind zone.

This model makes no claim to account for pulsed radiation, but interesting possibilities arise if the magnetic axis is oblique. The low-frequency magnetic dipole radiation then provides an additional means of accelerating particles to high energy as suggested by Gunn & Ostriker, and this process may be very important if NP 0532 is to supply the relativistic particle source in the Crab Nebula.

Space forbids a description of all the suggestions put forward to account for pulsed emission but many of them involve beamed radiation to produce a lighthouse effect. Few make sufficiently detailed predictions for comparison with the observations and many face difficulties with respect to 1) polarization and 2) the fact that, with few exceptions, pulsars radiate a single pulse during each complete rotation. It might, perhaps, be argued that for the bulk of the pulsars, which emit no interpulse, the observed period is actually one-half the rotation period, but the absence of odd-even effects in autocorrela-

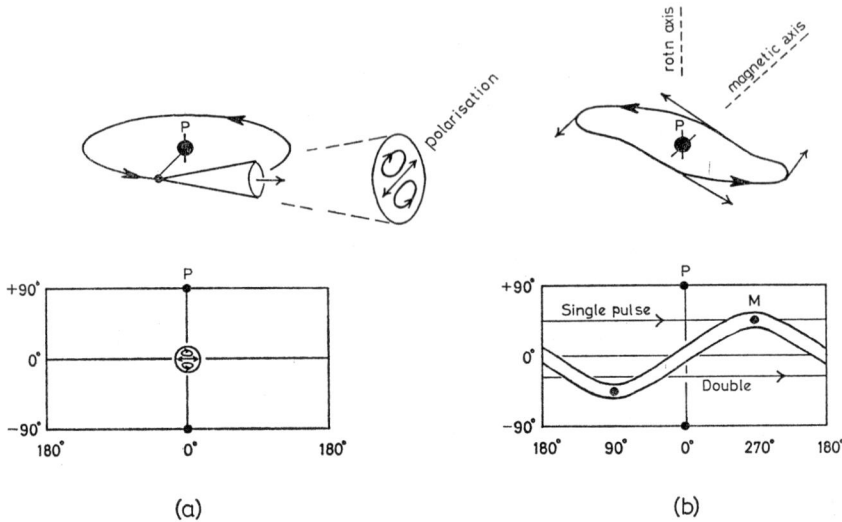

FIGURE 18. (a) The radiated beam for Gold's neutron-star model in a coordinate system stationary with respect to the source. (b) Possible radiated beam for a modified Gold model in which the magnetic axis M is inclined to the rotation axis P.

tion analyses, despite rapid pulse-to-pulse intensity variations intrinsic to the source, provides strong evidence that this is not the case (35, 52).

To illustrate the difficulties we first consider Gold's model (67). Gold made the fruitful suggestion that relativistic beaming of radiation from corotating charge-density fluctuations might account for the insensitivity of pulse shape with wavelength. To produce radio pulses in this way requires $\sim 10^8$-MeV electrons and the beamwidth will then depend upon the plasma distribution, although an idealized configuration in the form of a radial arm is required to give single pulses. The polarization is that of synchrotron radiation from a clump of charges moving in a circular orbit, which gives linear polarization in the plane of the orbit and elliptical polarization in opposite senses on either side (92) as shown in Figure 18. Such polarization is sometimes seen in individual pulses (34) but it does not appear to be typical (33).

Another model in which the beaming mechanism is specified is that of Ginzburg & Zaitsev (77) who show that induced scattering by plasma waves near the magnetic poles produces a beam in a plane perpendicular to the magnetic axis. A similar type of beam is given by Eastlund's model (72), which makes use of the synchrotron polar diagram emitted by particles gyrating in a corotating magnetosphere. The rocking of the magnetic equator for an oblique rotator then gives pulsed radiation in certain directions, but two unevenly spaced pulses are expected in the general case, rather than a single pulse. This difficulty, which has been discussed by Papagiannis (93), is

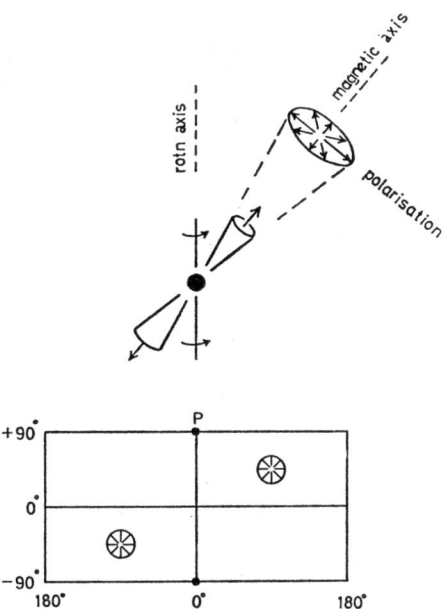

FIGURE 19. Radiated beam for particles ejected along field lines near the magnetic poles.

illustrated in Figure 18 where the coordinate system is a celestial sphere centered on the neutron star. If this sphere has poles defined by the rotation axis and rotates with the source, the radiation as viewed by a distant observer is obtained by moving along a line of constant latitude. Only if the observer's latitude is near that of one of the magnetic poles will a single pulse be received. This difficulty is a general one which afflicts other models besides those mentioned. It is, for example, also relevant to Gold's suggestion (67) that synchrotron radiation from charge-density fluctuations in a corotating magnetosphere might give pulses without the need for a unique plasma arm if the magnetic axis were sufficiently inclined.

A suggestion by Radhakrishnan et al (30) which accounts for both the swing of the polarization vector and single pulsed emission is to suppose that the radiation is beamed in the direction of the magnetic axis. This would arise if charged particles acquired sufficient acceleration while streaming outwards along curved field lines near the magnetic poles. Relativistic beaming of synchrotron radiation then gives a beam pattern as sketched in Figure 19. Obviously, a large obliquity is now necessary to produce interpulses and if these occur, the polarization vector will generally rotate in opposite senses in alternate pulses. To explain rotation in the same sense as in NP 0532 the magnetic axis must be inclined at approximately 90° to the rotation axis (32, 94). This implies that alternate pulses will have comparable magnitudes which agrees with observation. The almost complete linear polarization of PSR 0833 is somewhat higher than would be expected from synchrotron emission, but Radhakrishnan's model comes closer to explaining the observed features than any other. It also fits in with the conclusions of Goldreich & Julian (90) who predict a flux of particles from the poles. Clearly it may be complicated in a variety of ways, such as a bent magnetic axis to account for slightly asymmetrical interpulses, or a patchy effiux of plasma to explain irregular pulse shapes.

An attempt to explain both the radio and optical emission from NP 0532 has been made by Chiu & Canuto (75) who required vibration of the neutron star in order to establish population inversion of energy states in the polar regions. This idea has been criticized on theoretical grounds regarding the escape of the radiation (91) and it also predicts circular polarization of the optical radiation which is not observed.

Several suggestions have been made to explain the sudden change in period of PSR 0833. Changes of mass were invoked by Durney (95) and Feldman et al (96), while Greenstein & Cameron (97) have suggested that the outer shell of the star may rotate independently of the core and be accelerated occasionally by the onset of some hydrodynamic instability within. Ruderman (98), on the other hand, has considered that rigidity causes the neutron star to take up an equilibrium configuration discontinuously as spindown proceeds, with corresponding variations in the moment of inertia.

The long relaxation time for the recovery of the period of PSR 0833 has been taken as evidence for a superfluid core inside neutron stars by Baym et al (100). It is suggested that discontinuous change of angular velocity of the crust, caused by sudden alterations of shape (starquakes), couple rapidly with the electrons, via the magnetic field, but coupling with the neutron superfluid takes place slowly so that times of the order of a few years may be needed to re-establish effective solid-body rotation.

Although no model can yet be regarded as satisfactory, particularly concerning the radiation mechanism, there is reason to hope that a more thorough study along lines already opened may provide a solution. Certainly there is little to be gained from imaginative speculation which pays scant attention to the wealth of observational material now available.

LITERATURE CITED

1. Hewish, A., Bell, S. J., Pilkington, J. D. H. Scott, P. F. Collins, R. A. 1968, *Nature*, **217**, 709
2. Large, M. I., Vaughan, A. E., Wielebinski, R. 1968. *Nature*, **220**, 753
3. Staelin, D. H., Reifenstein, E. C. 1968, *Science*, **162**, 1481
4. Comella, J. M., Craft, H. D., Lovelace, R. V. E., Sutton, J. M., Tyler, G. L. 1969, *Nature*, **221**, 453
5. Cocke, W. J., Disney, M. J., Taylor, D. J. 1969, *Nature*, **221**, 525
6. Lynds, R., Maran, S. P., Trumbo, D. E. 1969, *Ap. J.*, **155**, L121
7. Lyne, A. G., Rickett, B. J. 1968, *Nature*, **218**, 326
8. Large, M. I., Vaughan, A. E., Wielebinski, R. 1969, *Nature*, **223**, 1249
9. Wielebinski, R., Vaughan, A. E., Large, M. I. 1969, *Nature*, **221**, 47
10. Cole, T. W. 1969, *Nature*, **221**, 29
11. Davies, J. G., Hunt, G. C., Smith, F. G. 1969, *Nature*, **221**, 27
12. Richards, D. W., Comella, J. M. 1969, *Nature*, **222**, 551
13. Mills, B. Y. 1969, *Nature*, **224**, 504
14. Hunt, G. C. 1969, *Nature*, **224**, 1005
15. Boynton, P. E., Groth, E. J., Partridge, R. B., Wilkinson, D. T. 1969, *Pulsar Conf. Rome, Dec. 18–20*
16. Richards, D. W., Pettengill, G. H., Roberts, J. A., Counselman, C. C. III, Rankin, J. M. 1969, *Pulsar Conf. Rome, Dec. 18–20*
17. Rankin, J. M., Comella, J. M., Craft, H. D., Richards, D. W., Campbell, D. B., Counselman, C. C. III 1969, *Pulsar Conf. Rome, Dec. 18–20*
18. Fishman, G. J., Harnden, F. R. Haymes, R. C. 1969, *Ap. J.*, **156**, L107
19. Campbell, D. B., Heiles, C., Rankin, J. M. 1970, *Nature*, **225**, 527
20. Drake, F. D. 1969, *URSI Gen. Assembly, Ottawa*
21. Goldstein, S. J., Meisel, D. D. 1969, *Nature*, **224**, 349
22. Radhakrishnan, V., Manchester, R. N. 1969, *Nature*, **222**, 228
23. Reichely, P., Downs, G. S. 1969, *IAU Circ. 2140*
24. Rickett, B. J., Lyne, A. G. 1968, *Nature*, **218**, 934
25. Bradt, H., Rappaport, S., Mayer, W., Nather, R. E., Warner, B., MacFarlane, M., Kristian, J. 1969, *Nature*, **222**, 728
26. Craft, H. D., Comella, J. M. 1968, *Nature*, **220**, 676
27. McCullough, P. M., Hamilton, P. A., Komesaroff, M. M., Cooke, D. J. 1969, *Proc. Astron. Soc. Aust.*, **1**, 225
28. Lyne, A. G., Smith, F. G. 1968, *Nature*, **218**, 124
29. Craft, H. D., Comella, J. M., Drake, F. D. 1968, *Nature*, **218**, 1122
30. Radhakrishnan, V., Komesaroff, M. M., Cooke, D. J., Morris, D. 1969, *Nature*, **221**, 443
31. Gardner, F. F., Whiteoak, J. B. 1969, *Nature*, **224**, 891
32. Wampler, E. J., Scargle, J. D., Miller, J. S. 1969, *Ap. J.*, **157**, L1
33. Ekers, R. D., Moffet, A. T., 1969, *Ap. J.*, **158**, L1
34. Clark, R. R., Smith, F. G. 1969, *Nature*, **221**, 724
35. Ekers, R. D., Moffet, A. T. 1968, *Nature*, **220**, 756
36. Drake, F. D., Craft, H. D. 1968, *Nature*, **220**, 231
37. Cooke, D. J. 1969, *Nature*, **224**, 469
38. Davidson, K., Terzian, Y. 1969, *Nature*, **221**, 729
39. Prentice, A. J. R., ter Haar, D. 1969, *Nature*, **222**, 964
40. Habing, H. J., Pottasch, S. R. 1968, *Nature*, **219**, 1137
41. Shuter, W. L. H., Venugopal, V. R., Mahoney, M. J. 1968, *Nature*, **220** 356
42. Grewing, M., Mebold, U., Rohlfs, K. 1969, *Nature*, **221**, 751
43. de Jager, G., Lyne, A. G., Pointon, L., Ponsonby, J. E. B. 1968, *Nature*, **220**, 128
44. Guélin, M., Guibert, J., Huchtmeier, W., Weliachew, L. 1969, *Nature*, **221**, 249
45. Gordon, C. P., Gordon, K. J., Shalloway, A. M. 1969, *Nature*, **222**, 129
46. Drake, F. D., Craft, H. D. 1968, *Science*, **160**, 758
47. Oke, J. B. 1969, *Ap. J.*, **156**, L49
48. Neugebauer, G., Becklin, E. E., Kristian, J., Leighton, R. B., Snellen, G., Westphal, J. A. 1969, *Ap. J.*, **156**, L115
49. Fritz, G., Henry, R. C., Meekins, J. F., Chubb, T. A., Friedman, H. 1969, *Science*, **164**, 709
50. Lyne, A. G., Rickett, B. J. 1968, *Nature*, **219**, 1339

51. Huguenin, G. R., Taylor, J. H. 1969, *Ap. Lett.*, **3,** 107
52. Taylor, J. H., Jura, M., Huguenin, G. R. 1969, *Nature*, **223,** 797
53. Rickett, B. J. 1969, *Nature*, **221,** 158
54. Huguenin, G. R., Taylor, J. H., Jura, M. 1969, *Ap. Lett.*, **4,** 71
55. Rickett, B. J. (To be published)
56. Lovelace, R. V. E., Craft, H. D. 1968, *Nature*, **220,** 875
57. Conklin, E. K., Howard, H. T., Craft, H. D., Comella, J. M. 1968, *Nature*, **219,** 1239
58. Scheuer, P. A. G. 1968, *Nature*, **218,** 920
59. Salpeter, E. E. 1969, *Nature*, 221, 31
60. Little, L. T. 1968, *Plan. Space Sci.*, **16,** 749
61. Drake, F. D. (To be published)
62. Slee, O. B., Komesaroff, M. M., McCullough, P, M. 1968, *Nature*, **219,** 342
63. Saslaw, W. C., Faulkner, J., Strittmatter, P. A. 1968, *Nature*, **217,** 1222
64. Ostriker, J. 1968, *Nature*, **217,** 1227
65. Hoyle, F., Narlikar, J. 1968, *Nature*, **218,** 123
66. Burbidge, G. R., Strittmatter, P. A. 1968, *Nature*, **218,** 433
67. Gold, T. 1968, *Nature*, **218,** 731
68. Israel, W. 1968, *Nature*, **218,** 755
69. Pacini, F. 1968, *Nature*, **219,** 145
70. Layzer, D. 1968, *Nature*, **220,** 247
71. McIlraith, A. H. 1968, *Nature*, **220,** 461
72. Eastlund, B. J. 1968, *Nature*, **220,** 1293
73. Black, D. C. 1969, *Nature*, **221,** 157
74. Gunn, J. E., Ostriker, J. P. 1969, *Nature*, **221,** 454
75. Chiu, H. Y., Canuto, V., Canuto, L. F. 1969, *Nature*, **221,** 529
76. Bertotti, B., Cavaliere, A., Pacini, F. 1969, *Nature*, **221,** 624
77. Ginzburg, V. L., Zaitsev, V. V. 1969, *Nature*, **222,** 230
78. Piddington, J. H. 1969, *Nature*, **222,** 965
79. Michel, F. C., Tucker, W. H. 1969, *Nature*, **223,** 227
80. Stothers, R. 1969, *Nature*, **223,** 279
81. Dyson, F. J. 1969, *Nature*, **223,** 486
82. Ostriker, J. P., Tassoul, J. L. 1968, *Nature*, **219,** 577
83. Faulkner, J., Gribbin, J. R. 1968, *Nature*, **218,** 734
84. Durney, B. R., Faulkner, J., Gribbin, J. R., Roxburgh, I. W. 1968, *Nature*, **219,** 20
85. Kristian, J. 1968, *Ap. J.*, **154,** L99
86. Thorne, K. S. Ipser, J. R. 1968, *Ap. J.*, **152,** L71
87. Gold, T. 1969, *Nature*, **221,** 25
88. Haymes, R. C. 1968, *Ap. J.*, **151,** L9
89. Ostriker, J. P., Gunn, J. E. 1969, *Nature*, **223,** 813
90. Goldreich, P., Julian, W. H. 1969, *Ap. J.*, **157,** 869
91. Roberts, J. A., Fahlman, G. G., 1969, *Nature*, **222,** 862
92. Ginsburg, V. L., Syrovatskii, S. I. 1965, *Ann. Astron. Ap.*, **3,** 297
93. Papagiannis, M. D. 1969, *Nature*, **222,** 1261
94. Böhm-Vitense, E. 1969, *Ap. J.*, **156,** L131
95. Durney, B. 1969, *Nature*, **222,** 1260
96. Feldman, B. A., Silk, J. I., Schwarz, R. A. 1969, *Nature*, **223,** 48
97. Greenstein, G. S., Cameron, A. G. W. 1969, *Nature*, **222,** 862
98. Ruderman, M. 1969, *Nature*, **223,** 597
99. Vila, S. C. 1969, *Nature*, **224,** 157
100. Baym, G., Pethick, C., Pines, D. D. 1969, *Nature*, **224,** 673
101. Lang, K. R., Rickett, B. J. 1970, *Nature*, **225,** 528
102. Wang, C. G., Rose, W. K., Schlenker, S. L. 1970, *Ap. J.*, **160,** 17

INTERNAL ROTATION OF THE SUN

R. H. DICKE

Joseph Henry Laboratories, Princeton University, Princeton, New Jersey

INTRODUCTION

The question of rotation in the deep solar interior is a controversial subject of considerable importance to relativity theory as well as to solar physics. The survival of Einstein's general relativistic theory of gravitation or the scalar-tensor theory may hinge respectively upon the absence or presence of rapid rotation in the deep solar interior. If a rapidly rotating solar core exists, with a rotational period under 2 days, it could be the source of angular momentum supplied to the solar wind, of internal mixing through the transport of matter along with angular momentum, and perhaps of solar activity leading to the sunspot cycle.

In prerelativity days the observed excess motion of Mercury's perihelion ($43''\pm0.4$ arc/century) (Clemence 1943, Duncombe 1958, Wayman 1966) was a mystery and led to several unsuccessful attempts to find a perturbation that could account for the effect (Leverrier 1859, Chazy 1928). The suggested sources include interplanetary material, Vulcan (a hypothetical and still undiscovered planet), and the flattened mass distribution of an oblate Sun (Newcomb 1897).

All of these suggestions must now be discarded. The interplanetary dirt and spare planet have not appeared and a solar gravitational quadrupole moment large enough to generate the full excess centennial motion of $43''$ of arc in Mercury's perihelion would also cause a $43''$ arc regression of the node on the plane defined by the Sun's equator, far too large to be allowed.

The urge to find a prosaic source for the excess motion of Mercury's perihelion disappeared with the appearance of Einstein's general relativistic theory of gravitation. This accounted for the full $43''$ arc motion as a relativistic effect. But with the recent increased interest in the scalar-tensor theory of gravitation (Jordan 1948, 1959; Thirry 1948; Bergmann 1948; Brans & Dicke 1961; Dicke 1962), the question of a possible nonrelativistic origin for part of the $43''$ motion is of interest. The scalar-tensor theory is a general relativistic theory for which the relativistic perihelion rotation of Mercury's orbit is

$$\frac{3\omega + 4}{3\omega + 6} \times 43'' \qquad 1.$$

where ω is the coupling constant of the Brans-Dicke (1961) form of the theory. This constant, ω, had been estimated on various grounds to fall in the range $4<\omega<7$ (Brans & Dicke 1961, Dicke & Peebles 1965, Dicke 1966).

If the "observed" excess motion, calculated from planetary perturbations only, is as accurate as claimed $43''0 \pm 0.4$ (Duncombe 1958), Einstein's theory is favored. But some additional perturbation, such as that of a flattened Sun, generating a motion of $4''$/century would favor the scalar-tensor theory with $\omega \sim 5$.

With these facts in mind the author suggested (Dicke 1964) that the Sun may have a distorted interior induced by a rapidly rotating core, the fossil remnant of the rotation of the young Sun. (Rapid internal rotation was also discussed by Roxburgh 1964, Plaskett 1965, Deutsch 1967.) It was assumed that in the density-stratified interior, below the convective zone, quasistable rotation of a core was possible with angular momentum diffusing to a thin shell of instability lying below the convective zone. This shell and the outer convective zone were assumed to have been rapidly rotating initially but slowed by a solar-wind torque. In collaboration with P. J. E. Peebles, I formulated a theory of the solar-wind torque along lines similar to the ideas of Schatzman (1959) and Cowling (1965) and used the resulting formulae in the 1964 paper to estimate the solar-wind torque. Equivalent formulae were independently derived by Modiesette (1967), Weber & Davis (1967), and Alfonso-Faus (1967).

The estimated torque 4×10^{30} dyne cm was based on early and preliminary measurements of the solar-wind flux and an estimated strength at the Sun of the magnetic field drawn out by the solar wind ($\sim 3/4$ G). Later, when observations were available, this field strength was found to be in reasonable agreement with observations of the field in the solar wind at the Earth, if a purely radial flow of the solar wind was assumed. It was also shown in the 1964 paper that the estimated solar-wind torque was in reasonable agreement with that derived from an approximate solution to the diffusion equation applied to the diffusion of angular momentum from the core to the outer slowly rotating shell.

One interesting aspect of stellar-wind braking concerns the old problem posed by the apparent break in the rotational distribution at special type F5. Stars much bluer than F5 are rapid rotators and old stars much redder than F5 are slow rotators. Schatzman (1959, 1962) pointed out that the transition between stars with deep radiative envelopes and those with deep subsurface convective zones occurs among the early F-type stars and he developed a theory of stellar braking using magnetic fields derived from jets or flares associated with such subsurface convective zones. Kraft (1967) noted that the connection between such subsurface convection and a stellar wind could explain why stellar-wind braking of young stars is limited to stars redder than F4.

It was noted in the 1964 paper that a solar gravitational quadrupole moment large enough to induce a centennial motion of $4''$ in Mercury's perihelion (needed for $\omega=5$) would also induce an oblateness of $\Delta r/r = (r_{eq} - r_p)/$ 5×10^{-5} in the Sun's atmosphere. This oblateness would be in addition to the 1×10^{-5} due to surface rotation. It was also noted that the 4" regression of the node on the solar equator expected from such a distorted Sun, when referred to the plane of the ecliptic, represents principally a centennial decrease of the inclination (0.21"). The observed residual in the rate of increase of the inclination $-0.12''\pm 0.18''$ (Clemence 1943) is to be compared with the above ($-0.21''$). (See later discussions of this question by Shapiro 1965, Audretsch et al 1967, Gilvarry & Sturrock 1967, and O'Connell 1968.)

In the spring of 1963 H. Hill, H. M. Goldenberg, and the author designed and built a new type of instrument designed to measure the solar oblateness. (See Figure 1.) This was installed in its own tiny observatory at Princeton and put into operation in the late summer of that year. The summers of 1964 and 1965 were used in studying and improving the instrument. The first high-quality measurements were made in the summer of 1966 and published in an abbreviated form (Dicke & Goldenberg 1967a). A full treatment will soon be given. These measurements showed the Sun to have the oblateness ($5\pm 0.7\times 10^{-5}$) expected if it possesses such a rapidly rotating core. Associated with an oblateness of 5×10^{-5} (if our interpretation is correct) is a perihelion rotation of 3.2" making the "observed" relativistic rotation $39.8''\pm 0.4$ consistent with the value 38.9" expected under the scalar-tensor theory with $\omega=5$. Unpublished measurements during the summer of 1967 gave the same oblateness with comparable precision.

The publication of our preliminary results was followed by a rash of criticisms, comments, and reinterpretations. Öpik (1967) and Ashbrook (1967) suggested that our observations may not have been as accurate as we thought. (See Dicke 1967a and 1967b for replies to these comments.) Roxburgh (1967a, 1967b), Cocke (1967a), Sturrock & Gilvarry (1967), and Durney & Roxburgh (1969) suggested that the excess solar oblateness did not imply a gravitational quadrupole moment. (See Dicke & Goldenberg 1967b and Dicke 1970a for discussions of these suggestions.) Howard et al (1967) and Goldreich & Schubert (1967a, 1967b) suggested that a rapidly rotating core was impossible because of spindown by Ekman pumping or a thermally driven instability respectively. For comments on these papers see Dicke (1967c, 1967d), McDonald & Dicke (1967), Colgate (1968), and Clark et al (1969). The instability argument of Goldreich & Schubert depends upon a model of the interior and is no more certain than the model.

In general, observations of the solar surface which have a bearing on the existence of a rapidly rotating core are more valuable than detailed calculations on the unknown solar interior. The important observations concern: the solar oblateness; the velocity fields, including rotation in the "seen layers" of the Sun; the magnetic fields in the "seen layers"; the abundance of lithium and beryllium; and the structure of the solar wind. In the absence of magnetic and velocity fields at the solar surface, the oblateness yields the gravitational quadrupole moment unambiguously. It must be emphasized

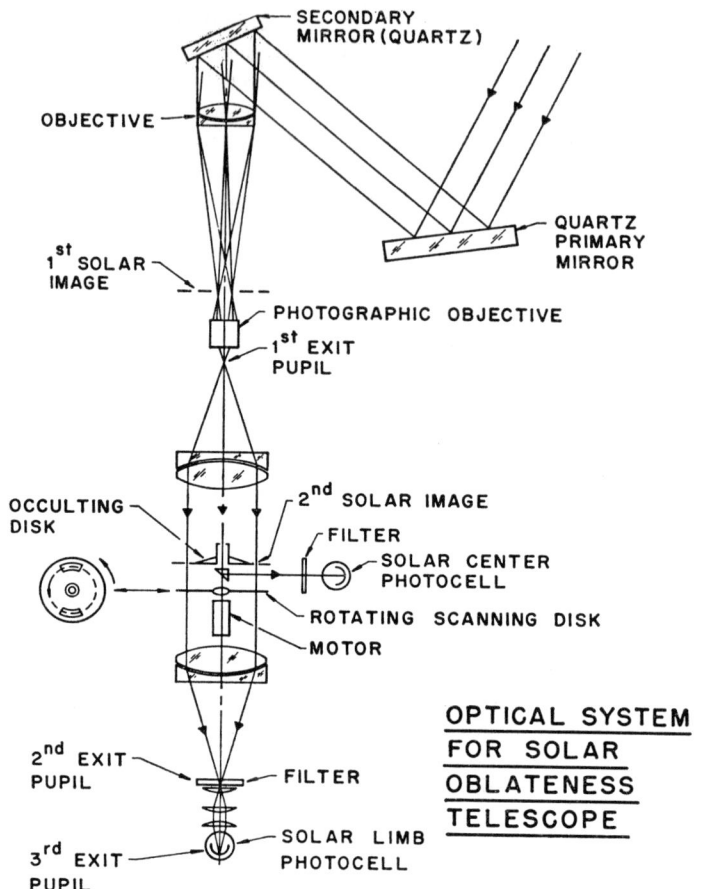

FIGURE 1. The optical system of the solar-oblateness telescope. An image of the Sun is projected on a stationary occulting disk that stops all the light except that from the outer 6.5″, 12.9″ and 19.2″ (arc). Light passing the occulting disk is "chopped" at ~100~/sec by a spinning disk. Light from the center of the Sun's disk provides a normalization signal. Calibration is carried out by replacing the circular occulting disk by one with a stepped edge.

that magnetic and velocity fields *must* be in the "seen layers" of the Sun, hence observable, if they are to affect this relationship. This will be discussed below.

The history of the solar system casts some light on the solar rotation problem. While the Sun's past may seem more hidden than its interior, with the assumption that it is a typical main-sequence star of 1 solar mass, observations of young solar-type stars are capable of showing the appearance of the Sun at the same age. We make this assumption.

Kraft (1967) has shown that stars of 1.2 solar masses in the Pleiades have surfaces rotating with $\langle V \rangle \sim 40$ km/sec, with the same angular velocity as that of the postulated rapidly rotating solar core. The extrapolation of the observation to G2 spectral-type stars in the Pleiades indicates that these solar-type stars may have an average surface velocity of ~ 10 km/sec. This represents an angular velocity only one-fourth that needed in the solar core. With the assumption that such stars possess solar-type rapidly rotating cores, stars redder than F5 probably arrive on the main sequence rotating differentially. For the Hyades, $\sim 5 \times 10^8$ years older, the rotation has decreased by a factor of 2. For very old stars the rotation is almost impercep-

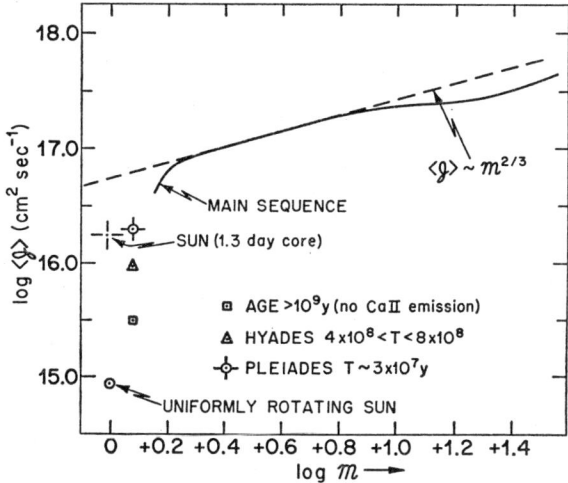

FIGURE 2. The logarithm of angular momentum/unit mass of stars various ages and masses vs log M; rigid rotation is assumed. For stars bluer than F_0 (more massive than log $M = 0.2$) the surface rotation is independent of age. This figure is based on Figure 17 of Kraft (1968).

tible. See Figure 2, based on Figure 17 of Kraft (1968). These observations suggest that the Sun was originally rotating with a 1–2 day period and that the solar wind has slowed the rotation of either the whole Sun or an outer shell.

A compelling argument in support of the contention that only an outer shell is slowed is provided by observations by Herbig (1965) and others of the abundances of lithium and beryllium in solar-type stars of various ages. (See survey article by Wallerstein & Conti 1969 for references.) Apparently solar-type stars arrive on the main sequence with abundances of both lithium and beryllium characteristic of chondritic meteorites. The lithium becomes depleted as the star ages, with a mean life of 7×10^8 years (Danziger 1969), beryllium *not* being depleted. Danziger (1967) has noted that the *e*-folding times for the decrease of lithium and rotational velocity are equal for young solar-type stars, which suggests that the two processes may be related.

The observations seem to imply that solar-type stars are mixed as deep as $r = 0.6$ but not as deep as 0.5 where beryllium is burned. But some type of mixing seems to be required if angular momentum is to be removed from the deep interior by the transport of material, for angular momentum will diffuse only $\sim r = 0.05$ during the life of the Sun.

Goldreich & Schubert (1967a) used this argument to conclude from the presence of lithium and beryllium in the Sun that the Sun arrived on the main sequence slowly rotating. If their argument were valid, it would imply that the late F's and early G's in the Pleiades must be slow rotators, for the old stars of these types are invariably slow rotators and they contain the normal amount of beryllium. Apparently the most reasonable interpretation to make of the observations is that the solar-wind torque slows only an outer shell (Dicke 1964) approximately 0.4 thick (Dicke 1970c).

Eighty-four percent of the moment of inertia of the Sun falls inside the radius $r = 0.6$. See Table 1. Hence, if the solar wind slows only an outer shell, in thickness only 0.4, the total angular momentum is substantially unaffected, and there is no great difference between the angular momentum per unit mass of solar-type stars and more massive stars. An enigma in old theories of the origin of the solar system has disappeared. These theories could not explain the slow rotation of the solar-type stars. The approximate equality of the angular momentum of the planetary system with the assumed initial total angular momentum of the solar system, an apparently fortuitous relation, compounded the difficulty, for it seemed necessary to find a mechanism to concentrate the angular momentum in the planets. Hoyle's (1960) theory of the solar system provides an effective means of transferring angular momentum out of the protosun through magnetic torque. But the observations of rapid rotation in young solar-type stars show that these stars initially have a great deal of angular momentum. Most of this remains in the deep interior if our picture is correct.

A difficult question concerns an internal magnetic field. One might expect a strong magnetic field to be trapped in the interior of the protosun. But with the alternating sunspot cycle, a 22-year period, the present solar field appears superficially to have its origin in the convective zone. It has been suggested that convective mixing during the Hayashi phase might convert the long-lived low magnetic modes to short-lived high modes that would decay after the cessation of convection in the interior but persist in the outer convective zone (Dicke 1964, Cowling 1965). To eliminate low modes by this mechanism sufficiently to permit internal differential rotation no longer seems feasible to me. At least two other possibilities remain. The internal magnetic field might be eliminated by a type of "beer foam process." If the internal field should become highly contorted during an initial convective phase, the radiative core might grow outward from the center by filling with field-free gas flowing along substantially field-free canals. Being free of the magnetic pressure this field-free gas would tend to be denser than its surroundings and settle to the center.

A second possibility seems more likely. It will be the subject of a future publication and is mentioned here only because of its relation to a rapidly rotating core. It is possible that the Sun arrives on the main sequence with its magnetic field oriented perpendicular to the rotation axis and cut off from the convective zone by a shell of differential rotation. Such an orientation might be expected because of an instability associated with the Hoyle (1960) magnetic braking of the protosun. For simplicity assume that the magnetic field is trapped in a dipolar configuration and links the central condensation with the outer solar nebula. The rapidly rotating protosun is flattened because of the tension—ρv^2 in the direction of fluid motion. As the energy of rotation is converted into toroidal magnetic energy, the negative motional tension is converted into the positive magnetic tension $B^2/4\pi$. When roughly half of the kinetic energy is lost the protosun becomes prolate and is probably unstable. It should then precess to the quasistable perpendicular position. This position may remain stable after the decay of the toroidal field. With a magnetic field in this perpendicular configuration, it penetrates only a few kilometers into the shell of differential rotation. Magnetic A stars might be exceptions where, for some reason, the shielding by differential rotation has not appeared, or been lost, exposing the strong internal field at the stars' surface.

The rapidly rotating core containing a magnetic field in the perpendicular orientation is capable of a torsional oscillation of high Q for which the north and south magnetic poles oscillate back and forth between the northern and southern hemispheres of the rapidly rotating core, and toroidal fields of opposite sign and alternating polarity are generated in the two hemispheres. Magnetic boyancy might cause this toroidal field to float up to the convective zone, providing an explanation of the sunspot cycle as an effect of this oscillation. With reasonable magnetic-field strengths, the period of this oscillation can be made to be 22 years. Owing to the stability of the frequency of the oscillating core, this model has implications that can be tested with observation of the sunspot cycle. These analyses have been carried out and will be reported elsewhere. (See Dicke 1970c for a brief discussion.) A torsional oscillation in the rapidly rotating core may have implications for the Goldreich-Schubert instability, for the θ component of velocity can be as great as 1 m per second, eliminating these slow-growing axial modes (Goldreich & Schubert 1967a).

The Solar Oblateness

In ancient times the Sun's oblateness was measured photographically (see discussion by Schaub 1938) and more accurately with the heliometer, a telescope with a split objective (e.g. the work of Schur & Ambronn 1895 at the Göttingen Observatory). Opposite limbs of the two solar images could be brought into contact by adjusting the two halves of the objective, whose separation provides a measure of the corresponding diameter.

Several possible difficulties with these measurements are apparent. The solar limb frequently has a width of 3″–5″ arc when the Sun is high in the sky, a necessary requirement if atmospheric refraction is to be manageable. But the anticipated oblateness represents a difference between equatorial and polar radii of only 0.05″, 1 percent of the "seeing" width. Owing to an expected small anisotropy in the seeing disk associated with anisotropy of the turbulence near the ground, the northern and southern limbs of the Sun might be more (or less) diffuse than the east and west limbs near the meridian, causing a systematic error. Furthermore, problems of personal bias are very difficult when the measured effects are so small relative to seeing widths. Also the heliometer may not have been free of systematic errors associated with gravitational distortion. These instruments required a 90° rotation of the objective system in the Earth's gravitational field. Any gravitational distortion would change with such a rotation.

The instrument designed by H. Hill, H. M. Goldenberg, and me incorporated a number of improvements. It is shown in Figure 1. Instead of measuring the position of the solar limb, the light flux was integrated from the edge of an occulting disk, a position near the limb, outward beyond the limb to an aperture stop a few tens of seconds of arc beyond the limb. Anisotropic seeing induced near the ground would be expected to spread the light but not change the flux. Anisotropy can still introduce some error because of the gradient in limb darkening at the edge of the occulting disk, but the effects are much reduced. The telescope was vertically mounted to avoid a change in gravitational distortion with rotation.

The problem of detecting the signal in noise was solved by measuring photoelectrically. A rapidly spinning wheel perforated by two apertures of different sizes at the ends of a diameter scanned the light flux passing the occulting disk. The photoelectric signals were analyzed electronically in an impersonal way, to measure the amplitudes of the sine and cosine terms of the second harmonic of the rotation frequency of the wheel, 1-min averages being recorded on a punched magnetic tape, the vertical telescope being rotated through 90° between the 1-min runs. The results recorded on the punched tape were analyzed by a computer. The scanning wheel also provided an error signal fed back to the main mirror to servolock the Sun's disk to the occulting disk, causing the Sun's disk to be accurately centered.

During a typical day of 6 hr the Sun's image rotated through 90° relative to the telescope. The sine and cosine amplitudes were combined linearly to give the north-south (or vertical) component of the oblateness $(\Delta r/r) \cos 2P$ and the northeast-southwest (or diagonal) component $(\Delta r/r) \sin 2P$ where P is the angular position of the Sun's rotational north pole measured eastward from the north point of the disk.

The two mirror cells were rotatable about the mirror axes, and the mirrors were cycled with a 2-day period through all four combinations of positions to permit the elimination of errors due to mirror astigmatism. The only astigmatic error not eliminated is the off-axis error associated with a slight curvature of the main mirror viewed obliquely. This error contributed to the vertical component of oblateness only; the diagonal component was unaffected. Except for the effect of this off-axis astigmatism, the instrument is believed to be free of significant systematic errors, and measurements of the diagonal component are believed to be reliable. The telescope aperture was about $2\frac{1}{2}$ inches and was stopped below 2″. The instrument probably has the largest ratio of pounds of electronics to pounds of telescope of any telescope in existence.

The instrument permitted checks for systematic errors and there was an accurate and reliable means of calibration repeated several times during each day. In addition, several internal checks of the data were possible. The contribution from atmospheric refraction to the measured diagonal component of the oblateness is usually large, ranging for September 1 from -4×10^{-4} at 9 AM to zero at noon to $+4 \times 10^{-4}$ at 3 PM. On July 16 the corresponding values are -2×10^{-4}, 0, and $+2 \times 10^{-4}$. This refractive contribution to oblateness is large, but it can be computed from atmospheric conditions measured in the observatory. For a laminar atmosphere, it is independent of conditions above the ground. After subtracting the computed values, the residuals are observed to be constant through the day (except for a small time-varying residual late in the summer when the Sun was low in the sky). On July 16 the residual in the diagonal component, $(\Delta r/r)$ sin $2P$, is approximately 7×10^{-6}, or only 3 percent of the refractive effect. That this small residual should be constant during the day indicated that the instrument has been correctly calibrated.

The most important internal check of the data is based on the change in orientation of the Sun's axis through the summer season. On July 7 the axis is in the north-south direction and the diagonal component should be zero. Also for any assumed constant oblateness along the rotation axis, the variation of the diagonal component with date (through the term sin $2P$) is predictable. For an oblique distortion axis the oblateness should vary with the solar-rotation period. This variation is not present to any marked degree. The change of the observed diagonal component with time through the summer of 1966 is shown in Figure 3. The curve is calculated with the assumption that the solar oblateness is equal to $\Delta r/r = (r_{eq} - r_p)/r = 5 \times 10^{-5}$. During 1967, the observational period was longer but the weather was substantially worse. The same oblateness was obtained with comparable precision.

One interesting interpretation of the data of Figure 3 is based on least-squares fits of the curve shown in Figure 3 to the data representing different amounts of exposed limb averaged in various ways. In permitting the curve to float up and down, the date for crossing the abscissa will vary and this change in crossing date can be interpreted as equivalent to an angle between

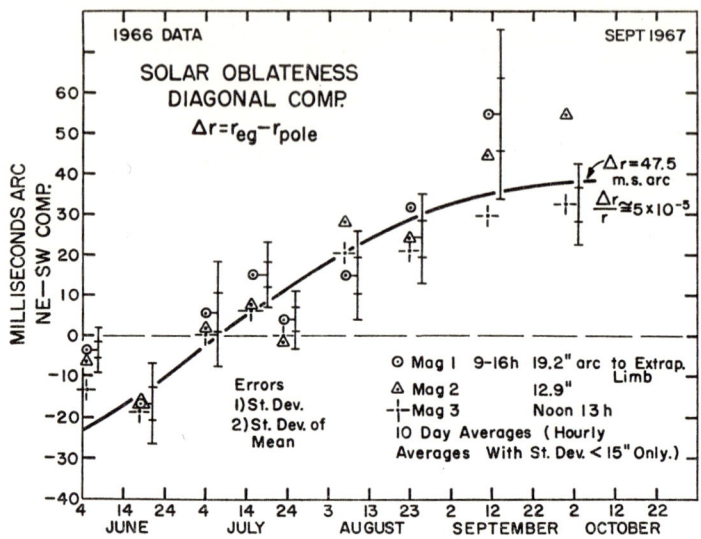

FIGURE 3. The solar oblateness, diagonal component, for 1966. Observations are averages over a 7-hr day. Points are ~10-day averages at three different amounts of exposed limb. Curve, calculated assuming $\Delta r/r = 5 \times 10^{-5}$.

the rotation axis and oblateness axis. Based on 15 different analyses of the data, the average crossing date is July (5.4 ± 3.4) whereas it should be July 7. This corresponds to the oblateness axis leading the rotational axis by $0°.7 \pm 1°.4$ as they rotate together counterclockwise on the sky through an angle of 40°. It is difficult to believe that these results are fortuitous, that instrumental and atmospheric effects would so conspire as to yield the curve of Figure 3 with a crossing date differing only a few days from July 7.

Measurements were made with three different distances from the edge of the occulting disk to the limb. This permits a separation of a signal due to the variation of brightness with latitude from the oblateness signal. The oblateness signal is proportional to the brightness of the photosphere at the edge of the occulting disk, but the intensity signal (associated with a variation with latitude of the photospheric brightness) is proportional to the integrated flux passing the occulting disk. For the edge of the occulting disk to be at the three distances 6.5″, 12.8″, and 19.1″ from the Sun's limb, the photospheric brightness at the edge obtained from a limb-darkening curve is proportional respectively to 0.380, 0.400, and 0.432. The integrated flux is respectively 3.1, 5.4, and 8.34 (Dicke 1970a). Measurements were made at these positions referred to the "extrapolated limb." With the measured values of edge brightness and light flux, the signals obtained at any two of the three positions are easily separated into the two parts.

Later measurements made with an annulus on the solar disk near the limb permitted a determination of oblateness from the limb-darkening effect and eliminated the chromosphere as a significant source of signal. The annulus technique was also used far from the limb to investigate the dependence of solar brightness upon latitude. None was found. Measurements of the solar oblateness were made with two broadband filters in the red and green and in 1967 these filters were frequently switched. No systematic dependence of oblateness on color was found. A few oblateness measurements were also made with an Hα filter as a separate check on the contribution to the oblateness signal from the chromosphere, and chromospheric lines on the disk.

One frequently raised question about the solar oblateness is concerned with solar activity and the effect of active patches, faculae, and sunspots on the measurements. There are two aspects to this question: (*a*) Is the shape of the solar surface distorted by the solar activity a distortion not mirrored in the gravitational quadrupole moment? (*b*) Does an active patch at the Sun's limb adversely affect the measurement of solar oblateness? The first question will be discussed in the next section. The answer to the second question is yes. A large sunspot lying on the limb can induce a 20–30 percent error in the results for that day, but this happened only infrequently in 1966 and 1967 when the Sun was reasonably quiet. Furthermore the error would be interpreted largely as an intensity signal and the absence of a significant signal of this type provides an internal check for the insignificance of this effect. The systematic error in average oblateness from this effect is negative in sign (i.e. prolate) but is too small to be significant.

The conclusions from these observations are:

1. The values of the solar oblateness during 1966 and 1967 were equal, and $\Delta r/r = 5 \times 10^{-5}$.

2. This photospheric oblateness was independent of color.

3. The contributions from the chromosphere and corona to the oblateness were unimportant.

4. The contribution from chromospheric lines on the solar disk was unimportant.

5. During 1966 and 1967 the effective temperature of the photosphere was remarkably free of variation with latitude. There was no convincing stationary dependence on latitude ($<3°K$).

Solar Oblateness and the Gravitational Quadrupole Moment

In this section it will be assumed that the solar oblateness has been measured to be 5×10^{-5}. The implication of this measurement for the existence of a quadrupole moment will be considered. This question has been discussed in detail (Dicke 1970a).

The direct and unambiguous relation between the oblateness of the Sun and the solar gravitational quadrupole moment under certain conditions hinges upon a functional relationship used by von Zeipel (1924). As used here it can be stated: In the *absence of magnetic and velocity fields in the "seen layers" of the Sun, surfaces of constant* P, ρ, T, *and* φ *(gravitational potential) coincide, i.e. pressure, density, and temperature can be considered to be functions of gravitational potential.* Also, as will be discussed below, for uniform rotation (or rotation on cylinders), these relations hold with φ replaced by φ plus a centrifugal potential. This functional relation was used by von Zeipel (1924) to derive his well-known paradox. Unlike von Zeipel we are here interested in the validity of these relations only in a limited region of the Sun, the part actually seen. If only a part of the solar surface is free of magnetic and velocity fields, the functional relation is applicable to this part.

The proof of this relation is trivial. In the absence of magnetic and velocity fields

$$0 = \nabla P + \rho \nabla \phi \qquad 2.$$

Thus the vector normal to a surface of constant P coincides with the normal to a surface of constant φ, which implies that two such surfaces coincide everywhere if they touch anywhere. Taking the curl of Equation 2 shows that surfaces of constant ρ and φ coincide. For a uniform composition, T is a function of P and ρ and surfaces of constant P, ρ, T and φ coincide; or P, ρ, and T are functions of φ. This result follows for any patch on the Sun's surface, simply connected or not, that is free of these fields.

Inasmuch as the gravitational potential has a simple, layered structure, the atmosphere must have the same layered structure in pressure, density, and temperature wherever the theorem applies. It has been verified that the effect on the shapes of surfaces of constant φ due to the presence of the gravitational quadrupole moment in question is so minor as to not affect noticeably this simple layered structure (Dicke 1970a). Hence, the limb-darkening curve would be substantially independent of latitude. Furthermore, an analysis of the factors affecting the position of the solar limb shows that, if such a quadrupole moment exists and von Zeipel's assumptions are valid, the position of the solar limb is determined by the position of a surface of constant density with an accuracy of ~ 3 m (Dicke 1970a). The expected brightness at the limb under these conditions should be free of any noticeable dependence on latitude. In summary, when von Zeipel's relations are applicable, and with the oblateness of a surface of constant gravitational potential at the Sun's surface being $\sim 10^{-5}$, the location of the limb of the Sun is determined to high precision by a surface of constant density, hence gravitational potential.

In an expansion of the gravitational potential outside the Sun in spherical harmonics the gravitational quadrupole moment is determined by the second zonal harmonic, falling off inversely as the cube of the distance from the Sun's center. The oblateness of surface of constant gravitational potential at the Sun's surface is given by

$$\left(\frac{\Delta r}{r}\right)_o = \frac{\varphi_2}{rg} = \frac{3}{2} J_2 \qquad 3.$$

where

$$\varphi_2 = \frac{3}{2} J_2 \frac{GM_0}{r^3} r_0^2$$

is the coefficient of the term $(\frac{1}{3} - \cos^2 \theta)$ in the expansion. The limb temperature is expected to be remarkably uniform and the variation of disk brightness with latitude should be less than .01 percent (Dicke 1970a).

The limb occurs at an optical depth of ~ 0.004 and a density of $\sim 2 \times 10^{-8}$. One might think that a very strong magnetic or turbulent velocity field just below this layer would affect the oblateness of the limb, but, as shown above, such is not the case. Similarly one might think that the tension of the strong magnetic field of a sunspot would depress the level of the photosphere over a large area surrounding the sunspot, but, as shown above, such a strong field cannot affect the height of the photosphere anywhere except at the location of the magnetic field. Also, the only significant effect of the sunspot on the measurement is induced by the darkening of the sunspot, not the change in level (Dicke 1970b).

As was noted above, von Zeipel's functional relations include the effects of rotation whenever the rotation of the surface layers is on cylinders, i.e. the angular velocity ω is a function only of distance from the rotation axis. If, and only if, purely rotational motion is on cylinders, the "centrifugal force" term is derivable from a potential and the inertial term can be added to Equation 2 by including a centrifugal potential. Then the equation is still valid with φ replaced by

$$\Phi = \varphi - \int_0^{r \sin \theta} \omega r \sin \theta d(r \sin \theta) \qquad 4.$$

Including the effects of rotation of the Sun's surface, the oblateness of a surface of constant ρ is also the oblateness of a surface of constant Φ. From this oblateness, the contribution of the centrifugal potential (8×10^{-6}) must be subtracted before the oblateness of the surface of constant φ is obtained.

Only one thing more is needed for a complete description of the connection between the observed solar oblateness and the gravitational quadrupole moment. When magnetic and velocity fields exist in the "seen layers" of the solar surface, von Zeipel's relations are not exactly satisfied and the effects of these fields must be included. Such fields may contribute generally different amounts to the oblateness of surfaces of constant density and pressure. The surface rotation discussed above is a special case for which these two contributions are equal and the temperature hence constant on the surface.

The surface fields usually induce different oblateness in density and pressure surfaces, and the contribution to the oblateness of a constant density surface is usually accompanied by a variation of brightness with latitude. Only for a carefully selected stress distribution in the surface layers is the surface brightness independent of latitude.

For an arbitrary distribution of magnetic and velocity surface stresses these contributions to the oblateness of constant density and pressure surfaces are known (Dicke 1970a). They are conveniently expressed in terms of integrals over the surface of a set of basic stresses. These are: P_f, the field pressure; S_r, the radial shear; S_t, the transverse shear; and S_m, the meridional shear.

$$P_f = \frac{1}{8\pi} B^2 + \frac{1}{3} \rho v^2$$

$$S_r = -\frac{1}{4\pi}[2B_r^2 - B_\theta^2 - B_\varphi^2] + \rho[2v_r^2 - v_\theta^2 - v_\varphi^2]$$

$$S_t = -\frac{1}{8\pi}(B_\theta^2 - B_\varphi^2) + \frac{1}{2}\rho(v_\theta^2 - v_\varphi^2)$$

$$S_m = -\frac{1}{4\pi} B_r B_\theta + \rho v_r v_\theta$$

5.

In Legendre polynomial expansions of ρ and P over a surface of constant φ, $\delta\rho$ and δP, the coefficients of the second Legendre polynomial ($\frac{1}{3}-\cos^2\theta$), are given by

$$\delta P = \frac{45}{8}\int_0^\pi (P_f - \tfrac{1}{6}S_r)(\cos^2\theta - \tfrac{1}{3})\sin\theta\,d\theta$$

$$-\frac{15}{8}\int S_t \sin^3\theta\,d\theta - \frac{15}{8} r^{-2}\frac{d}{dr} r^3 \int S_m \cos\theta \sin^2\theta\,d\theta$$

6.

and

$$g\delta\rho = \frac{45}{16}\frac{1}{r^2}\frac{d}{dr}r^2 \int S_r(\cos^2\theta - \tfrac{1}{6})\sin\theta\,d\theta$$

$$+\frac{15}{8}\frac{d}{dr}\int S_t \sin^3\theta\,d\theta + \frac{15}{8}\left[\frac{4}{r} + r^{-1}\frac{d^2}{dr^2}r^2\right]\int S_m \cos\theta \sin^2\theta\,d\theta$$

7.

δP and $\delta\rho$ can also be interpreted as contributions to the equatorial excesses of pressure and density on surfaces of constant φ. The contributions to the oblatenesses of surfaces of constant pressure and density induced by these fields are

$$\left(\frac{\Delta r}{r}\right)_P = \frac{\delta P}{r g \rho}$$

8.

and
$$\left(\frac{\Delta r}{r}\right)_\rho = \frac{\lambda_\rho}{r}\frac{\delta\rho}{\rho} \qquad 9.$$

where λ_ρ is the density scale height and g is the gravitational acceleration. The temperature excess at the equator on a surface of constant density is

$$\delta T = -[(\Delta r/r)_P - (\Delta r/r)_\rho]\mu g r R^{-1} \qquad 10.$$

where μ and R are molecular weight, and gas constant respectively.

The oblateness of the Sun is observed at the limb at an optical depth of ~ 0.004, but the whole of the Sun is observed to be remarkably free of a variation of brightness with latitude. This implies that any acceptable distribution of surface fields must generate equal oblateness in the observed surfaces of constant ρ and P, and this oblateness must be substantially independent of optical depth over the range 0.004 to 1. Thus,

$$\left(\frac{\Delta r}{r}\right)_\rho = \left(\frac{\Delta r}{r}\right)_P = \text{const} \qquad 11.$$

The condition 11 in turn implies that such a distribution of surface fields induces a force per unit volume of the form $-\rho\nabla W$, where W is some scalar function of the polar angle θ and is only weakly dependent on r. A variation of W with latitude by an amount δW generates an outward displacement of a surface of constant ρ, P, and T by an amount

$$\delta r = -\delta W/g \qquad 12.$$

In order that the surface fields induce an oblateness $(\Delta r/r)_f$,

$$W = gr\left(\frac{\Delta r}{r}\right)_f (\cos^2\theta - \tfrac{1}{3}) \qquad 13.$$

Equations 6 and 7 can be solved subject to the constraint 11 to give the three independent solutions. The first is

$$P_f - \tfrac{1}{6}S_r = \tfrac{3}{2}P_f = \rho W \qquad 14.$$

with W given by Equation 13 and with S_t and S_m zero. Note that these equations refer to second Legendre polynomials. Other nonzero terms could be present. The second solution is

$$S_t = -\frac{1}{2}\rho g r\left(\frac{\Delta r}{r}\right)_f \sin^2\theta \qquad 15.$$

with $P_t = S = S_m = 0$ and W given by Equation 13. The third is

$$S_m = \rho g r\left(\frac{\Delta r}{r}\right)_f \left(\frac{\lambda_\rho}{r}\right)\sin^2\theta \qquad 16.$$

with $P_f = S = S_t = 0$ together with Equation 13. Any linear combination of Equations 14–16 is also suitable.

Roxburgh (1967a), Cocke (1967a), and Sturrock & Gilvarry (1967) suggested that the observed excess oblateness was due respectively to variation with radius of the angular velocity, meridional currents acting against turbulent viscous forces, and magnetic fields. Equations 6, 7, and 14–16 have been used to analyze these suggestions. Roxburgh (1967a) suggested that Coriolis forces acting on the convective zone induce a heat-transfer imbalance that in turn causes a strong increase of the angular velocity with radius. The stresses associated with convective transport can be ignored since they occur below the optical depth of 0.004 of the limb. At the limb, only the rotation is significant. From equations similar to 6 and 7 it was shown that Roxburgh's assumed rotational distribution reduces, rather than increases, the oblateness, and a large variation of brightness with latitude is generated, contrary to the observations (Dicke & Goldenberg 1967b). (See later rediscussion by Roxburgh 1967b.) In similar fashion, through the use of Equations 14–16 it was shown that meridional currents and magnetic fields strong enough to generate the excess oblateness without generating a variation of brightness with latitude were incompatible with the observations (Dicke 1970a). It is concluded that the observed surface stresses do not generate the observed excess oblateness.

The great uniformity in surface brightness is probably no accident. Any inequality between the radiation rate at the surface and the rate of heat transport to the surface from the interior would generate circulation currents that would transport magnetic fields over the Sun's surface to redistribute surface stresses. Probably surface stress distributions are automatically adjusted by this feedback mechanism until the stress distribution is sufficiently uniform for the surface to be uniformly bright (except as sunspots where the magnetic field is strong enough to inhibit the convective transport of heat from below). Whatever the mechanism, the observations show the random-velocity field and the background magnetic field to be independent of latitude. For the weak-background magnetic fields of the quiet Sun, this uniformity is strikingly shown by Livingston's (1966) pictures. An analysis of one of his pictures shows no systematic variation of the background field with latitude (Dicke 1970a).

In summary: The oblateness of the undisturbed Sun's surface is very nearly that of a surface of constant density. After subtracting the contribution from surface rotation, the remainder represents the oblateness of a surface of constant gravitational potential. This oblateness in turn uniquely determines the gravitational quadrupole moment. Magnetic and velocity fields in the "seen layers" could change these conclusions, but no such fields are found capable of seriously affecting these relations.

The Spindown Problem and Other Questions of Stability

It has been claimed (Howard, Moore & Spiegel 1967) that the Sun probably could not have a rapidly rotating core because of loss of angular momentum by dynamically driven circulation currents associated with the formation of an Ekman layer, i.e. the spindown effect. It has also been claimed that because of an instability due to a thermally driven turbulence, associated with a thermal diffusivity large compared with that of angular momentum, a rapidly rotating core is precluded (Goldreich & Schubert 1967a,b; Fricke 1968). This firm position of Goldreich & Schubert was later modified somewhat (Goldreich & Schubert 1968) after Colgate (1968) showed that a compositional gradient in the zone of differential rotation could stabilize a rapidly rotating core.

The spindown effect is easily seen in a cup of tea where a rotation of the stirred tea ceases in a time short compared with the diffusion time. The rapid slowing is due to pumping of the tea through a thin (Ekman) layer at the bottom of the cup.

In attempting to relate this phenomenon to the solar interior it is essential to understand the significance of an important difference between a cup of tea and the solar interior. The density of tea is constant whereas the solar medium is compressible. Furthermore, there is an important difference between the outer convective zone of the Sun and its radiative core. In the convective zone, pressure and density are functionally connected through the adiabatic condition, but not in the radiative core. If we neglect viscous forces, purely rotational motion is on cylinders in the convective zone. This is seen by noting that for $P = P(\rho)$ and for purely rotational motion with the angular velocity ω,

$$\nabla P + \rho \nabla \varphi + \rho \omega \times (\omega \times r) = 0 \qquad 17.$$

and, dividing through by ρ,

$$\nabla \int_0^P \frac{1}{\rho} dP + \nabla \varphi + \omega \times (\omega \times r) = 0 \qquad 18.$$

Equation 18 is valid also for the incompressible tea. From 18, $\omega \times (\omega \times r)$ must be the gradient of a scalar, and normal to circular cylinders.

$$\omega \times (\omega \times r) = -\nabla \int \omega^2 r \sin \theta \, d(r \sin \theta) \qquad 19.$$

Thus, ω is constant on circular cylinders.

In the stirred cup of tea, because of the boundary condition on the bottom of the cup, purely rotational motion on cylinders is impossible. The inclusion of the viscous-force term significantly affects the motion in a thin layer (Ekman) at the cup bottom and permits the boundary condition to be satisfied. Owing to the reduction of the centrifugal-force density in this thin layer, fluid is propelled inward, pumping the whole content of the cup through the Ekman layer where viscous dissipation is great. This causes a slowing of the rotation in a time $\sim a/(\omega \nu)^{1/2}$ where a is the cup dimension, ω is the angular velocity of the fluid, and ν is the kinetic viscosity of the fluid. This is much less than the diffusion time a^2/ν.

Spindown would be expected in the convective zone of the Sun but other complications are associated with the functional relation between pressure and density. Angular momentum per unit mass could not increase inward toward the rotation axis because of the Kelvin-Helmholtz instability. Also, for an appreciable variation of angular velocity, turbulence would be excited, the Reynolds criterion being easily satisfied.

In the presence of turbulence the spindown phenomenon becomes modified, a phenomenological turbulent viscosity playing a role similar to molecular viscosity. The presence at the Sun's surface of differential rotation in spite of the enormous turbulent-viscous force has long been something of a mystery. To drive this differential rotation requires a large torque. The best candidate for this force seems to be the viscous force itself, the extra component added through anisotropic turbulence providing the driving force (Kippenhahn 1963, Cocke 1967b).

Just below the convective zone the temperature gradient is nearly adiabatic, but a few thousand kilometers deeper the temperature gradient has greatly decreased and the gas is strongly density stratified. The appropriate criterion for stability against turbulence for an angular-velocity gradient normal to constant-density surfaces is not that of Reynolds, but rather that of Richardson

$$\left(\frac{r}{\omega}\frac{d\omega}{dr}\right) \leq \frac{4}{\gamma}\left(\frac{g}{\omega^2}\right)\left[(\gamma - 1)\frac{d}{dr}\ln \rho - \frac{d}{dr}\ln T\right] \qquad 20.$$

This stability criterion holds only for rotation on spheres $\omega = \omega(r)$. From the Weymann (1957) solar model the values of $+r/\omega \, d\omega/dr$ for equality in Equation 20 are $-90, -1230, -1730$ at $r = 0.84, 0.76, 0.64$ respectively. If ω is not constant on spherical surfaces, i.e. depends upon θ, turbulence should be excited. It might be expected that this turbulence would eliminate the dependence of ω on θ.

Pressure and density are uncoupled in the radiative core. The functional relation between P and ρ, which leads to Equation 18 and forces purely rotational motion to be on cylinders, is relaxed. For an arbitrary choice of $\omega(r, \theta)$ the dependences of both P and ρ on θ are separately determined by the function $\omega(r, \theta)$ (Dicke 1967c). Choosing the θ component of Equation 16, defined to lie in a surface of constant gravitational potential φ, and integrating this equation over the surface yield the dependence of P on θ over this surface. By taking the curl of Equation 16 a similar integration can be carried out for ρ. Thus the θ dependence of P and ρ (hence of T if the mean molecular weight is constant) are determined by the rotational distribution.

Purely rotational motion (no spindown) is possible in the density-stratified solar interior, if the temperature distribution is appropriate, but the adopted rotational distribution need not be stable. The importance of density stratification on spindown has been discussed several times from different viewpoints by Holton (1965), Pedlosky (1967, 1969), Dicke (1967c), Holton & Stone (1968), Sakurai (1969a,b), and Clark et al (1969).

The effect of density stratification on spindown was exhibited experimentally (McDonald & Dicke 1967). A density-stratified fluid was established in corotation with a steadily rotating cylindrical dish. The rotational rate of the dish could be changed by a fractionally large amount without inducing spindown, if the change was made slowly in very small steps. If the angular velocity were changed discontinuously by as much as 1 percent, spindown would occur through a series of complex events, starting with the excitation of gravity waves, followed by mixing in two layers and separate spindown of each of the mixed layers. The sudden change in the angular velocity of the dish imposes on the density-stratified fluid a rotational distribution that, for purely rotational motion, is incompatible with the actual density distribution.

It is concluded that dynamically driven spindown currents do not occur in the density-stratified solar interior. The time scale $\sim 10^{10}$ years for slowing the solar rotation by the solar wind is extremely long compared with the rotation period, and the inertial effects of the circulation currents that maintain the correct density distribution are negligible. Thus there is adequate time for the solar temperature to automatically adjust itself to satisfy the dynamical requirements of a purely rotational motion. While dynamically driven Ekman-type currents probably do not exist, Eddington-Sweet thermally driven circulation currents should occur, unless there is a gradient in molecular weight in the zone of differential rotation. In general, the dual requirements of the rotational distribution, on pressure and density, lead either to a variation of molecular weight or else to a temperature distribution that is incompatible with the requirements of heat balance. The velocities of Eddington-Sweet currents associated with differential rotation can be orders of magnitude greater than the more familiar thermally driven currents associated with uniform rotation (Schwarzschild 1958). It is a common mistake to apply the time scale associated with Eddington-Sweet currents under uniform rotation to situations with differential rotation.

The instability discussed by Goldreich & Schubert (1967a, 1967b) and Fricke (1968) takes place through the development of axially symmetric angular-velocity variations on spherical surfaces. Thin toruses, ~ 1 km thick, are continuously generated and destroyed, moving upward and downward and transporting angular momentum. It was noted (Dicke 1967b) that the theory of this instability assumed the absence of a magnetic field in the deep interior of the Sun. Goldreich & Schubert had noted that a negligibly small θ component of velocity was required. Clark et al (1969) have proposed that the oscillating motion of internal gravity waves driven by turbulence in the convective zone could provide the θ velocity component that would stabilize the flow. Fricke (1969) has investigated the effects of magnetic fields on the instability. He finds that a strong toroidal magnetic field ($\sim 10^5$ G) in the zone of differential rotation can stabilize the rotation if the field strength increases outwardly. Colgate (1968) had shown that this instability could be eliminated by the existence of a molecular weight gradient in the shell of differential rotation (also see Goldreich & Schubert 1968). The required gradient in the mean molecular weight is slight and could be established by the production of helium in the core if a sufficiently rapid means of mixing the core were available. The gradient in mean molecular weight necessary to stabilize the core satisfies (Goldreich & Schubert 1968),

$$\frac{rd \ln \mu}{dr} < 2 \left(\frac{\omega^2 r}{g} \right) \frac{1}{r\omega} \frac{d}{dr} (r^2 \omega) \qquad 21.$$

For the model of a rotating core to be discussed below, the right side of Equation 21 is roughly 6×10^{-2} and the fractional increase in mean molecular weight μ in the core over that of the exterior need be only 3×10^{-3}. Roughly 5×10^7 years of nuclear burning with the products mixed uniformly through the core would be required to increase μ by this amount in the core.

The ordinary Eddington-Sweet thermally driven currents associated with a rigidly rotating core are too slow to mix the core, even if the core rotates as rapidly as we postulate (Schwarzschild 1958). But differential rotation in the core, or a strong poloidal magnetic field buried in the core in the perpendicular rotator configurations could greatly increase the circulation rate, by several orders of magnitude. Mixing by thermally driven currents might occur for a few hundred million years but then be choked by the accumulated molecular weight gradients. If this happened the evolutionary tracks in the H-R diagram might be only slightly modified.

If the Goldreich-Schubert-Fricke instability does occur, what is the limiting velocity distribution, assuming an initially uniform and rapidly rotating Sun slowed by the solar wind? This instability is very effective at transporting angular momentum as long as the angular momentum per unit mass increases inward toward the rotation axis. But ordinary viscosity-driven turbulence would be expected to develop if ω were a function of θ. Thus the quasistable limiting distribution would be expected to be of the form $\omega \sim r^{-2}$ below the convective zone with ω constant in that zone except for the above-mentioned differential rotation generated perhaps by anisotropic turbulence in the differentially rotating zone. The long-term stability of the distribution below the convective zone depends upon the effectiveness of thermally driven currents, upon whether or not they have been choked by gradients in molecular weight, and possibly upon other complications such as an internal magnetic field.

One possible distribution, particularly interesting because it can be tested observationally, is a rapidly rotating core inside a shell of differential rotation (of thickness $\delta\, r/r_0 \sim 0.05$) through which the angular momentum leaks by molecular diffusion. Outside of this to the convective zone is a thick shell through which the angular momentum is transported by the Goldreich-Schubert process and in which $\omega \sim r^{-2}$. Outside the shell is the convective zone.

This model differs from the one first proposed (Dicke 1964) in that the zone of molecular diffusion could lie substantially deeper. The observation that lithium, but not beryllium, is depleted with time suggests that the outer radius of the zone of molecular diffusion may fall below $r = 0.58$, where ^{7}Li is quickly burned, but outside 0.5 if the Weymann (1965) solar model is correct. This new model will be discussed in some detail below.

To summarize, it is the lack of observations of the deep solar interior that makes conclusions about instabilities uncertain. Because of strong density stratification below the convective zone and the mild nature of braking by the solar wind, dynamically driven spindown currents probably do not exist, but fairly rapid thermally driven circulation currents in the zone of differential rotation are possible, though very easily choked by gradients in mean molecular weight. The Goldreich-Schubert-Fricke instability is easily inhibited by such a molecular weight gradient or by oscillatory motion in the θ direction. Density stratification stabilizes purely rotational motion on spheres; and it is concluded that ω is a function of r if a stable rapidly rotating core exists. Of more importance than these theoretical arguments concerning the deep interior are the observations of the solar surface.

The Solar-Wind Torque

While the structure of the solar wind is not directly of concern to us, observations of the solar wind can provide a measure of the solar-wind torque, and this is of importance to the problem of internal solar rotation. If we make the questionable assumption that the solar wind blows substantially radially out to the vicinity of Venus and the Earth, measurements of solar-wind flux carried out with the Mariner space probes, and of the magnetic-field strengths in the wind, permit an evaluation of the solar-wind torque. The assumption of radial flow when the Sun is quiet may be questionable in the light of the appearance of the corona.

The solar-wind torque density on the solar surface at the equator is (Dicke 1964, Modiesette 1967, Weber & Davis 1967, Alfonso-Faus 1967)

$$K = Jr^2\omega_0 \qquad 22.$$

where J is the mass flux density at the solar surface, ω_0 is the angular velocity, and r is a "critical radius" for which

$$\rho v^2 = \frac{1}{4\pi} B^2 \qquad 23.$$

namely, the radius at which v, the radial component of the solar-wind velocity, equals the Alfvén velocity calculated from B, the radial component of the magnetic field.

The magnetic field is trapped in the wind and B falls off inversely as the square of the radius. Equation 23 can be written as

$$\rho = 4\pi J^2/B_0^2 \qquad 24.$$

where $B_0 = (r/r_0)^2 B$ is the strength of the trapped magnetic field referred to the solar surface. This is a particularly interesting way to express the field, for the magnetic field at the solar surface cannot be too strong or trapping is impossible. The physical properties of the chromosphere and lower corona limit the strength of the trapped field. Before the strength of the interplanetary field was measured, B_0 had been estimated to be .75 G (Dicke 1964). This estimate was based on the assumption that the field strength would lie near its upper limits.

Measurements of the magnetic-field strength at 1 au with the Mariner II space probe gave an rms value for the radial component of roughly $\sim 3.5 \times 10^{-5}$ G, $B^0 \sim 1.4$ G (Coleman 1966). If we make the simplified assumption that near the Sun's surface substantially cylindrical magnetic flux tubes are stretched out from the solar surface by the solar wind, the magnetic pressure $1/8\pi\, B^2$ cannot exceed the gas pressure outside these tubes. From Allen's model of the solar corona at the equator, the rough upper limit for B_0 takes on the values 0.8, 0.7, and 0.5 from gas pressures at $r = 1.01$, 1.1, and 1.4 respectively. The concentration of the magnetic field toward the equatorial plane may have increased the field strength at the Earth's radius.

The mass flow in the solar wind is determined by the rate of heating of the corona. This heating is believed to be caused by acoustic noise generated by turbulence in the convective zone. It would be expected to be more or less constant in time, depending upon the relative importance of magnetohydrodynamic waves in coupling the corona to the convective zone. There are observational reasons for believing that young solar-type stars are more active magnetically than older stars (Wilson 1966).

The magnitude of the surface density of the mass flux in the solar wind, from the Mariner II space probe, is $J \simeq 1.7 \times 10^{-11}$ g/cm² sec (Neugebauer & Snyder 1966). From Equation 24, combining the above results obtained with the space probe with Allen's (1963) model of the corona gives $r = 20\, r_0$ for the critical radius. From Equation 22 $K = 9 \times 10^7$ dyne/cm. For the total solar-wind torque

$$\frac{8\pi}{3} r_0^2 K = 3.8 \times 10^{30} \text{ dyne/cm} \qquad 25.$$

One should not be misled by the apparent precision of this number which may be uncertain by a factor of 2. This precision and similar apparently accurate numbers below are introduced only to make the arithmetic well defined.

The torque (Equation 25) is proportional to the angular velocity of the solar surface. With the assumptions that the magnetic field B_0 lies near its maximum value and that the solar-wind strength J has been reasonably constant, the torque density per unit angular momentum, K/ω_s, would be reasonably constant over the life of the Sun. But the greater magnetic activity in young solar-type stars would be expected to increase the torque. If magnetic coupling to the corona provides the dominant means of heating the corona, the solar-wind flux in the young Sun could have been substantially greater.

The Evolution of the Rapidly Rotating Core in the Sun

In this section the picture developed above will be adopted as a working hypothesis, and a quantitative history of the Sun's rotation will be developed. The radius of the core will be assumed to be $r_c = 0.54$, permitting rapid burning at the core boundary of lithium but not of beryllium. (See Table 2.) It will be assumed that the solar core, of radius 0.54, is rotating uniformly with the angular velocity $\omega = 20\ \omega_0$ needed to generate the quadrupole moment associated with the solar oblateness ($\omega_0 = 2.87 \times 10^{-6}$ sec^{-1}). The angular velocity of $20\ \omega_0$ corresponds to a 1.27-day period; at a core radius of 0.8, an angular velocity of $15\ \omega_0$ is needed (McDonald 1969).

It is impossible to transport angular momentum from the outer bounds of such a core to the bottom of the convective zone at $r_v = 0.86$ (Weymann 1957) by molecular diffusion. It will be assumed that the thermally driven turbulence discussed by Goldreich & Schubert (1967b, 1968) permits the transport by turbulent diffusion in the range $r_c < r < r_v$, a thin shell of molecular diffusion limiting the flow of angular momentum from the core. As discussed above, thermal turbulence is easily inhibited. It will be assumed that the core boundary is stabilized, probably by a molecular weight gradient or by oscillatory motion in the θ direction.

Thermal turbulence is so effective when it occurs that the angular velocity gradient cannot appreciably exceed the threshold value $r(d\omega/dr)/\omega = -2$. This gradient will be assumed, or $\omega r^2 =$ const in this shell. As discussed above, it will be assumed that ordinary mechanically driven turbulence keeps the angular-velocity gradient parallel to the density gradient, i.e. angular velocity is a function only of r.

It will be assumed that the Sun is a typical star. Thus solar history might be illuminated by Kraft's (1967) observations of rotation in young stars. He finds that the surface rotation $20\ \omega_0$ of stars of $1.2\ M_0$ in the Pleiades, 3×10^7 years old, has dropped to $10\ \omega_0$ in the Hyades, 5×10^8 years old. Observations are missing for G2 stars but extrapolation curves suggest angular velocities as low as $5\ \omega_0$ and $2.5\ \omega_0$ respectively for the Pleiades and Hyades. If the above picture is correct, solar-type stars, with their deep convective envelopes, arrive on the main sequence either with a substantial amount of differential rotation already present or with a strong stellar wind acting to slow the outer shell in a time as short as 3×10^7 years.

We assume that the angular velocity is substantially constant at the surface value ω_s in the convective zone (down to $r_v = 0.86$) and that for a fully developed turbulent zone, $\omega = (r_v/r)^2 \omega_s$ for $r_c < r < r_v$. A stellar-wind equatorial torque density $K_s = 10^8 (\omega_s/\omega_0)$ dyne/cm (substantially the same as that observed for the solar wind) must act for 1.63×10^9 years on an initially uniformly rotating star to develop the thermal turbulent zone down to the core radius $r_c = 0.54$. At this time $\omega_s = (r_c/r_v)^2 \omega_c = 0.394 \omega_c$. Subesquent slowing of the fully developed shell, together with the convective zone, occurs with an e-folding time of 6.05×10^9 years. This neglects the (initially small) contribution to the solar-wind torque from angular momentum leaking out of the core. See Table 1.

TABLE 1. The slowing of rotation of a shell by a solar wind of equatorial torque density 10^8 dyne/cm

r_c	I_s/I	T (year) (rigid rotation)	T (year)
0.86	0.0122	0.173×10^9	0.173×10^9
0.78	0.034	0.48	0.51
0.70	0.074	1.04	1.28
0.62	0.140	1.98	2.88
0.54	0.241	3.40	6.05

Decay time in years and fractional moment of inertia as functions of the inner radius of a uniformly rotating outer solar shell; it is assumed that only the shell is slowed by an equatorial solar-wind torque density of 10^8 dyne/cm. The decay time for the whole Sun rotating rigidly is 14.1×10^9 years. In the last column the decay time is calculated for a rigidly rotating connective zone and a differentially rotating inner zone, with $\omega = (r_v/r)^2 \omega_s$ for $r_c < r < r_v = 0.85$.

To account for the factor of 2 decrease of ω_s from the Pleiades to the Hyades in 5×10^8 years would require that the solar-wind torque density $K_s = 8.35 \times 10^8 (\omega_s/\omega_0)$. The solar-wind torque may have been even greater in the first 3×10^7 years. If the torque density were as great as $K_s = 150 \times 10^8 (\omega_s/\omega_0)$, the Sun could initially have been uniformly rotating on the main sequence. For reasons discussed above, a torque this great seems unlikely, and a strong torque during the late Hayashi phase seems more likely. It will be assumed that the Sun arrived on the main sequence with the surface rotating with an angular velocity $\omega_s \sim 5\omega_0$.

The solar-wind torque density K_s can be decomposed into two parts, K_t and K_d. Here $K_t = -6.05 \times 10^8 (\dot{\omega}_s/\omega_0)$ represents the contribution from the deceleration of the outer shell ($\dot{\omega}_s$ means time derivative with 10^9 years as the unit of time) and K_d is the contribution from the loss of angular momentum from the core. K_d is evaluated by solving the diffusion equation

$$\frac{\partial}{\partial r}\left(\rho\nu r^4 \frac{\partial \omega}{\partial r}\right) = \rho r^4 \frac{\partial \omega}{\partial t} \qquad 26.$$

as a boundary-value problem, assuming that $(r_v/r_c)^2\omega_s$, the angular velocity at the core boundary, is known as a function of the time. If $\rho\nu r^4$ varies slowly enough through the shell of molecular diffusion (thickness ~ 0.05), a good approximation is obtained by replacing it by its (constant) value at $r_c = 0.54$. At this point ν, 10 percent of the kinematic viscosity 14.3 cm^2/sec is due to radiation transport and the remaining is due to transport by ions. In this approximation, for $r < r_c$

$$\omega(r,t) = -(r_v/r_c)^2 \int_{-\infty}^{t} (d\omega_s/d\tau)$$
$$\cdot \{\text{erf}\,[(r_c - r)/\sqrt{2\nu(t-\tau)}] - 1\}d\tau + \omega_c \qquad 27.$$

From Equation 27 the radial derivative at r_c is calculated, giving

$$K_d = -r_0^{-2} r_c^4 \rho_c \nu_c (\partial\omega/\partial r)_c \qquad 28.$$

as the core's contribution to the solar-wind torque density. If ω_s decreases exponentially to zero with a decay constant λ,

$$K_d = r_0^2 (r_c/r_0)^2 \rho_c \omega_c [\nu_c/\pi t]^{1/2} \sqrt{\lambda t}\, W(\lambda t) \qquad 29.$$

where

$$W(x) = e^{-x} \int_0^x e^y y^{-1/2} dy \qquad 30.$$

The function

$$D(x) = x^{1/2} W(x) \sim 1.3x(1 - \tfrac{1}{4}x) \qquad x < 2$$
$$\sim 1 + 1.2/(x+2) \qquad x > 2$$

Substituting numerical values gives

$$K_d = 1.38 \times 10^8\, t^{-1/2} D(\lambda t)\ \text{dyne/cm} \qquad 31.$$

where t is in units of 10^9 years. For $\lambda t > 0.5$, K_d varies slowly with time. The above formalism is easily generalized to cover the case

$$\omega_s(t) = \int_0^\infty A(\lambda) e^{-\lambda t} d\lambda$$

for which $D(\lambda t)$ in Equation 31 is to be replaced by

$$\bar{D} = \int A D d\lambda \Big/ \int A d\lambda \qquad 32.$$

The time dependence of the surface rotation is obtained as a solution of the differential equation $K_s = K_t + K_d$, namely

$$N(t) \times 10^8 (\omega_s/\omega_0) = -6.05 \times 10^8 (\dot\omega_s/\omega_0) + K_d(t) \qquad 33.$$

where $N(t)$ is the ratio of the solar-wind torque density to surface angular velocity, expressed in units of $10^8 \omega_0$ g/sec characterizing the present solar wind. The general solution to this equation is easily written, if we assume that $N(t)$ and $K_d(t)$ are known. For sufficiently large value of N and slow variation of K_d/N, ω_s falls until $|K_t| \ll K_d$. These conditions seem to be approximately satisfied for N, varying in such a way as to give a satisfactory account of the surface rotations adopted for the Pleiades and Hyades, and of that observed in the Sun. As a result, it is to be expected that the present value of the solar-wind torque is approximately equal to

$$K_d \sim 0.8 \times 10^8 \text{ dyne/cm} \qquad 34.$$

in satisfactory agreement with the observations of the solar wind.

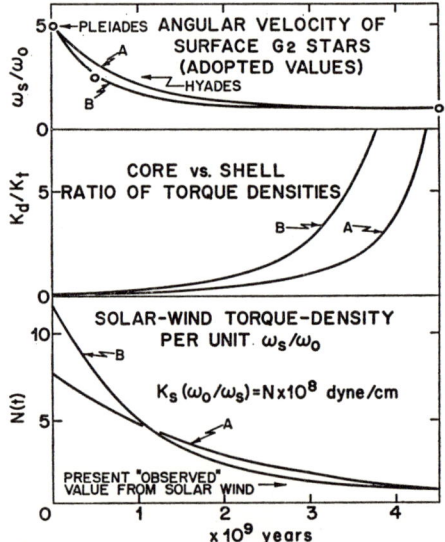

FIGURE 4. The slowing of the Sun's rotation assuming: A, a solar-wind torque with N decreasing exponentially in time; B, N decreasing exponentially to a constant value.

Two variations of N with time seem to be particularly interesting. The results of numerical integrations with these choices are plotted in Figure 4. For the curves A of Figure 4, N is assumed to fall exponentially with time, and the three parameters characterizing N and the initial value of ω_s are adjusted to give a satisfactory account of the three "known" values of ω_s. For the curves B, N is the sum of a constant and an exponential. The mean life of the exponential is arbitrarily taken to be $\sim 10^9$ years. This term is introduced to represent the decay of the initially strong solar activity. The decay with time of magnetic activity in solar-type stars, as exhibited by Ca II emission (Wilson 1966), might be due to the decay of short-lived magnetic modes originally trapped in the Sun. As noted above, increased magnetic activity could result in a stronger solar wind, particularly if magnetic coupling to the corona is the primary source of coronal heating. The initial values of the two terms are adjusted to give surface rotations agreeing with the "observations."

It should be remarked that the integrations A and B require the present solar-wind torque density to be $\sim 0.85 \times 10^8$ dyne/cm. This is in satisfactory agreement with the observations of the solar wind.

LITHIUM DEPLETION IN SOLAR-TYPE STARS

The depletion of lithium in solar-type stars can be related to the deceleration of their surface rotations if the model developed above is reasonably correct. In the thermal-turbulent zone the diffusivities of angular mo-

TABLE 2. Fractional depth r_b/r_o at which burning takes place with indicated mean life

Mean life	3×10^6 years	3×10^7 years	3×10^8 years
Li^6	.57	.60	.63
Li^7	.51	.55	.58
Be^9	.42	.45	.47

Based on Fowler, W. A., Caughlan, G. R., Zimmerman, B. A. 1967, *Ann. Rev. Astron. Ap.*, **5**, 525 and Weymann's (1957) solar model.

mentum and lithium are equal. But this diffusivity is determined by two requirements, that the angular momentum flux be correct and that ωr^2 be constant in this zone, with ω independent of latitude.

Integrating Equation 26 from r_c to r and using the relation

$$r^2 \omega = r_v^2 \omega_s \qquad 35.$$

gives

$$\rho \nu r = \frac{1}{2\omega_s} \left(\frac{r_0}{r_v}\right)^2 \left[K_d + K_t \frac{M(r) - M(c)}{M(v) - M(c)} \right] \qquad 36.$$

In Equation 36 $M(r)$, $M(c)$, and $M(v)$ are stellar masses inside the designated radii; the minor contribution from the convective zone has been omitted from K_t (as defined in the paragraph above Equation 26).

The diffusion of lithium is controlled by the equation

$$\frac{\partial}{\partial r}\left(\rho v r^2 \frac{\partial F}{\partial r}\right) = \rho r^2 \frac{\partial F}{\partial t} \qquad 37.$$

where F represents the fractional abundance of ^{7}Li or ^{6}Li (by mass or number).

The solution to Equation 37 is eased by the simplifying assumption that the zone of burning has a sharp boundary. This requires $F=0$ as a condition on the boundary. There is also a condition to be satisfied at the inner boundary of the convective zone. This is determined by the requirement that the radial derivative of F, which determines the flow of lithium from the convective zone, be proportional to the time derivative of F which gives the loss of lithium in the convective zone.

A normal solution to Equation 37, with Λ as the decay constant, satisfies the eigenvalue equation

$$\frac{\partial}{\partial r}\left(\rho v r^2 \frac{\partial F}{\partial r}\right) + \Lambda \rho r^2 F = 0 \qquad 38.$$

This is to be integrated subject to the above-described boundary conditions. Equation 36 is first substituted in 38. If the zone of burning is $r < r_c$, the normal solutions depend upon the parameter (K_d/K_t). For the lowest mode, we shall need the slope-to-value ratio $r(\partial F/\partial r)/F$ evaluated at r_v (as a function of K_d/K_t). This is given in Table 3.

All higher normal modes decay rapidly, an order of magnitude faster for the second mode, and the subsequent fractional decay rate is that of the lowest mode, providing K_d/K_t varies slowly with time.

TABLE 3. The slope/value ratio of F versus K_d/K_t

K_d/K_t	0.02	0.05	0.10	0.2	0.4	0.8	1.6	3.2	6.4
$r\frac{dF}{dr}/F$	0.074	0.096	0.118	0.150	0.193	0.234	0.273	0.304	0.319

The decay rate of the lowest mode is conveniently expressed in terms of the solar-wind torque. Integrating Equation 38 from r_v to r_0, the solar surface, gives

$$-\left(\rho v r^2 \frac{\partial F}{\partial r}\right)_v = \int \rho r^2 \dot{F}_v dr = \frac{1}{4\pi} \dot{F}_v [M(0) - M(v)] \qquad 39.$$

Substituting Equation 36 gives

$$\Lambda = 2\pi \frac{K_s}{\omega_s}\left(\frac{r_0}{r_v}\right)^2 \left[\frac{r \partial F/\partial r}{F}\right]_v \frac{1}{M(0) - M(v)} \qquad 40.$$

Taking K_s/ω_s and K_v/K_t from Figure 4 and using Table 2 gives Λ as a function of time. This permits the integration of $\dot{F} = -\Lambda F$. The resulting curves A and B are plotted in Figure 5 which is based on Figure 2 of Danziger (1969). Note that these curves contain no adjustable constants.

It should be emphasized that the above integration is based on the assumption that the boundary of lithium burning is sharp and that it occurs at the bottom of the zone of turbulent diffusion. This may be reasonable, if we assume that the outward-moving boundary is initially somewhat below

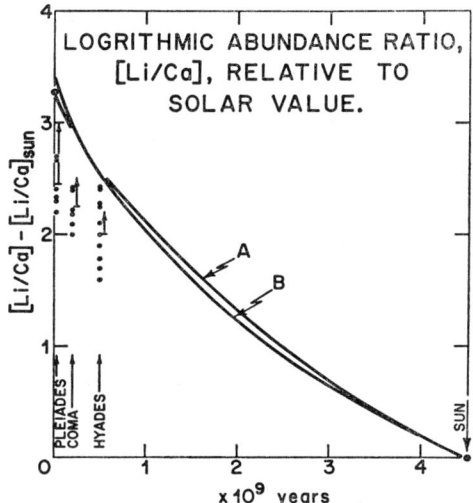

FIGURE 5. The depletion of lithium; the same turbulence viscosity as that associated with the transport of angular momentum is assumed (curves A & B of Figure 4). The turbulence is assumed to be driven *in part* by the thermal effect of Goldreich & Schubert (1967b). The plotted points represent individual stars (Danziger 1969). The three arrows are Danziger's corrections for "curve-of-growth effects."

r_c. Two corrections tending slightly to lower the upper ends of curves A and B of Figure 5 have been omitted. The first is caused by the delay in arrival of the outward-moving boundary of lithium burning at r_c, possibly a time $\sim 10^8$ years. The second is caused by the delay in the start of depletion caused by the necessity of the decay of the higher decay modes of F, a time $\sim 10^8$ years.

If the boundary of burning extends somewhat outside r_c, the upper ends of the curves are raised slightly, but explicit integrations have not been carried out. If the boundary of lithium burning occurs at least $0.1\ r_0$ above r_c, the right side of Equation 36 is approximately independent of r, which simplifies the solution of Equation 38. This case has previously been discussed by Dicke (1970c), assuming that the zone of burning occurs at

$r = 0.58$, for the adopted radii of the convective zone $r_v = 0.86$, 0.78, and 0.70. For the "observed" present solar-wind torque-ratio density the mean decay times for ^{7}Li are found to be roughly the same as Danziger's value of 7×10^8 years. But the decay times are almost a factor of 2 too small for a solar-wind torque density adequate to slow the outer shell with $r_c = 0.54$.

SUMMARY

The types of observations having a bearing on the rotation of the deep solar interior are:
1. The observation of the solar oblateness of $(r_{eq} - r_p)/r = 5 \times 10^{-5}$.
2. The independence of latitude of the solar photospheric brightness.
3. The magnetic- and velocity-field distributions in the photosphere.
4. The structure of the solar wind, suggesting a present solar-wind torque density, at the equator, of 10^8 dyne/cm.
5. The rotation observed in F- and G-type stars in young clusters.
6. The lithium and beryllium abundance in the Sun and in solar-type stars of young clusters.

In the light of the uniformity of solar photospheric brightness, the solar oblateness seems to require a gravitational quadrupole moment sufficient to advance the perihelion of Mercury's orbit by $\sim 4''$/century.

A reasonable account of all of the above observations can be given by assuming:
1. That the Sun is a typical star of 1 solar mass.
2. That it possesses a core of radius ~ 0.55 rotating at an angular velocity 20 times as great as the Sun's surface.
3. That it arrived on the main sequence with the surface rotation already slowed to ~ 5 times the present surface value, but with the core rotating uniformly at 20 times the present surface rate.
4. That the young Sun had a ratio of solar-wind torque to surface angular velocity roughly an order of magnitude greater than its present value (an increase probably associated with the increased magnetic activity of the young Sun).
5. That this enhanced torque decayed with time, either quickly or more slowly.

It should be explicitly stated that if the overall picture is qualitatively correct, the radius of the core is rather accurately fixed by the requirement that it fall outside 0.5 where beryllium is burned, but inside 0.58 where ^{7}Li is burned [expressed in terms of Weymann's (1957) solar model as given by Schwarzschild (1958, see p. 259)].

The radius of the rapidly rotating core is fixed by the requirement that the outer parts of the Sun be mixed down to $r = 0.58$ (to destroy lithium) but not to 0.5 (to avoid destroying beryllium). The rotation of the core is fixed by the observation of the solar oblateness, assuming that the oblateness implies a quadrupole moment due to a rapidly rotating solar interior. The past slowing of surface rotation in the Sun is crudely fixed by observations on young stellar clusters, assuming that the Sun is a typical star. To slow the surface rotation by this amount yields a present value for the solar-wind torque in agreement with observations of the solar wind, but only if the Sun possesses such a rapidly rotating core. The present value of the rate of loss of angular momentum from the core is substantially independent of the time scale for slowing the young Sun. Without any adjustable parameters the depletion of lithium in solar-type stars is determined by the loss of angular momentum, if the model is correct. The resulting losses are found to be in reasonable agreement with observations of lithium in young clusters and the Sun.

LITERATURE CITED

Alfonso-Faus, A. 1967, *J. Geophys. Res.*, **72**, 5576
Allen, C. W. 1963, *Astrophysical Quantities* (2nd ed., London: Athlone Press)
Ashbrook, J. 1967, *Sky Telescope*, **34**, 229
Audretsch, J., Dehnen, H., Hönl, H. 1967, *Ap. J. Lett.*, **150**, L127
Bergmann, P. 1948, *Ann. Math.*, **49**, 255
Brans, C., Dicke, R. H. 1961, *Phys. Rev.*, **124**, 925
Chazy, J. 1928, *La théorie de la relativité et de la mécanique céleste* (Paris: Gauthier-Villars)
Clark, A., Thomas, J. H., Clark, P. A. 1969, *Science*, **164**, 290
Clemence, G. M. 1943, *Astron. Pap., Am. Ephem.*, **11**, 1
Cocke, W. J. 1967a, *Phys. Rev. Lett.*, **19**, 609
Cocke, W. J. 1967b, *Ap. J.*, **150**, 1041
Coleman, P. J., Jr. 1966, *J. Geophys. Res.*, **71**, 5509
Colgate, S. A. 1968, *Ap. J. Lett.*, **153**, L81
Cowling, T. G. 1965, In *Stellar Structure*, Chap. 8 (Aller, L. H., McLaughlin, D. B., Eds., Univ. Chicago Press)
Danziger, I. J. 1967, *Ap. J.*, **150**, 733
Danziger, I. J. 1969, *Ap. Lett.*, **3**, 115
Deutsch, A. J. 1967, *Science*, **156**, 236
Dicke, R. H. 1962, *Phys. Rev.*, **125**, 2163
Dicke, R. H. 1964, *Nature*, **202**, 432
Dicke, R. H., Peebles, P. J. E. 1965, *Space Sci. Rev.*, **4**, 419
Dicke, R. H. 1966, *Stellar Evolution*, 319 (Stein, R. F., Cameron, A. G. W., Eds., New York: Plenum Press)
Dicke, R. H. 1967a, *Int. Astron. J.*, **8**, 29
Dicke, R. H. 1967b, *Sky Telescope*, **34**, 371
Dicke, R. H. 1967c, *Ap. J.*, **149**, L121
Dicke, R. H. 1967d, *Science*, **157**, 960
Dicke, R. H., Goldenberg, H. M. 1967a, *Phys. Rev. Lett.*, **18**, 313
Dickie, R. H., Goldenberg, H. M. 1967b, *Nature*, **214**, 1294
Dicke, R. H. 1970a, *Ap. J.*, **159**, No. 1
Dicke, R. H. 1970b, *Ap J.*, **159**, No. 1
Dicke, R. H. 1970c, In *IAU Colloq. No. 4, Stellar Rotation* (Slettebak, A., Ed., Holland: Reidel)
Duncombe, R. L. 1958, *Astron. Pap., Am. Ephem.*, **16**, 1
Durney, B. R., Roxburgh, I. W. 1969, *Nature*, **221**, 646
Fricke, K., 1968, *Z. Ap.*, **68**, 317
Fricke, K. 1969, *Astron. Ap.*, **1**, 338
Gilvarry, J. J., Sturrock, P. A. 1967, *Nature*, **216**, 1283
Goldreich, P., Schubert, G. 1967a, *Science*, **156**, 1101
Goldreich, P., Schubert, G. 1967b, *Ap. J.*, **150**, 571
Goldreich, P., Schubert, G. 1968, *Ap. J.*, **154**, 1005
Herbig, G. H. 1965, *Ap. J.*, **141**, 588
Holton, J. R. 1965, *J. Atmos. Sci.*, **22**, 402
Holton, J. R., Stone, P. H. 1968, *J. Fluid Mech.*, **33**, 127
Howard, L. N., Moore, D. W., Spiegel, E. A. 1967, *Nature*, **241**, 1297
Hoyle, F. 1960, *Quart. J. Roy. Astron. Soc.*, **1**, 28
Jordan, P. 1948, *Astron. Nachr.*, **276**, 1955
Jordan, P. 1959, *Schwerkraft und Weltall* (Braunschweig: Vieweg)
Kippenhahn, R. 1963, *Ap. J.*, **137**, 564
Kraft, R. 1967, *Ap. J.*, **150**, 551
Kraft, R. 1968, In *Stellar Astronomy*, 2 (Chiu, H. Y., Warasila, R., Remo, J., Eds., New York: Gordon & Breach)
Leverrier, U. J. 1959, *Ann. Obs. Paris*, **5**, 104
Livingston, W. C. 1966, *Sci. Am.*, **215**, 107
McDonald, B. E., Dicke, R. H. 1967, *Science*, **158**, 1562
McDonald, B. E. 1969 (Private communication)
Modiesette, J. L. 1967, *J. Geophys. Res.*, **72**, 1521

Neugebauer, M., Snyder, C. W. 1966, *J. Geophys.*, **71**, 4469
Newcomb, S. 1897, *Suppl. Am. Ephem. Naut. Alm.*
O'Connell, R. F. 1968, *Ap. J. Lett.*, **152**, L11
Öpik, E. J. 1967, *Int. Astron. J.*, **8**, 29
Pedlosky, J. 1967, *J. Fluid Mech.*, **28**, 463
Pedlosky, J. 1969, *J. Fluid Mech.*, **36**, 401
Plaskett, H. H. 1965, *Observatory*, **85**, 178
Roxburgh, I. W. 1964, *Icarus*, **3**, 92
Roxburgh, I. W. 1967a, *Nature*, **213**, 1077
Roxburgh, I. W. 1967b, *Nature*, **216**, 1286
Sakurai, T. 1969a, *J. Phys. Soc. Jap.*, **26**, 840
Sakurai, T. 1969b, *J. Fluid Mech.*, **37**, 689
Schatzman, E. 1959, *IAU Symp. No. 10*, 129 (Greenstein, J. L., Ed.)
Schatzman, E. 1962, *Ann. Ap.*, **25**, 18
Schaub, W. 1938, *Astron. Nachr.*, **265**, 161
Schur, W., Ambronn, L. 1895, *Astronomische Mittheilungen*, Part 4 (Göttingen Obs.). Also see 1905, Part 7, for discussion of results
Schwarzschild, M. 1958, *Structure and Evolution of the Sun*, 175–84 (Princeton Univ. Press)
Shapiro, I. I. 1965, *Icarus*, **4**, 549
Sturrock, P. A., Gilvarry, J. J. 1967, *Nature*, **216**, 1280
Thirry, Y. R. 1948, *C. R. Acad. Sci.*, **226**, 216
von Zeipel, H. 1924, *MNRAS*, **84**, 665
Wallerstein, G., Conti, P. S. 1969, *Ann. Rev. Astron. Ap.*, **7**, 99
Wayman, P. A. 1966, *Quart. J. Roy Astron. Soc.*, **7**, June
Weber, E. J., Davis, L., Jr. 1967, *Ap. J.*, **148**, 217
Weymann, R. 1957, *Ap. J.*, **126**, 208,
Wilson, D. C. 1966, *Science*, **151**, 1487

EXCITATION AND IONIZATION BY ELECTRON IMPACT

OLEG BELY

Observatoire de Nice

HENRI VAN REGEMORTER

Observatoire de Paris

In this review we survey the principles of different theoretical methods for calculating excitation and ionization cross sections and we discuss the reliability and the accuracy of the various approximations.

At the present time, although it is possible in many cases to obtain estimates of cross section, accurate calculations are still uncertain and a great deal of investigation remains to be done. On the experimental side also, more work must be done to obtain accurate absolute cross sections.

One must keep in mind that even in the simple case of the $1s$-$2p$ excitation in H there is no exact solution of the problem, the work involved in accurate calculations being considerable. Moreover, it can be dangerous to apply the same method to different cases: one method can be appropriate in one case and useless in another case. When applied to a wrong case, an elaborate quantum-mechanical calculation can be worse than a more approximate method giving a quick estimation.

Many recent review articles deal with the theoretical work on the electron-impact excitation of atomic systems [Massey 1956 (1), Seaton 1958 (2), Seaton 1962 (3), Peterkop & Veldre 1966 (4), Burke 1968 (5), Moiseiwitsch & Smith 1968 (6)]. In the last paper, experimental methods and results are also discussed, as in review papers by Fite 1962 (7) and Heddle & Keesing 1968 (8).

Electron collisions with positive ions have been discussed by Seaton 1968 (9). Three recent review articles concerned with excitation of forbidden lines in planetary nebulae (10–12) include up-to-date reference lists.

Two recent review articles deal with ionization processes: one on the experimental situation by Kieffer & Dunn 1966 (13), the other on theoretical studies by Rudge 1969 (14).

Many other references can be found in the IAU reports for Commission 14—see the 1964 and the 1967 reports—as well as in the systematic collection of data provided by the Information Center of the Joint Institute of Laboratory Astrophysics (15, 16) and by the Oak Ridge Atomic and Molecular Information Center (17).

The role of collisions in astrophysical plasma is reviewed by Branscomb (18) and by Pagel (19).

We shall consider first the problem of excitation of atoms and positive ions by electron impact and give a critical survey of the theory, comparing the different approximate methods with each other and with the few available experiments.

EXCITATION

1. INTRODUCTION

The rate of collisional excitation.—With N_e electrons per cubic centimeter, having a normalized velocity distribution $f(v)$, the probability per unit time of a collisionally induced transition $n' \to n$ with $E_{n'} > E_n$ is

$$\alpha_{nn'} N_e = N_e \int_0^\infty v_{n'} f(v_{n'}) Q(n' \to n) dv_{n'} \qquad 1.$$

where the cross section Q has the dimensions of a surface. Cross sections are frequently given in units of $\Pi a_0^2 = 8.806 \ 10^{-17}$ cm².

If one assumes a Maxwellian distribution for $f(v)$, the rates of deexcitation $\alpha_{n'n}$ and of excitation $\alpha_{nn'}$ are

$$\alpha_{n'n} = \frac{8.63 \ 10^{-6}}{\omega_{n'}} T_e^{-1/2} \int_0^\infty \Omega(nn') \exp\left(-\frac{mv_{n'}}{2kT}\right) d\left(\frac{mv_{n'}^2}{2kT}\right) \qquad 2.$$

where the collision strength $\Omega(n', n)$ is a dimensionless number related to the cross section:

$$Q(n' \to n) = \frac{1}{\omega_{n'}} \frac{1}{k_{n'}} \Omega(n,n') \Pi a_0^2 \qquad 3.$$

It is convenient to use the collision strength $\Omega(nn')$, which is symmetric, $\Omega(n' \to n) = \Omega(n \to n')$, because of the symmetry of the quantum-mechanical problem. This symmetry insures detailed balancing,

$$\alpha_{n'n} = \frac{\omega_n}{\omega_{n'}} \alpha_{nn'} \exp\frac{E_{n'} - E_n}{kT} \qquad 4.$$

which insures the Boltzmann relation between level populations when collisional processes are much more important than radiative processes. Here ω_n and $\omega_{n'}$ are the statistical weights.

The total energy of the system atom+electron, which is conserved during the collision, is given in atomic units by

$$E = \tfrac{1}{2} k_n^2 + E_n = \tfrac{1}{2} k_{n'}^2 + E_{n'}$$

It is important to stress that forbidden transition cross sections $Q(n' \to n)$

EXCITATION AND IONIZATION BY ELECTRON IMPACT   331

can be of the same order as those for optically allowed transition. Only when $n' = n$ can the former be neglected.

One must also keep in mind that the rate of excitation by collisions with other particles like protons or neutral atoms can be important, in particular for transitions of very small energy separation.

From classical to quantum theory.—In a semiclassical theory, the impinging electron is a classical charged particle that produces a variable field in the neighborhood of the atom. Using quantum perturbation theory one calculates the probability P of an induced transition. This method, as we shall see below, makes no allowance for the change in kinetic energy when the transition occurs, so it is valid only if the kinetic energy of the electron is much greater than the transition energy.

Let R be the impact parameter, the distance of closest approach between the electron and the atom if the electron were undeflected. Obviously,

$$\sum_{n' \neq n} P_{nn'}(R) \leq 1 \quad \text{for all } R \qquad 5.$$

The cross section for the $n \to n'$ transition is simply

$$Q[n \to n'] = \int_0^\infty P_{nn'}(R) 2\Pi R dR \qquad 6.$$

Let v_n and L_n be the velocity and the angular momentum of the incident electron at large distance. Classically,

$$L_n = m v_n R \qquad 7.$$

In quantum theory $L_n = \hbar \sqrt{l_n(l_n+1)}$ with l_n integral. Then

$$2RdR = (2l_n + 1) \frac{1}{k_n^2} \quad \text{with } k_n = \frac{m v_n}{\hbar}$$

Replacing the integral 6 by a sum ($\delta \bar{l}_n = 1$) gives

$$Q_{n \to n'} = \frac{\Pi}{k_n^2} \sum_{l_n=0}^\infty (2l_n + 1) P_{nn'}(R_{l_n}) \qquad 8.$$

Introducing the collision strength Ω

$$\Omega(n - n') = \sum_{l_n} \Omega_{l_n}(n - n') \quad \text{with } \Omega_{l_n}(n - n') = (2l_n + 1) P_{nn'}(R_{l_n}) \qquad 9.$$

one obtains expression 3 for the cross section.

In quantum theory the decomposition of the wavefunctions into partial waves of angular momentum l corresponds to the different impact parameter R. The conservation condition 5 corresponds to

$$\Omega_l(n - n') \leq (2l + 1)$$

which is very useful to give an upper bound of a cross section when an approximate method gives an overestimation of the cross section.

As it is obvious from 7, for a small value of v_n, a large value of L can arise only for very large values of R, but such remote encounters are not likely to produce transitions. Therefore for low-energy collisions, only a few values of l are important. This is why a semiclassical treatment is not valid in most cases at low energies.

In formula 6 the inelastic cross section is related to the probability of excitation from level n to level n'. In quantum theory the cross section is defined as the ratio of the flux of particles of energy $\frac{1}{2}k_n^2$, scattered in all directions over the flux of the incident particles of energy $\frac{1}{2}k_{n'}^2$ across a unit area perpendicular to their direction. This ratio is related to the asymptotic behavior of the wavefunctions representing the system atom+electron.

2. Quantum Theory

The atomic eigenfunction expansion method.—In most of the calculations in electron atom collisions, one expands the total wavefunction in terms of the unperturbed atomic eigenfunction ϕ_n.

$$\Psi(\mathbf{r}_1\mathbf{r}_2) = A \sum_n \phi_n(\mathbf{r}_1)\chi_n(\mathbf{r}_2) \qquad 10.$$

where A is an antisymmetrizing operator.

In practice the ϕ_n are the best available atomic wavefunctions, and one takes only a few terms in the expansion in order to be able to solve the problem on a computer. Here, for simplicity, ϕ_n is the wavefunction of the valence electron of an atom with only one electron outside a closed shell and we are neglecting the distortion of the atomic core.

Putting Equation 1 into the Schroedinger equation for the total system system atom+electron

$$[H - E]\Psi(\mathbf{r}_1\mathbf{r}_2) = 0 \qquad 11.$$

and taking account of the Schroedinger equation for the atom only, one obtains the integrodifferential equations for the perturbing electron wavefunction

$$[\nabla^2 + k_n^2]\chi_n(\mathbf{r}_2) = 2 \sum_{n'} (V_{nn'} - W_{nn'})\chi_{n'}(\mathbf{r}_2) \qquad 12.$$

where $V_{nn'}$ and $W_{nn'}$ are the interaction potential and the exchange operator.

If the incoming electron impinges upon a neutral atom in state 1, the asymptotic behavior takes the form

$$\chi_n(r_2) \xrightarrow[r_2 \to \infty]{} \exp(i\mathbf{k}_1 \cdot \mathbf{r})\delta_{1n} + r_2^{-1} \exp(ik_n r)f_{1n}(\theta\phi) \qquad 13.$$

where the scattering amplitude $f_{1n}(\theta\phi)$ is given for the polar angles with the direction of the incident electron.

The cross section is given in terms of the scattering amplitude by

$$Q_{1\to n} = \frac{k_n}{k_1} \int\int |f_{1n}(\theta\phi)|^2 \sin\theta d\theta d\phi \qquad 14.$$

The partial-wave analysis.—For simplicity we suppose that all the levels n considered in expansion 10 can be excited—all the $k_n^2 \geq 0$; if this is not the case, resonances are obtained on solving the coupled equations 12, as we shall discuss later on.

In all accurate calculations, in order to solve this system of equations, it is necessary to use a partial-wave treatment, solving the differential equations for each value of the total angular momentum and of the total spin. In fact, the atomic state is characterized by the quantum numbers $nLMSM_S$ and the incoming electron by $klmsm_s$. In the coupled representation nLS $lsL^TS^TM^TM_S{}^T$, the total angular momentum $L^T = L+l$ and the total spin $S^T = S+s$ are conserved; the system being invariant by rotation, a channel of the total system will be defined by the set $nLlS\frac{1}{2}L^TS^T$.

For each value of S^T and L^T, the radial parts of the wavefunction $\chi_n(r_2)$ of the colliding electron are then solutions of a set of coupled equations

$$\left[\frac{d^2}{dr^2} + k_n^2 - \frac{l(l+1)}{r^2}\right] F_\nu(r_2) = 2\sum_{\nu'}[V_{\nu\nu'} - W_{\nu\nu'}]F_{\nu'}(r_2) \qquad 15.$$

where ν stands for the quantum numbers $nLSl$.

The detailed expressions of the direct and exchange potentials V and W have been given by Percival & Seaton (20) for hydrogen, and for one-electron atoms ($S = \frac{1}{2}$) when neglecting the core distortion, and have been calculated recently for complex atoms (21, 22). These calculations involve the use of Racah algebra techniques, and the angular coefficients can be calculated with a computer. In a recent alternative approach (23) all the algebraic reductions are made by a computer, which applies the more elementary techniques given by Condon & Shortley (24).

The cross section is now related to the asymptotic forms of the radial functions F given in terms of the elements of the reactance matrix $\mathbf{R}$ or of the scattering matrix $\mathbf{S}$. The $\mathbf{R}$ matrix is real and symmetric. The S matrix is symmetric and unitary

$$\mathbf{SS}^* = 1 \qquad \mathbf{S} = \tilde{\mathbf{S}} \qquad 16.$$

This is equivalent to the conservation theorem 5 and to the reprocity condition $\Omega(i\to j) = \Omega(j\to i)$.

The cross section is conveniently written in terms of the transmission matrix $\mathbf{T}$ related to $\mathbf{S}$ and $\mathbf{R}$ by

$$\mathbf{T} = 1 - \mathbf{S} = \frac{-2i\mathbf{R}}{1 - i\mathbf{R}} \qquad 17.$$

For inelastic cross section $T_{ij} = S_{ij}$.

The total cross section for a transition between two spectral terms is given by

$$Q[\alpha LS - \alpha' L'S'] = \frac{\Pi}{k_{nL}^2} \frac{1}{(2S+1)(2L+1)}$$
$$\cdot \frac{1}{2} \sum_{l'L^T S^T} (2S^T+1)(2L^T+1) \qquad 18.$$
$$\cdot |\langle \alpha Ll L^T S^T | T | \alpha' L'l' L^T S^T \rangle|^2$$

α designating the configuration; i.e. all the quantum numbers needed for a unique specification of the state.

The close-coupling approximation.—In the close-coupling approximation one retains a few atomic states in the expansions 10 and solves a finite set of coupled integrodifferential equations using numerical techniques.

Since the first calculation on excitation of $1s$–$2s$ in H neglecting the coupling with all other states by Bransden & McKee in 1956 (25), this method has been used a great deal. In the table of results below are listed the elements and the transitions for which these accurate close-coupling calculations have been done. Many papers are concerned with the simple cases H+e and He$^+$+e, the purpose being to obtain accurate elastic cross sections for elastic scattering by the ground state. For this problem the inclusion of a few excited states in expansion 10 may yield good results. In the calculation of an excitation cross section it is necessary to take into account at least the first excited state above the upper level of the transition.

The close-coupling approximation will be good if the coupling with all the states that are neglected is very weak. This is the case in elastic scattering by alkali atoms in which the coupling with the resonance state is very strong compared to the coupling with the other excited states (the polarizability of an alkali is mainly given by the resonance-state contribution). In inelastic scattering, this is the case when a few levels are close in energy, strongly coupled together, and very weakly coupled to other distant levels that are neglected. For example the transition between terms of the ground-state configurations of some atoms of the type p^n (26, 27).

If the close-coupling method has been very successful in predicting the positions and width of resonances in elastic scattering just below the first excitation threshold (28), it has apparently been less successful for calculating the excitation cross sections of H and He$^+$, showing a lack of convergence with respect to the addition of more atomic states into the trial wavefunction expansion.

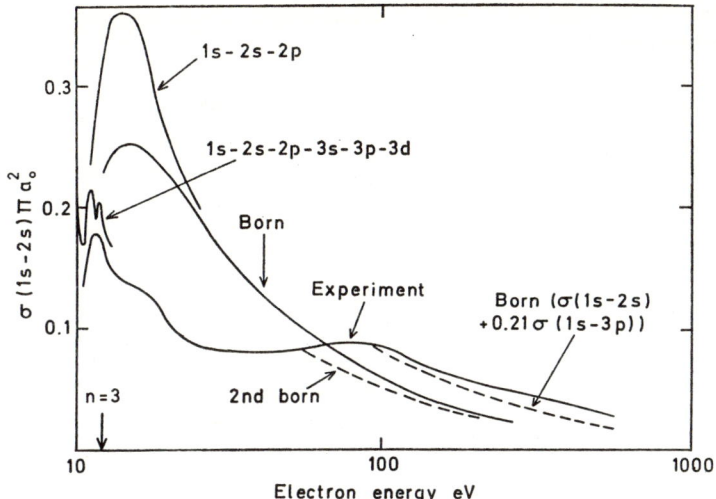

FIGURE 1. Close-coupling calculations for excitation of 1s-2s in H, including three states and six states in expansion 10. See (5) and (32).

As shown in Figure 1, good agreement is obtained for energies below the $n=4$ threshold when all the $n=3$ states are included in the wavefunction expansion. The experimental curve is not normalized to the Born approximation at high energies, because of cascade effects. The very delicate question of the comparison between theory and experiment is thoroughly discussed in two recent papers (30, 31).

At high energies the second Born approximation must be good but in the intermediate region between 12.7 eV ($n=4$) and 150 eV a good theory remains to be found, the close-coupling method becoming intractable.

The convergence of the close-coupling method and the way to improve it have been discussed in a series of papers by Burke and co-workers (30, 32, 33) and in three recent review articles by Burke (5, 28) and Smith (29).

In fact, the close-coupling method suffers from two defects at low energies. If one neglects coupling with the higher states and with the continuum, the asymptotic behavior of the effective potential does not always include the polarization potential properly. At the same time, at short distances the strong interaction between the atomic electron and the colliding electron is not adequately described.

The method can be improved by taking full account of the polarizabilities of both the initial and the final states of the transition studied (34, 35). On the other hand, correlation terms have been added to the usual eigenfunction expansion method analogous to the correlation terms introduced by Hylleraas for representing the short-range interaction of the two electrons in He. This has been done by Burke and co-workers (32), but the mathematical difficulty of solving the coupled integrodifferential equations increases rapidly with the number of states involved.

3. QUANTAL APPROXIMATIONS

The Born approximation.—At high energies, the Born approximation is valid. When the incident kinetic energy is large compared to the interaction energy, the wavefunction solution of Equation 12 may be approximated by a plane wave and the cross section is proportional to a squared matrix element of the form

$$|\langle \Psi_{n'}^+ | V | \Psi_n \rangle|^2 \qquad \qquad 19.$$

with $\psi_n = \phi_n F_n$ and $\psi_{n'} = \phi_{n'} F_{n'}$ where F_n and $F_{n'}$ are plane waves. For collision with positive ions, one should take F_n and $F_{n'}$ to be Coulomb waves because of the distortion by the Coulomb field. This is the so-called Coulomb-Born approximation.

In electron scattering the Born approximation fails when the energy decreases below a few hundred electron volts. For inelastic scattering, close collisions are less important and the domain of validity of the Born approximation is larger.

The Born approximation can be considered as a first-order approximation. One can think of higher approximation in the so-called Born series in which the exact scattering amplitude is expanded in powers of the interaction. The improvement with the second Born approximation is not very great at low energies. In fact when the momentum transfer in the vicinity of the target is not small compared to the incident momentum, one cannot expect good convergence of the Born series. At low energies the incident electron spends more time near the target where complicated effects of distortion, exchange, and coupling can take place.

In some cases the Born approximation can be improved by taking account of the distortion of the wavefunction by the static atomic field—retaining the potential V_{nn} and $V_{n'n'}$ in the two equations 12 corresponding to the two states involved in transition $n \rightarrow n'$. This is the distorted-wave approximation. On the other hand, exchange is included in the so-called Born-Oppenheimer approximation, in which the exchange amplitude is estimated to first order, but at low energy this very rough estimation of exchange gives worse results than if one ignores exchange.

It is surprising that taking account of the electron indistinguishability should give worse results than ignoring indistinguishability, but as Ochkur explained (36), this shows how bad a first-order approximation can be. To correct the Born-Oppenheimer approximation, Ochkur suggested that one should calculate the exchange terms including at low energies only the terms of the exchange amplitude that are important at high energies (36). The so-called Ochkur approximation, which is not rigorously based, has been modified by Rudge (37, 38) and can give completely unreliable results, in particular for transitions between excited states and between fine-structure levels of the same atomic configuration (39). This method cannot be applied to positive ions.

The main defect of the Born approximation, particularly for optically allowed transition for which the linestrength is large, is the assumption of weak coupling—i.e. the $V_{nn'}$ of Equations 12 are small compared to k^2. In fact the single scattering $n \rightarrow n'$ is combined with multiple scattering of the type $n \rightarrow n'' \rightarrow n$ or n' and of higher-order interactions.

For improving the first Born approximation in this respect, remembering the failure of the higher-order Born approximation which cannot properly take into account this strong coupling effect, it is necessary to use the partial-wave analysis—see below the Born II and Coulomb-Born II approximation. As a consequence of the Coulomb potential, approximate methods are more accurate for positive ions than for neutrals, and as we shall see below, good results can be obtained with the Coulomb-Born approximation for positive ions.

For neutrals new calculations have been done using the Born approximation—see tables pp. 350, 360—and a very optimistic report has been given by Vainshtein & Sobelman (40) giving the Born expression for allowed transitions and the Ochkur expression for the intercombinational transitions ($\Delta S \neq 0$). The last has to be used with caution for reasons given above.

Vainshtein, Presnyakov & Sobelman (41) have introduced a method that is not rigorously established, in which the repulsion between the atomic electron and the incoming electron is taken into account exactly but the interaction of the incoming electron and the atomic core is approximated. This method has been discussed by different authors (42–45) and its use is still questionable.

The unitarized Born approximation.—At low energies, i.e. kT of the order of ΔE_{ij}, the excitation energy or smaller, the breakdown of the Born approximation is due to the importance of close collisions. As we explained above, at small kinetic energies, the contribution of the partial waves corresponding to an angular momentum l becomes very large. The corresponding classical orbits are in fact penetrating the target and the incident particles spend a long time in the interaction region. Therefore one cannot assume as in the Born approximation that the trajectory of the colliding electron is undistorted by the atomic field.

For these close collisions, the interactions are very strong; one cannot neglect the different coupling interaction potentials in Equation 12, which is to say that the different terms of the **R** matrix in 17 are not small compared to one.

In fact, within the framework of a partial-wave analysis, the Born approximation is obtained by setting

$$\mathbf{T} = -2i\mathbf{R}_B \qquad 20.$$

instead of using the exact relation 17, where the matrix elements of R are calculated using plane waves (Born approximation B I) or Coulomb waves in the case of positive ions (Coulomb-Born approximation CB I).

For close strong collisions, the unitary condition $SS^+ = 1$ is not satisfied. That is to say, the conservation conditions 5 are not satisfied and the cross section is overestimated.

In the unitarized Born approximation proposed by Seaton (46), the **R** matrix is still calculated in the same way, i.e. $\mathbf{R} = \mathbf{R}_B$, but the exact relation 17 is used

$$\mathbf{T} = -\frac{2i\mathbf{R}_B}{1 - i\mathbf{R}_B} \qquad 21.$$

which insures that the conservation conditions are automatically satisfied

$$P_{ij} = |T_{ij}|^2 = \frac{\Omega_{ij}}{2l+1} \leq 1 \qquad 22.$$

In this way, through the relation between **T** and **R**, some allowance for strong coupling is made and the method is an important improvement for strong allowed transition (47). This is called the Born II approximation.

Within the unitarized Born approximation as within the semiclassical approach which insures conservation conditions for small impact parameter (see below), our inability to calculate the close collisions is obvious, but the upper limit of the contribution of the close collisions cannot be surpassed!

A few calculations have been done with the Born II approximation for neutrals (47, 48) and the corresponding Coulomb-Born II method for positive ions has been extensively used. For strong resonance transitions like $4s-4p$ in Ca^+ and $3s-3p$ in Mg^+ the Coulomb-Born II results (3, 49, 50) are only 20 percent higher than elaborate close-coupling calculations (51).

The peculiar case of positive ions.—Because of the ion field the distance of closest approach R_c of the perturbing electron when its velocity tends to zero remains finite for all finite values of its angular momentum

$$R_c = l^2[2me^2z]^{-1} \qquad 23.$$

where z is the ionic charge. Consequently the collision strength Ω has finite contributions for all values of l even at threshold; Ω is finite at threshold.

Compared to the neutrals, for which at low energies the cross section is mainly given by low values of l, much higher values of l are involved in the calculation of positive ions, and in consequence the accuracy of the approximate methods at low energies will be better in the case of positive-ion excitation. It is clear in the study of highly ionized ions that the presence in the Schroedinger equation of a large Coulomb long-range interaction of the form zr^{-1} permits considering the other interactions as small perturbations, as is done in the Coulomb-Born approximation I.

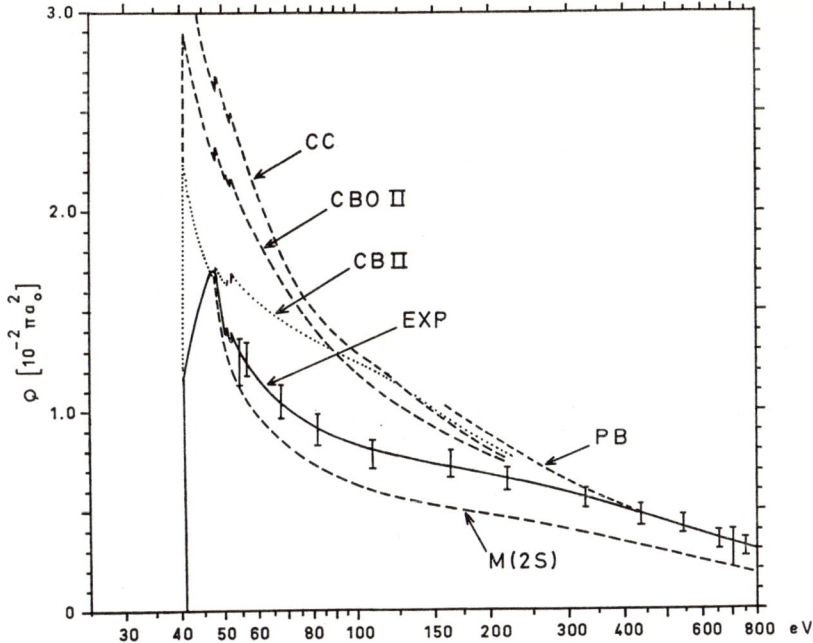

FIGURE 2. Excitation cross section Q ($1S$–$2S$) in He$^+$. The experimental curve $M(2S)$ is renormalized to the Born results PB (1) at high energies. See (53) and (110) where recent calculations including correlation are compared to two experiments.

Because close collisions are less important, one can neglect the possibility of exchange in the case of optically allowed transitions; and when these transitions are strong, the coupling is taken into account with the unitarized Coulomb-Born approximation II. For ions more than three times ionized these approximations I and II give the same results.

When exchange cannot be neglected, for example for excitation of intercombination lines, with a change of the atomic spin during the collision, a method proposed by Bely (39, 52) derived from the Ochkur approximation (36) can be applied to positive ions. This Coulomb-exchange I approximation can also be generalized as the unitarized Coulomb-Born approximation in order to satisfy the flux conservation Coulomb-exchange II.

In Figure 2 the different approximations applied to the calculation of the $1s$–$2s$ cross section in He$^+$ by electron impact are compared to an experiment by Dance, Harrison & Smith (53) using the crossbeam method. The Coulomb-Born I method (55) gives better results than the close-coupling method (54). At low energies the experimental curve is below the theoretical curves, a fact difficult to explain when considering the very good agreement of the close-coupling method in the calculation of the positions and the widths of the autoionized states of He compared to the experimental data of Madden & Codling (56).

If the Coulomb-Born is good for He$^+$ one can expect that it will give better results for more ionized atoms of the same isoelectronic sequence. The systematic study of some transitions along the isoelectronic series has been performed by many authors—see table of results. At high energy, where the Coulomb-Born approximation is valid, it may be shown that $Z^4 Q(i \rightarrow j)$ does not depend on the charge Z (57).

On the other hand, one disadvantage in the study of highly ionized atoms, which we shall discuss when discussing the methods of calculation of excitation between terms of a p^n configuration, comes from configuration interaction which is important as well as all the departures from LS coupling. This is a consequence of the orbital angular momentum degeneracy when $Z \rightarrow \infty$. At the same time the contribution of resonances in a low-energy $i \rightarrow j$ cross section is more important when Z increases.

All these peculiar features of positive-ion excitation are discussed in two review papers, one by Bely (58) and the other by Seaton (9).

The Bethe approximation. The $\bar{g}$ empirical formula.—Calculations within the Born approximation and within most of the different approximations discussed above that require a partial-wave analysis are still very long. As is well known, at high energy a further simplification is valid: the distant encounters are really the most important and the colliding electron remains outside the atom most of the time. This is the Bethe approximation for which the excitation cross section Q is given by the very simple expression

$$Q(i \rightarrow j) = \frac{8\Pi}{\sqrt{3}} \frac{1}{k_i^2} \frac{I_H}{E_j - E_i} f(ji) g \Pi a_0^2 \qquad 24.$$

in terms of the atomic oscillator strength and the Gaunt factor g proportional to the probability of an induced free-free transition of the colliding electron.

By comparing the form 24 with some experimental data and a few calculations available in 1961, Van Regemorter (59) has derived values of an effective $\bar{g}$ for optically allowed transitions. Particularly simple is the case of positive ions for which excitation cross sections are finite at threshold and $\bar{g}$ can be taken to be a constant ~ 0.2 at low energies.

This $\bar{g}$ approximation has been extensively used by plasma physicists and astrophysicists, but at the same time new calculations have shown that this very convenient procedure must be used with caution.

For positive ions, many calculations using the Coulomb-Born I and II methods (21, 60, 61) have shown that $\bar{g}$ is generally larger than 0.2 for transitions corresponding to identical principal quantum numbers ($n = n'$) and smaller than 0.2 when $n \neq n'$, but the Gaunt factor $\bar{g}$ is also an increasing function of the ion charge Z. Straightforward application of the $\bar{g}$ empirical formula may give a considerable error for the cross sections and the excitation rates.

4. Semiclassical and Classical Theories

The impact-parameter method for permitted transitions.—Much better results can be obtained with the semiclassical impact-parameter method introduced for atom-electron collisions by Seaton (62) following a procedure extensively used in nuclear excitation problems (63).

Assuming a classical path for the colliding electron, one can calculate the probability of excitation $P_{ij}(R)$ for impact parameter R using perturbation theory equivalent to the Born approximation. The dipole approximation is assumed and in fact the colliding electron remains outside the target—this is equivalent to the Bethe approximation.

The cross section is given by an expression analogous to expression 6 with appropriate lower cutoff, because for a small impact parameter corresponding to low values of the angular momentum the probability $P_{ij}(R)$ is in fact $\gg 1$ and the conservation condition 5 is not satisfied. On the other hand, exchange of energy between the perturber and the target is neglected if we assume a classical path for the former. Consequently the reciprocity condition

$$\omega_i P_{ij}(R) = \omega_j P_{ji}(R) \qquad 25.$$

the ω being the statistical weights, is not satisfied. When the kinetic energy of the perturber is not much bigger than ΔE_{ij} one has to use symmetrized expressions (62, 63).

The use of two cutoffs is discussed by Seaton. The weak-coupling cutoff is the effective radius of the atom in the lower state of the transition, operating mainly when the transition is weak. In most cases, for strong allowed transitions, the cutoff R_1 is such that $P_{ij}(R_1) = \frac{1}{2}$.

For neutral atoms, the classical path is a straight line. The cross section is given in terms of two functions tabulated in (62). For positive ions the corresponding function has been derived from the repulsive case (63) by Burgess (64) and in different articles on the application of the impact-parameter method to the problem of line broadening by charged particles (65, 66).

The results are much better with the impact-parameter approximation because the method avoids the overestimation of the close collisions obtained with the Born approximation at low energy. In fact, exactly as in the unitarized Born approximation II and Coulomb-Born II this overestimation is avoided by virtue of the conservation condition; i.e. $P_{ij}(R) \leq 1$ for all values of R. As we said previously this is a way to obtain reasonable results without knowing what is really going on during a close strong interaction. Indeed, when the coupling is strong the impact-parameter method gives roughly the same results as the Born II and Coulomb-Born II approximation which involve longer calculations (21, 58).

The method is particularly appropriate for evaluating the excitation cross section for the $n \to n+1$ optical transitions in hydrogen. The cross section is simply given by the expression

$$Q(n \to n+1) = \chi_{n+1,n} \Pi r_n^2 \qquad 26.$$

as a function of the mean radius r_n and of a quantity which is tabulated by Saraph (67). The impact method has been applied for transitions in H of the type $nl \to nl'$ ($l' = l \pm 1$), which are important in recombination spectra (68) and for transition between quasidegenerate levels in line-broadening theory (69).

The generalization of the impact-parameter method to quadrupole transitions (70) cannot be reliable. The results are too sensitive to the cutoff because close collisions are important.

In conclusion, when properly applied—so that the conservation condition and detailed balancing are satisfied—the impact-parameter method affords a simple way of obtaining reliable results, particularly for strong allowed transitions from the ground state or between $n \to n+1$ levels. Compared to the Coulomb-Born approximation—which itself is not very accurate—semiclassical results should be reliable to within 50%. The impact-parameter method requires only a knowledge of the oscillator strengths f and is very easy to use for neutral (62) and positive ions (64, 65).

Pure classical approach.—In this approach the collision between the electron and the atom is treated as a binary electron-electron encounter. The transfer of energy from the incident electron to the bound electron is computed as if the two electrons are free. This energy transfer must be large compared to the binding energy of the atomic electron and the method is more suitable for ionizing collisions and pure exchange collisions than for direct excitation.

In this method the electron-electron interaction is treated exactly, including exchange (which is treated as a quantum-mechanical collision between identical particles) and taking account of the interference between direct and exchange scattering (64, 71). On the other hand, different assumptions are made for the velocities of the atomic electron by Thomson (72,) Gryzinski (73), and Stabler (74).

With the inclusion of exchange, A. Burgess (64) is using this classical approach at low energies—where close strong binary encounters are important, and the semiclassical impact-parameter method at high energies, where the pure classical theory breaks down and does not have the good log E/E behavior. The classical method appears to be a powerful approach for improving the treatment of close collisions for which all the approximate methods reviewed up to now are useless.

Approximate classical theories are discussed in a recent report by

A. Burgess & I. C. Percival (75). All the improvements since the pioneer works of Thomson and Gryzinski are thoroughly discussed in this paper and must be taken into account in any calculation.

Classical methods are particularly suitable for ionization and in this case the exact solution of the classical problem has been given by Percival (75). For excitation there are some ambiguities in choosing the final energy band. For excitation from level one to level n, the transfer of energy ΔE_{1n} lies in the range

$$E_n - E_i \leq \Delta E \leq E_{n+1} - E_1 \qquad 27.$$

but all angular momenta of the final state are included in this cross section. It is possible to estimate the relative probability of excitation of a definite angular momentum state (71).

All the useful formulae are given by Burgess (64, 71) and by Burgess & Percival (75). Some applications for excitation of high excited states have been made (76) and for $n \to n+1$ transitions with high values of n (77, 78). For this last case, the extensively used formula of Gryzinski (73), like the Born approximation, overestimates the cross section when compared to the semiclassical calculation of Saraph (67), which must be good in this case. More recently an approximate formula given by Percival & Richards (79) that is valid for transitions between some highly excited states of hydrogen and in good agreement with the calculation of Saraph.

For excitation from the ground state to any level in hydrogen and in hydrogenlike ions, I. C. Percival (80), applying classical arguments, gives recommended cross sections in terms of scaling factors and of the experimental cross section for the $1 \to 2$ transition. Approximate formulae for different other processes are also discussed in this paper.

Lastly it must be emphasized that all classical formulae are more empirical for electron–positive ion collisions, for which it is necessary to introduce a semiempirical additional factor to account for the focusing effect of the long-range Coulomb field of the ion (64).

5. The Influence of Resonances in Inelastic Scattering

In discussing the close-coupling method we have said that the neglect of the higher states in the wavefunction expansion leads to neglect of various distortion effects. It can also lead to neglect of resonance effects in the cross section. If the atom is in its ground state i and is excited to level j by an incoming electron whose energy is too small to excite the next level k, one says that the channels corresponding to i and j are open and the channels k are closed. When one takes account of these channels k in solving the coupled equations 12, the new coupling effects give rise to resonances in the cross section $i \to j$.

In fact, below the inelastic threshold of excitation of level k, the incoming electron and the target can form a compound atom (or ion) in states α^* that are not stationary and have a finite width. For energies corresponding to these pseudobound states, the colliding electron spends a long time near the target before being scattered.

The cross section $i \to j$ is increased by the indirect excitation $i \to \alpha^* \to j$ through the "level" α^*. This two-step excitation at a definite energy is a combination of a capture of the electron by the target in level i in a doubly excited state and autoionization, which leaves the target in level j plus an electron of different energy. This excitation is illustrated in Figure 3.

When the widths of the resonances are very small compared to the electron energy spread or when the resonances do not appear in the near-threshold region within the energy range of astrophysical interest, they

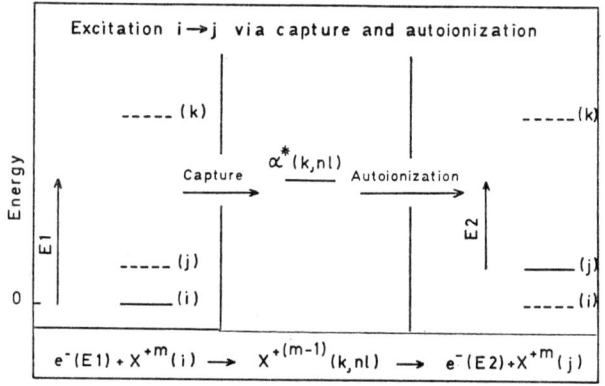

FIGURE 3. The two-step excitation via capture and autoionization.

should not strongly influence the excitation rates. But the width of a resonance can be large (>0.1 eV) and the distance between them very small. Their total contribution can be important, in particular in forbidden-line excitation for which the direct excitation $i \to j$ is generally less strong.

Calculations by Bely & Petrini (81) have shown that in Ca^+ the cross section $4s$–$3d$ is multiplied by a factor 2 in the energy range of $3d$–$4p$. These effects are particularly strong in the case of highly ionized positive ions. According to Petrini the cross section $\Omega(^2P_{1/2} - {}^2P_{3/2})$ in Fe XIV is multiplied by a factor 8 for energies smaller than 40 eV (82). In these calculations the contribution of the resonances is estimated in a very simple way by using a formula given by Gailitis (83) and Coulomb-Born techniques. The results are equivalent within 20% to the solution of the coupled equations.

The problem of resonance in electron scattering has been reviewed recently by Smith (29) and Burke (29). These articles deal mainly with elastic scattering and inelastic scattering by H and He^+.

The problem of resonances in cross section for excitation of forbidden lines, in a complex atom of ground configuration p^n, has been discussed very

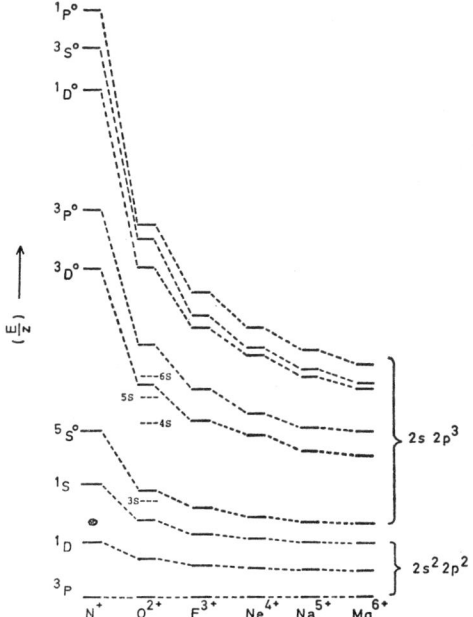

FIGURE 4. Energy differences divided by Z in the carbon isoelectronic sequence. For O^{+2} is shown only the $2s2p^3ns$ of $O^{+2}+e$ resonances states converging on the $2s2p^3 {}^3P^0$ state of O^{+2}.

recently by Seaton (9). In this case resonances due to the coupling with some closed channels do appear in the near-threshold region of astrophysical interest.

It is important to be able to predict the position of the resonances. If for example we are interested in the transitions between the sublevels of configuration $2s^22p^2$ in the carbon isoelectronic sequence, resonances may or may not occur because of the coupling with the $2s2p^33s$ configuration. As one sees from Figure 4 some of the levels of the $2s2p^33s$ configuration lie above the ground state $2s^22p^{2\,3}P$ for ions like $N^+(Z=1)$ and $O^{2+}(Z=2)$. On the contrary, for larger values of Z the $2s2p^33s$ states lie below the ground state and are true bound states. In the first case the $2s2p^33s$ states are resonance states and in this case the coupling of the $2s^22p^2$ open channels and the $2s2p^3$ closed channels in the collision problem gives a resonance structure.

For $Z>2$ this does not mean that the cross section will not have a resonance structure, because many other states of type $2s2p^3nl$ may give rise to resonances. As a matter of fact, as discussed by Seaton (9) for highly ionized members, as for positive ions in the solar corona, there are many series of resonances. Inelastic cross sections are averaged over resonances and Seaton (84) showed that resonant and nonresonant contributions to excitation rates are of comparable magnitude.

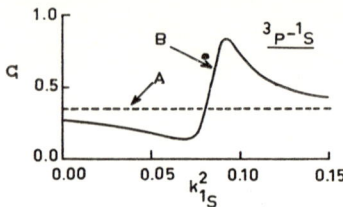
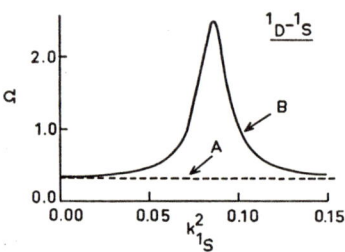

FIGURE 5. Forbidden transitions in the ground configuration $2s^22p^2$ of O^{2+}. Curve A, without coupling with $2s2p^3$. Curve B, with coupling. See (9), (12), and (89).

Many calculations for excitation of forbidden lines in O^{+2}, done in the past by Seaton and co-workers (9, 85–88), have been redone recently, taking into account the coupling with configuration other than the ground configuration. In Figure 5 are shown the results for two transitions of O^{+2}. The old results are shown as broken lines (89).

On the other hand, when all the possible resonance states are far above the excitation threshold of the forbidden transition $i \rightarrow j$ the influence of the closed channels in the near-threshold region is in fact equivalent to an additional polarization potential. This is the way to improve the close-coupling calculations for forbidden lines in the ground states of neutrals like C, N, O (26, 27).

6. Transition Between Fine-Structure Levels

In many astrophysical applications it is necessary to know the excitation cross sections between fine-structure levels, in particular, for forbidden-line excitation in planetary nebulae and in the solar corona.

Even when LS coupling is valid, calculation of excitation cross sections between fine-structure levels requires the use of a few transformation coefficients that can be found in the literature on the theory of angular momentum.

In fact, assuming LS coupling, the **S** matrix is usually calculated in a representation of total angular momentum and total spin $\mathbf{L}^T\mathbf{S}^T$, which are conserved during the collision. The collision strength between spectral terms

EXCITATION AND IONIZATION BY ELECTRON IMPACT   347

characterized by the quantum number αLS, where L and S are the orbital angular momentum and spin and where α represents the internal configuration of the state, is given by

$$\Omega[\alpha LS - \alpha'S'L'] = \frac{1}{2} \sum_{l'S^TL^T} (2L^T + 1)(2S^T + 1) \qquad 28.$$
$$\cdot | S(\alpha LSlS^TL^T, \alpha'L'S'l'S^TL^T) |^2$$

To obtain

$$\Omega[\alpha LSJ - \alpha'S'L'J'] = \frac{1}{2} \sum_{l'jj'J^T} (2J^T + 1) \qquad 29.$$
$$\cdot | S(\alpha LSJljJ^T, \alpha'L'S'J'l'j'J^T) |^2$$

it is necessary to go from the $LSl\frac{1}{2}L^TS^T$ representation to the $LSJljJ^T$ representation using the transformation

$$S[(\alpha LSJljJ^T, \alpha'L'S'J'l'j'J^T)]$$

$$= \sum_{S^TL^T} A \begin{Bmatrix} L & S & J \\ l & \frac{1}{2} & j \\ L^T & S^T & J^T \end{Bmatrix} S(\alpha LSlS^TL^T, \alpha'L'S'l'S^TL^T) A \begin{Bmatrix} L' & S' & J' \\ l' & \frac{1}{2} & j' \\ L^T & S^T & J^T \end{Bmatrix} \qquad 30.$$

The transformation coefficients A are the usual transformation coefficient between LS and jj coupling. They are related to the Wigner $9j$ symbol (Racah W) and given in the literature (90).

When the transition studied has the same spin before and after the collision ($S = S'$) and when only one multipole is important for computing the cross section ($\lambda = 1$ in the dipolar case), the transformation formulae are simpler.

In particular, when the Born approximation is valid (or the Coulomb-Born for positive ions), remembering the relation 3 between Q and Ω, one has

$$Q[\alpha LSJ \rightarrow \alpha'L'SJ']$$
$$= (2L + 1)(2J' + 1)W^2(LL'JJ'\lambda S)Q[\alpha LS \rightarrow \alpha'L'S] \qquad 31.$$

where W is the Racah coefficient, a formula that remains valid when configuration mixing is allowed for the states of the ion (91). Formula 31 is valid with the semiclassical dipole approximation and the approximation derived from the Bethe approximation in which the cross section is proportional to the oscillator strength.

Using the orthogonality properties of the A and the W coefficient, one can show that when exchange can be neglected ($S = S'$)

$$\sum_{J'} Q[\alpha LSJ \rightarrow \alpha'L'SJ'] = Q[\alpha LS \rightarrow \alpha'L'S] \qquad 32.$$

In the very special case of $L=0$ $L'=1$ ($s \to p$ transition)

$$Q[\alpha LSJ \to \alpha'L'SJ'] = \frac{2J'+1}{(2S+1)(2L'+1)} Q[\alpha LS \to \alpha'L'S] \qquad 33.$$

For a large number of levels the approximation of LS coupling is inadequate. Then, the coupling is of an intermediate type and formula 30 is not valid (91). This is particularly important in the study of transitions between terms of the same configuration. In fact, when LS coupling is violated, intercombinational transitions become possible not only by exchange but also by the mixing of states of different wavefunctions. Most of the recent calculations on nebular and coronal forbidden transitions within the p^n configuration are made in intermediate coupling. In addition, particularly in the case of highly ionized atoms, configuration interaction is very important, for example in the study of the green line of Fe^{+13} $3s^2 3p(^2P_{3/2} \to {}^2P_{1/2})$ (92).

7. Forbidden Transitions Within Ground Configuration of the Type p^n

These transitions are of very great importance in gaseous nebulae, the solar corona, and quasars, and consequently have been very carefully studied.

During recent years, calculations (87, 88, 93, 94) to improve the preliminary calculation of Seaton (85) have been based on the exact-resonance approximation, i.e. the neglect of the difference of energy between the fine-structure levels, or based on the quantum-defect extrapolation method as for the $p_{3/2}$-$p_{1/2}$ transition in C^+ and Si^+ (95).

Calculations using improved forms of the exact-resonance and distorted-wave approximations give results similar to the exact solutions of coupled equations of the close-coupling approximation (26) corrected in (27). Until recently the main approximation was the neglect of coupling with states of other configuration—see the discussion on resonances. In Seaton's approach variational techniques are used to obtain approximate solution of the coupled equations. Once the reactance matrix R is calculated for an ion X^m it is possible to extrapolate it to negative energies and to calculate the bound states of ions X^{m-1}. As a final step, parameters can be introduced into the expression of the matrix R and adjusted in order to have the best agreement with observed bound states. The adjusted matrices can then be used to obtain improved collision cross sections. Intermediate coupling has to be used.

All this work has been done for transitions between the $2p^n$ and $3p^n$ configurations by Seaton's group at University College London and by Czyzak & Krueger (96). The details of the method used are described by Saraph, Seaton & Sheming (97) and are discussed in three recent valuable reviews (11, 12, 97).

In very recent papers (11, 12, 89) it is shown that resonances can produce an important modification of the cross sections—see Figure 5. Resonance

EXCITATION AND IONIZATION BY ELECTRON IMPACT 349

structures are calculated by using methods of the quantum-defect theory.

The use of quantum-defect theory (99, 100) for the analysis of resonance structure is fully discussed by Seaton (84, 89).

The method just discussed, like the close-coupling method, implies long calculations. A more approximate method is the Coulomb-exchange approximation, which allows for exchange (52). The method is probably valid in a few cases; calculations have been done for the ^3P-^1D transition in the silicon and carbon series and the results agree with Seaton & Czyzak's results, excluding resonance effects.

8. Results for Electron-Impact Excitation

A few recent results are listed in Table 1. More references can be found in other review articles cited in the introduction. In particular, all available results on forbidden transitions within the p^n configurations are given elsewhere (11, 12, 97) and the experimental results on excitation functions can be found in a very recent review (8).

For close-coupling calculations, the states included in expansion 10 are given. Excitation cross sections are usually calculated for most of the transitions between these states.

IONIZATION
9. A General Remark

Ionization theory is much more difficult than excitation theory. However, as we shall see, the results given by the various approximations are in better statistical agreement with the experimental data than for excitation. While it is often difficult to obtain a 50% accuracy for excitation cross sections, this accuracy is generally reached in the ionization cross-section calculations. Also, simple empirical formulae for excitation are uncertain while results within a factor of 2 or better can be obtained very quickly for ionization. These agreeable features of the ionization problem may perhaps be explained in the following way.

If we consider the excitation cross section $dQ_l(\epsilon)$ of a narrow energy band $d\epsilon$ in the l continuum of an atom, it is possible, after an appropriate normalization, to show that $dQ_l(\epsilon)$ may be expressed as an excitation cross section $Q(1 \rightarrow nl)$. The ionization cross section being written as

$$Q(E) = \sum_l \int_0^{(E-I)} (dQ_l/d\epsilon) d\epsilon \qquad 34.$$

we can see that ionization corresponds in fact to the summation of a great number of excitation cross sections (up to a dozen sometimes).

It turns out that although each $dQ_l(\epsilon)$, or excitation cross section, is difficult to approximate, the sum 34 can be obtained more accurately because of cancellations. Also the integration over $d\epsilon$ seems to work in the same way. However, when a contribution to the ionization cross section comes from

TABLE 1. Excitation cross sections

Element	Transitions	Method	Reference
HYDROGEN			
H	$1s, 2s, 2p, 3s, 3p, 3d$	Close coupling	32
H	$1s, 2s, 2p$	Close coupling+correlations	33
H	$1s, 2s, 2p$	Close coupling and experimental	30
H	$1s-2s, 1s-2p$	Experimental	101
H	$1s-2p$	Experimental	102
H	$1s-2s$	Experimental	103
H	$1s-2s$	Experimental	104
H	$1s-nl, 2s-nl, 3s-nl$	Born	105
H	$1s-ns, 1s-np$	Born	106
H	$n's-ns, n's-np, n's-nd$	Born	107
H	$n \to n+1$	Impact parameter	67
H	General	Approximate methods	80
H	$1s-2s$	Approximate methods	108
HYDROGENLIKE IONS			
All	$1s-2s, 1s-2p$	Coulomb-Born II, D.W.	55
All	General	Approximate methods	80
He$^+$	$1s-2s$	Experimental	53
He$^+$	$1s, 2s, 2p$	Close coupling	54
He$^+$	$1s, 2s, 2p$	Close coupling	109
He$^+$	$1s, 2s, 2p$	Close coupling+correlations	110
He$^+$	$1s, 2s, 2p, 3s, 3p, 3d$	Close coupling	111
HELIUM[a]			
He	$1^1S, 2^1S, 2^3S$	Close coupling	115
He	$1^1S, 2^1S, 2^3S, 2^1P, 2^3P$	Close coupling	116
He	$1^1S \to 2^1S, 2^3S, 2^1P, 2^3P$	Experimental	117
He	$2^1S \to 2^3S$	Experimental	118
He	$1^1S \to 2^1S, 2^3S, 2^1P, 2^3P$	Experimental	119
He	$1^1S \to 2^3S, 2^3P$	Experimental	120
He	$1^1S \to 2^1S, 2^3S, 2^1P, 2^3P$	Experimental	121
He	$1^1S \to 2^3S, 2^3P$	Ochkur	122
He	$1^1S \to 2^1S, 2^3S, 2^1P$	Born	123
HELIUMLIKE IONS			
All	1^1S-2^1S	Coulomb-Born	124
Li$^+$, C^{+4}	1^1S-2^3S	Coulomb-Born-Oppenheimer	125
C^{+4}, O^{+6}	$1^1S-2^1P, 2^3P$	Experimental	126
O^{+6}	$1^1S-2^1P, 2^3P$	Experimental	127
O^{+6}	$1^1S-2^1P, 2^3P$	Coulomb exchange	128
LITHIUM AND LITHIUMLIKE IONS			
Li	$2s-2p$	Close coupling	129
Li	$2s-2p$	Close coupling	130
Li	$2s-2p$	Experimental	131

TABLE 1. (Continued)

Element	Transitions	Method	Reference
All	$2s \to ns, np, nd$	Coulomb-Born	60, 132
Be^+	$2s-2p$	Close coupling	133, 134
N^{+4}	$2s-2p$	Close coupling	135
N^{+4}	$2s \to 2p, 3s, 3p, 3d$	Experimental	136
	OTHER ALKALILIKE IONS		
All	$3s-3p$	Coulomb-Born	21
Mg^+	$3s, 3p, 3d$	Close coupling	51
Ca^+	$4s, 4p, 3d$	Close coupling	51
Ca^+	$4s, 4p, 3d$	Coulomb-Born	50
Ca^+	$4s-3d$	Coulomb-Born	81
Ca^+	$4s-4p$	Experimental	137
Sr^+	$5s-5p$	Experimental	137
Ba^+	$6s-6p$	Experimental	137, 138
Fe^{+15}	$3s-3p, 3p-3d$	Coulomb-Born	139
Fe^{+7}	Many	Coulomb-Born	140

a Excitation functions of many lines have been measured recently (112–114). More references are given in a review article by Heddle & Keesing (8) and in (6).

autoionizing levels (see Chapter 7), one again faces the excitation problem and the subsequent uncertainties.

10. CLASSICAL THEORIES

The first classical approach of the ionization problem is due to Thomson (72). Neglecting the interaction between the ionizing electron and the nucleus and supposing the atomic electron at rest, he derived a simple formula valid for a target having ξ electrons with binding energy I. The expression for the ionization cross section is

$$Q = 4\xi(I_H/I)^2 X^{-1}(1 - X^{-1})\Pi a_0^2 \qquad 35.$$

where $X = EI^{-1}$ is the reduced ionizing energy and I_H is the ionization energy of hydrogen. If one introduces a reduced cross section $Q^R(X)$ defined by

$$Q^R(X) = (I/I_H)^2 \xi^{-1} Q \qquad 36.$$

one finds that this expression is equal to $4X^{-1}(1-X^{-1})$ and is *a universal function* in Thomson's theory. This is an important result because it suggests that it is possible to predict an approximate ionization cross section given either by Thomson's theory or by a reduction of all available experimental and theoretical data. This point will be discussed in detail below.

Of course the asymptotic behavior of Q in this simple theory is wrong; quantum mechanics requires $Q \sim \log(E)/E$ at high energy.

In 1959 Gryzinski (73) tried to improve the classical theory by assigning an initial velocity to the atomic electron. He approximated the relative velocity $|V_1-V_2|$ of the two electrons by its averaged value $(V_1^2+V_2^2)^{1/2}$ and derived a fairly simple expression for the ionization cross section.

Later on, Ochkur & Petrunkin (141) and Stabler (74) removed the approximation on the speeds made by Gryzinski. However, the predictions of these new theories were not much better than Thomson's in the medium- and low-energy regions and, of course, did not agree with the high-energy quantum-mechanical behavior. In further works Gryzinski (142) sought the best velocity distribution for the atomic electron in order to improve the agreement with experimental results. No very satisfactory answer was found.

Interestingly enough, Thomson's theory gives a linear threshold law for the ionization cross section, in agreement with the latest quantum-mechanical derivations, while the modified theories do not give a linear law. Experimentally, the linear law is well verified except perhaps in the near vicinity of the threshold. Discussions about the threshold law are still going on but from a practical point of view a linear dependence is fully satisfactory (see Chapter 11).

In recent years a more sophisticated classical approach has been used by Percival's group (Abrines & Percival 145, Percival & Valentine 146, Abrines, et al 147). Having shown that the classical and the quantum-mechanical speed distribution for the hydrogen atom are the same (Percival 80), they were able to solve the exact classical problem numerically. The agreement with the experimental data is satisfactory in the low- and medium-energy range ($X \leq 10$). At high energies $Q \sim E^{-1}$. However, this theory can only be regarded as a test for classical theories and not as a tool for practical calculations since the results cited above require a tremendous amount of computation without providing the complete answer one would expect from such long calculations. Classical investigations have also been performed by Kingston (148) and Mapleton (149).

Modified classical theories.—In 1963 and 1964 Burgess (64, 71) improved Thomson's theory by feeding in some quantum-mechanical properties. In particular, he introduced exchange between the two electrons in a way that agrees with the recent quantal improvements (equality between the direct and the exchange scattering amplitudes). The atomic electron was also allowed to have an initial kinetic energy. A similar approach was also used by Vriens (150). The improvement over Thomson's theory is obvious in the low- and medium-energy range where exchange is supposed to play a role. However, once again, at high energies this approach breaks down. Then Burgess combined the preceding theory in the low- and medium-energy range with the correct high-energy results given by the impact-parameter method. This method, which was discussed in the excitation part of this

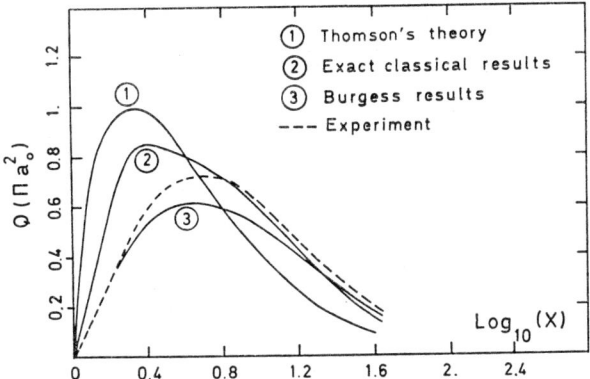

FIGURE 6. Classical and semiclassical ionization cross section for H(1s) compared to the experimental curve obtained by averaging the results given in (203) and (204).

paper (see Chapter 4), gives the right energy behavior, provided that the oscillator strength f (here from bound to free states) is known. There is no modification of the general conclusions when ionization is considered in place of excitation. The oscillator strength f (bound-free) is generally obtained from photoionization data or from direct calculations.

In Figure 6 are plotted the results given by theories of Thomson (72), Abrines et al (147), and Burgess (64). Experimental data are given for comparison.

11. QUANTUM THEORY

It is not our purpose here to review in detail the theory of ionization. A full analysis can be found in a series of papers by Peterkop (151–155), in a paper by Rudge & Seaton (156), and in two review articles by Veldre (157) and Rudge (14). A critical analysis of experimental data is available in a review paper by Kieffer & Dunn (13). We wish to sketch the most important features due to recent quantum-mechanical improvements of the ionization theory.

Asymptotic charges.—The ionization problem is, from a theoretical point of view, much more difficult than the excitation problem. One difficulty is to find the asymptotic fields in which the ejected and the scattered electrons are moving. In the excitation problem the target constitutes, *before* and *after* the collision, a *bound* system. Consequently, the colliding electron, for large values of r, is in a well-defined asymptotic field. It is a neutral field for a neutral atom, a Coulomb field for a positive ion. A lot of theoretically safe approximations can be done on these grounds (Born, Born-Coulomb, distorted waves, etc. . . .).

But after an ionizing collision, two electrons are moving away: the ejected and the scattered electron. After the collision the target is no longer a bound system and the fields in which the two electrons move are not so simple.

For simplicity we shall consider only the ionization of atomic hydrogen. Similar conclusions are valid for complex systems. If we denote by $\mathbf{k}$ and $\mathbf{k'}$, respectively, the momenta of the ejected and scattered electrons and if we look at the problem from a classical point of view we shall have, a long time after the collision

$$\mathbf{r} = \mathbf{k}t \qquad \mathbf{r'} = \mathbf{k'}t \qquad\qquad 37.$$

Calling z and z' the asymptotic charges that each electron would see if they were treated independently, the potential energy will be

$$V = -z/r - z'/r' \qquad\qquad 38.$$

On the other hand this energy must also be equal to

$$V = -1/r - 1/r' + 1/|\mathbf{r} - \mathbf{r'}| \qquad\qquad 39.$$

Equating 38 and 39 and using 37, we get

$$z/k + z'/k' = 1/k + 1/k' - 1/|\mathbf{k} - \mathbf{k'}| \qquad\qquad 40.$$

In other words, if one makes an assumption about the charge seen by one of the electrons, the charge seen by the other one will be given by Equation 40. One must also notice that this relation is angle-dependent. For practical uses a spherical average of Equation 40 is often made and is written

$$z/k + z'/k' = 1/k + 1/k' - 1/k_> \qquad\qquad 41.$$

where $k_>$ is equal to max (k, k').

Direct and exchange amplitudes.—As in the excitation problem we can define a direct scattering amplitude $f(\mathbf{k}, \mathbf{k'})$ and an exchange amplitude $g(\mathbf{k}, \mathbf{k'})$. Because two free electrons moving away after the collision are indistinguishable, the probability that the ejected electron has momentum $\mathbf{k}$ and the scattered electron momentum $\mathbf{k'}$ is the same as for scattering with $\mathbf{k}$ and ejection with $\mathbf{k'}$. It follows that $|f(\mathbf{k}, \mathbf{k'})|$ and $|g(\mathbf{k'}, \mathbf{k})|$ should be equal. In fact quantum theory gives

$$g(\mathbf{k'}, \mathbf{k}) = f(\mathbf{k}, \mathbf{k'}) \qquad\qquad 42.$$

if a correct choice of phase is made for the wavefunctions. By computing $f(\mathbf{k}, \mathbf{k'})$ through simple approximations and using Equation 42 it is possible to obtain good exchange approximations.

Integral expression for the scattering amplitude.—As in any scattering problem, the amplitudes of scattering are defined through the asymptotic forms of the wavefunctions (see the chapter on excitation). The main difference between ionization and excitation lies in the fact that for ionization the asymptotic forms are taken with respect to r and r' while for excitation only one electron is going away. This greatly complicates the problem, and an alternative way must be found for getting the scattering amplitudes if practical applications are intended. This was done by Peterkop and Rudge & Seaton in the papers cited above. Denoting by $\psi(r,r')$ the exact wavefunction of the (H+e) system before the collision (with appropriate asymptotic form), we may write the scattering amplitude $f(\mathbf{k},\mathbf{k'})$ as an integral

$$f(\mathbf{k}, \mathbf{k'}) = - (2\Pi)^{-5/2} \exp{(i\Delta)} \int \Psi(\mathbf{r}, \mathbf{r'})[H - E]\phi(z, -\mathbf{k}, \mathbf{r})$$
$$\cdot \phi(z, -\mathbf{k'}, \mathbf{r'}) d\mathbf{r} d\mathbf{r'}$$

43.

where H is the Hamiltonian of the system, E its total energy, Δ a phase factor, and the ϕ Coulomb wavefunctions corresponding to charges z and z' and momenta $-\mathbf{k}$ and $-\mathbf{k'}$; z and z' are related by Equation 40. This expression is the theoretical basis for any good approximation.

Threshold law.—Knowledge of the ionization-threshold law is important for two reasons. First, it makes possible to calibrate the electron energy in experimental work by extrapolating the cross section towards the ionization limit (known with accuracy from spectroscopic data). Second, in many practical problems the temperatures are such that only the low-energy part of the cross section is needed for computing the rates.

From expression 43 Rudge & Seaton (156) derived a linear-threshold law, apparently in agreement with most of the experimental data. However, the derivation of this law is not, mathematically speaking, satisfactory.

A recent experiment by McGowan & Clarke (159) seems to prove that the law is indeed not linear in the near-threshold region. However, the experimental cross section becomes linear for energies above 13.7 eV (i.e. 0.1 eV above threshold).

The threshold law predicted earlier by Wannier (160), $Q \propto (E-I)^{1.127}$, based on statistical mechanics, also disagrees with the preceding experimental results.

Personally we think that the linear law, even if not fully proved, must be not far from the truth. Many experimental data and all reasonable approximations prove that for practical needs the linear law is good.

Supposing both free electrons moving in the $z=1$ Coulomb field, Geltman (161) also derived a linear-threshold law. Hereafter we shall consider the linear law to be the exact one.

High-energy behavior.—For high-energy incident electrons the interaction takes a simple form and the cross section can be written as

$$Q = A \log E/E + B/E \qquad 44.$$

which is the Bethe form (Bethe 162).

A is simply related to the optical properties of the target and is accurately known for hydrogenic systems. The B term in Equation 44 comes from the short-range part of the interaction and may also be evaluated. Omidvar (163) gives analytical expressions for A and B for hydrogenic systems in their ground states. For complex atoms these quantities are difficult to compute with a good accuracy. However, using Pekeris two-electron wavefunctions, Inokuti & Kim (164) recently gave accurate A and B values for H^-.

Experimental values of A for complex atoms can be deduced from ionization and photoionization measurements (see for example Seaton 165). An extensive compilation of photoabsorption data is given by Kieffer (166).

12. Quantal Approximations

Here also we shall consider for simplicity the ionization of the hydrogen atom. Similar methods exist for complex atoms.

Born approximations: Born (a) and Born (b).—The wavefunction of the whole system before the collision is written as a product of a plane wave (describing the colliding electron) by the wavefunction of the bound electron. The charge seen by the ejected electron is taken to be $z = 1$ while the scattered one moves in a neutral field ($z' = 0$). In other words, the ejected electron completely screens the nucleus to the scattered one. Equation 42 is applied for the scattering amplitude. The threshold law given by this approximation is $(E-I)^{3/2}$ for neutral atoms and linear for positive ions. The incorrect threshold law obtained for neutral atoms is physically explained by the assumption made about z and z'. In fact, as follows from Equation 40, a complete screening of the nucleus never takes place. The cross section is obtained by integrating $f_B(\mathbf{k},\mathbf{k}')|^2$ over all possible energies and directions of the free electrons.

$$Q[\text{Born (a)}] = (\Pi k_0)^{-1} \int_0^{E-I} kk' d(\tfrac{1}{2}k^2) d\hat{k} d\hat{k}' \,|\, f_B(\mathbf{k}, \mathbf{k}')\,|^2 \qquad 45.$$

where k_0 is the initial momentum of the colliding electron. However, the neglect of exchange in this approximation is not consistent with Equation 42 and a better approximation is obtained when the upper integrand in 45 is taken to be $(E-I)/2$. This approximation is called Born (b) and is extensively used now. Calculations using Born (b) were performed for H ($1s$) (Rudge & Seaton (156) and for He^+ ($1s$) (Rudge & Schwartz (167) and compared with experimental data. An accuracy of 50% may be expected from these calculations, provided the wavefunction describing the bound state of the target is good.

Born-Oppenheimer approximations.—This approximation is obtained by

EXCITATION AND IONIZATION BY ELECTRON IMPACT 357

including exchange in a way similar to the excitation case, i.e. by antisymmetrizing the free electrons. As in the excitation case the results given by the Born-Oppenheimer approximation are often too large because of the lack of orthogonality of the wavefunctions describing the system (see for example Schiff 168). The results are better for highly charged ions.

Born-exchange approximations.—The relation 42 between direct and exchange amplitudes is used for computing $g_B(\mathbf{k},\mathbf{k}')$ from $f_B(\mathbf{k},\mathbf{k}')$. Since exact wavefunctions are not used for the relation between f and g, different choices of this phase factor lead to different results. However, they do not disagree by more than 20%. The agreement between theory and experiment is good for H (1s) (Geltman, Rudge & Seaton 169 and Peterkop 52) and perfect for He^+ (1s) (Rudge & Schwartz 167).

Other approximations.—In the preceding types of approximation the charge relation 40 is violated and the initial stage of the (e+atom) system is roughly approximated. It is possible to improve the last point by putting better initial wavefunctions. For example a good description of the system (e+H) should be given by the close-coupling wavefunction (see Chapter 2). Although the corresponding calculations are very complex and long, the results are no better than the Born (b) results (Veldre & Vilkalns 170, Burke & Taylor 171). The final states can also be improved by using 40 or its spherical averaged form 41.

Here again the calculations are tedious and no decisive improvement over Born (b) and Born-exchange results is obtained (Rudge & Schwartz 167). In fact z and z' are not uniquely defined by 40 and when approximate wavefunctions $\psi(\mathbf{r},\mathbf{r}')$ are used (which is always the case), the accuracy of the results depends on the choice of one of the z's.

Approximations using the Ochkur approach (Ochkur 172, Prasad 173) were also proposed with results comparable to Born-exchange.

The Racah algebra for computing Born (a) and (b) approximations in the $nLSJ$ scheme for complex atoms and ions is given by Bely & Schwartz (174).

In conclusion we can say that the Born (b) and Born-exchange are the best available approximations up to now. Figure 7 compares some theoretical results for H (1s) with experimental data.

For positive ions all the methods studied here give a linear-threshold law. As for the excitation case, all the approximations studied above will be called Coulomb-Born (a), (b), E, etc. . . when positive ions are considered.

13. IONIZATION VIA EXCITATION AND AUTOIONIZATION

The effect of autoionization on ionization was first pointed out by Fox, Hickam & Kjeldaas (175). Since then many features observed experimentally in ionization cross sections have been explained through this mechanism (see for example Kaneko 176 and Brink 177). More recently Peart & Dolder (178) observed (the first time for an ion), this effect in the ionization of Ba⁺ by electron impact. From a theoretical point of view the autoionization effect is simple. Some bound states of the target always lie above the ionization limit (this is of course not true for hydrogenic systems). Most of the states due to the excitation of an internal electron, i.e. an electron belonging to an inner shell, are above the ionization limit. For example the LiI state $1s2s^2$ lies far above the ionization limit, i.e. this state is in an energy region where one should find only continuum states of LiI. The excitation of the continuum leads to the usual ionization process. Now if the $1s2s^2$ state is excited, two phenomena can take place. First the bound state can interact with the adjacent continuum and give a free lithium state. This

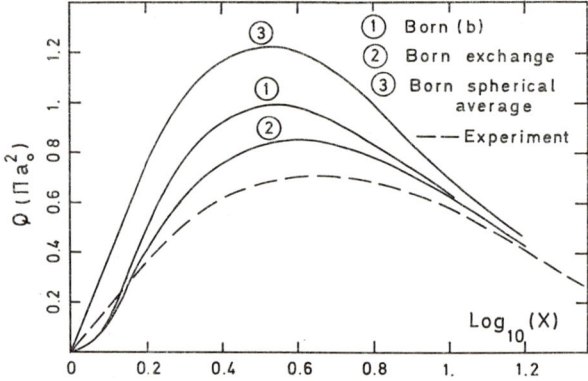

FIGURE 7. Quantum-mechanical ionization cross sections compared to the experimental results for H(1s).

is called autoionization. The former double process, excitation followed by autoionization, is finally equivalent to a normal ionization to which it adds up. The second mechanism is simply the radiative decay of the $1s2s^2$ state into another bound state lying below the ionization threshold. These two mechanisms are illustrated on Figure 8. If we call, respectively, A_a and A_r the autoionization and the radiative transition probabilities, the total cross section is written

$$Q(\text{total}) = Q(\text{ionization}) + \sum_i Q_i(\text{exc}) A_a(i)/(A_a(i) + A_r(i)) \quad 46.$$

where the summation is extended over all target bound states lying above the ionization limit.

For neutral atoms we have generally $A_a \simeq 10^{13}$ sec⁻¹ and $A_r \simeq 10^8$ sec⁻¹ so the former formula reduces to

$$Q(\text{total}) = Q(\text{ionization}) + \sum_i Q_i(\text{excitation}) \quad 47.$$

which is used in astrophysical applications.

However, we must keep in mind two important features of A_a and A_r. The dipole transition probability A_r varies like Z^4 along an isoelectronic sequence. On the other hand, it is possible to show that the autoionization probability A_r is about constant along a sequence.

Putting these characteristics together, we can easily convince ourselves that for highly ionized positive ions ($Z \geq 10$) one should be careful in applying Equation 47. In ions like Fe^{+13} and Fe^{+14} the branching ratio $A_a/(A_a + A_r)$ could be close to 1. A diminution of the autoionization contribution would result.

The autoionization effect can vary suddenly along the sequence of different stages of ionization of a given element. For example in Fe^{+8} to Fe^{+15}

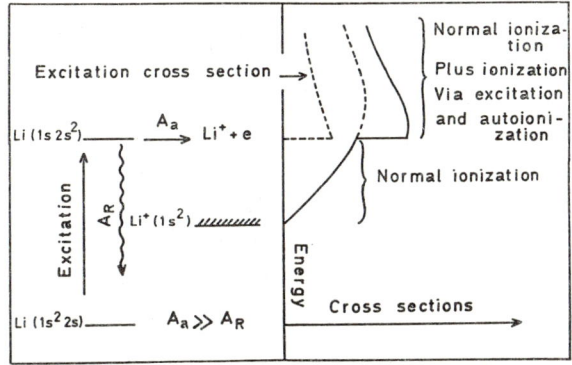

FIGURE 8. Ionization via excitation and autoionization.

(ground configurations of the type $3s^q\ 3p^r$), the autoionizing levels are mainly due to the internal $n=2$ shell excitation. The autoionization effect is large compared to the ionization rate (Bely 179) (a factor of 5 for Fe^{+15} at $T = 2 \times 10^{6\circ}K$). But in Fe^{+16} the autoionizing levels are due to the excitation of the $n=1$ shell, since the $n=2$ shell leads to normal states located below the ionization threshold. The excitation energy corresponding to $n=1$ being large for usual coronal temperatures, no substantial modification of the ionization rates will be obtained.

Also, note that the autoionization rates have a temperature dependence $T^{-1/2} \exp(-\Delta E/kT)$, like excitation rates, while ionization rates vary like $T^{1/2} \exp(-I/kT)$ in the usual temperature range ($kT \leq I$). It turns out that no simple extrapolation of the autoionization rates is possible from the internal-shell ionization rates.

As far as calculations are concerned, we have to face two problems. First, we must find which levels of the target lie above the ionization limit. The best available techniques must be used, in particular if the level considered is close to the ionization limit. Once the levels and the wavefunctions are obtained, excitation cross sections must be computed (and sometimes autoionization and radiative probabilities).

TABLE 2. Calculation of ionization cross section

Target	Method	Reference
HYDROGENIC SYSTEMS		
$H(1s, 2s)$	Born exchange	156, 167
$H(n=1, 10)$	Born (a)	202
$H(1s, 2s, 2p)$	Born-Ochkur	173
$H(n=1, \infty)$	Empirical	80, 201
$H(1s)$	Experimental	203, 204
$He^+(1s, 2s)$	Coulomb-Born exchange	167
$He^+(2p, 3p, 4p, 3d, 4d)$	Born (a)	205
$He^+(1s)$	Experimental	206
$C^{+5}(1s, 2s, 2p, 3d)$	Coulomb-Born (a)	207
Hydrogenic ions $(n=1, \infty)$	Empirical	80, 201
Hydrogenic ions	Coulomb-Born	167
HELIUMLIKE SYSTEMS		
H^-	Coulomb-Born (a)	194
H^-	Experimental	191, 192
He	Born exchange	208
He	Experimental	209
Li^+	Coulomb-born (b)	210
Li^+	Experimental	211, 212
C^{+4}	Coulomb-Born (a)	207
O^{+6}	Coulomb-Born (a)	213
OTHER SYSTEMS		
Li	Born exchange	214
Li	Experimental	183
O^{+5}	Coulomb-Born-Oppenheimer	215
Ca^{+17}	Coulomb-Born (a)	213
Be	Born exchange	214
O^{+4}	Coulomb-Born-Oppenheimer	216
B	Born (b)	217
C	Born (b)	217
N^+	Experimental	218
N	Born (b)	217
N	Experimental	219
O	Born (b)	217
O	Experimental	220
O^-	Experimental	192
F	Born (b)	217
Ne^+	Experimental	221
Ne	Born (b)	217
Ne	Experimental	209
Na^+	Experimental	222, 223
Fe^{+16}	Coulomb-Born (O)	213
Na	Born (b)	214

TABLE 2. (Continued)

Target	Method	Reference
Na	Experimental	183
Mg^+	Coulomb-Born (b)	210
Mg^+	Experimental	184
Fe^{+15}	Coulomb-Born exchange	224
Mg	Born exchange	214
Al	Born (b)	217
Fe^{+13}	Coulomb-Born (b)	174
Si	Born (b)	217
P	Born (b)	217
S	Born (b)	217
Cl	Born (b)	217
A	Born (b)	217
A	Experimental	209
K	Experimental	183
K^+	Experimental	222, 223
Kr	Experimental	209
Rb	Experimental	183
Sr^+	Coulomb-Born (b)	181
Sr	Experimental	227
Y^+	Coulomb-Born (b)	181
Xe	Experimental	219
Cs	Experimental	182, 226
Ba^+	Coulomb-Born (b)	181
Ba	Experimental	228
Hg	Experimental	225
Fe^{+18} to Fe^{+24}	Coulomb-Born (b)	199

Second, we must insist that all cross sections should be considered, those corresponding to the permitted transitions as well as those corresponding to the forbidden ones. Recent calculations (see Chapter 3) have fully demonstrated that when $n \neq n'$ the dipole cross section is no longer the dominant one. The use of the $\bar{g}$ approximation (valid only for dipole transitions) could lead then to erroneous results.

The accuracy obtained in computing the autoionization contribution is comparable to the accuracy obtained when computing excitation cross sections: 30% if careful calculations are performed, a factor of 2 with simpler methods.

Recent calculations using Born II and Coulomb-Born II approximations were done for sodiumlike series (Bely 180)[1]; for Mg^+ (Moores & Nussbaumer 210); for Ba^+, Sr^+, Yr^+ (Bely & Schwartz 181); and for CsI, RbI, StrI, and BaI (Dubau 182). Comparison with experimental data is possible for Na (McFarland & Kinney 183), Mg^+ (Martin, Peart & Dolder 184), and

[1] Extrapolated cross sections were used in that reference.

Ba$^+$ (Peart & Dolder 178). In Ba$^+$ the agreement is satisfactory as far as the position and the magnitude of the autoionization process are concerned. For CsI, RbI, StrI, and BaI the comparison with existing experimental data shows fair agreement (Dubau 182). In Mg$^+$ the results obtained by Moores & Nussbaumer (210) are in fair agreement with the experimental results. These authors also pointed out that the extrapolated results obtained by Bely (180) are overestimated for the first elements of the sodiumlike ions, in particular for Na and Mg$^+$. No apparent structure is observed in the sodium-ionization cross section.

After the theoretical review of this subject in a paper by Goldberg, Dupree & Allen (185), one must say that a great deal of work still has to be done, in particular on the branching-ratio aspect of this question.

14. Ionization of Negative Ions

The ionization cross section of the most important astrophysical negative ion, H$^-$, was first estimated by Page (186) who found a very large cross section. Several other approximate calculations showed a large spread of the results (Geltman 187, Rudge 188, Smirnov & Chibisov 189, Rogalski 190).

Experimental results of Dance, Harrison & Rundel (191) and of Tisone & Branscomb (192) indicated that the ionization cross section of H$^-$ was much smaller that most of the theoretical predictions. Recently, calculations by McDowell & Williamson (193) and Bely & Schwartz (194) gave satisfactory agreement with experimental data.

Because of the repulsive Coulomb field between the colliding electron and H$^-$, the cross section is not a linear function in the threshold-energy region but decreases exponentially when the incident energy approaches the ionization limit from above (Rudge 188). Experimental results are also available for O$^-$ (Tisone & Branscomb 192).

15. Empirical Formulae

Contrary to the excitation case, it is not very difficult to obtain a quick and reliable estimate of an ionization cross section in the energy range $E/I \lesssim 10$ (we are dealing here only with normal ionization, i.e. we disregard autoionization). As we have seen, the classical Thomson theory suggests that the reduced ionization cross section defined by Equation 36 should be about the same for any species. In practice one ascertains that almost all available theoretical and experimental data lead to reduced cross sections that are, within a factor of 2, equal in the above-cited energy range.

Most of the empirical formulae are based upon this important result. Figure 9 shows some experimental reduced ionization cross sections. The reduced cross sections will start to differ more strongly in the high-energy region because only one multipole is there responsible for the cross sections (dipole interaction). The cross section then approaches Bethe's form $A \log E/E$.

Consequently the reduced cross sections will depend more on the nature

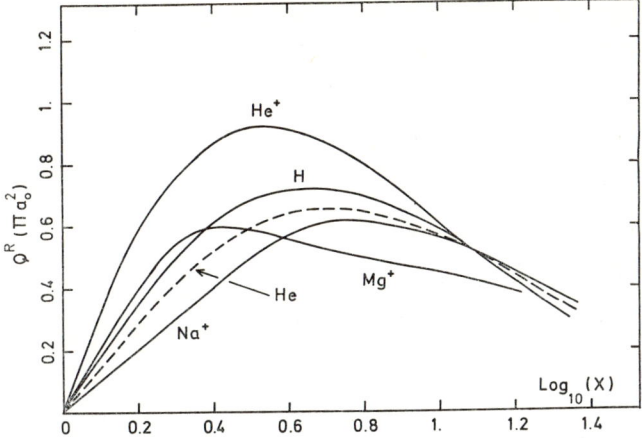

FIGURE 9. Some experimental reduced ionization cross sections.

of the target than do those in low- and medium-energy range where many multipoles as well as short-range forces make the Q^R more nearly independent of the target species.

Elwert (195) was the first to apply the reduced cross-section technique to give simple empirical formulae. Since then a number of formulae have been proposed (Drawin 196, Burgess & Seaton 197, Lötz 198).

One of the simplest is that of Burgess & Seaton (197). Approximating the reduced cross section by a straight line, whose slope is deduced from available results, they find for the ionization rate (valid if $kT \leq I$)

$$g = 2.0\xi I(\text{eV})^{-2} T(^\circ\text{K})^{1/2} 10^{-5040 I(\text{eV})/T(^\circ\text{K})} (\text{sec}^{-1}\ \text{cm}^3) \qquad 48.$$

This result is particularly good for highly charged positive ions. A theoretical study by Bely & Schwartz (199) of the iron coronal ions Fe^{+8} to Fe^{+24} gave results equal to those given by Equation 48 within 30%!

Very recently Lötz (198) presented an extensive and convenient tabular compilation of results.

Vriens (200) has given semiempirical formulae for the ionization of inert gases.

All these approximate expressions give, statistically, a 50% accuracy and may also be used for the inner-shell ionization cross sections, provided the ionization energy is known.

Special attention was devoted to the hydrogenic systems by Percival (80). He has given formulae that represent the average cross section for ionization from excited states of hydrogen and hydrogenic positive ions. There is no restriction on the energy range of validity. This careful analysis yields the reduced cross sections

$$Q_1{}^R(X) = (1.19 \ln (X) + 5.26)(X - 1)/(X^2 + 1.67X + 3.57)$$
$$\text{for } n = 1 \qquad 49.$$

and

$$Q_n{}^R(X) = (1.28 n^{-1} \ln (X n^{-2}) + 6.67)/(X^2 + 1.67 X + 3.57)$$
$$\text{for } n \geq 2 \qquad 50.$$

For hydrogenic ions Percival writes

$$Q_n{}^Z(X) = [1 + 2.3/((1 - Z^{-1})^2 + 2(X - 1)^2)] Q_n{}^R(X) \qquad 51.$$

Percival's results must be averaged numerically for giving the ionization rates.

LITERATURE CITED

1. Massey, H. S. W. 1956, *Handb. Phys.*, **36** (2), 307 (Springer-Verlag)
2. Seaton, M. J. 1958, *Rev. Mod. Phys.*, **30**, 979–91
3. Seaton, M. J. 1962, in *Atomic and Molecular Processes*, 374–420 (Bates, D. R., Ed., Academic)
4. Peterkop, R., Veldre, V. 1966, *Advan. At. Mol. Phys.*, **2**, 264
5. Burke, P. G. 1968, in *Int. Conf. At. Phys.*, 265–94 (Plenum)
6. Moiseiwitsch, B. L., Smith, S. J. 1968, *Rev. Mod. Phys.*, **40**, 238–53
7. Fite, W. L. 1962, in *Atomic and Molecular Processes*, 421–92 (Bates, D. R., Ed., Academic)
8. Heddle, D. W. O., Keesing, R. G. W. 1968, *Advan. At. Mol. Phys.*, **4**, 267–98.
9. Seaton, M. J. 1968, in *Int. Conf. At. Phys.*, 295–319 (Plenum)
10. Seaton, M. J. 1968, in *Planetary Nebulae. Symp.*, 129–37 (Osterbrock, D. R., O'Dell, C. R., Eds., Reidel)
11. Seaton, M. J. 1968, in *Advan. At. Mol. Phys.*, **4**, 331–80
12. Seaton, M. J. 1969, *Mem. Soc. Sci. Liège*, **17**, 45–55
13. Kieffer L. J., Dunn, G. H. 1966, *Rev. Mod. Phys.*, **38**, 1; *JILA Rep. 51*
14. Rudge, M. R. 1968, *Rev. Mod. Phys.*, **40**, 564
15. Kieffer, L. J. 1966, *Low Energy Electron Collision. JILA Rep. 2, 4 with addendum; JILA Rep. 30*
16. Kieffer, L. J. 1969, *Low Energy Electron Collision, Part 2. Line and Level Excitation. JILA Rep. 7*
17. *Bibliog. At. Mol. Processes* (At. Mol. Processes Inf. Cent. Oak Ridge Nat. Lab.)
18. Branscomb, L. M. 1968, *Invited Pap. Leningrad Conf.*, 12, Boulder, Colo. (JILA)
19. Pagel, B. E. J. 1968, *Proc. Roy. Soc. A*, **306**, 91–118
20. Percival, I. C., Seaton, M. J. 1957, *Proc. Cambridge Phil. Soc.*, **53**, 654
21. Bely, O., Tully, J., Van Regemorter, H. 1963, *Ann. Phys.*, **8**, 303
22. Bely, F., Lan, Vo Ky 1968, *C. R. Acad Sci.*, **276**, 533
23. Eissner, W., Nussbaumer, H. 1969, *J. Phys. B*, **2**, 1028–43
24. Condon, E. U., Shortley, G. H. 1935, *Theory At. Spectra* (Cambridge)
25. Bransden, B. H., McKee, J. S. C. 1956, *Proc. Phys. Soc. A*, **69**, 422
26. Smith, K., Henry, R. J. W., Burke, P. G. 1967, *Phys. Rev.*, **157**, 51
27. Henry, R. J. W., Burke, P. G., Sinfailam, A. L. 1969, *Phys. Rev.*, **178**, 218–25
28. Burke, P. G. 1968, in *Advan. At. Mol. Phys.*, **4**, 173–217
29. Smith, K. 1966, *Rep. Progr. Phys.*, **29**, (2) 373
30. Burke, P. G., Taylor, A. J., Ormonde, S. 1967, *Proc. Phys. Soc.*, **92**, 345
31. Fite, W. L., Kaupilla, W. E., Ott, W. R. 1963, *Phys. Rev. Lett.*, **20**, 409
32. Burke, P. G., Ormonde, S., Whitaker, W. 1967, *Proc. Phys. Soc.*, **92**, 319

33. Burke, P. G., Taylor, A. J. 1967, *Proc. Phys. Soc.*, **92**, 336
34. Damburg, R. J., Karule, E. 1967, *Proc. Phys. Soc.*, **90**, 637
35. Damburg, R. J., Geltman, S. 1968, *Phys. Rev. Lett.*, **20**, 485
36. Ochkur, V. I. 1963, *JETP*, **45**, 734
37. Rudge, M. 1965, *Proc. Phys. Soc.*, **85**, 607
38. Rudge, M. 1965, *Proc. Phys. Soc.*, **86**, 763
39. Bely, O. 1966, *Proc. Phys. Soc.*, **88**, 587–95
40. Vainshtein, L. A., Sobelman, I. I. 1967, *Lebedev Rep. 66*, 1–32 (UKAEA Transl., Culham Lab: Born-approximation excitation cross sections for atoms by electrons)
41. Vainshtein, L. A., Presnyakov, L., Sobelman, I.I. 1964, *JETP*, **18**, 1383
42. Crothers, D., McCarroll, R. 1965, *Proc. Phys. Soc.*, **86**, 753
43. Omidvar, K. 1966, *NASA Rep. X641 66 245*
44. Omidvar, K. 1967, *Phys. Rev. Lett.*, **18**, 153
45. Crothers, D. 1967, *Proc. Phys. Soc.*, **86**, 753
46. Seaton, M. J. 1961, *Proc. Phys. Soc.*, **77**, 174
47. Burke, P. G., Seaton, M. J. 1961, *Proc. Phys. Soc.*, **77**, 199
48. Somerville, W. B. 1963, *Proc. Phys. Soc.*, **82**, 446
49. Van Regemorter, H. 1960, *MNRAS*, **121**, 213–31
50. Petrini, D. 1965, *C. R. Acad. Sci.*, **260**, 4929–32
51. Burke, P. G., Moores, D. 1968, *Proc. Phys. Soc. B*, **1**, 575–85
52. Bely, O. 1967, *Nuevo Cimento*, **49**, 66–86
53. Dance, D., Harrison, M., Smith A. 1966, *Proc. Roy. Soc. A*, **290**, 74–93
54. Burke, P. G., Mc Vicar, D., Smith, K. 1964, *Proc. Phys. Soc.*, **83**, 397
55. Burgess, A., Hummer, D. G., Tully, J. 1969 (To be published in *Phil. Trans. Roy. Soc.*)
56. Madden, R. P., Codling, K. 1965, *Ap. J.*, **141**, 364
57. Burgess, A. 1961, *Mem. Soc. Roy. Sci. Liège*, **4**, 299
58. Bely, O. 1966, *JILA Rep. 89* (Boulder, Colo.)
59. Van Regemorter, H. 1962, *Ap. J.*, **136**, 906
60. Bely, O. 1966, *Ann. Ap.*, **29**, 131
61. Blaha, M. 1969, *Ap. J.*, **157**, 473–77
62. Seaton, M. J. 1962, *Proc. Phys. Soc.*, **79**, 1105
63. Alder, K., Bohr, A., Huus, T., Mottelson, B., Winther, A. 1956, *Rev. Mod. Phys.*, **28**, 432–542
64. Burgess, A. 1964, *Culham Conf. At. Collisions, AERE Rep. 4818*, 63
65. Feautrier, N. 1968, *Ann. Ap.*, **31**, 305–9
66. Sahal-Brechot, S. 1969, *Astron. Ap.*, **1**, 91–123
67. Saraph, H. E. 1964, *Proc. Phys. Soc.*, **83**, 763
68. Pengelly, R. M., Seaton, M. J. 1964, *MNRAS*, **127**, 165
69. Feautrier, N. Praderie, F., Van Regemorter, H. 1967, *Ann. Ap.*, 45–54
70. Stauffer, A., McDowell, M. R. C. 1965, *Proc. Phys. Soc.*, **85**, 61
71. Burgess, A. 1963, *Proc. Intern. Conf. Electron At. Collisions*, 237 (North-Holland)
72. Thomson, J. J. 1912, *Phil. Mag.*, **23**, 449,
73. Gryzinski, M., 1959, *Phys. Rev.*, **115**, 374
74. Stabler, R. C. 1964, *Phys. Rev.*, **A133**, 1268
75. Burgess, A., Percival, I. C. 1968, in *Advan. At. Mol. Phys.*, **4**, 109–41
76. Kingston, A. E. 1964, *Phys. Rev.*, **A135** 1529
77. Kingston, A. E., Lauer, I. E. 1966, *Proc. Phys. Soc.*, **87**, 399
78. Kingston, A. E., Lauer, I. E. 1966, *Proc. Phys. Soc.*, **88**, 597
79. Percival, I. C., Richards, D. 1970, *Ap. Lett.*, **4**, 235
80. Percival, I. C. 1966, *Nucl. Fusion*, **6**, 182
81. Bely, O., Petrini, D. 1966, *Phys. Lett.*, **23**, 442–43
82. Petrini, D. 1969, *Astron. Ap.*, **1**, 139
83. Gailitis, M. 1963, *JETP*, **17**, 1328–32
84. Seaton, M. J. 1969, *J. Phys. B*, **2**, 5–11
85. Seaton, M. J. 1953, *Proc. Roy. Soc. A*, **218**, 400
86. Seaton, M. J. 1955, *Proc. Phys. Soc.*, **68**, 457
87. Seaton, M. J. Saraph, H. E., Shemming, J. 1966, *Proc. Phys. Soc.*, **89**, 27
88. Czyzak, S. J., Krueger, T. K. 1967, *Proc. Phys. Soc.*, **90**, 623
89. Eissner, W., Nussbaumer, H., Saraph, H. E., Seaton, M. J. 1969, *J. Phys. B*, **2**, 341–55
90. Edmonds, A. R. 1957, *Angular Momentum in Quantum Mech.* (Princeton Univ. Press)
91. Bely, O. 1967, *Nuovo Cimento*, **49**, 87–102
92. Petrini, D. 1967, *C. R. Acad. Sci.* **264**, 411

93. Blaha, M. 1968, *Ann. Ap.*, **31**, 311
94. Blaha, M. 1969, *Astron. Ap.*, **1**, 42–43
95. Osterbrock, D. E. 1965, *Ap. J.*, **142**, 1423
96. Czyzak, S. J., Krueger, T. K., Saraph, H. E., Shemming, J. 1967, *Proc. Phys. Soc.*, **92**, 1146
97. Saraph, H. E., Seaton, M. J., Shemming, J. 1969, *Phil. Trans. Roy. Soc. A*, **264**, 77–105
98. Czyzak, S. J., Krueger, T. K., Martins, P., Saraph, H. E., Seaton, M. J., Shemming, J. 1968, in *Planetary Nebudae, IAU Symp. 34*, 138–42
99. Seaton, M. J. 1958, *MNRAS*, **118**, 504–18
100. Seaton, M. J. 1966, *Proc. Phys. Soc.*, **88**, 801–14
101. Stebbings, R. F., Fite, W. L., Hummer, D. G., Brackmann, R. T. 1960, *Phys. Rev.*, **119**, 1939; 1961, *Phys. Rev.*, **124**, 2051
102. Chamberlain, G., Smith, S., Heddle, D.W.O. 1964, *Phys. Rev. Lett.*, **12**, 647–49
103. Lichten, W., Schultz, S. 1959, *Phys. Rev.*, **116**, 1132–39
104. Hils, D., Kleinpoppen, H., Koschmieder, H. 1966, *Proc. Phys. Soc.*, **89**, 35–40
105. Omidvar, K. 1965, *Phys. Rev.*, **140**, A38
106. Veldre, V., Rabik, L. L. 1966, *Opt. Spectrosc.*, **19**, 265
107. Vainshtein, L. A. 1965, *Opt. Spectrosc.*, **18**, 538
108. Thruhlar, D. G., Cartwright, D. C., Kupperman, A. 1968, *Phys. Rev.*, **175**, 113
109. McCarroll, R. 1964, *Proc. Phys. Soc.*, **83**, 409–17
110. Burke, P. G., Taylor, A. J. 1969, *J. Phys. B*, **2**, 44–51
111. Ormonde, S., Whitaker, W., Lipsky, L. 1967, *Phys. Rev. Lett.*, **19**, 116
112. Heddle, D. W. O., Keesing, R. G. W. 1967, *Proc. Phys. Soc. A*, **299**, 212
113. Zapesochnyi, I. P., Feltsan, P. V. 1965, *Ukr. Fiz. Zh.*, **10**, 1187
114. Zapesochnyi, I. P., Shpenik, O. B. 1966, *JETP*, **23**, 592
115. Marriott, R. 1966, *Proc. Phys. Soc.*, **87**, 407
116. Burke, P. G., Ormonde, S., Cooper, J. W. 1969, *Phys. Rev.*, **183**, 245–64
117. Schulz, G. J., Fox, R. E. 1957, *Phys. Rev.*, **106**, 1179
118. Phelps, A. V. 1955, *Phys. Rev.*, **99**, 1307
119. Holt, H. K., Krotkov, R. 1966, *Phys. Rev.* **144**, 82
120. Pichanick, F. M. T., Simpson, J. A. 1968, *Phys. Rev.*, **168**, 1, 64
121. Ehrhardt, H., Langhans, L., Linder, F. 1968, *Z. Phys.*, **214**, 179
122. Ochkur, V. I., Bratsev, V. F. 1966, *Sov. Astron.*, **9**, 797
123. Vriens, L., Simpson, J. A., Mielczarec, S. R. 1968, *Phys. Rev.*, **165**, 7
124. Sural, D. P., Sil, N. C. 1966, *Proc. Phys. Soc.*, **87**, 201
125. Beigman, I. L., Vainshtein, L. A. 1967, *JETP*, **25**, 119–23
126. Kunze, H. J., Gabriel, A. H., Griem, H. R. 1968, *Phys. Rev.*, **165**, 267–76
127. Elton, L. R. B., Koppendorfer, W. 1967, *Phys. Rev.*, **160**, 194
128. Bely, O. 1968. *Phys. Lett.*, **26A**, 408–9
129. Burke, P. G., Taylor, A. J. 1969, *Proc. Phys. Soc. B*, **2**, 869–78
130. Karule, E. M., Peterkop, R. 1965, in *Atomic Collisions* (Veldre, Y. Ya., Ed., Chicago: SLA Transl. Cent.)
131. Perel, J., Englander, P., Bederson, B. 1962, *Phys. Rev.*, **128**, 1148–54
132. Bely, O. 1966, *Ann. Ap.*, **29**, 683–87
133. Moores, D. 1966, *Proc. Phys. Soc.*, **87**, 843–59
134. Moores, D. 1967, *Proc. Phys. Soc.*, **88**, 830–41
135. Burke, P. G. Tait, J. H., Lewis, B. A. 1966, *Proc. Phys. Soc.*, **87**, 209
136. Boland, B. C., Jahoda, F. C., Jones, T. J. L., McWhirter, R. W. P. 1968, *2nd Int. Conf. Vac. UV Spectrosc. Maryland*
137. Lee, A. R., Carleton, N. P. (To be publ. 1969)
138. Bacon, F. M., Hooper, J. W. 1969, *Phys. Rev.*, **177** (Feb.)
139. Krueger, T. K., Czyzak, S. F. 1965, *Mem. Roy. Astron. Soc.*, **69**, 145
140. Czyzak, S. F., Krueger, T. K. 1966, *Ap. J.*, **144**, 381
141. Ochkur, V. I., Petrunkin, A. M. 1963, *Opt. Spectrosc.*, **14**, 245
142. Gryzinski, M., 1965, *Phys. Rev.*, **A138**, 305
143. Gryzinski, M. 1965, *Phys. Rev.*, **A138**, 322
144. Gryzinski, M. 1965, *Phys. Rev.*, **A138**, 336
145. Abrines, R., Percival, I. C., 1966, *Proc. Phys. Soc.*, **88**, 861
146. Percival I. C., Valentine, N. A. 1966, *Proc. Phys. Soc.*, **88**, 885
147. Abrines, R., Percival, I. C., Valentine, N. A. 1966, *Proc. Phys. Soc.*, **89**, 515
148. Kingston, A. E. 1964, *Phys. Rev.*, **135**, A1537
149. Mapleton, R. A. 1966, *Proc. Phys. Soc.*, **87**, 219

150. Vriens, L. 1966, *Phys. Rev.*, **141**, 88
151. Peterkop, R. K. 1960, *Izv. Akad. Nauk Latv. SSR*, **9**, 79
152. Peterkop, R. K. 1962, *Sov. Phys. JETP*, **14**, 1377
153. Peterkop, R. K. 1962, *Opt. Spectrosc.*, **13**, 87
154. Peterkop, R. K. 1963, *Sov. Phys. JETP*, **16**, 442
155. Peterkop, R. K., 1963, *Sov. Phys. Dokl.*, **27**, 987
156. Rudge, M.R.H., Seaton, M. J. 1965, *Proc. Roy. Soc. A*, **283**, 262
157. Veldre, V. 1965, *Riga Rep., Transl. TT 66–1239* (Chicago: John Crerar Library)
158. Rudge, M. R. H., Seaton, M. J. 1964. *Proc. Phys. Soc.*, **85**, 607
159. McGowan, J. W., Clarke, E. M. 1968, *Phys. Rev.*, **167**, 43
160. Wannier, G. M. 1953, *Phys. Rev.*, **90**, 817
161. Geltman, S. 1956, *Phys. Rev.*, **102**, 171
162. Bethe, H. A. 1930, *Ann. Phys.*, **5**, 325
163. Omidvar, K. 1969, *Phys. Rev.*, **177**, 212
164. Inokuti, M., Yong-Ki-Kim 1968, *Phys. Rev.*, **173**, 154
165. Seaton, M. J. 1954, *Phys. Rev.*, **113**, 814
166. Kieffer, L. J., 1968, *JILA Inf. Cent. Rep. 5*
167. Rudge, M. R. H., Schwartz, S. B. 1966a, *Proc. Phys. Soc.*, **88**, 563
168. Schiff, L. I. 1952, *Quantum Mechanics* (New York: McGraw Hill)
169. Geltman, S., Rudge, M. R. H., Seaton, M. J., 1963, *Proc. Phys. Soc.*, **81**, 375
170. Veldre, V., Vinkalns, I. 1963, in *At. Collision* (Akad. Nauk Latv. SSR)
171. Burke, P. G., Taylor, A. J. 1965, *Proc. Roy. Soc. A*, **287**, 105
172. Ochkur, V. I. 1965, *Sov. Phys. JETP*, **20**, 1175
173. Prasad, S. S. 1965, *Proc. Phys. Soc.*, **85**, 57
174. Bely, O., Schwartz, S. B. 1969, *Astron. Ap.*, **1**, 281
175. Fox, R., Hickam, W., Kjeldaas, T. 1953, *Phys. Rev.*, **89**, 555
176. Kaneko, Y. 1961, *J. Phys. Soc. Jap.*, **11**, 2288
177. Brink, G. 1962, *Phys. Rev.*, **127**, 1204
178. Peart, B., Dolder, K. T. 1968a, *J. Phys. B*, **1**, 872
179. Bely, O. 1967, *Ann. Ap.*, **30**, 953
180. Bely, O. 1968, *J, Phys. B*, **1**, 23
181. Bely, O., Schwartz, S. B. (To be publ. 1970)
182. Dubau, J. 1969., *Troisième Cycle* (Unpubl. thesis, Paris)
183. McFarland, R. H., McKinney, J. D. 1964, *Phys. Rev.*, **137**, 1058
184. Martin, S. O., Peart, B., Dolder, K. T. 1968, *J. Phys. B*, **1**, 537
185. Goldberg, L. Dupree, A., Allen, J. 1965, *Ann. Ap.*, **28**, 589
186. Pagel, B. 1959, *MNRAS*, **119**, 609
187. Geltman, S. 1960, *Proc. Phys. Soc.*, **75**, 67
188. Rudge, M. R. H. 1964, *Proc. Phys. Soc.*, **83**, 419
189. Smirnov, B. M., Chibisov, M. I. 1966. *JETP Sov. Phys.*, **22**, 585
190. Rogalski, M. 1966, *Acta Phys. Polon.*, **29**, 15
191. Dance, D. F., Harrison, M. F., Rundel, R. D. 1967, *Proc. Roy. Soc. A*, **299**, 525
192. Tisone, G., Branscomb, L. 1968, *Phys. Rev.*, **170**, 169
193. McDowell, M. R. C., Williamson, J. H. 1963, *Phys. Lett.*, **4**, 159
194. Bely, O., Schwartz, S. B. 1969, *J. Phys. B*, **1**, 159
195. Elwert, G. 1952, *Z. Naturforsch.*, **7a**, 432
196. Drawin, H. W. 1961, *Z. Phys.*, **164**, 513
197. Burgess, A., Seaton, M. J. 1964. *MNRAS*, **127**, 355
198. Lötz, W., 1967, *Ap. J. Suppl.*, **14**, 207
199. Bely, O., Schwartz, S. B. (To be publ. 1970)
200. Vriens, L. 1965, *Physica*, **31**, 385
201. Percival, I. C. 1965, *Culham Rep. CLM-P87*
202. Omidvar, K., 1965, *Phys. Rev.*, **A140**, A26
203. Fite, W. L., Brackmann, R. T. 1958, *Phys. Rev.*, **112**, 1141
204. Rothe, E. W., Marino, L. L., Neynaber, R. H., Trujillo, S. M. 1962, *Phys. Rev.*, **125**, 582
205. Dalgarno, A., McDowell, M. R. C. 1966, *The Airglow and Aurorae* (Armstrong, B., Dalgarno, A., Eds., New York: Pergamon)
206. Dolder, K. T., Harrison, M. F. A., Thonemann, P. C. 1961, *Proc. Roy. Soc. A*, **264**, 367
207. Beigman, I. L., Vainshtein, L. A. 1968, *Sov. Astron.*, **11**, 712
208. Sloan, I. H., 1965, *Proc. Phys. Soc.*, **85**, 435
209. Schram, B. L., De Heer, F. J., Van Der Wiel, M. J., Kistemaker, J. 1964, *Physica*, **31**, 94 [Many other refs. in (13)]
210. Moores, D., Nussbaumer, H. (To be publ. 1970 in *J. Phys. B*)

211. Lineberger, W. C., Hooper, J. W., McDaniel, E. W. 1966, *Phys. Rev.*, **141**, 151
212. Peart, B., Dolder, K. T. 1968, *J. Phys. B*, **1**, 872
213. Beigman, I. L., Vainshtein, L. A. 1968, *Lebedev Inst. Rep. 104* (Moscow)
214. Peach, G. 1966, *Proc. Phys. Soc.*, **87**, 381
215. Trefftz, E. 1963, *Proc. Phys. Soc. A*, **271**, 379
216. Malik, F. B., Trefftz, E. 1961, *Z. Naturforsch.*, **169**, 583
217. Peach, G. 1968, *J. Phys. B*, **1**, 1088
218. Harrison, M. F. A., Dolder, K. T., Thonemann, P. C. 1963, *Proc. Phys. Soc.*, **82**, 368
219. Peterson, J. R. in *Atomic Collision Processes* (McDowell, M. R. C., Ed., Amsterdam: North-Holland)
220. Fite, W. L., Brackmann, R. T. 1959, *Phys. Rev.*, **113**, 815 [Other refs. in (13)]
221. Dolder, K. T., Harrison, M. F. A., Thonemann, P. C., 1963, *Proc. Phys. Soc. A*, **274**, 546
222. Hooper, J. W., Lineberger, W. C., Bacon, F. M. 1966, *Phys. Rev.*, **141**, 165
223. Peart, B., Dolder, K. T. 1968, *J. Phys. B*, **1**, 240
224. Rudge, M. R. H., Schwartz, S. B. 1966b, *Proc. Phys. Soc.*, **88**, 579
225. Liska, J. N. 1934, *Phys. Rev.*, **37**, 808 [Other refs. in (13)]
226. Korchevoï, Y., Pronski, A. 1967, *Sov. Phys. JETP*, **24**, 1089
227. Ziesel, J. P., Abouaf, R. 1967, *J. Phys. Chem.*, **64**, 695
228. Ziesel, J. P., Abouaf, R. 1967, *J. Phys. Chem.*, **64**, 702

Copyright 1970. All rights reserved

THE NUCLEI OF GALAXIES

G. R. BURBIDGE
University of California, San Diego

1. INTRODUCTION

To the observer, a nucleus can properly be defined qualitatively as a small region of high luminosity usually, but not always, symmetrically placed in the center of a galaxy. From the theoretical standpoint, dynamical considerations lead to the general conclusion that in all galaxies with symmetry there will be a concentration of mass toward the center. The studies and analyses of the rotations of galaxies with symmetry show that the central densities are always much greater than the mean densities.

The very central region, often called the nucleus of a galaxy, is frequently highly conspicuous because it is often highly luminous when contrasted with its surroundings. But the relative luminosity of the nucleus is not necessarily a direct measure of the mass concentration in the nucleus, since very different mass-to-light ratios are possible in different circumstances.

In this article I shall first review the observational evidence concerning the nuclei of galaxies. Later I shall discuss the possible origins and evolution of galactic nuclei and the ways in which their activity bears on general galactic evolution and on cosmological problems.

The term "nucleus" cannot be *precisely* defined from an observational standpoint. Direct photographs often show a bright center in a galaxy which may be described as a stellar or semistellar nucleus. The size of the object defined as the nucleus is a function of the distance of the object, if it is measured by normal optical methods where the Earth's atmosphere is a limiting factor. The smallest size that can be detected is limited by the atmospheric seeing to about 1/2″. This corresponds to about 1.5 pc in M31, about 40 pc at the distance of the Virgo cluster, and about 2000 pc for an object at a cosmological redshift z of 0.1.[1]

The nucleus of our own Galaxy cannot be seen in optical wavelengths but, as will be described, it has been investigated extensively at radio wavelengths and in the infrared. Since we are situated only about 10 kpc from this nucleus, much smaller-scale structures can be investigated directly than is the case for other galaxies.

[1] We shall assume that the distance to the Virgo cluster is 15 Mpc and that the Hubble constant $H = 75$ km/sec Mpc^{-1} though there are now some indications that it may be smaller.

Two techniques are now being exploited to measure small-scale structures. For work at radio wavelengths intercontinental interferometric methods and scintillation techniques enable structures as small as 10^{-3} sec of arc to be looked at. Thus, the radio astronomer is able to find and study nuclear structure with sizes of the order of light years or less in galaxies as far away as the Virgo cluster or even the Perseus cluster. For technical reasons we are very far from using these methods in optical wavelengths for sources as faint as the nuclei of galaxies.

The other method, which leads to circumstantial evidence strongly suggesting that very small-scale structures are present in the nuclei of galaxies and related objects, is the study of flux variations in optical, radio, and less certainly in infrared wavelengths. As will be discussed, large variations have been found to occur in times ranging from years to days in the nuclei of galaxies and related objects. Unless relativistic expansion of the regions is invoked, and this is a possibility in some circumstances, these variations show that there are structures as small as $R \lesssim c\tau$, where τ is the period of variation. This gives upper limits for R ranging from $\sim 10^{15}$ to $\sim 10^{18}$ cm.

Much can be learnt about galactic nuclei by studying their stellar compositions. Concentrations of mass in the nuclear region can exist (a) in the form of normal stars radiating energy through thermonuclear processes; (b) in the form of evolved stars such as white dwarfs and neutron stars, or in collapsed matter. Concentrations of light in galactic nuclei can lead to their becoming conspicuous if the light-to-mass ratio is very high, or if there is a large mass concentration involving ordinary stars.

In most galaxies we have little information on the mass distribution in the nuclear region, though in some cases an upper limit to the mass within a small volume can be set. However, large optical luminosities can arise because:

1. There is a large population of high-luminosity stars, including giant and supergiants in the nucleus.

2. There is a cloud of ionized gas which emits large fluxes in lines and continuum. This gas may be energetically maintained by the high-luminosity stars in the nuclear region, or it may be excited by other energy sources.

3. There is a nonthermal energy source in the nuclear region. Such sources are now known to exist in many different situations in galaxies. As will be discussed later, it is generally accepted that the radiation which is seen is generated by the synchrotron process (coherent or incoherent) and possibly by Compton scattering. This requires that large fluxes of relativistic particles, magnetic flux, and radiation fields with very large energy content be present. The characteristic frequencies at which the radiation is emitted by these mechanisms are determined by the energies of the particles and the magnetic fields, while some constraints are placed by the properties of the environment in which these mechanisms are working. So far the bulk of the observations have been made in the optical wavelength range 3000–7000 Å, and in radio frequencies between $\sim 10^7$ and 10^{10} Hz. Observations of a few

objects in the infrared, mostly between about 1 and 20 μ, have shown that nuclei are often powerful emitters in this frequency range also. This radiation must be due ultimately to the same physical processes that give rise to the optical and radio flux. Such radiation may be generated directly, but a case can also be made for its being nonthermal optical flux which is absorbed by dust and re-emitted at long wavelengths. Thus, the presence and even the production of dust in galactic nuclei will be discussed.

On the theoretical side it is necessary to ask how a nucleus forms, how it evolves in time, and in what form and at what rate energy will be released in the evolution. Finally, it is necessary to ask how important galactic nuclei are, as far as the overall properties of the extragalactic universe are concerned.

2. OBSERVATIONS OF THE STELLAR AND GASEOUS COMPONENTS IN NORMAL GALACTIC NUCLEI

In the majority of galaxies, when we observe the central regions in optical wavelengths, we are studying the stellar population. Baade, in his fundamental work on stellar populations, studied in detail the central part of M31 and the dwarf ellipticals in the local group. He pointed out that a central "stellar" nucleus was a feature of most galaxies. He described the central nucleus of M31 as a flattened system with dimensions of $2.5'' \times 1.5''$ (1), and he considered this typical of many spiral systems. It is generally understood that in spirals there are two systems, a central spherical system and a disk that exhibits spiral structure. Baade argued that the central system could be quite large (he gave as an example NGC 4594) or it could be so small that it finally appears as a semistellar point. He pointed out that a very short exposure would be required to show the semistellar nucleus of M101, for example.

In the case of the elliptical galaxies there is a smooth light distribution following the law originally found by Hubble (2) to be of the form $I = I_0(r/a+1)^{-2}$ where r is the distance from the center and a is a parameter which varies from galaxy to galaxy. However, the observers have remarked that many ellipticals have very bright semistellar nuclei. A good example of this is the nucleus of M32 with a size of about $0\rlap{.}''8 \pm 0.1$ (3).

We have already remarked that the visibility of such nuclei is a function of the distance of the galaxy, the relative luminosity of a cylinder with a diameter $\sim 1''$ of arc through the galaxy, compared with the surrounding luminosity, and the scale and exposure time of the plate. The important questions from the standpoint of the physics of the galaxies are what kinds of stars are present in the nuclei, what is the state of the gas, and what are the dynamical characteristics of each. We consider the two components in turn.

Stellar Populations

The types of stars that are present can be investigated by measuring the energy distribution, using *UBV* photometry or scanner techniques, and by obtaining slit spectra of the central regions. The spectroscopic methods have been developed by Morgan and his collaborators (4–10), Spinrad (11–13), Deutsch (14), Wood (15), and McClure & van den Bergh (16). Morgan has developed a spectral classification scheme for galaxies which can be related to the original Hubble classification scheme based on the forms of galaxies. A preliminary correlation of form with spectral type is evident from the classification made by Humason in Humason, Mayall & Sandage (17). The classification depends on one of the criteria used by Hubble in his original classification scheme—the degree of central concentration of luminosity—and this is related in detail by Morgan and Mayall to the spectral class of the central part of the galaxy based on spectra of classification dispersion ($\sim$100 Å/mm) taken in the blue-violet wavelength region. From the integrated spectra they are therefore making an estimate of the types of star which contribute to the bulk of the light between about 3500 and 5000 Å. Spinrad's approach is rather different. He is investigating the central parts of galactic nuclei at lower dispersion, extending into the yellow-red spectral range, $\lambda\lambda 4400$–6800 Å. Both groups have been concerned only with the brighter galaxies lying at distances no greater than about 15 Mpc.

It should be stated at the outset that the stellar populations that these authors deduce to be present are typical in general of a much larger *volume* than the volumes that we shall be considering in discussing the properties of the nuclei as far as nonthermal radiation and violent activity is concerned. Also, the volumes being investigated are sometimes much smaller for local-group galaxies than they are for galaxies at the distance of the center of the Virgo cluster. A typical angular size over which spectra and scanner observations of the type used by Morgan and Spinrad are obtained is about 5" of arc or greater, and the majority of the galaxies are at distances $\lesssim$15 Mpc. Thus, despite the rapid falloff in the star density in the centers of galaxies according to the Hubble law, the bulk of the light will be coming from a region outside the innermost nucleus with an angular size $\leq 1"$. Thus, if the stellar population is quite different in the innermost core, it will not be detected unless the stars give rise to characteristic spectral features which dominate in some part of the spectrum. Bearing this in mind, the results are as follows. Morgan has shown that:

1. Irregular galaxies contain large numbers of hot early-type stars and H II regions. Such bright comparatively nearby galaxies as NGC 4449 and NGC 4214 show many small bright regions—large H II regions—throughout with some concentration in the inner part. Many spirals have a large number of bright H II regions in the center. Irregular systems which fall into this category have no well-defined nucleus and thus, in a sense, they are irrelevant to this discussion. However, the type of stellar population seen in these systems is at one end of a sequence, and the next stage is what Morgan calls the galaxies with intermediate stellar population, and these do have well-defined nuclear regions.

2. These are the galaxies of Morgan's classes f and fg which contain stars

in their nuclear region classified F8 to K0. They often show [O II] λ3727 indicating that ionized gas is present. Well-known galaxies which are classified by population in this way are NGC 4303 and NGC 4321 (F8), NGC 4501 (G5), and NGC 4594 (K0). The nuclear regions range in luminosity from the comparatively faint nucleus of NGC 4303 through the increasing nuclear luminosity of NGC 4321 and NGC 4501 to the very large nucleus of NGC 4594. Morgan deduced that the mean spectral types of the nuclear regions of the majority of the bright spirals were rather similar and contain large numbers of stars of types F to K.

3. The Morgan system then leads on to the galaxies whose central regions have what he has called *the amorphous population*. He puts into this class the centrally condensed spirals like M31 and M81, and the bright elliptical galaxies. The integrated spectra have types near K0 in the blue and violet regions, near K5 in the green, and M in the red. M31 has been compared with the more distant galaxies by using spectra covering a very large angular size in M31. The conclusion is that galaxies of this type contain large numbers of giant stars with spectral types in the range G8-K3, and the M giants make a significant contribution to the red spectral region as deduced from the Ti O bands. According to Morgan the bright elliptical galaxies have similar stellar populations, though the higher-velocity dispersion of the stars in their central regions means that the lines appear to be weaker because of Doppler broadening.

Since it is not possible to make optical observations of the central region of our Galaxy directly, comparison of the stellar population in the center of our Galaxy with that in other galaxies is difficult. However, the work of Nassau & Blanco (18), who have classified stars in the relatively unobscured region near the globular cluster NGC 6522, which is at a projected distance of no more than 700 pc from the galactic center, has shown that the light is coming mainly from M5-M9 giants. These observations in the red suggest to Morgan a great similarity between the population in the nuclear region of our own Galaxy, and that in M31 and in elliptical galaxies in general.

The studies of Morgan and his colleagues are aimed at classification schemes which relate form to stellar population in the central regions of galaxies, and they have discussed the way in which the spectral characteristics, and hence the spectral types, change as one moves outward from the center.

Spinrad and his colleagues have been interested not in a classification, but in the problem of identifying as far as possible all of the different types of stars which contribute to the light emitted from the centers of galaxies. Spinrad's first investigation (12) was concerned with the nuclear regions of 32 galaxies with Hubble types Sb, Sa, S0, and ellipticals with various degrees of flattening. The sample ranged from M31 and M32 to galaxies in the Coma cluster. He concluded that in addition to the late-type giant stars, which contribute a large amount of light in the centers of galaxies of intermediate and amorphous type, a large population of dwarfs is often present. Spinrad reached this conclusion after he found that the Na I D lines were exceedingly strong in some galaxies in his sample, in particular, M31, M81, and NGC 4594, which are all amorphous spirals in Morgan's nomenclature, and after he concluded that these lines must arise from dwarf stars. To reach the latter conclusion it was necessary to show that the Na I D lines could not come from giant stars nor could they have an interstellar origin. From considering the equivalent widths of the D lines in stars on the normal giant branch, he showed that they would give rise to D lines which would only be marginally detected at the dispersion used by him. Taking reasonable parameters for the interstellar gas in the nuclei of other galaxies, it was again possible to show that the Na I D lines were not likely to arise there. This does not exclude the possibility that in particular objects the interstellar contribution could be large. This left only dwarf stars where strong Na I lines are found.

Deutsch (14) confirmed the results of Spinrad. He pointed out that the D lines are intrinsically very strong in many galaxies in the Virgo cluster. Moreover, he showed a correlation between redshifts and D-line strength in Spinrad's results, which is explained by the fact that airglow emission lines of Na I tend to fill in the galactic absorption at small redshifts, and only when the redshift is large enough will this effect be removed. Thus the estimates of D-line absorption and hence the contribution of dwarfs were underestimated by Spinrad.

These investigations were followed up by Wood (15) and Spinrad (13) who attempted to make composite models for the stellar populations in a range of galaxies. Earlier work of this type was carried out by Baum (19). Wood used narrowband photoelectric photometry in 12 colors—7 continuum colors in the range $\lambda 3400-\lambda 7300$, and 5 line indices including Na D, Mgb, Ti O, and Hα. Observations were made of 20 galaxies, many of them of the same galaxies studied by Spinrad and Morgan, and a large number of galactic field stars ranging in type from B7 to M6 were used. The relative number of stars in each class which would give a good fit to the observations was determined. Some of Wood's results are shown in Table 1. The synthesis was successful for the elliptical and S0 galaxies, and a good correlation was found between the absolute visual magnitude M_0 and the mass-to-light ratios, with the luminous galaxies containing relatively larger numbers of dwarfs. However, the total mass of these stars that can completely account for the light is only about 1/3 to 1/5 of the total mass of the galaxies obtained by dynamical methods (45). This leads to the conclusion that there may be much hidden mass in these systems. This nonluminous mass may be distributed throughout the galaxy, or it may be concentrated in the nuclear region. We shall return to this question in a later section.

Spinrad (13) and Wood (15) considered in some detail a similar model for the central part of M31. Spinrad's results are shown in Table 2. Again they found a large population of dwarfs with $M_0 = +5$ to $+12$.

The differences between the luminosity function for stars in the solar neighborhood and in galactic or globular star clusters, and those required to

TABLE 1. Models for elliptical galaxies: Comparison of nucleus of NGC 4406 with entire galaxies, according to Wood (15)

Star type	NGC 4406 nucleus	NCG 4406 E3	NGC 4552 E0	NGC 4621 E5
F V	5.00 (3)	3.63 (7)	9.54 (6)	—
G V	5.00 (4)	3.63 (8)	1.59 (8)	1.36 (8)
Early K V	6.00 (5)	4.35 (9)	2.38 (9)	3.39 (9)
Late K V	3.40 (6)	2.46 (10)	7.94 (9)	6.33 (9)
Early M V	3.00 (7)	2.18 (11)	9.54 (10)	2.71 (10)
Late M V	3.00 (7)	2.18 (11)	5.16 (10)	1.13 (11)
G IV	6.00 (4)	4.35 (8)	3.18 (8)	3.16 (8)
A–F III	1.10 (3)	7.98 (6)	7.94 (5)	2.60 (6)
G–K III	6.00 (3)	4.35 (7)	2.38 (7)	2.26 (7)
Late K III	2.00 (3)	7.25 (5)	3.42 (6)	9.03 (6)
Late M III	8.00 (1)	7.25 (3)	2.78 (5)	1.92 (5)
Baade giant	2.50 (1)	—	—	—
O–B horizontal branch	4.20 (2)	3.05 (6)	—	2.60 (6)
$(M/L)_{pv}$	12.5	14.2	10.1	7.4
Total mass	—	1.4 (11)	5.3 (10)	3.9 (10)

Numbers in parentheses are powers of 10 by which values are to be multiplied.

match the integrated spectra of Sa, Sb, and elliptical galaxies, are very great and indicate that the processes of star formation in these different locations were very different.

Also, there may be differences in chemical composition of the stars between different galaxies, and between the nuclear regions and outer parts of some systems, and these composition differences may account for some part of the difference between the integrated spectra. Three effects determine the integrated spectra: the relative numbers of stars of different types, the element abundances in these stars, and the stellar velocity dispersion, and it is quite difficult to disentangle them. McClure & van den Bergh (16) and McClure (20), using five-color observations together with the photometry of Wood (15), studied this question. They found that the CN λ4150 feature is very strong in the semistellar nucleus of M31 and they concluded that the stars in this region may have a much higher abundance of the heavier elements than the population further out. Unpublished work of Spinrad and Taylor also indicates an increase in the H/metal ratio with increasing distance from the center of M31. However, some of the spectral features used are highly sensitive to stellar luminosity, so that real abundance differences are not yet proved.

McClure & van den Bergh (16) also found a similar effect involving CN in NGC 4472. This could mean that the nuclear region of this galaxy is more abundant in heavier elements than the envelope. It also means that the cen-

TABLE 2. Preliminary model for the nucleus of M31, according to Spinrad (13)

Star type	V light (%)	Mass (%)	$\mathfrak{M}/\mathfrak{M}_\odot$	$L/L_\odot$	No. of stars
Glob. cl.	2.37	0.14	1.0×10^5	1.0×10^5	1.0 cluster
B star	2.47	0.00	5.0	1.0×10^3	1.04×10^2
Early A	0.00	0.00	2.0	4.0×10	0
Late A	0.00	0.00	1.7	1.6×10	0
Late F	0.00	0.00	1.2	2.5	0
Early G V	4.74	0.28	1.0	1.0	2.0×10^5
Late G V	4.74	0.42	0.9	0.6	3.33×10^5
K0 V–K2 V	4.08	0.43	0.7	0.4	4.3×10^5
Middle K V	6.64	0.79	0.6	0.3	9.33×10^5
Late K V	8.54	2.55	0.5	0.1	3.6×10^6
Early M V	6.64	15.87	0.4	1.0×10^{-2}	2.8×10^7
Middle M V	6.64	79.33	0.2	1.0×10^{-3}	2.8×10^8
K subgiants	6.64	0.08	1.2	6.0	4.67×10^4
G8 III	25.62	0.08	2.0	4.0×10	2.7×10^4
K0 III–K2 III	6.64	0.01	2.0	1.0×10^2	2.8×10^3
Middle K III	9.49	0.01	2.0	1.0×10^2	4.0×10^3
Late K–M0 III	2.85	0.00	2.0	1.0×10^2	1.2×10^3
Early M III	1.90	0.00	2.0	1.0×10^2	8.0×10^2
Middle M III	0.00	0.00	2.0	1.0×10^2	0
Late M III	0.00	0.00	2.0	4.0×10	0

Computed mass-to-light ratio = 16.7

tral parts tend to be bluer than the outer envelopes, so that changes in color discussed by Hodge (21) and Tifft (22, 23) may be due to differences in composition as well as to differences in stellar population.

The attempts, using low-dispersion spectra, to sort out the stellar populations in the central parts of galaxies can be summarized as follows. The small bright nuclei found in many late-type spirals are due largely to the presence of comparatively young massive stars. This is also the case for the small bright regions often described as nuclei of irregular galaxies, though in many irregulars similar stellar populations apparently exist throughout the galaxy. In the spiral galaxies with small bright nuclei there is sometimes no direct evidence for lower-mass older stars in the nuclei. However, such stars are present in the disk population and probably form a substratum also in the nucleus; the difficulty in detecting them is presumably due to the fact that the smaller number of high-luminosity stars and the gas they excite overwhelm them.

In spiral galaxies where the nuclear region is more pronounced, the bulk of the light comes from older low-mass stars, much of the light in the visible and ultraviolet region coming from G-K-M giants, a main-sequence contribution, and an increasing population of low-mass dwarfs. In galaxies with well-defined spiral arms, the galactic center population is rather different from that which gives the bulk of the light in the arms where O and B stars predominate, but the substratum of stars in the disk is largely made up of low-mass stars.

In spiral galaxies of types Sa and S0 and in the ellipticals, there is a smooth light distribution and the stars in the central region are probably representative of the whole galaxy. The stellar population is of the type given in Tables 1 and 2. However, very small structures can often be seen in the centers of both elliptical galaxies and spirals from types Sb on. These are the true nuclei. Often these nuclei can be detected because they have colors which are different from the main amorphous body of the galaxy and the presence of ionized gas is very often an important factor.

In addition to the spectroscopic investigations, many photoelectric and scanner observations of the energy distribution in galaxies have been made, including those by Stebbins & Whitford (24, 25), Pettit (26), de Vaucouleurs (27), Tifft (22), and Oke & Sandage (28), but they have been mostly concerned with the energy distribution of the total luminosity. Some of the most accurate work using photographic methods is that of Holmberg (29). Far fewer data are available for the nuclear regions of galaxies. We shall discuss the data on some individual peculiar objects later. Apart from this, the only investigations in which different regions of galaxies are sampled for a large number of systems are those by Tifft (30, 31). He restricted his investigations to a sample of Shapley-Ames galaxies north of $-16°$ and all at approximately the same distance (members of the Virgo cluster). Observations were made through a series of diaphragms with diameter ranging from $10''$ to $65''$. Both "normal" ellipticals and spirals show the same trend of gradual reddening toward the nucleus. This is presumably due to stellar population differences or abundance differences, or both, and to interstellar reddening in some cases. However, some spirals and ellipticals have small blue nuclei superimposed on normal outerparts. Tifft also believed that he found both spirals and ellipticals with abnormally red nuclei. He concluded that the nuclear reddening or bluing appeared to be an unresolved nuclear effect at the smallest scale available ($10''$ or about 750 pc at the distance of the central part of the Virgo cluster). He found that blue nuclei appeared to be surrounded by a red zone and red nuclei by a blue zone, compared with normal systems. For ellipticals he found that the red nuclei showed steep light gradients with stellar or near-stellar nuclei averaging about $3''$. The nuclei of normal ellipticals and those with significant bluing had larger diameters, $\sim 6''$, and less-steep light gradients.

While Tifft's results are still highly qualitative, his descriptions of nuclei do strongly suggest that there are a variety of forms for the light distribution and colors in the central parts of many systems. It is also interesting that he concluded that the galaxies with anomalously red or blue nuclei, as distinct from those with no measurable color differences between $10''$ and $\sim 60''$ (the range of diaphragms used), tended to be concentrated in a region with diameter of about 600 kpc in the center of the Virgo cluster.

It appears that the nuclear effects are not due to the presence of interstellar matter, either dust or gas (since they are not correlated with the presence of emission lines). It must be assumed therefore that there are differences in the stellar population or composition between the nuclear regions and the outer parts, or that other radiating sources are present.

In addition to this survey work, Miller & Prendergast (32, 33) found a small degree of reddening in the nuclear region of the elliptical galaxy NGC 3379. They found a constant value of $B-V=0.96$ for $r \geq 20''$ from the center, but near $r=0$, $B-V=1.12$. They argued that this difference was more likely to be due to a different stellar population in the nucleus than to the effects of dust at the center.

Ionized Gas in the Nuclei of Normal Galaxies

The most extensive spectroscopic investigations of normal galaxies remain those of Humason and Mayall, made in the period between approximately 1925 and 1955 in connection with the measurement of the redshift-apparent-magnitude relation and the expansion of the Universe. From these data published by Mayall (34) and by Humason, Mayall & Sandage (17), the proportion of galaxies containing ionized gas was first estimated. Since the bulk of the spectra were taken in the blue, the primary indicator for the presence of ionized gas was [O II] $\lambda 3727$. As was discussed in the previous section, the irregular galaxies and spirals of later type contain a conspicuous component of high-luminosity early-type stars, both in the nuclear regions and in the spiral arms. Associated with them, and ionized by them, is uncondensed gas, so these galaxies tend to show comparatively strong emission, both in the nuclear region and outside. An extensive discussion of the situation was given by Mayall (35). As one goes successively from these objects to the early-type spirals, S0 and elliptical galaxies, the work of Humason and Mayall showed that the frequency of emission from ionized gas (in their case the [O II] doublet) becomes rarer. Mayall (35) found that [O II] $\lambda 3727$ was only visible in about 14–18% of all of the elliptical galaxies that had been studied adequately. For ellipticals and early-type spirals, emission when present is usually found only in the nuclear region of the galaxies.

These investigations of the general characteristics and frequency with which emission features indicating the presence of ionized gas are found in galaxies of different types were followed by the work of Burbidge & Burbidge (36, 37) in the red spectral region, using Hα and [N II]$\lambda\lambda 6583, 6548$ as primary indicators for the presence of ionized gas. For the spiral and irregular galaxies they confirmed and extended the early investigations of Mayall, showing that the irregular galaxies have the highest proportion of ionized gas, which is spread throughout their volume, and that this proportion decreases steadily through the barred spirals of types SBb and SBc and normal spirals of types Sc, Sbc, Sb, and Sa. The nuclei of many Sc and Sbc galaxies appear to contain *less* ionized gas than do the spiral-arm regions. The most striking observational result is that between the outer parts and the nuclear regions in many of the spiral galaxies, the ratio of the strength of Hα to [N II]$\lambda 6583$

decreases from values ~3 to values of 1 to 0.1. Among the spirals the reversal of the intensity ratio is more common among the Sb and Sa galaxies with large prominent nuclei, but it also appears in some Sc galaxies with small nuclei.

For the elliptical, S0, and Sa galaxies, Burbidge & Burbidge (37) studied a larger sample of galaxies (85 in all) in which Humason had indicated that [O II]λ3727 is present in the nucleus. Usually in the elliptical and S0 galaxies only [N II]λ6583 is seen. Among the galaxies showing emission lines in the red, the intensity ratio Hα/[N II]λ6583 is $\lesssim 1$ in all of the ellipticals studied, in 81% of the S0 galaxies, and in 55% of the spirals. The early investigation had shown that it was $\gtrsim 3$ in all of the irregulars.

Two possible explanations were offered for this effect. The most plausible was thought to be that in the nuclear regions where the K giants provide most of the radiation the gas is heated to electron temperatures of 10000–20000°K (38) and the energy fed into the gas has a kinetic origin (stellar winds, etc). Alternatively, it was suggested that the N$^+$/H$^+$ ratio was higher in the nuclei of galaxies than in the outer parts. In addition, Morgan & Osterbrock (10) suggested that the changes in the [N II]/Hα ratio are due to the state of ionization in nitrogen and not to temperature or abundance effects.

The question was reconsidered by Peimbert (40) who studied in detail the emission in the nuclei of M51 and M81. In these nuclei not only Hα, [N II]$\lambda\lambda$6548, 6583, and [O II]λ3727 are seen, but also [O I]λ6300, [O III]$\lambda\lambda$5007, 4959, Hβ, and Hγ. Using quantitative measures of the line intensities, and particularly the lines from three stages of ionization of oxygen, Peimbert concluded that collisions by thermal electrons were not responsible for the ionization. This, coupled with the data on the Balmer decrement, led him to the idea that radiative excitation was responsible. He investigated the possibility that blackbody radiation from hot stars was responsible for the excitation, reaching the conclusion that such radiation could be supplied by only ~50 stars with R~10 $R_\odot$ and T_{eff}~40000 degrees, or ~10^4 horizontal-branch stars with R~0.7 $R_\odot$. A comparatively weak ultraviolet nonthermal source could also explain the observations. Peimbert also derived masses of ionized gas in the range $3 \times 10^3 - 8 \times 10^5$ $M_\odot$ in the nucleus of M51 and $2 \times 10^4 - 10^5$ $M_\odot$ in that of M81 and showed that large density fluctuations are probably present. The presence of [O I]λ6300 means that much neutral gas is also present. He concluded that the N/H ratio might well be 2 to 6 times higher than the normal (solar) abundance ratio, and that this probably explains the high [N II]/Hα ratio in the nuclear region. Alternative explanations—either higher electron temperatures or lower ionization—would give [O II] lines much stronger than observed in these objects. This is the only quantitative study of the line intensities and the ionized gas in nearby spirals.

We pointed out earlier that the frequency of occurrence of emission lines in the nuclei of normal S0 and elliptical galaxies is comparatively low, and there are very few ellipticals or S0 galaxies in which emission lines other than the comparatively low-excitation Hα, [N II], and [O II] lines are seen. Minkowski & Osterbrock (41) discussed the problem of estimating the electron density in ellipticals in which [O II]λ3727 is present in the nucleus, using the fact that the intensity ratio of the two components of the doublet is proportional to the electron density. They applied the method to NGC 1052, which is an elliptical with abnormally strong nuclear emission (they also measured [O III]λ5007) and Osterbrock (42) studied NGC 4278, which is an E1 galaxy showing Hα, Hβ, [N II]λλ6548, 6583, [O II]λ3727, [O III]λ5007, and [Ne III]λ3869 in the nucleus. They found that the mass of ionized gas responsible for such emission is at most 10^6 $M_\odot$ and concluded that the ionization is caused either by ultraviolet radiation from stars or by the dissipation of kinetic energy associated with the stars and gas in the nuclei of these ellipticals.

To summarize, ionized gas is found in the nuclear regions of all types of galaxy. It is always present in late-type spirals and irregular systems, but as one goes along the Hubble sequence it is less frequently detected. The amount of gas present is in general not known, but in the few galaxies in which mass estimates have been made it has been shown that comparatively small masses ($\lesssim 10^6\, M_\odot$), compared with the total mass in the nucleus (Section 3), can explain the observed emission-line intensities. The mechanisms of excitation in irregular and Sc spirals on the one hand, and ellipticals and S0s on the other, are likely to be different. In the systems with many high-luminosity O and B stars and much uncondensed matter the ionization in the nuclear regions, as elsewhere, is likely to be ultraviolet radiation from the hot main-sequence stars. However, in the ellipticals and S0 and in nuclei of Sb galaxies such as M51 and M81, the ionization and continued energy input is likely to be due to the high kinetic energy of the stars and gas ejected from them (see Section 3) and also possibly due to ultraviolet radiation from horizontal-branch stars or other sources. The differences in the intensity ratio Hα/[N II] between the nuclei and outer parts may be due to abundance effects or to higher electron temperature but more probably the former.

NEUTRAL GAS IN THE NUCLEI OF NORMAL GALAXIES

For the majority of the galaxies we have no direct information concerning the presence of neutral gas in the nuclei, though the presence of dust shows indirectly that much comparatively cool matter must often be present. Only for our own Galaxy, M31, and other nearby spiral and irregular galaxies have studies using 21-cm techniques been made, and they will be discussed individually in the later sections. The problem is, first, one of detection of 21-cm radiation from an external galaxy. Very few galaxies at a distance greater than that of the center of the Virgo cluster have so far been detected as 21-cm sources. Secondly, the distribution of neutral atomic hydrogen and the question of how much neutral gas is concentrated in nuclei can be investigated only in systems with angular sizes large compared with the antenna beam size, and this has severely restricted what can be done.

Attempts to detect neutral gas in nearby ellipticals have not been successful. A limit has been set for M32 by Wentzel & van Woerden (43) showing that $M_H \lesssim 2.5 \times 10^7 \, M_\odot$. Other ellipticals have been observed (44), but no neutral hydrogen has been detected.

3. THE DYNAMICS OF STARS AND GAS IN NORMAL GALACTIC NUCLEI

Studies of the spectra of stars and gas making up the nuclei of galaxies not only give information about the composition of the nuclear regions, but measurement of the shifts and the broadening of lines give information about the motions of stars and gas. Many investigations have been made to measure the rotations of spiral and irregular galaxies, and a number of studies have been made to measure the random motions of the stars and the gas in the nuclei of systems that are not flattened. In both cases the original idea was to use the results to determine the masses of galaxies from rotation curves or by the virial theorem (45). Most of the observations have been made in the last decade, and in the course of the systematic investigation of many galaxies it became clear that in the highly flattened systems, not only is there rapid rotation, but noncircular motions are frequently present in the nuclei, and in some cases in the outer parts also. Only a few elliptical galaxies have been studied at dispersions high enough so that the random motions of the stars in the nuclear region have been measured with reasonable accuracy. These random motions range from about 100 km/sec $= \langle v_r^2 \rangle^{1/2}$ in M32 to about 490 km/sec in M87 (39). Here v_r is the radial-velocity component.

In addition to these investigations, the nuclei of a number of comparatively nearby galaxies have now been studied, using high resolution and moderate to high spectroscopic resolution. We consider these in turn as they give the best information available at present for a variety of different types of galaxy.

M31.—A high-resolution spectroscopic study of the nucleus of M31 was carried out by Lallemand, Duchesne & Walker (46), using the Lallemand image tube at the coudé focus of the Lick 120-inch telescope. They found that the image of the nucleus was about 4".4, corresponding to a diameter of about 15 pc. They found that the absorption lines Ca II H and K were inclined. On the assumption that the inclination is due to rotation of the nucleus, they determined a rough rotation curve, which reaches a maximum at a distance of about 2" (6 pc) from the center and is roughly linear. The position angle of the slit was chosen to lie very close to the position angle of the major axis of M31 (the image rotator was not used for technical reasons), and correcting the observed velocity from the relation $V_c = V_o \sec 15° \sec \theta$

where sec 15° represents the correction to the plane of M31, and θ is the difference between the position angle of the slit and that of the major axis, they obtained a maximum rotational velocity of 87 km/sec. From this they determined the mass of the nucleus to be about 1.3×10^7 $M_\odot$. The tilt of the absorption lines decreased beyond about $2''.2$ (6.5 pc) from the center and this indicated that there was a minimum in the rotation curve beyond about $6''$ (18 pc) from the center. The luminosity of the nucleus was estimated from the work of Hubble (47), and in addition Lallemand et al made photoelectric measures. From these they obtained a mass-to-light ration in the blue of 3–6.

A detailed study of the luminosity profile of the nucleus of M31 was made by Kinman (48). He showed that the luminosity distribution could be fitted to a model involving 33 homogeneous spheroids. He showed that if the stellar orbits are circular, the theoretical rotation curve for the nucleus cannot be fitted with the observed one. Neither can the assumption of isotropic or ellipsoidal velocity dispersions fit both the luminosity and rotation data. However, he concluded that an isotropic velocity dispersion might exist in the nucleus giving a value of M/L of about 50. He suggested that the orbits might be mildly eccentric.

A study of the energy distribution in the central parts of M31 and extending into the near infrared (2.2 μ) was made by Sandage, Becklin & Neugebauer (49). They showed that there is a pronounced variation of $U-B$ across the central $\pm 40''$ of M31, a variation not found in $B-V$ or $V-R$. This change in color is similar to that found by Tifft, discussed earlier, and de Vaucouleurs (27), for more distant galaxies, but the size of the region is very much less.

It has been suggested, both from these data and from the dynamical arguments given earlier, that the nuclear region is really separate from the remainder of the galaxy—that it is somewhat like a gigantic globular cluster with a stellar population somewhat different from that in the outer parts, and that stars in this region have orbits which confine them there.

Sandage et al compared the central brightness of M31 at 2.2 μ with the comparable nuclear region in our Galaxy (over a dimension of ± 13 pc) and found that for M31 it is 2.2×10^{-28} W/m² Hz⁻¹, while for our Galaxy it is about 2.4 times greater. The intensity profiles over the central ± 400 pc are similar. We turn next to the evidence for noncircular motions.

An important question is whether Lallemand et al were really measuring rotation when they measured the tilt of the absorption lines in the nucleus. A gradient in a spectrum line can be interpreted as due either to rotation, or to radial motion inward or outward. While it is not likely that motion other than rotation is present in the stellar component, a check on this would be to obtain spectra along the minor axis where, if only circular motions are present, there should be no tilt of the lines. This has not been done.

Münch (50) has investigated the motion of gas in the nuclear region by looking for the [O II]λ3737 doublet at coudé scale and dispersion (67Å/mm) with the Hale telescope. He obtained spectra both along the minor axis and in other position angles. The doublet was found to be clearly resolved, showing that the velocity dispersion in the ionized gas is considerably smaller than that of the nuclear stars (the velocity dispersion in the line of sight of the stars is about 220 km/sec). In the very center it was not possible to see the lines against the stellar continuum, but a little further outside, major departures from circular motion were found. They range from ~100 km/sec at 300 pc in the south following side to −30 km/sec at 200 pc from the center in the north preceding side. Münch concluded that the observations were best explained as due to expansion in all directions but confined to the equatorial plane. He concluded that the gas flow from the center might amount to ~1 $M_\odot$/year.

An extensive study of the velocity field in M31 was recently completed by Rubin & Ford (51). They measured emission lines produced in the ionized gas, and found a maximum in the plot of rotational velocity against distance from the center at about 600 pc and then a steep drop, followed by an increase again. This form of the curve appears to indicate that an annulus of ionized gas with low angular momentum is present in this region and that this is matter which has been ejected from the center.

M32.—Although M32 and NGC 205 are dwarf systems, they are the only ellipticals close enough so that their nuclei can be investigated with good scale and high dispersion. Walker (52) studied the nucleus of M32 in the same way as was done for M31 (46). Again Walker found a tilt in the absorption features in a position angle close to the major axis, which he interpreted as due to rotation. The curve shows an approximately linear increase out to a distance of about 2″.5 (7.5 pc) from the center and then a decrease to a distance of about 9″ (27 pc). The maximum velocity measured was about 65 km/sec. This gave a mass for the central nucleus of about $10^7 M_\odot$, a central density ~$3-6\times 10^3 M_\odot/pc^3$, and a mass-to-light ratio (photographic) ~$1-2$. Walker concluded that, since the nucleus shows rapid rotation which decreases sharply outside the bright semistellar center, the nucleus is dynamically separate from the main body of the galaxy. However, as for M31, observations in more position angles are required to prove that only rotation is present. If pure rotation is all that is present, the similarity between the masses and dynamical properties of M31 and M32, despite their very different overall properties, is of considerable interest.

There is no evidence for any gas, either neutral or ionized, in the nucleus or elsewhere in M32.

M33.—Since this galaxy is the nearest Sc system, it is important to investigate its nucleus in as much detail as possible. Walker (53) has done this by the method used for M31 and M32. The nuclear region consists of a compact cluster of blue stars with a diameter of about 5″ ($\simeq 17$ pc). The spectral type was given as A7 by Mayall & Aller (54) and the color is $B-V = +0.65$

(55). Walker confirmed that the spectral type is A5–A7. He could only see Ca II K and the hydrogen lines of his spectra, and they showed no tilt, so he concluded that the rotation was less than $\sim$10 km/sec with respect to the center. He pointed out that this might suggest either that there was no rotation, or that angular size of the nucleus was smaller than the seeing disk. At the coudé focus the nucleus did not appear much larger ($1\frac{1}{2}''$) than the seeing disk (1''). In an extensive study of the velocity field in M33 by Carranza et al (56), they found the rotation in the nuclear region to be no greater than 10 km/sec per 100 pc, and found no noncircular motions there.

M81.—Münch (57) pointed out that the nucleus has an angular size of about 5'' ($\simeq$80 pc). The integrated spectral type is K. It contains gas as shown by the presence of [O II] λ3727 doublet (which is resolved) and [Ne III] λ3869. The lines are tilted. From the intensity ratio of the [OII] doublet Münch found that the electron density was about 10^3 cm^{-3}, so that if it is homogeneous, the mass of ionized gas is about $10^4 M_\odot$. The random motions of the gas $\lesssim$75 km/sec, but from the widths of the absorption features Münch concluded that the random motions of the stars were more than twice this. A more detailed study of the ionized gas was carried out by Peimbert (40); his results were described in the previous section.

M51.—The nuclear region of M51 is complex. Inside about 15'' from the center there appear to be many H II regions, together with a number of thin dust lanes (58, 59). There is a semistellar nucleus with a diameter of about $2''.7$ ($\approx$55 pc) (59). The excitation conditions in the gas and estimates of its mass have been discussed by Peimbert (40), and have been described in the previous section. No detailed information is available on the stellar component of the nucleus, though classification of a spectrum by Burbidge gives a type of G8 III plus a considerable contribution of main-sequence F. By analyzing a large number of spectra taken at many different position angles through the nucleus and, in general, out to a distance of about 600 pc from the center, Burbidge & Burbidge (59) showed that complex patterns of noncircular motion are present in the ionized gas. The size of the motions is $\sim$100 km/sec in the line of sight. While interpretation is very difficult, it appears that gas is being ejected from the nucleus asymmetrically in the form of jets or streams that are present in some quadrants in the plane of the galaxy, but also below and above the plane. As Peimbert has shown, very considerable masses of gas are present in the nucleus. Thus, large kinetic energies are present in ejected gas, even in this comparatively "normal" nuclear region. Recent observations by Carranza, Crillon & Monnet (60) have confirmed the presence of large noncircular motions in the nuclear region.

NGC 253.—This comparatively nearby Sc galaxy has been studied by Burbidge, Burbidge & Prendergast (61) and by Demoulin & Burbidge (62).

There is a bright region in the center that is small and elongated. However, there is no well-defined semistellar nucleus. Instead there is a group of intense H II regions. Emission lines from the ionized gas in the nucleus on spectra taken at many different position angles within 5″–10″ of the geometrical center (i.e. within about 50–100 pc from the center) showed a velocity of approach of about 120 km/sec with respect to the systemic velocity, suggesting that gas was flowing out of the center. The most probable geometrical interpretation of the measures just outside this region was that a cone of gas emerges from the nucleus, out of the equatorial plane at an angle intermediate between the axial and radial directions. From direct photographs it is obvious that much dust is present in the central region. There is evidence for noncircular motions also in the spiral arms, and these are likely to be related to the outflow from the nucleus. The galaxy is a weak radio source of the core-halo type. The core component centered about the nucleus has dimensions 0.7′×0.7′ with a power of 2.5 fu at 1425 MHz. The radio source has been studied by Fomalont (63) by Gardner, Morris & Whiteoak (64), and by Ekers (65).

NGC 4939.—Recent observations of the central part of this galaxy (66) have shown quite unexpectedly that very close to the nucleus, and close to the minor axis, a large and presumably massive cloud of ionized gas appears to be moving outward with a velocity in the line of sight of 700 km/sec relative to the center. This again is preliminary evidence for large-scale activity in the nuclear region.

NGC 1097 and *NGC 1365.*—The dynamics of the ionized gas in the nuclei of these two southern barred spirals have been investigated by Burbidge & Burbidge (67) and Burbidge, Burbidge & Prendergast (68). Morgan (4) pointed out that they both have hot spots (highly luminous H II regions) in their nuclei. Since they are at distances of about 20 Mpc, nuclear structure extending over $\pm 10″$ of arc from the center has dimensions ± 1 kpc. Both nuclei show very large tilts in the emission lines Hα and [N II] $\lambda 6583$, and these were first interpreted as due to rotation alone. In this case it amounted to about 550 km/sec between one side and the other in NGC 1097, and 810 km/sec in NGC 1365. This led to the conclusion that large masses must be present in the nuclear regions. However, when a more detailed study of NGC 1365 was made, by taking spectra across the nucleus in many different position angles, it was shown that noncircular motions must be present. It was further shown that a velocity field that is a linear function of position— this includes solid-body rotation, pure axisymmetric expansion, linear shear, or a combination of all three—was ruled out by the observations. It was concluded that an enormous variety of streaming motions and z motions might be compatible with the observations.

In some barred spirals, but not in these galaxies, some evidence suggests that matter may be flowing along the bar and off the ends. There may be, in general, outflow from the center, as is suggested by the observations of NGC 1365.

NGC 3504.—The rotation of this galaxy, which is intermediate in type between SBb and Sb, was studied by Burbidge, Burbidge & Prendergast (69). The nucleus is very bright and well defined with a diameter of just over 10″ (about 1 kpc). The emission lines are strong in the nucleus; $H\alpha$, $H\beta$, $H\gamma$, [S II] $\lambda\lambda 6716, 6731$, [O III] $\lambda 5007$, and [N II] $\lambda\lambda 6548, 6583$ are seen and are only visible over a diameter of about 5″, and there is a sharp drop in brightness outside this region. The rotation curve obtained by measuring these lines is very steep, and the mass out to about 600 pc turns out to be $2.5 \times 10^9 M_\odot$. The visible matter in the nuclear region largely consists of high-luminosity stars and hot gas.

NGC 4245.—This is a bright Sc galaxy. While no detailed modern studies of the nuclear region have been made, nearly 50 years ago Lampland (70) discovered secular changes in the form and structure in its nuclear region and small changes in the brightness of the nucleus. Walker (71) re-examined Lampland's plates and made new observations. He confirmed the variations reported by Lampland and concluded that a stellar object about 2″.5 (200 pc) from the center may be responsible for some of the changes. Walker suggested that this object might have some similarities to a quasistellar object.

NGC 3310.—The nuclear region of this Sb galaxy has been studied by Walker & Chincarini (72) who published very good pictures of the nuclear structure. The nucleus is very bright with a diameter of only about 2″ (≈ 130 pc). The spectrum contains the Ca II K line and the Balmer series in absorption, together with [O II] $\lambda 3727$, [Ne III] $\lambda 3638$, [O III] $\lambda\lambda 4959, 5007$, and the Balmer series in emission. Velocity gradients are seen, but Walker & Chincarini were not able to decide whether the velocity field could be explained as due simply to rotation, or whether noncircular motions are present. However, the center of rotation of the galaxy does not appear to coincide with the nucleus as defined by the light distribution.

This concludes the discussion of the velocity field and dynamics of gas and stars in the few nuclei in comparatively normal galaxies, some quite close to us, which have been investigated. In nearly all cases there is some evidence either that there is a small dense nucleus, dynamically separate and possibly rotating rapidly, or that gas and dust appear to be flowing out of the nuclear region. And the galaxies discussed here are a pseudorandom sample, investigated either because they are in the vicinity of our Galaxy and thus can be studied with good scale and resolution, or for other reasons. The sample has not been chosen with nuclear activity in mind. In addition to the galaxies described here, many have been found to have strong narrow tilted emission lines in their nuclei. In view of the evidence given here, it appears that in many cases these tilts may be due to gas outflow as well as to rotation.

4. THE NUCLEUS OF OUR GALAXY

While this is the nearest galactic nucleus, it is not possible to investigate it in optical wavelengths, and investigations have been restricted to radio, infrared, millimeter, and γ-ray wavelengths. No X-ray source has been found at the galactic center. The most important studies are those made at radio and infrared wavelengths. We now summarize what has been deduced from these observations.

CONTINUUM OBSERVATIONS

The galactic nuclear region contains both a powerful continuum radio source, and 21-cm and OH-line radiations. Investigations using the lines show a complex velocity field in the neutral gas. We discuss first the radio continuum data.

The continuum source was shown to be discrete by Piddington & Minnett (73) who named it Sagittarius A. Further investigations by Haddock, Mayer & Sloanaker (74), Drake (75), Pariiskii (76), and Biraud, Lequeux & LeRoux (77) showed that the strong component has a size of about 3', or about 10 pc at a distance of 10 kpc. There are also several weaker components in this region. The early work was surveyed by Burke (78), but the most detailed recent investigations are due to Downes & Maxwell (79), Maxwell & Taylor (80), and Thompson, Riddle & Lang (81) [see also the review by Lequeux (82)]. The central structure appears to have a form as follows. There is a core source whose size is not known. This has a spectral index $P(\nu) \propto \nu^{-\alpha}$, with $\alpha \simeq 0.25$, i.e. it is a very flat source embedded in a larger region with a spectral index $\alpha = 0.7$. This source is remarkably smooth, and the half-power diameter is $2' \times 3'$. The lunar occultation data (80, 81) show little significant structure down to sizes of 20" ($\sim$1 pc). In the region surrounding Sgr A there are a number of other sources with angular diameters of 5' to 10'. These all appear to have spectral indices indicating that they are thermal sources arising in H II regions. Downes & Maxwell have estimated that about $10^6 M_\odot$ of ionized hydrogen must be present within about 1° or 100 pc of the center. Mezger & Höglund (82) and Reifenstein, Wilson, Burke & Mezger (84) have detected the hydrogen recombination line 109α in several of these sources, thus confirming that they are H II regions. Cameron (86) has shown that the H II regions found in this way are arranged in much the same way as the H II regions, called hot spots by Morgan, found in the nuclear regions of many external galaxies. Evidence for jets of continuum radiation at a wavelength of 20 cm, lying at angles to the plane and apparently originating from the center, has been given by Kerr & Sinclair (85).

The energetics of the synchrotron source with a spectral index of 0.7 have been investigated by Downes & Maxwell. They have made the usual equipartition calculations and conclude that $5\times10^{-5}<H<5\times10^{-4}$ G and 3×10^{48} ergs $<E_{tot}<3\times10^{50}$ ergs. The equipartition condition may not be fulfilled. If it is, this means that the electrons have energies of order a few GeV, and half-lives $\sim10^4$–10^6 years.

Observations in the infrared and millimeter wavelength range have now added considerably to the information available from continuum radio studies. A measurement at 3.3 mm (87) suggests that the source has a size $<1'.6$ (<5 pc) at this wavelength, and the flux falls below the extrapolation of the spectrum at longer wavelengths.

Measurements of the center region in the range 1.5–1500 μ have been made by Becklin & Neugebauer (88, 89) and by Low, Kleinmann, Forbes & Aumann (90) while Hoffmann & Frederick (91) made a measurement near 100 μ with a detector with comparatively poor angular resolution. Becklin & Neugebauer (88) established that the radiation between 1.5 and 4 μ originates in a source whose position and size agree very well with that of Sgr A. The diameter of this dominant source is about 5'. Within it and centered on it they found a pointlike source. There also appeared to be a more extended background and some discrete extended sources. A large part of the radiation from 1 to 3.4 μ is very likely to be due to starlight which has been reddened by dust and to very cool stars radiating directly. The distribution is similar to that found in M31 (see previous discussion). However, the dominant source coincident with Sgr A is a very powerful source in the far infrared, and it appears likely that either it is due to a very dense dust cloud irradiated by a central ultraviolet source of nonthermal origin, or else it is nonthermal radiation emitted directly by the synchrotron process. The spectrum is similar to that found in the nuclei of Seyfert galaxies, but the power emitted is much less. The flux radiated between 5 and 25 μ appears to come from a source $\sim15''$ (0.7 pc) in diameter, and the power emitted is $6\times10^{39}-3\times10^{40}$ ergs/sec. The flux radiated between 40 and 350 μ comes from a region less than 10 pc in diameter and the total luminosity in this wavelength range is about 3×10^{41} ergs/sec. Earlier, Hoffmann & Frederick had detected a much more extended source at 100 μ, with an extension along the galactic plane of 6.5° and a total flux of 2.7×10^{42} ergs/sec. Low et al did not see this extended component; they also attempted to measure the flux at 1250 μ and, while they did not detect it, a limit could be set.

The infrared measurements of the central component, taken all together, suggest that within 0.7 pc a powerful nonthermal source is operating. The limit at 1250 μ requires that at some wavelength >100 μ, a maximum is reached, and the flux drops thereafter. The most plausible view is that this is due to synchrotron self-absorption. Since the source is resolved at a size of 0.7 pc, application of the well-known self-absorption condition leads to the conclusion that if the source were a single coherent object, its magnetic field must have a strength in excess of 10^{20} G (92). Since this is quite unreasonable, it suggests that the source is made up of a number of very small components

(90, 92), each with a strong ($\sim$100 G) magnetic field situated in a magnetosphere around it, and each generating particles continuously. A similar situation may be operating in galactic nuclei in general. If we compare the infrared flux emitted in the central source in our Galaxy with that emitted by typical Seyfert nuclei in comparable wavelengths (cf Section 4), we see that the Galactic nucleus is really a miniature Seyfert nucleus operating at a power level $\sim 10^{-2}$–10^{-3} of a classical Seyfert nucleus. As far as the radio properties of Sgr A are concerned, a similar situation prevails. These results strongly suggest that the Galactic nucleus may be emitting optical and ultraviolet synchrotron radiation at a power level $\sim 10^{38}$ ergs/sec.

In quite a different energy range, γ radiation with photon energies $\gtrsim 100$ MeV has been detected from the galactic plane in the direction toward the center (93). The published intensity is 5×10^{-4} photon/cm^2 sec^{-1} rad^{-1}, though recalibration suggests that the flux may be a factor $\sim$3 lower. No more than a small fraction of this flux can come from the immediate vicinity of the galactic nucleus. It appears to come from a rather large volume in the disk with a size of many kiloparsecs and to be generated by π^0 decays following p-p collisions, or electron bremsstrahlung, or to come from a number of widely distributed unresolved discrete γ-ray sources, which might be highly condensed objects generating relativistic particles.

Line Radiation

From the time of its discovery the 21-cm line has been a powerful tool in studies of galactic spiral structure and in investigations of the central region of the Galaxy. Some of the early work was surveyed by Burke (78). Work by the Dutch group (94–96) showed that large expansion motions are present in the central region. Rougoor (96) suggested that the data were best explained in terms of a flattened rotating disk with a radius of about 750 pc, and a rotational speed of about 200 km/sec at its outer edge. Much further away, at about 4 kpc from the center, is the so-called expanding arm moving outward at about 50 km/sec, and there is also another arm beyond the center with an expansion velocity of about 135 km/sec.

More detailed studies (97–99) show considerable evidence for motions outward, at angles to the plane of the Galaxy, in the form both of isolated features that are presumably clouds, and of more extended structures. The total mass of neutral hydrogen lying within the central region has been estimated to be about $\gtrsim 5 \times 10^7 M_\odot$ within about 4 kpc (96, 99), and a significant fraction of this, perhaps 70%, is moving with velocities $\gtrsim 50$ km/sec (line-of-sight component) relative to the local circular velocity field. This means that the total kinetic energy of ejected gas in the central region $\gtrsim 10^{54}$ ergs.

A large body of evidence now suggests that gas is flowing into the Galaxy in the form of high-velocity clouds at high galactic latitudes (100–103). This may be intergalactic gas (102), more distant neutral clouds, or gas originally ejected from the center or from vast explosions in the plane.

Studies of the neutral gas in the region of the center have also been made,

390   BURBIDGE

using the OH lines (cf 104, 105). There is no easy way of comparing OH features with neutral hydrogen data, and from these features there appears to be no evidence for the rapidly rotating disk found from the 21-cm measurements. However, Robinson & McGee (104) have pointed out that radial motions found using the OH lines are at least as important as the rotational components. Presumably OH absorption features seen against Sgr A arise in dense regions where molecules are plentiful, and the velocity structure of these regions is quite different from that found in the lower-density clouds of atomic hydrogen.

Gravitational Radiation

Recently Weber (106, 107) claimed to have detected gravitational waves at a frequency of 1660 Hz that he believes are radiated from the galactic nucleus. If this is correct, the rate of mass loss in this frequency band is about 0.2 $M_\odot$/year. His bandwidth is only 0.017 Hz. Since it would be reasonable to assume that the radiation is emitted over a broadband, his results suggest mass-loss rates $\sim 10^2$–$10^3 M_\odot$/year. While his results are doubted by many [see the experimental difficulties raised by Beron & Hofstadter (106)], various speculations about their implications have been published (108, 109). If correct, the result is fantastically important since it means that, unless matter is continuously created in a galaxy throughout its life, the mass may be reduced drastically in a time comparable with, or short compared with, a Hubble time, if the mass-loss rate remains at this level. The orbits of gas and stars about the center will tend to expand, and Field, Sciama & Rees (108, 109) have discussed the compatibility of this result with observation.

As will be seen as we discuss the observations of other galaxies with active nuclei, the nucleus of our Galaxy is apparently very similar in its nuclear properties to many active galaxies, though the scale of activity is lower than it is in the more spectacular cases. The implication of this general result will be discussed in the theoretical sections. However, in view of the similarity of the nucleus of our Galaxy to the nuclei of other systems, we can predict that it will eventually be found to be variable in the high-frequency radio range, and possibly in the infrared as well.

5. GALAXIES WITH HIGHLY ACTIVE OR EXPLOSIVE NUCLEI

There are now known to be many galaxies in which violent activity is present in their nuclear regions. The first general discussion of the observations that pointed to this from radio and optical data was given by Burbidge, Burbidge & Sandage (110). We now have further evidence from infrared measurements, and more objects are known. We discuss the data on the best-studied systems.

M82.—This is the nearest galaxy in which violent activity is seen. Its distance is 3.2 Mpc, so that $1'' \approx 16$ pc. The original suggestion that a large-scale outburst had taken place in M82 came from the work of Lynds &

Sandage (111) following the discovery of the small, weak radio source in this galaxy. The dimensions of this source are about $45'' \times 45''$ (112). Lynds & Sandage showed that a large amount of gas had been ejected above and below the plane and that the outburst must have started about 10^6 years ago. They and Sandage & Miller (113) also found evidence for a very high degree of optical polarization in the filaments, which led them to conclude that the outburst had given rise to optical synchrotron radiation. Further studies of the outer filaments have been made by Sandage & Visvanathan (114). Elvius (115, 116) has also made polarization measures, but has argued against this interpretation. Extensive studies of the velocity field in the exploding cones of gas were made by Burbidge, Burbidge & Rubin (117). Since the galaxy is filled with dust, it is not possible to detect nuclear structure at ordinary optical wavelengths. However, Solinger (118) suggested that the matter which has been ejected, and the polarized optical radiation, result from the presence of a nucleus of Seyfert type which is obscured by dust. Solinger attributed the polarization of the filaments to electron scattering of optical flux from the nucleus. Elvius tried to attribute it to dust. However, the optical synchrotron explanation still appears the most plausible. In view of Solinger's suggestion, van den Bergh (119), Bertola, D'Odorico, Ford & Rubin (120), and Raff (121) have all tried to investigate the central region by photographing it in the near infrared. Optical spectra are also available from Burbidge et al (117) and Bertola et al (120). From this work it was found that in two respects the nuclear region is not of the classical Seyfert type. There is no evidence for a stellar nucleus nor is there any evidence for very broad emission lines. However, a small condensation with an approximate size of $3'' \times 8''$ on a 7500 Å plate was found by Bertola et al to become more pronounced going from the blue plate to the 10,000 Å plate. They concluded that this feature may be the nucleus or part of it. It coincides with the center of the explosion determined by Lynds & Sandage (111) and the center of the magnetic field derived from the general pattern of electric vectors obtained from the polarization measures of Elvius. Peimbert & Spinrad (122) have done accurate photometry of the nuclear region as defined by Burbidge et al (117). They concluded that interstellar absorption of 4.2 mag at Hβ is present. By studying the intensities of a number of emission lines they also concluded that the ionization is likely to be due to ultraviolet radiation. They found that the total luminosity within $10'' = 32$ pc is comparable to that of Seyfert nuclei, and the Hα emission is 6.9×10^{41} ergs/sec. The nuclear region is also a very strong infrared source (123). That there is a substantial difference in the size of the Balmer jump between the nucleus and the regions outside is probably due to differences in stellar population in the sense that the nuclear region has an earlier spectral type (earlier than B1).

Fairly large departures from circular motion of the ionized gas are found throughout the galaxy (117), and it is believed that the disturbance giving rise to these originated in the nucleus.

All of the evidence points to a violent outburst in the nucleus of M82 which was active as long as 2×10^6 years ago. However, the tremendous amount of dust in the galaxy obscures the central part and makes detailed investigation of the nuclear region in optical wavelengths very difficult.

M87.—This galaxy is the nearest giant elliptical (at a distance of 15 Mpc) in which direct evidence of violent nuclear activity is found. The evidence is of several different kinds: 1. The existence of the optical jet first discovered by Curtis (124). 2. The discovery that M87 is a powerful nonthermal radio source. 3. The physical conditions in the ionized gas investigated, using the [O II] $\lambda 3727$ line. 4. The discovery that the galaxy contains a powerful X-ray source.

The sequence of events that led to the realization that very violent activity was taking place in the nucleus of M87 was as follows. Bolton, Stanley & Slee (125) first identified M87 with a powerful nonthermal radio source which was then studied optically by Baade & Minkowski (126). Shklovsky (127) proposed that the optical jet was due to synchrotron radiation, and he predicted that it would show a high degree of linear polarization. Baade (128), using photographic methods, found that the light was highly polarized. This result was confirmed by Hiltner (129) who made photoelectric measurements. A detailed theoretical investigation of the properties of the jet was then made by Burbidge (130) who showed that the energy content of relativistic particles and magnetic field in the jet must be very high, and that if the mean magnetic field is $\sim 10^{-3}$–10^{-4} G the electrons must have energies $\sim 10^{11}$–10^{12} eV. This means that their lifetimes must be very short ($\lesssim 1000$ years) so that they could not travel from the nucleus to the end of the jet (a distance of about 2000 pc) without dissipating their initial energies. [See also the later investigations of Bless (131).] This analysis led to the suggestion that either the particles are continuously accelerated in the jet, or they are secondary particles generated either in proton-proton collisions in the jet or in collisions between protons and antiprotons. Detailed studies of these possibilities were made by Felten (132) and by Felten, Arp & Lynds (133). In the meanwhile, further studies of the structure of the optical jet had been made (134, 133), and Arp (135) had reported the discovery of a counterjet. Moreover, an X-ray source had been found which was identified with M87 (136, 137, 138, 139). It is not known whether the X rays arise in the nucleus, in the jet, or in a more extended volume, nor whether they are generated by the synchrotron process or by Compton scattering, or are bremsstrahlung. However, the X-ray luminosity between ~ 2 and ~ 500 keV is $\sim 10^{43}$ ergs/sec, compared with about 5×10^{41} ergs/sec for the optical jet. The possibility that the X rays are generated in very small volumes by the synchrotron process or Compton scattering (140) means that continuous generation of high-energy electrons in strong magnetic fields is indicated. The lifetimes of synchrotron electrons are proportional to $\nu_c^{-1/2}$ so that, for a synchrotron X-ray source, the problems already encountered in trying to explain the optical jet (130) are magnified.

Very recently Cohen et al (141) and Hogg, McDonald, Conway & Wade (142) have shown that very small high-frequency radio sources exist in M87. One has a size of no more than 2.5 light months and is within 1" (75 pc) of the optical center of M87. This, together with the evidence for small optical structures in the jet (133, 134) and the existence of an X-ray source, suggests that a string of highly condensed regions able to generate relativistic particles are distributed along the jet and possibly the counterjet, and lead to its being detected as a powerful nonthermal source. In principle, it is not possible to decide whether such activity comes from outside and is due to infall, or is generated at the nucleus and is due to matter being thrown out, but there is a high probability that the latter is correct. The much larger halo of radio emission, originally discovered by Bolton et al (125), has also presumably arisen in the nucleus.

M87 is therefore an excellent example of a galaxy whose nucleus has generated a tremendous amount of nonthermal activity. There is also the possibility that it has ejected coherent clouds of plasma identified as radio sources far from M87, but still in the Virgo cluster. 3C 272.1 associated with M84 is in a direction with exactly the same position angle (290°) as the jet (143).

Not only is all of this activity present in the continuous radiation, but there is evidence of excitation and peculiar motions of ionized gas in the nucleus of M87. Humason and Minkowski (144, 145) found [O II] $\lambda 3727$ in the nucleus, and they showed that the emission line was broad and asymmetrical and has a velocity of 225 km/sec toward us relative to the recession velocity of the stars in the nucleus. Osterbrock (42) studied the profile of the [O II] doublet on higher dispersion. He showed that it has two components, one with the same redshift as the stellar absorption, and a second weaker component, with a velocity of 900 km/sec toward us with respect to the center. The age and velocity of the jet were inferred from this emission-line measure by Shklovsky (146, 147) who discussed various theoretical aspects relating to the maintenance of the jet. Further observations using the Lallemande camera at coudé dispersion were carried out by Walker & Hayes (148). They did not confirm the emission-line structure described by Osterbrock. They concluded that the nucleus contains a number of clouds of ionized gas both within the seeing disk (<75 pc from the center) and outside, to distances $\sim$300 pc, moving with velocities of several hundred km/sec, and similar to those which they had discovered in Seyfert nuclei. These clouds have a net outward motion toward us of about 200 km/sec, and random motions with a line-of-sight component, derived from the broadening of about 450 km/sec. The latter value is comparable with the line-of-sight component of the random motions of the stars ($\sim$490 km/sec). They could not see a component with a feature indicating motions as large as 900 km/sec relative to the nucleus. Since they found it hard to accept the idea that the profile had changed in the $\sim$5 years between the two sets of observations, the possibility that the apparent differences were due to instrumental effects was explored. The results were inconclusive.

Since very small sizes and variations in continuum flux with time are known to be commonplace in the nuclei of radio galaxies, we may ask whether such apparent changes could be real. In this case it is not likely since it would imply that a very large flux of forbidden-line radiation comes out of a region with a size $\lesssim 5$ light years, and density arguments can probably be used to rule this out. However, despite the uncertainties, it is clear that gas is being ejected from the nucleus of this galaxy.

M87 is the nearest of the powerful radio galaxies with active nuclei which can be clearly seen and investigated. NGC 5128 (Centaurus A), while it is closer than M87, cannot be included in this discussion because its nuclear region is totally obscured by dust as far as observation in optical wavelengths is concerned, and no infrared observations have yet been made.

In Section 7 we shall describe the general properties of the radio galaxies and the way in which these properties are intimately connected with galactic nuclei.

NGC 1052.—This is a bright elliptical galaxy (E3) in which Mayall (149) first identified strong [O II] $\lambda 3727$ emission. In Section 3 we discussed the low incidence of emission lines in general in ellipticals, and described how Minkowski & Osterbrock (41) investigated the physical conditions in the ionized gas in this galaxy, which has unusually strong emission. The emission lines are broad and indicate motions of 600–900 km/sec. While Osterbrock & Minkowski attempted to explain the excitation of ionized gas by processes involving stars alone, studies of the radio flux from this galaxy show that violent activity is going on in its nucleus.

Heeschen (150) investigated the radio emission from NGC 1052 at high frequencies (at wavelengths of 2, 6, and 11 cm) and showed that the bulk of the emission comes from a region $<2''$ (200 pc). The radio spectrum has positive curvature and is of the type found for small-diameter sources in QSOs and strong radio galaxies. The object is one of the brightest radio sources at 2 cm though in total flux it is a weak radio source (total power $\sim 10^{39}$–10^{40} ergs/sec). Wills (151) has shown that it was detected at much longer wavelengths (178 MHz) and has concluded that a two-component model is required to explain the radio source. It appears most probable that both radio and optical activity stems from a very small nonthermal source in the nucleus.

NGC 4278.—This is also a bright elliptical with an abnormally strong emission-line spectrum in its nucleus (42), and we discussed the physical conditions in the gas in Section 3.

Radio observations by Heeschen (150), Lang & Terzian (152), and de Jong (153) show this to be a very bright, small, high-frequency source with a size less than $2'' = 85$ pc at 2700 MHz (150). The radio spectrum reaches a maximum at about 1000 MHz and decreases at lower frequencies. Wills (151) has suggested that this turnover is due to thermal absorption in the ionized gas cloud investigated by Osterbrock.

In this galaxy also, the optical and radio observations taken together suggest that violent activity is occurring in the nuclear region.

NUCLEI OF SEYFERT GALAXIES

In 1943 Seyfert (154) described a class of galaxy whose nuclei set them apart from all other systems known at that time. The features characterizing these objects had been first discovered by Fath (155) and Slipher (156) (discovery of emission lines and their widths in NGC 1068), and by Campbell & Moore (157) (emission in NGC 4151), by Hubble (158) (NGC 4051, 4151, 1068), and by Mayall (159) (NGC 4151, 3516, and 7469). The nuclei are starlike in appearance, and have a high-excitation emission-line spectra, while the lines are exceedingly broad. Seyfert listed 12 galaxies, thought at that time to have these characteristics. They were all comparatively close systems with the exception of NGC 1275, with redshifts $z < 0.01$. Much later it was pointed out that 4 of the galaxies originally classified in this way by Seyfert, NGC 4258, 3077, 2782, 6814, do not fulfill the Seyfert criteria and should not be included (161). Since the 8 galaxies originally identified as having Seyfert nuclei are likely to be nearly all of this type in the Shapley Ames catalogue (containing about 1500 bright galaxies), and allowing for some incompleteness in the survey, it was concluded that these objects form about 1–2% of spiral galaxies. Following the work of Seyfert, little attention was paid to these objects until 1959 when NGC 1068 was reinvestigated by Burbidge, Burbidge & Prendergast (162), and theoretical speculations on Seyfert nuclei were made by Woltjer (163). In recent years many investigations of the classical Seyfert nuclei have been made, and a number of new galaxies with Seyfert nuclei have been discovered. In 1968 a conference on Seyfert galaxies and related subjects was held, and the proceedings contain much new information (164). Twenty-five galaxies with Seyfert characteristics have been discovered. In the following discussion we summarize the information available on each system.

NGC 1068.—This is one of the brightest and best studied of the Seyfert galaxies. The distribution of mass in the central region was studied by Burbidge, Burbidge & Prendergast (162) who showed that the total mass within a radius of 2000 pc is about $2.6 \times 10^{10} M_\odot$. By considering the changes in the shape of the rotation curve which would be caused by a concentrated mass in the nuclear region (over a radius $\sim$500 pc), it was shown that the mass in the nucleus could not be much more than about $3 \times 10^9 M_\odot$. NGC 1068 does not appear to have such a concentrated nucleus as some of the other Seyfert galaxies. Walker (165), using the coudé image-tube spectrogram at Lick, investigated the line profiles with high resolution. He concluded that there are a number of discrete clouds of material having high internal turbulent motions. The velocities of these clouds as a whole are up to 600 km/sec relative to the center; their diameters are ~200–350 pc and their masses are ~10^6–$10^7 M_\odot$. Though it is not certain whether these clouds have velocities in excess of the escape velocity, there is a strong presumption that they have, and are being ejected from the nucleus.

Studies of the line spectrum have been made by Seyfert in his original paper, by Osterbrock & Parker (166), and by Dibai & Pronik (167). The emission lines measured are [Ne V] λλ3346, 3425, [O II] λλ3727, 7320, 7330, [Ne III] λλ3869, 3968, [S II] λλ4068, 4076, 6717, 6731, [O III] λ4363, 4959, 5007, He IIλ4686, [Fe VII] λλ5158, 5276, 5721, 6086, [N I] λ5199, [N II] λλ5755, 6548, 6583, He I λ5876, [O I] λλ6300, 6363, [S III] λ6310, [Ar III] λ7136, the Balmer series, and Ca II K in absorption. The mean total linewidth of the emission lines is about 2900 km/sec (154). Ionization and excitation mechanisms will be discussed as they apply to all of the data of Seyfert nuclei following the discussion of individual objects.

Spectral scans suggest that the ultraviolet flux from NGC 1068 is greater than that found in normal spirals (168, 169). Polarization in the visual part of the spectrum was measured by Dibai & Shakhovskoi (170), by Walker (165), and by Hagen-Thorn & Dombrovski (171). Visvanathan & Oke (169) concluded that the optical continuum of NGC 1068 had two components, one which comes from a normal stellar component and the hot gas, and a second which is of nonthermal origin, and is presumably synchrotron radiation. The polarization is high in the ultraviolet and gradually decreases toward longer wavelengths. Since the amount of polarization decreases as the diaphragm size is increased, it appears that the intrinsic polarization is confined to a region $\lesssim 10''$ in diameter. Polarization ~5% in the U band is indicated (172).

A large flux of infrared radiation out to a wavelength of about 20 μ has been detected (173, 174). This infrared flux is either of thermal origin indicating the presence of large amounts of dust in the nucleus, or it is synchrotron radiation (175, 92). Rapid variability of the infrared flux at 2.2 μ has been reported by Pacholczyk & Weymann (176). Rapid variations in the infrared make a thermal dust cloud origin for this radiation very difficult to accept (92).

It has been shown by Bash (177) that 75% of the flux at 11 cm comes from a component with a half-brightness diameter of 11".5 while the remainder comes from a source <1."5 which scintillates (178). The scintillating source is responsible for 10% of the flux at long wavelengths and is contained within a linear size of about 16 pc. At millimeter wavelengths the source is powerful and has varied significantly in the range 1–10 fu in a year (179).

NGC 1275.—As well as being a classical Seyfert galaxy, this object is also a strong radio source (Perseus A, see Table 4). The optical spectrum has been studied by Seyfert (154), Minkowski (126), Burbidge & Burbidge (180), and Dibai & Pronik (181). The emission-line spectrum has two components. In the Seyfert nucleus the emission lines seen are [O II] λ3727, Ne III λλ3869, 3968, [S II] λλ4068, 4076, 6717, 6731, [O III] λλ4363, 4959, 5007, [O I] λλ6300, 6363, and the Balmer series. No quantitative measurements of linewidths have been given but they correspond to several thousand km/sec if they are interpreted as Doppler widths. All of the classical Seyfert galaxies are spirals, but NGC 1275, partly because it is much further away, has been harder to classify. It is obviously not a regular spiral, but it has a bright central part about 8 kpc in diameter, outside which are absorbing clouds and extended structures, some of which are broad. Other features show filamentary structures. Since the galaxy is the brightest member of the Perseus cluster, which is largely made up of ellipticals, it has been classified as such, but Minkowski (182) has shown that this is not correct. A spectrum taken 2" (700 pc) north of the nucleus is of type A. Studies of the velocity field in the gas outside the Seyfert nucleus (183, 180) show that from immediately outside the Seyfert nucleus to distances $\sim$30" (10 kpc) from it, gas moving with very high velocities $\sim$3000 km/sec with respect to the center is present. It appears that this gas has been ejected from the nucleus, and a very large part of the outer structure of filaments, etc is included. The emission-line spectrum of this rapidly moving gas shows comparatively sharp lines, Hα, [N II] λλ6548, 6583, [O II] λ3727.

Recently Lynds (160) took direct photographs of NGC 1275 through Hα interference filters with widths of approximately 55 Å centered on both systems of velocity. Each corresponds to a velocity spread of 2500 km/sec. These pictures show the enormous extent of the filamentary structure which has been generated in the explosion, and confirm the form of the velocity field found by spectroscopic investigations. The photograph taken through the filter centered at λ6694A, showing the low-velocity system, shows that NGC 1275 resembles the Crab Nebula on a vastly greater scale.

The continuum energy distribution indicates the presence of a rather steep nonthermal spectrum (184). Optical polarization was measured by Walker (165) who found that it amounted to 5.1% with a 13".6 diaphragm in the ultraviolet, decreasing toward longer wavelengths, by Hagen-Thorn & Dombrowski (171), and by Dibai & Shakhovskoi (170). Low & Kleinmann (174) showed that the flux increases very steeply in the infrared.

The radio source that emanates from NGC 1275 is very complex. Studies carried out by Ryle & Windram (185) strongly suggest that this galaxy is the origin of a radio source distribution which is spread throughout the Perseus cluster over distances $\sim$1 Mpc, and which has been ejected over a time during which the clouds of gas were ejected ($\sim 10^6$–10^7 years). Comparatively compact plasma clouds or secondary sources appear to have been ejected.

The radio source centered on the galaxy is also complex and appears to have at least three components (186). The largest has a size of about 80 kpc and is typical of the structure seen in many strong radio galaxies. The radio spectrum of this component is normal. The second component reaches a maximum flux density near 800 MHz and has a size between 0".1 and 0".03

or between 8 and 26 pc (187). If the low-frequency cutoff in the radio spectrum is due to synchrotron self-absorption, and the source is 8 pc in diameter, the minimum magnetic-field strength is about 10^{-2} G. The third component is optically thick at centimeter wavelengths and has a size smaller than 10^{-3} sec or <0.25 pc (141). Both of these latter sources are variable in time, and in the case of the very small source the rate of change is such that the source does not have to expand relativistically, i.e. the size is compatible with the simple relation $R \lesssim c\tau$, where τ is the period over which a significant variation occurred.

NGC 1566.—This Sc galaxy was studied by Sérsic (188). It was found to be a Seyfert galaxy by de Vaucouleurs & de Vaucouleurs (189, 190). The spectrum has Balmer lines with widths 3000–3500 km/sec. Other emission lines are [O III] $\lambda\lambda 4959$ and 5007.

NGC 3227.—This classical Seyfert galaxy was listed by Seyfert (154), but not studied by him. It has been investigated by Dibai & Pronik (191) and by Rubin & Ford (192). The line spectrum contains [O II] $\lambda 7330$, [S II] $\lambda\lambda 6717$, 6731, [N II] $\lambda\lambda 6548$, 6583, [O I] $\lambda\lambda 6300$, 6363, [N I] $\lambda 5198$, [O III] $\lambda\lambda 4959$, 5007, [O III] $\lambda 4363$, and the Balmer series. Also, absorption lines including the Na I D lines, $\lambda 4174$, Ca I $\lambda 4227$, the G band, Mg I $\lambda 5176$, and other features characteristic of a solar-type spectrum are seen. Widths ~ 3000 km/sec are found for the [O III] lines. Rubin & Ford noted a sharply defined nucleus with discrete blobs of Hα emission at its boundaries. They found that the very broad lines are confined to the nucleus and the emission lines outside are very sharp. This is also the case in NGC 1068. They were able to measure the velocity field out to several kiloparsecs. In the nucleus, within about 200 pc of the center, there are discrete clouds moving with velocities that spread over several thousand km/sec. As was also found in NGC 1068, the mean redshift found for these clouds is less than the redshift of the center of mass of the galaxy, which is obtained from the rotation curve for the outer parts. Thus the whole system of clouds in the nucleus appears to be expanding outward with a velocity of ~ 150 km/sec. From the rotation curve the total mass within 620 pc of the center is estimated to be $\sim 3 \times 10^9 M_\odot$, and out to 3900 pc the mass is $\sim 2.6 \times 10^{10} M_\odot$. These values are again very similar to those found for NGC 1068. Since the velocity of escape from the center is only ~ 400 km/sec, and velocities of thousands of km/sec are found for the discrete clouds, it is concluded that matter is continuously being ejected from the nucleus.

Polarization of about 1% is found with a diaphragm size of 26″ (171). The galaxy has been detected as a weak radio source by Wade (193) and the radio emission appears to be confined to the Seyfert nucleus and its immediate environs.

NGC 3516.—This galaxy was originally studied by Seyfert, and more recently it has been investigated by Dibai & Pronik (191) and by Andrillat

& Souffrin (194). Seyfert only observed the Balmer series together with [O III] λλ4959, 5007. He gave widths of 1400 km/sec for the [O III] lines and 8500 km/sec for the hydrogen lines. The more recent investigations have led to a very important conclusion, that the line spectrum and probably the continuum also have varied at some times within the last 25 years. The lines that can now be seen include [O II] λ3727, [Ne V] λλ3346, 3426, [Ne III] λ3863, together with the [O III] lines. The Balmer lines appear to have almost disappeared. The intensities of all of the forbidden lines have markedly increased. Andrillat & Souffrin have interpreted this change as being due to the injection from a central source of comparatively dense clouds with masses $\sim 100 M_\odot$ at very high speeds, ~ 1000 km/sec. The clouds then expand. It is also argued that an increased flux of ultraviolet photons, presumably of nonthermal origin, has given rise to enhanced ionization.

No optical polarization can be measured, nor has the galaxy been detected at radio frequencies.

NGC 4051.—This galaxy was investigated by Seyfert and more recently by Dibai & Pronik (191). The line spectrum contains [Ne III] λ3869, [O III] λλ4363, 4959, 5007, He II λ4686, [O I] λλ6300, 6363, [N II] λλ6548, 6583, [S II] λλ6717, 6731, and the Balmer series. The widths of the lines, according to Seyfert, are ~ 1200 km/sec for the [O III] lines and ~ 3600 km/sec for the Balmer lines. The galaxy has been detected in the near infrared (176), and it is a weak radio source with a size less than about 100 pc (193).

NGC 4151.—This is probably the best-studied Seyfert galaxy. Its structure outside the nucleus is very faint, and it is clear that if it were somewhat fainter, or further away, it would be impossible to distinguish it from a quasistellar object. The object was studied by Seyfert, who identified a large number of lines—it has the richest spectrum of any known Seyfert nucleus, and a detailed study of the spectrum has been made by Oke & Sargent (195). We reproduce in Table 3 a table from their paper, giving the lines and their measured intensities compared with the spectrum of the galactic planetary nebula NGC 7027. The appearance of the coronal lines [Fe X] 6374, [Fe VII] 6085, 5721, and [Fe XIV] 5303 is particularly interesting. Ford, Purgathofer & Rubin (196), working in the near infrared, found that [S III] λλ9069, 9532 were also very prominent. Oke & Sargent stressed that their studies show the existence of a smooth continuum with no underlying absorption lines. This continuum is largely of nonthermal origin. Seyfert gives widths for the forbidden lines ~ 1000 km/sec, and ~ 7500 km/sec for the Balmer lines.

From an analysis of relative intensities, Oke & Sargent concluded that the emission-line spectrum arises in a region with $\overline{N}_e \simeq 5000$ cm^{-3} and $\overline{T}_e \simeq 20000°$K. The total mass of this gas is about $2 \times 10^5 M_\odot$; it fills only about 1/40 of the volume of the nucleus and appears to be in the form of clouds or filaments with random motions ~ 450 km/sec.

Anderson & Kraft (197) showed that the faint absorption feature be-

TABLE 3. Emission lines in nucleus of NGC 4151: Identifications, intensities, and comparison with the planetary nebula NGC 7027, according to Oke & Sargent (195)

λ(Å)	Identification	Equivalent width (Å)	Flux at source (units of 10^{40} ergs/sec)	Strength relative to $H\beta = 100$ NGC 4151	NGC 7027
10830.2	He I	163.5	5.98	81	87
10049.4	Pζ	30.0:	1.19:	16:	5
7329.9 } 7330.7	[O II]	11.0	0.62	8	32
6731.3	[S II]	33.3	2.08	28 }	6
6717.0	[S II]	27.7	1.75	24	
6583.6	[N II]	29.0	1.83	25	90
6562.8	Hα wings	360.0	22.75	307 }	290
	Hα core	34.0	2.15	29	
6548.1	[N II]	6.2	0.39	5	30
6374.5	[Fe X]	2.8:	0.18	2:	—
6363.9	[O I]	4.7	0.31	4	6
6300.2	[O I]	20.6	1.36	18	20
6085.3	[Fe VII]	7.6	0.52	7	—
5875.6	He I	4.4	0.31	4	11
5754.8	[N II]	3.0:	0.22:	3:	8
5720.9	[Fe VII]	4.4	0.32	4	—
5303.6	[Fe XIV]	1.0:	0.08	1:	—
5006.8	[O III]	188.0	15.80	214	1460
4959.9	[O III]	62.0	5.23	70	480
4861.3	Hβ wings	72.0	6.07	82 }	100
	Hβ core	16.0	1.35	18	
4799.5	[Fe III]	1.0:	0.10:	1:	—
4740.3	[A IV]	1.7	0.15	2	10
4711.4	[A IV]	1.7	0.15	2	8
4685.7	He II	21.7	1.88	25	46
4658.1	[Fe III]	5.9	0.51	7	—
4471.5	He I	1.0:	0.10:	1:	4
4363.2	[O III]	5.4	0.48	7	26
4340.5	Hγ wings	21.4	1.92	26 }	47
	Hγ core	5.6	0.50	7	
4243.0	?	0.5:	0.04:	1:	—

TABLE 3. (Continued)

λ(Å)	Identification	Equivalent width (Å)	Flux at source (units of 10^{40} ergs/sec)	Strength relative to $H\beta = 100$	
				NGC 4151	NGC 7027
4228.0	?	0.5:	0.04:	1:	—
4101.7	Hδ wings	5.4	0.51	7	26
	Hδ core	3.7	0.35	5	
4076.2	[S II]	2.3	0.22	3	16
4068.6	[S II]	3.5	0.34	5	
3970.1	Hε	10.3	1.03	14	52
3968.5	[Ne III]				
3889.1	Hζ	4.3	0.47	6	20
3888.6	He I				
3869.7	[Ne III]	19.3	2.12	29	120
3728.9	[O II]	28.7	3.75	51	35
3726.2	[O II]				
3425.8	[Ne V]	14.5	2.46	33	130

tween the strong emission lines [Ne III]λ3869 and He I+Hζ λ3889, first discovered by Wilson, is made up of three sharp absorption features, which they identify as three components of He I λ3889 with velocities of -280, -550, and -840 km/sec with respect to the emission lines. They also find weak absorptions in the Balmer lines at two of these three velocities. They have therefore concluded that three discrete shells of gas with these velocities are being ejected from the nucleus. They have then calculated the mass loss from the nucleus, and find that it amounts to between 10 and 1000 $M_\odot$ per year. Cromwell & Weymann (439) have suggested that other observations indicate that hydrogen absorption lines are not always present. They have therefore concluded that the ejection is sporadic. This means that the average mass-loss rate calculated by Anderson & Kraft may be too high. A revised rate of about 1 M_0 per year may be indicated. Clearly more detailed studies are required. The higher mass-loss rates would mean that in times $\sim 10^8$ years suggested by the frequency of the Seyfert phenomenon, masses comparable to the total masses of the galaxies might be ejected.

Danielson, Savage & Schwarzschild (198) measured, from above the Earth's atmosphere, an upper limit to the half-intensity angular diameter of the nucleus of NGC 4151 to be 0."18, or about 10 pc. This refers to the nonthermal continuum radiation, and comparison with the analysis of Oke & Sargent shows that this source must be significantly smaller than the region giving rise to the emission-line spectrum.

The nucleus of NGC 4151 appears to be variable in light and the fluctuations are due to changes in the continuum nonthermal source and not in the emission-line region. The time scale for changes is of order of a year (199, 200). The continuum is also known to increase steeply into the infrared (174, 176, 201) and it may also be variable at a wavelength near 10 μ on a time scale of order of a year. There is every indication therefore that the continuum source radiating in the frequency range from 10^{15}–10^{13} Hz has a size smaller than a light year or so.

Optical polarization has been measured by Walker (165), Hagen-Thorn & Dombrovski (171), and Kruszewski (172).

The galaxy has been detected as a weak radio source, and an unresolved component with a size less than 1."8 (90 pc) is present in the nucleus (193).

NGC 5548.—The nucleus of this galaxy has been studied by Dibai, Esipov & Pronik (202). The lines seen are [O II] λ3727, [Ne III] $\lambda\lambda$3868, 3967, [O III] $\lambda\lambda$4363, 4959, 5007, [N II] $\lambda\lambda$6548, 6583, and the Balmer series. The total widths are $\sim$1000 km/sec for the forbidden lines and $\sim$5000 km/sec for the Balmer lines. Bardin, Chopinet & Duflot-Augarde (203) made a spectrophotometric study of the nucleus and concluded that it was variable and that the electron temperature had varied. Optical polarization amounting to 1.9% in the uv is present (165). The galaxy has not been detected as a radio source.

NGC 7469.—This galaxy was originally discussed by Seyfert. The emission-line spectrum contains [O II] λ3727, [O III] $\lambda\lambda$4959, 5007, and the Balmer series. The linewidths are $\sim$1000 km/sec for the forbidden lines and $\sim$5000 km/sec for the Balmer lines (162). Since the galaxy has a redshift of 0.017, the maximum diameter of the nucleus, limited by seeing, is 1" = 460 pc. Burbidge et al (162) measured the rotation of the galaxy and concluded that the mass within about 3600 pc of the center was $1.1 \times 10^{10} M_\odot$. They concluded that the escape velocity was significantly less than the velocities of the clouds giving rise to the line emission if the random velocities are measured by the extent of the line broadening. There is a weak radio source in the central region with a size less than about 1 kpc (193).

III Zwicky-2.—This object was identified as a compact galaxy by Zwicky and his colleagues (204), and was investigated by Arp (205). He identified [O II] λ3727, [O III] $\lambda\lambda$4959, 5007, [Ne III] $\lambda\lambda$3869, 3968, and the Balmer series. The redshift is 0.089. The hydrogen linewidths are about 7000 km/sec. This object clearly has the typical Seyfert characteristics in its spectrum, but it is difficult to decide what sort of galaxy, if any, surrounds the nucleus.

Zwicky pointed out that it was only slightly nonstellar in appearance, and Arp showed that there are only a few faint wisps surrounding it, in contrast to the situation in the classical Seyfert galaxies where a galaxy, usually a spiral, is clearly visible. III Zw-2 differs from a QSO such as 3C 48, which has similar wisps, only in having a much smaller redshift (see Section 6). Because III Zw-2 has a much greater redshift than that of a classical Seyfert galaxy, it could well be argued that the outer parts of a small galaxy in which this very bright nucleus is embedded could simply be too faint to be detected.

The object is a weak radio source with a rather flat spectrum.

3C 120 = II Zw-14 = PKS 0430+05 = 4C 05.20.—This object was identified by Clarke, Bolton & Shimmins (206) as a radio source and independently by Zwicky as a compact galaxy. It was observed spectroscopically by Burbidge (207), Sargent (208), and Arp (205). It is clearly a Seyfert galaxy since it has a bright core with a dimaeter of 5″, which contains a starlike nucleus, while the total extent of the whole galaxy appears to be about 40″ on the Palomar Sky Atlas. The spectrum contains [Ne V] $\lambda\lambda 3345, 3426$, [O II] $\lambda 3727$, [Ne III] $\lambda\lambda 3869, 3968$, He II $\lambda 4686$, [O III] $\lambda 4959, 5007$, He I $\lambda 5876$, [N II] $\lambda\lambda 6548, 6583$, and the Balmer series. The redshift is 0.033. The Balmer lines have widths ~ 3300 km/sec.

Continuum measurements were made by Oke, Sargent, Neugebauer & Becklin (210). The object has shown rapid variations in optical wavelengths amounting to factors of 2 in months (209), and rapid variations involving a number of successive outbursts have been discovered at centimeter wavelengths (186).

3C 120 is an exceedingly powerful infrared source (174). The bulk of the energy observed so far is emitted between about 5 and 20 μ, and this luminosity amounts to about 10^{46} ergs/sec.

Zw 0039.5+4003.—This galaxy was first identified by Zwicky and it has recently been studied by Zwicky et al (211). It has a nucleus of about 1″ diameter with a halo extent of about 3″. Its line spectrum contains [Ne V] $\lambda\lambda 3345, 3426$, [O II] $\lambda 3727$, [Ne III] $\lambda\lambda 3869, 3968$, [O III] $\lambda\lambda 4363, 4959, 5007$, He I $\lambda 3889$, and the Balmer series in emission. The redshift is 0.1026. The Balmer lines have widths ~ 5000 km/sec while the forbidden lines have widths ~ 850 km/sec. The continuum is similar to that of 3C 120 and steepens into the near infrared (1.65 μ). The object has a variable nucleus, which may almost disappear at times. No radio flux has been detected from the object.

I Zw 0051+12.—This object was discovered by Zwicky and investigated by Sargent (212). It is classified as a Seyfert nucleus because it has a starlike nucleus and emission lines with widths at half-intensity ~ 3000 km/sec. However, its emission-line spectrum differs from the other objects in this class because the Balmer series and a large number of permitted lines of Fe II

TABLE 4. Spectroscopic properties of N galaxies identified with radio sources

Object	z	Lines
MSH 05-43 (Pictor A)	.0342	[O II]λ3727, [Ne III]λ3869, Hδ, Hγ, Hβ, [O III]$\lambda\lambda$4363, 4959, 5007
3C 371	.0508	[O II]λ3727, [O III]$\lambda\lambda$4959, 5007, Ca II H and K abs
3C 445	.0568	[Ne V]λ3426, [O II]λ3727, [Ne III]$\lambda\lambda$3869, 3968, [SII]$\lambda\lambda$4068, 4076, Hδ, Hγ, Hβ, [O III]$\lambda\lambda$4363, 4959, 5007. No abs
3C 390.3[a]	.0569	[Ne V]λ3426, [O II]λ3727, [Ne III]$\lambda\lambda$3869, 3968, Hδ, Hγ, Hβ, [O III]$\lambda\lambda$4363, 4959, 5007, Hα. No abs
PKS 0521-36	.061	[Ne V]$\lambda\lambda$3346, 3426, [O II]λ3727, [Ne III]$\lambda\lambda$3869, 3968, Hδ, Hγ, [O III]$\lambda\lambda$4959, 5007
3C 227[b]	.0855	[Ne V]λ3426, [O II]λ3727, [Ne III]$\lambda\lambda$3869, 3968, Hγ, Hβ, [O III]$\lambda\lambda$4363, 4959, 5007, He I λ5876, Hα. No abs
PKS 1417−19	.1192	[O II]λ3727, [Ne III]λ3869, [O III]$\lambda\lambda$4363, 5007, Hγ, Hβ, Hα. No abs
PKS 2300−18	.129	[Ne III]λ3869, Hγ, Hβ, [O III]$\lambda\lambda$4363, 4959, 5007
PKS 1340+05	.1333	[O II]λ3727, Ca II H and K abs
PKS 2349−01	.174	[O II]λ3727, Hδ, Hγ, Hβ, [O III]$\lambda\lambda$4363, 4959, 5007
3C 234	.1846	[Ne V]$\lambda\lambda$3346, 3426, [O II]λ3727, [Ne III]$\lambda\lambda$3869, 3968 Hζ, Hδ, Hγ, [O III]$\lambda\lambda$4363, 4959, 5007, He II $\lambda\lambda$3203, 4686, Hβ. No abs
3C 287.1	.2156	[Ne V]λ3426, [O II]λ3727, [Ne III]$\lambda\lambda$3869, 3968, Hβ, [O III]$\lambda\lambda$4959, 5007
3C 17	.2201	[O II]λ3727, [Ne III]λ3869, Hβ, [O III]$\lambda\lambda$4959, 5007
3C 459	.2205	[Ne V]λ3426, [O II]λ3727, [Ne III]λ3869, [O III]$\lambda\lambda$4959, 5007. Weak em; no abs
3C 171	.2387	[O II]λ3727, [Ne III]$\lambda\lambda$3869, 3968, [O III]$\lambda\lambda$4363, 4959, 5007, Hβ
3C 79	.2561	[Ne V]$\lambda\lambda$3346, 3426, [O II]λ3727, [Ne III]$\lambda\lambda$3869, 3968, Hγ, [O III]$\lambda\lambda$4363, 4959, 5007, He II 4686, Hβ
3C 109	.3056	[Ne V]λ3426, [O II]λ3727, [Ne III]$\lambda\lambda$3869, 3968, Hγ, Hβ, [O III]$\lambda\lambda$4363, 4959, 5007
3C 177	No published z	No emission lines, Ca II H, K, and G band abs

[a] Balmer emission lines very broad, with two maxima.
[b] Balmer emission lines double.

are all that are seen. The strongest Fe II line is λ5018.4 which is 2/3 as strong as Hβ. The object has an optical energy distribution similar to that of a QSO. In 3C 273, the best-studied QSO, less prominent Fe II lines are also seen (213), though they are not dominant. The object has a redshift $z=0.061$, so that if it is at a cosmological distance, it has an absolute magnitude ($M_v = -23.2$). In this case it is much brighter than a classical Seyfert nucleus. The presence of Fe II emission and the absence of [Fe II] radiation means that the electron density $N_e > 10^6$ cm^{-3}. It is not clear whether the Fe/H abundance is normal, though Sargent assumed that it was. The stellar nucleus is surrounded by some faint nebulosity, which can be resolved into two faint armlike structures terminating in two faint stars 4 or 5 mag fainter than the central object. This object has not been detected to be a radio source.

Markarian 9.—This galaxy was identified by Markarian (see Section 8). Its spectrum has been discussed by Khachikian (214) and Weedman & Khachikian (215). It shows [O III] $\lambda\lambda 5007, 4959$, [O II] $\lambda 3727$, He II $\lambda 4686$, and the Balmer series. The Balmer lines are very broad, $\gtrsim 6000$ km/sec, while the forbidden lines are narrower. The redshift is 0.038.

Markarian 10.—This galaxy is similar to Markarian 9. It has broad hydrogen lines with widths ~ 6000 km/sec and comparatively narrow forbidden lines. The redshift is 0.029.

Markarian 34.—This was identified as a Seyfert galaxy by Weedman & Khachikian (215). It shows [O II] $\lambda 3727$, [Ne III] $\lambda 3869$, He II $\lambda 4686$, [O III] $\lambda\lambda 4959, 5007$, and the Balmer series. The width of Hα is ~ 1500 km/sec, and the widths of the forbidden lines are less. The redshift is 0.0507.

Markarian 42.—This was identified by Weedman & Khachikian (215). It shows [O II] $\lambda 3727$, [O III] $\lambda\lambda 4959, 5007$, and the Balmer series. The Balmer lines have widths ~ 2000 km/sec. The redshift is 0.024.

Markarian 50.—This was identified as a Seyfert galaxy by Sargent (216). Its spectrum is more like that of NGC 4151 than NGC 1068. The redshift is 0.023.

I Zw 1535+55.—This object was discovered by Zwicky and investigated by de Veny & Lynds (436). The object has a redshift $z = 0.0386$. Its angular diameter is $<2''$. The emission-line spectrum contains [O III] $\lambda\lambda 4959, 5007$, [Ne III] $\lambda 3869$, [S II] $\lambda 4069$, He II $\lambda 4686$, He I $\lambda 5876$, the Balmer series, and a number of Fe II lines, including Fe II $\lambda\lambda 4177, 4297, 4924$, and 5169. The presence of Fe II lines makes the object rather similar to I Zw 0051+12 just discussed (212).

Markarian 69.—This object was identified as a Seyfert galaxy by Sargent (216). Its spectrum is similar to that of NGC 4151. Its redshift is 0.076.

Zw II 2130+09.—This compact galaxy was investigated by Fairall (217) and Sargent (218). It shows [Ne III] $\lambda 3869$, [O III] $\lambda\lambda 5007, 4959$, and the Balmer series in emission. The hydrogen lines have widths ~ 3000 km/sec. The redshift $z = 0.061$.

VV 144 = II Zw 1122+54.—This object from Vorontsov-Velyaminov's catalogue (later independently listed by Zwicky) (cf Section 8) was first found to have Seyfert characteristics by Burbidge & Burbidge (219), and was further studied by Sargent (218). In the spectrum one sees the broad Balmer lines (widths ~3000 km/sec) together with [O III] λλ4959, 5007, and [O II] λ3727. The object has a redshift of 0.021. A bright jet extends from the nucleus in which the Seyfert characteristics are found.

VV 150 = I Zw 26.—This object from Vorontsov-Velyaminov's catalogue was first found to have Seyfert characteristics by Burbidge & Burbidge (220). It has further been studied by Sargent (221). Emission lines of hydrogen and [O II] λ3727, [O III] λλ4959, 5007, and [Ne III] λ3869 are seen. The total widths of the hydrogen lines are ~1000 km/sec. The galaxy is one in a chain and its redshift is 0.027.

NGC 3783.—This galaxy has been stated to have Seyfert characteristics by Page.

This concludes our discussion of the observed properties of known Seyfert galaxies. Before describing other related classes of extragalactic objects which have excited nuclei, we shall next briefly describe the infrared properties of some Seyfert galaxies and other systems. Kleinmann & Low (440) have recently given a list of galaxies in which they have been able to detect infrared flux out to a wavelength of about 25 μ. Since there are great similarities between the shapes of the infrared spectra of the nucleus of our Galaxy and these external galaxies in the wavelength range out to 25 μ, they have extrapolated their measurements beyond 25 μ using the observations out to 300 μ which have been made for the Galactic center. Very high luminosities are then obtained. We show these luminosities in Table 10, at the end of the article, together with more conservative estimates obtained by supposing that the infrared flux is cut off (quite unrealistically from a physical standpoint) beyond 25 μ.

N GALAXIES

These galaxies will not be described individually. They were first classified according to form by Morgan (4), who defined them as "systems having small brilliant nuclei superposed on a considerably fainter background." Later in that paper he classified NGC 4051 and 4151, two classical Seyfert galaxies, as N systems. In the paper on radio galaxies by Matthews, Morgan & Schmidt (222) they were defined as "galaxies having brilliant starlike nuclei containing most of the luminosity of the system. A faint nebulous envelope of small visible extent is observed." It was also pointed out in that paper that they may be related to the compact galaxies discovered by Zwicky.

It is therefore clear that, as far as form is concerned, "N galaxy" and "Seyfert galaxy" are equivalent descriptions. However, the majority of the N galaxies investigated have first been identified as strong radio sources. All of those found in this way are listed in Table 5 (Section 7); the majority of them have much larger redshifts than the galaxies with starlike nuclei originally identified by Seyfert. Descriptions of the spectra of these objects have been given by Schmidt (223), Sandage (224), and Burbidge (207). We list their properties in Table 4. The majority of them identified with strong radio sources show the same emission-line spectrum as that found in the characteristic Seyfert nuclei, though in general the emission lines are not as broad (225, 226). In most of these systems there is no evidence for absorption features due to stars. However, some N galaxies have been identified as strong radio sources, but do not show a strong emission-line spectrum: e.g. 3C 177 (226) having only the normal absorption lines seen in an elliptical galaxy; 3C 371 (227, 228) showing only rather weak emission and absorption features. Some N galaxies have been shown to be variable in light (e.g. 3C 371, 3C 109), and this variability is due to the nonthermal continuum radiation and not to the line spectrum. Of course, as in other variable sources, this indicates the presence of a very small component with dimensions of light years or less.

Sandage (228) has shown that N galaxies studied by him have UBV colors that restrict them to a small region in a $(U-B)$, $(B-V)$ diagram, as is the case also for QSOs, and there has been some tendency therefore to use such a color criterion to partially define such objects.

6. QUASISTELLAR OBJECTS

The properties of these objects have been discussed extensively elsewhere (229-232), We discuss here briefly only the relationship between them and the nuclei of galaxies.

QSOs are defined as a class of objects which have a starlike appearance on direct plates and have very large redshifts in their spectra. Their spectra contain broad emission lines which are weaker relative to the continuum than is the case in Seyfert nuclei and N systems, and they sometimes have sharp absorption lines. They all show a strong and often variable nonthermal continuum. They also resemble the nuclei of galaxies described earlier in that they often are very powerful radio sources, and infrared emitters. What criteria separate them from N galaxies or Seyfert nuclei?

In general, as far as morphology is concerned, one would say that they are really starlike with *no* surrounding structure of any kind. However, exceptions can be found. For example, 3C 48, the first QSO identified, has faint wisps surrounding it, which would make some want to redefine it as an N galaxy.

As far as redshifts are concerned, it is usually found that $z_{Seyfert} < z_N < z_{QSO}$. However, exceptions are known. QSOs with redshifts as small as 0.06 have been found, while Seyfert galaxies with redshifts $z \sim 0.1$, and N galaxies with $z \sim 0.3$ have also been identified.

B264 was originally defined as a QSO by Braccesi, Lynds & Sandage

TABLE 5. Optical properties of radio galaxies

Source	Optical object	z	Type	Spectroscopic[a] character	Radio flux[b] (flux units)
3C 231	M82	.001	Irr	s.e. A	13.0
MSH 13–42 Centaurus A	NGC 5128	.00157	DE 3	w.e. A	8700 (85.5)
3C 71	NGC 1068	.00377	Seyfert	s.e. A	13.5
MSH 03–31 Fornax A	NGC 1316	.00577	D3–4	w.e. A	249 (408)
3C 272.1	NGC 4374	.0029	E2 (cl)	w.e. A	18.0
3C 270	NGC 4261	.00697	ED 3 (cl)	A	44
3C 274 Virgo A	NGC 4486	.0041	E2 (cl)	w.e. A	970
PKS 0722−09		.0073	Sc	s.e. A	1.4 (1410)
4C 25.35	NGC 3689	.0088		w.e. A	2.8
4C 17.52	NGC 3801	.0105		w.e.	3.4
3C 278	NGC 4782–83	.0143	E double	A	42 (159)
4C 03.01	NGC 193	.0145		A	3.4
4C 39.11	NGC 1233	.0163		w.e.	6.8
3C 31	NGC 383	.0170	DE 3	A	15.5
3C 40	NGC 545–7	.0180	D4 (cl)	A	24
3C 83.1	NGC 1265 ?	.0181	ED 3–4 (cl)		28
3C 84	NGC 1275	.0181	ED Seyfert (cl)	s.e.	58
3C 449		.0181		A	13.5
4C 34.09	NGC 1167	.0203		w.e.	5.0
3C 264	NGC 3862	.0206	DE 1	A	24
4C 39.12		.0209		A	2.7
3C 66		.0215	ED 2	A	33
3C 296	IC 5532	.0237	E4	A	12.5
3C 255		.0238	S pec	w.e. A	16.5
3C 75		.0241	Double	A	23
4C 31.42		.0241		w.e.	4.1
4C −01.32		.0249		w.e.	2.1
3C 442	NGC 7236–7	.0262	E double	w.e. A	20
3C 78	NGC 1218	.0289	DE 3	w.e. A	15.0
3C 465	NCG 7720	.0301	D4 (cl?)	w.e. A	35
3C 88		.0302	D4	w.e. A	16.0
3C 338	NGC 6166	.0303	D4 (cl)	w.e. A	41
3C 98		.0306	ED 3	s.e. A	41

[a] w.e.—weak emission, meaning that one or more of the lines [O II] λ3727, Hα, [N II] λ6548, 6583 is present.
    s.e.—strong emission, meaning that lines other than those listed above are present.
    A—absorption lines are present.

[b] Values given are generally at a frequency of 178 Mc/s. Where this is not the case the frequency at which the flux has been measured is given in brackets.

TABLE 5. (Continued)

Source	Optical object	z	Type	Spectroscopic[a] character	Radio flux[b] (flux units)
3C 353		.0307	D2	w.e. A	203
3C 76.1		.0326	DE 3	A	9.5
3C 455		.0331	D4	A	13.0
3C 120		.0333	Seyfert	s.e.	11.5 (159)
MSH 05-43	Pictor A	.0342	ND 1	s.e.	570 (85.5)
3C 317		.0351	D4 (cl)	s.e. A	43
4C 56.16		.0356		w.e.	4.2
3C 305		.0416	E or Sa pec	s.e. A	13.5
3C 29		.0450	ED 2	A	15.0
3C 293		.0454	D5 ?	w.e.	12.0
3C 371		.0508	N	w.e.	9.5
3C 218 Hydra A		.0530	D2 double (cl)	w.e.	210 (159)
3C 310		.0543	Double	w.e. A	51
PKS 0634−20		.056	E	s.e.	7.0 (1410)
3C 390.3		.0569	N	s.e.	44
3C 405 Cygnus A		.0570	D3 (cl)	s.e.	8100
3C 445		.0568	N1	s.e.	23
3C 382		.0586	D3 ?	s.e.	20
3C 403		.059	DE 3-4		26
3C 192		.0596	E0	s.e.	19.5
3C 33		.0600	DE 4	s.e.	49
PKS 0521−36		.061	N	s.e.	37 (408)
PKS 0349−27		.066	E	s.e.	5.2 (1410)
PKS 1131+21		.066	E	s.e.	0.7 (1410)
3C 15		.0733	D1		14.5
MHS 12+04 A		.0756	E	w.e. A	30 (85.5)
MSH 12+04 B,C		.0771	D	A	
3C 285		.0797	D ?	w.e.	10.5
3C 198		.0809	D4 (cl)	s.e.	16.5
3C 452		.0820	ED 1	s.e.	49
MSH 23-112		.0825	D5	s.e.	30 (85.5)
3C 227		.0855	N1	s.e.	28
3C 277.3 Coma A		.0857	D2	s.e. A	11.5
3C 388		.0917	D3 ?	w.e.	22
3C 236		.0988	DE 4	w.e.	9.0
3C 433		.1025	D4 ? (double)	s.e.	52
3C 327		.1041	DE 3-4	s.e.	40
3C 223.1		.1075	E5	s.e.	9.5
3C 315		.1086	Double	s.e.	17.5
PKS 1417−19		.1192	N	s.e.	5.1 (408)
3C 135		.1270	DE ? (cl)	s.e.	16.0

TABLE 5. (Continued)

Source	Optical object	z	Type	Spectroscopic[a] character	Radio flux (flux units)
PKS 1340+05		.1333	N	w.e. A	4.9 (408)
3C 223		.1370	E (cl)	s.e.	14.5
3C 348 Hercules A		.1540	D4 (cl)	w.e. (1 line)	325
3C 381		.1614	ND ?	s.e.	12.5
PKS 2349−01		.174	N	s.e.	1.6 (1410)
3C 219		.1745	D5 (cl)	s.e.	44
3C 234		.1846	N1	s.e.	29
3C 28		.1959	(cl)	w.e.	14.0
3C 26		.2106		s.e.	8.0 (159)
3C 436		.2154		s.e.	15.0
3C 287.1		.2156	N	s.e.	11.5
3C 17		.2201	N	s.e.	21
3C 459		.2205	N	s.e.	22
3C 456		.2337		s.e.	13.0
3C 171		.2387	N	s.e.	23
3C 79		.2561	N	s.e.	24
3C 109		.3056	N	s.e.	19.5
3C 295		.461		w.e. (1 line)	73

(233). It lies in a cluster of galaxies. However, studies of its line spectrum, its form, and its energy distribution (234, 235) have led to its redefinition as an N system. Thus there is some overlap and some confusion because, depending on one's primary criterion, there may not be agreement on the classification of an object into Seyfert, N, and QSO categories. However, we are describing one general class of phenomena, embracing all three of these classes of object, when we find an object which (a) is very compact, with or without a faint halo; (b) shows a strong emission-line spectrum with rather broad lines; (c) shows a strong nonthermal component variable in time; (d) may be a measurable nonthermal radio source; (e) may be a powerful infrared emitter.

In the case of a classical (nearby) Seyfert galaxy we know that these are the properties of a galactic nucleus, since the galaxy has been classified as a spiral, and direct observation of the outer parts shows that it is indeed a stellar system similar to our own. In the case of N galaxies at much greater redshifts, we see structure outside the nucleus, and there is a presumption by some that a normal galaxy may be present, but we *do not know*. In the case of a QSO we have no idea as to whether or not we are looking at the nucleus of a galaxy.

Clearly, the definitions are strongly influenced by the relative distances of the objects. If the redshifts in all cases are directly measures of distances, it is indeed very difficult to intercompare classical Seyfert nuclei with N galaxies and QSOs, since the light that is seen and integrated in the starlike nuclei comes from very different volumes. This may explain some of the differences in the spectra and the relative levels of continua and emission lines. We know from the variability that much of the continuum radiation in each case comes from a very small volume not greater than a few light years in diameter. However, suppose we compare the spectra of the nuclei of two identical galaxies whose distances from us are in the ratio 1:10, both having tiny identical nonthermal continuum sources. In the first case we shall be looking at a spectrum made up of two components, a nonthermal continuum flux I_{NT} plus a thermal contribution from hot gas clouds consisting of an emission-line spectrum and a thermal continuum $I_L + I_T$. We assume that in both cases the flux is observed from an angular size of 1″. Since $(I_L + I_T)$, but not I_{NT}, is proportional to the volume integrated, the total flux from the nearer object will be $(I_{NT} + I_L + I_T)$, and that from the further object will be $(I_{NT} + I_L' + I_T')$, where $(I_L' + I_T') \approx 1000 \, (I_L + I_T)$. Thus the differences between the apparent emission-line strengths, or widths, will be measured in some sense by the ratios $I_L/(I_{NT} + I_T)$ and $I_L'/(I_{NT} + I_T')$, and the spectra of the "nuclei" of the objects may look rather different. Thus, if one wishes to intercompare systems using the kinds of criteria developed to define these nuclei, one must compare objects at *comparable* distances only. If it turns out that the redshifts of QSO and perhaps some N systems are not measures of distance, intercomparison may be easier, since they all are likely to be fairly close.

If the redshifts of Seyfert nuclei, N galaxies, and QSOs are all entirely cosmological in origin, on average there is a progression of optical luminosity with the Seyfert galaxies being intrinsically the faintest objects, while the QSOs are the brightest. However, the very large scatter in the redshift-apparent-magnitude relation for QSOs (cf 232) means that there is a wide range in intrinsic luminosities of the QSOs. A similar progression exists for the radio luminosities though again there is a large scatter, and it must also be remembered that the bulk of the QSOs are not strong enough even to be identified as radio sources at present.

The realization that there is such a continuity if all of the redshifts are tacitly assumed to have a cosmological origin led some (228) to argue that this was evidence for the cosmological redshift hypothesis for QSOs. However, it is fairly obvious that if the QSOs or even the N systems have large intrinsic redshift components, so that QSOs and Seyfert nuclei are at comparable distances, there is again continuity (236). Since the objects all belong to a class of phenomena with a narrower range in optical luminosities, i.e. the intrinsic luminosities are much more nearly the same, the main difference between QSOs and the other objects is then the nature of the redshift itself.

In any case the observed properties of Seyfert nuclei, N galaxies, and QSOs are so similar that the physics of all of these nuclei must be very similar, in the sense that the energy-generating mechanisms are likely to be the same, and the differences only those of degree. Thus in discussing the physics of the regions giving rise to the line spectra and continua in Seyfert nuclei and N galaxies, much of the argument can be taken over for QSOs. For the line spectra there are some differences in the analysis because the large redshifts make the lines in the ultraviolet observable.

The only other major difference comes when one uses estimates of size of very small structures obtained from flux variations. This method gives sizes independent of distance. Thus, if two objects at very different distance, say a Seyfert nucleus and a QSO assumed to have a cosmological redshift, are measured to have the same upper limit to the core size from variations assuming nonrelativistic motion, and if they have the same apparent brightness, very different radiation densities will be deduced, and consequently the inverse Compton effect will appear to be more important in the distant object (236). Also, if an angular size of the variable component has been measured directly, comparison of the linear size with the size deduced from variations ($R \lesssim c\tau$) will give very different apparent expansion speeds, depending on the distance of the object. For example, these quantities have been observed at high radio frequencies in NGC 1275 (Section 5) and in 3C 273 (237). The time scales for variations and the angular sizes in these two objects are similar, and yet because it is assumed that 3C 273 has a cosmological redshift, and is therefore some ten times further away than NGC 1275, the expansion velocity in NGC 1275 is only a fraction of the velocity of light, while for 3C 273 it appears to be about $3c$. Hence Gubbay et al have claimed that the 3C 273 result demonstrates that there is a relativistic effect present and the source is coherently expanding at relativistic speed. The alternative explanation is that 3C 273 is no further away than NGC 1275.

7. RADIO GALAXIES

In Section 5 we described, object by object, the evidence that violent activity is taking place in the nuclear regions of many galaxies, using the data available from the whole electromagnetic spectrum. Historically the discovery of nonthermal radio emission from extragalactic objects was one of the first indications that violent activity is taking place in galactic nuclei (110). The large energies that must be present in particles and field to explain such powerful sources have often been discussed (110, 130, 238). In the case of the strong sources, minimum energies $\gtrsim 10^{61}$ ergs must have been released in the nuclei (238). Since current theory suggests that wherever a nonthermal radio source is found, violent activity in the sense of the very rapid release of very large amounts of gravitational or nuclear energy has occurred, it is necessary in an article on galactic nuclei to survey briefly the data on radio galaxies. It is convenient to divide the discussion under the headings of "strong radio galaxies" and "normal radio galaxies."

Strong Radio Galaxies

The strong radio galaxies can be defined as those which have luminosities $\gtrsim 10^{41}$ ergs/sec in the frequency range $10^8 - 10^{10}$ Hz. In general, the radio sources have sizes much greater than those of the optical galaxies. A survey of their properties has been given by Moffet (239). In some cases high-resolution studies show a central radio component within the confines of the optical galaxy, but very frequently it is found that the radio source is a double system with the optical galaxy approximately placed symmetrically between the two components, while in some cases the radio emission originates in a halo centered on the optical galaxy. High-resolution studies now show that often the extended sources have complex structures.

While very large numbers of extragalactic radio sources have been discovered, only a small fraction of them have been identified with optical objects. In Table 5 we list all of those identified with galaxies for which redshifts have been obtained, together with the types of galaxy with which they have been identified, and a simple classification of their line spectra.

It can be seen from this table that the bulk of the identifications of the more luminous radio sources are with giant ellipticals, some classified as D-type systems. Because the majority of these galaxies are at great distances compared with those described in the previous sections, the linear sizes associated with their "nuclei" range between 200 pc and several kiloparsecs. Thus, when we discuss their nuclear properties, we are discussing regions which cannot be compared with the fine detail seen in sources like M87; we are looking only at very gross structures. It can also be seen that there is a progression in spectral characteristics and morphological types of the optical galaxies as the redshift increases, so that galaxies of all types have been found to be radio sources at small redshifts while, as the redshift increases, sources with higher luminosities are identified and the galaxies are then restricted more and more to giant ellipticals, or N systems with very strong optical emission-line characteristics.

To estimate the frequency of outbursts which give rise to strong radio sources, the only general method available is to compare the total number of radio galaxies in a given volume to the total number of galaxies. If it is assumed that in all galaxies, assumed to have the same age, say 10^{10} years, there is an equal chance of a nuclear event giving rise to a radio source greater than a given strength, then if x is the number of radio galaxies and y is the total number of galaxies, the duration of an outburst in a source is $\sim 10^{10} (x/y)$ years. This type of argument led to the early conclusion that, since Seyfert nuclei are present in $\sim 1-2\%$ of the spiral galaxies in the Shapley-Ames catalogue, the total duration of a Seyfert outburst must be about 10^8 years. When this argument was initially applied to the radio sources, not many radio galaxies were identified, and the assumption had to be made that all types of galaxy were equally likely to become strong radio sources. It was then found that only about 1 in 10^3 or 10^4 galaxies was a strong radio source, and consequently the time scale was estimated to be $\sim 10^6 - 10^7$ years (240). This crude estimate was then thought to be compatible with the dynamical time scale for the expansion to the large dimensions frequently seen; this time scale t might be approximately $R/c < t < R/v_A$, where v_A is the Alfvén speed in a hypothetical circumgalactic medium.

With the realization that the strong radio galaxies were largely identified with giant ellipticals with $M_v \leq -20$, Schmidt (241) made a careful study, restricting himself to sources of this type alone, so that the sample was rather small. He concluded that about 1 in 10 of the giant ellipticals is a strong source. These results were confirmed by Rogstad & Ekers (242). This argument suggests therefore that among these galaxies the harmonic mean lifetime for the outburst is $\sim 10^9$ years. This is very long when compared with the dynamical time based on the estimates made from a characteristic distance divided by a velocity. It also means that the total radiated energy for a powerful source with $L_r \simeq 10^{44}$ ergs/sec is about 3×10^{60} ergs.

The simple picture of the generation of a powerful extended radio source was originally that a cloud containing relativistic particles and magnetic flux expands out of the nucleus. There are many complications and some severe difficulties associated with such models, as follows.

Repeated activity in the nuclei of galaxies is known to occur. The variability of the very small high-frequency radio sources seen in some nuclei strongly suggests that successive bursts of relativistic particles are being ejected outward (243). Also, the forms of the radio structures in many extended sources indicate that more than one outburst has occurred (244, 245). Accepting that repeated activity does occur, van der Laan & Perola (246) concluded that the only model of conventional type, i.e. expansion of clouds from the nucleus, which is compatible with the spectral data, requires noncumulative continuous ejection for a large fraction of the life of a galaxy.

Many authors have attempted to develop models in which it is argued that clouds of relativistic particles are ejected from the nucleus of a galaxy and expand outward at relativistic speeds, the only confinement being due to an intergalactic medium (247–251). In these schemes confinement is achieved either by an intergalactic magnetic field, or by the ram pressure of the intergalactic medium. The difficulties with these models arise in part because there is no independent and unambiguous evidence for an intergalactic medium at all, and certainly not one with the density or the magnetic-field strength required. The values are unreasonably large: $\rho_{ig} = 10^{-27}$ g/cm^3 in the model of Mills & Sturrock to explain the structure in Cygnus A by the ram pressure, using the magnetic-field and intergalactic magnetic-field values $\sim 10^{-4} - 10^{-5}$ G in a realistic model of the Gold type (252). Nor does it appear that components with linear size r very small compared with the distance from the nucleus R ($r/R \simeq 10^{-2} - 10^{-3}$) can be explained by these schemes.

The alternative proposal (252), made originally because of the difficulties with Gold's model and because very compact radio components ($<0\rlap{.}''1$) had been found in large sources far from the centers from which they were ejected, is that the confinement is gravitational in origin, i.e., coherent objects are ejected and these are the secondary sources of particles. If this type of model is correct the double nature of many radio sources is probably due to the breakup of a massive object and the ejection of fragments (cf Section 11). The lifetime of the source is determined first by the period in which ejection of fragments takes place and by the spectrum of ejection velocities, and secondarily, by the period over which these fragments eject particles. It is no longer a simple matter of estimating the lifetime from a size divided by some typical expansion velocity.

So far, in determining what proportion of galaxies and which types are likely to generate strong radio sources, it has been assumed that the optical form of the galaxy is not affected by the outburst from the nucleus. How can we test this assumption? How sure are we that galaxies of the type that give rise to strong radio sources are generating the bulk of their luminosity by thermonuclear processes in stars? Could a significant fraction of their luminosity be due to radiation from a hot gas cloud or a nonthermal continuum generated in the explosion, rather than from stars? In some cases we know that the nuclear emission has this origin. This question has been studied by Grewing, Demoulin & Burbidge (253), who concluded from the only optical data at present available, namely redshifts and physical conditions obtained from spectra of the nuclear region, and integrated UBV colors, that such possibilities cannot be ruled out in many cases.

When the optical luminosity or form of the galaxy is changed by the outburst, so that the galaxy is given a quite different classification, estimates of the frequency of outburst made by the methods described earlier are meaningless.

We turn now to a brief review of the optical spectra of the strong radio galaxies. As can be seen from Table 5 they show: absorption lines only, of the type seen in normal ellipticals; absorption together with [O II] $\lambda 3727$, [N II] $\lambda\lambda 6548, 6583$, and H$\alpha$, again a situation seen in a fraction of normal galaxies; very strong emission. Some of the galaxies in Table 5 are Seyfert galaxies or N galaxies whose spectra have been described in Section 5. However, many of the high-excitation lines seen in those galaxies are also seen in radio galaxies not classified as Seyfert or N systems (207, 223, 224), The best example of such a galaxy is 3C 405 (Cygnus A) whose spectrum contains [Ne V] $\lambda 3426$, [Ne III] $\lambda 3869, 3968$, [O II] $\lambda 3727$, [O III] $\lambda\lambda 4363, 4959, 5007$, He II $\lambda 4686$, and the Balmer series. In this object the outbursts have generated a cloud of hot gas with an extent of many kiloparsecs. If there is a normal galaxy underlying this cloud it is totally overwhelmed by the strong emission.

While there is a tendency for the stronger radio galaxies to show nuclear emission more frequently than the weak sources, some well-known very powerful radio galaxies, for example 3C 218 (Hydra A), show only normal absorption lines in their nuclei.

If a single explosive event gave rise to optical excitation in the nuclear region and to an extended radio source, it is likely that the time scale for decay of the optical excitation is very much shorter than that for the radio source. Thus, the fact that optical activity appears very often with a powerful radio source also suggests that continuous nuclear activity is the rule rather than the exception.

Normal Radio Galaxies

The strong radio galaxies were first found because the radio astronomers discovered sources, and then identified them with optical objects. The other approach, to search for radio emission from known galaxies, has led to the discovery that nonthermal radio emission is commonly radiated by normal spiral galaxies, and less commonly by ellipticals. Surveys have been conducted by Heeschen & Wade (254), de Jong (255, 153), Cameron & Glanfield (256), Tovmassian (257, 258), Rogstad & Ekers (242), and Lang & Terzian (259). The power levels of such sources lie in the range $\sim 10^{38} - 10^{41}$ ergs/sec. To what extent is this radiation another manifestation of nuclear activity? The answer to this question is not completely clear, but the following points appear to be of importance.

1. Most of the radio flux is coming from a region whose angular extent is sometimes equal to, but usually smaller than, the angular size of the optical galaxy. It is also now clear that radio halos are not common. As was discussed in Section 5, Seyfert galaxies and other systems known to have very active nuclei have nuclear radio sources that generate most of the radiation. However, high-resolution studies are beginning to show that galaxies with comparatively normal nuclei in optical wavelengths sometimes have very powerful nuclear components. For example, Wade (193) has shown that 40% of the flux at 11 cm from M81 comes from a region $<2''$ (<30 pc) in diameter centered on the nucleus. On the other hand, such a source is certainly not present in M31.

2. The radio spectra of the spirals show what appears to be a break between about 200 and 1200 MHz, so that as one goes to higher frequencies the slope of the spectrum steepens. Lang & Terzian (259) have shown that the change in the spectral index is 0.8 ± 0.4. This effect is present in the spectrum of our own Galaxy (260, 261) and also in the spectrum of M31 (262). The most plausible interpretation of this effect and one suggested by de Jong and by Lang & Terzian is that the electrons are being generated largely in recurrent bursts in the galactic nucleus, that the bursts are frequent enough to maintain most spiral galaxies as weak sources, but that they are infrequent enough so that the aging effect leading to steepening of the spectra has time to take place between bursts. The conditions in the nuclei of normal radio galaxies, e.g. our Galaxy and NGC 253, are certainly not incompatible with this idea. Of course this implies that the bulk of the radio electrons arise in the nucleus of the galaxy, and not in supernova remnants in spiral arms, as has sometimes been supposed.

3. Elliptical galaxies are not as frequently found to be radio sources as are the spirals. However, when they are detected as normal or weak radio sources, they either tend to have normal straight radio spectra with spectral indices $\alpha \approx 0.9$ with a source size comparable to that of the optical galaxy, or tend to have peculiarities of the type found in NCG 4278 and NGC 1052 (Section 5), which indicate nuclear activity (263). Heeschen (263) has pointed

NUCLEI OF GALAXIES 417

out that the occurrence of such peculiar radio spectra and very small nuclear emission cores is relatively common in ellipticals.

4. Tovmassian (257) found in his studies of many of the galaxies in the lists of Markarian (Section 8) that the radio flux from such nuclei markedly exceeds that from normal spirals. Wade (193) makes the same point for Seyfert galaxies discussed individually in Section 1.

To summarize: the available data strengthen the earlier conclusion that nuclear activity is responsible for a large part of all of the radio emission from galaxies with radio luminosities ranging all the way from 10^{38} to 10^{45} ergs/sec. In apparently normal spiral galaxies some relativistic electrons are undoubtedly generated by supernova remnants, but this may be a comparatively minor effect. In both weak and strong sources, nuclear activity over a rather long time scale is indicated.

8. SURVEYS WHICH HAVE LED TO THE DISCOVERY OF OBJECTS WITH PECULIAR NUCLEI

A number of surveys have led to the discovery of galaxies with nuclei that are conspicuous in various ways. Two methods have been used. In one method the form of the galaxy is used as the primary criterion. In the other the selection is made on the basis of the energy distribution (crudely, the color) of the nucleus. To make the discussion complete we briefly describe these investigations.

Form Classification

Morgan's approach—As described earlier (4), Morgan classified galaxies from large-scale plates and therefore restricted himself to bright galaxies, mostly contained in the Shapley-Ames catalogue. More than 600 galaxies were classified. His category of galaxies with peculiar or complex nuclei includes N galaxies (NGC 4051, for example) and also the galaxies with multiple hot spots (like NGC 5248, 1808, 4321, and 3351). He was therefore drawing attention, not only to the type of nucleus discussed in Section 5, but also to objects whose nuclei are made up of large numbers of bright H II regions which may be excited by stars. In such galaxies with hot spots there is at present no evidence of nonthermal energy generation.

The classification of Sersic & Pastoriza.—Sersic & Pastoriza (264–266) picked out bright galaxies which have "peculiar nuclei" in the sense that "there is a luminosity profile in which there is a change in the slope, and some structure due to the existence of high excitation gas clouds around the area of the true nucleus of the galaxy." Since they used large-scale plates taken at the Mount Wilson and Cordoba Observatories, there is some overlap with Morgan's work. Some of their examples are NGC 613 where the nuclear region itself resembles a small spiral, the barred spirals NGC 1097 and 1365, discussed previously, NGC 1808 (one of the galaxies with hot-spot nuclei), NGC 5236, and NGC 2997. Among the galaxies brighter than $M_v = 11$, they found 20 galaxies with peculiar nuclei out of 136 galaxies having orientations such that their nuclei can be clearly classified. They found no correlation between the incidence of such peculiarities and galactic types among the spirals.

The classification of Vorontsov-Velyaminov.—Vorontzov-Velyaminov and his collaborators (267–270) produced an *Atlas of Interacting Galaxies* and a Morphological Catalogue of Galaxies from studies of the Palomar Sky Atlas. Arp (271) published an atlas of large-scale photographs of many of these objects. Vorontsov-Velyaminov's main purpose was not to concentrate on nuclei, but spectroscopic investigations of systems first pointed out by him (e.g. VV 144 and VV 150) led to the realization that some of them are objects with remarkable nuclei.

Compact galaxies classified by Zwicky.—Zwicky (204) defined compact galaxies as follows: "We call compact galaxies those which can just be distinguished from stars on plates taken with the Palomar 48-inch Schmidt telescope and which have diameters of 2″ to 5″." Earlier it is stated "We shall call a galaxy moderately compact if its image on photographs taken with the 18-inch Palomar Schmidt can just barely be distinguished from stars of the same apparent brightness. These systems have diameters of about 5″ to 10″." Many of them are listed in the *Catalogue of Galaxies and Clusters of Galaxies* (272–277).

Zwicky privately circulated seven lists, containing about 2000 objects, entitled Compact Galaxies and Compact Parts of Galaxies; Eruptive and Posteruptive Galaxies. The classification of objects in this way has helped to pick out galaxies whose nuclei appear to be very prominent and often dominant. However, there has developed some confusion in nomenclature, because in their catalogues of galaxies and clusters, Zwicky and his collaborators have referred to galaxies as being compact, very compact, or extremely compact, whereas according to the definitions given by Zwicky above, they should all be moderately compact. Also, Sargent (220) has pointed out that many of the 2000 objects in Zwicky's list do not appear in his catalogues, and many of the objects described as compact in the catalogues are not in his lists. Fairall (278) compiled a list of compact galaxies in areas which do not overlap with Zwicky. He concentrated on small and rather blue objects, and listed about 80 objects.

The only extensive spectroscopic investigation of compact galaxies is that due to Sargent (220) who studied low-dispersion spectra of about 140 Zwicky objects. He also measured colors of some and obtained large-scale photographs of about 30. Needless to say, he found that they are a heterogeneous collection: 10 have Seyfert spectra and are individually described in the previous discussion, 33 have sharp emission lines, 20 have both absorption and emission lines, 60 have absorption lines only, and 3 have continuous spectra. The remainder are galactic stars, while one turned out to be a planetary nebula. Since the mean redshift is 0.033, on average the sample contains objects at an average distance of greater than 100 Mpc, and the mean absolute magnitude is -20.2, while the range in absolute magnitude is from -14.9 to -23.2, the brightest being Seyfert objects (assumed to have cosmological redshifts). For the galaxies for which large-scale photographs were available, Sargent showed that the objects were indeed compact in the sense of having an abnormally high luminosity per square parsec; the values range from ~ 10 to greater than 1000 $L_\odot$ per square parsec compared with values for normal galaxies of about 1 $L_\odot$ per square parsec.

The UBV colors and the line spectra are related. Most of the Zwicky galaxies populate the same region of the $U-B$, $B-V$ diagram as the normal galaxies. The exceptions are all compact systems with emission lines. Many of these, particularly the objects with broad emission lines, are bluer in $U-B$ than any normal galaxy. They are N or Seyfert galaxies and they have similar colors to QSOs. This is also true of some of the objects with sharp emission lines.

Sargent also showed that objects with H, K, and the G band in absorption are all redder than $U-B = +0.1$. Objects with early-type absorption lines and emission lines populate the range from $-0.15 < U-B < 0.2$. Objects with emission lines only have $U-B < -0.15$. Clearly, Zwicky surveys have led to the discovery of abnormal types of galaxy which are often dominated by very bright nuclei. However, Sargent's work has shown that investigations in which observations of the line spectrum and energy distribution are made are essential to give a rough idea of what one is looking at.

Classification According to Color

The discovery of extragalactic objects whose nuclei play a dominant role has also been made, using the color of the object as the primary criterion.

The method of Haro.—Haro (279) used filters and the objective prism attached to the Tonantzintla Schmidt telescope to pick out objects with strong ultraviolet fluxes. Many of these objects turn out to be galaxies with strong-emission-line nuclei, some with Seyfert characteristics (cf 280). About one third have jets associated with the nuclei (281).

The method of Markarian.—Markarian and his colleagues (282–285) investigated galaxies which have abnormal nuclei in the sense that they have a large ultraviolet excess or strong emission lines, and no indication of a normal stellar population. These investigations are similar to the earlier one of Haro (279). Markarian (282) gave a list of 40 bright galaxies with these characteristics and, more recently (285), he published a further list of 70 fainter galaxies observed with a 1.5° objective prism and 40-inch Schmidt at the Byurakan Observatory. He divided the nuclei into two types: those which, from their energy distribution, are similar to QSOs, and those in which the energy distribution appears to arise from H II regions and high-luminosity stars of the type seen in spiral arms.

Studies of individual objects picked out by Markarian were made by Khachikian and Weedman, by Sargent, and others, and are described in Section 5. They confirmed that a number of the Markarian objects have Seyfert nuclei, while many others show narrow strong emission lines and early-type stellar spectra.

The method of Sandage & Luyten used to discover quasistellar objects.— This section would not be complete without some mention of the methods used to discover QSOs independent of radio astronomy. In this case it is necessary to distinguish them from stars. They must therefore be isolated first from their energy distribution alone, and then confirmed as extragalactic objects by obtaining spectra. The method has been described in detail by Burbidge & Burbidge (229). Sandage (286) first developed it, using as a basis the lists of blue stellar objects found by Luyten, and by Haro and his colleagues. Sandage & Luyten (288, 287) concluded by surveying small areas of the sky that the density of such objects down to about 22^m may amount to some 200 objects per square degree. Braccesi (289) and Braccesi, Lynds & Sandage (233) showed that such objects can be picked out from ultraviolet stellar objects, many of which are white dwarfs, by choosing those objects with excess radiation in the near infrared. Not all of the objects studied in this way have turned out to be completely stellar, and in some cases they have later been reclassified as N galaxies (e.g. B264).

9. PHYSICAL CONDITIONS IN SEYFERT NUCLEI AND N GALAXIES

REGIONS GIVING RISE TO THE LINE SPECTRUM

From the large amount of data now available on the line spectra of some of the brightest Seyfert nuclei and N galaxies, it has been possible to obtain information about the excitation and ionization conditions in the gas, the composition, and the possible mechanisms of line broadening.

References for studies of individual Seyfert galaxies were given in the previous section. In addition, theoretical studies have been carried out by Woltjer (163), Williams & Weymann (290), Souffrin (291, 292), Shklovsky (293), Kaneko & Ohtani (294), Osterbrock (295–298), Nussbaumer & Osterbrock (437), and Weymann (438). We summarize the results of these studies.

The electron temperature is obtained from the [O III] line ratio I(4959 +5007)/I(4363) and the [N II] ratio I(6548+6583)/I(5577), the latter being less certain, in part because [N II] $\lambda\lambda 6548, 6583$ are blended with Hα. Typical electron temperatures obtained by this method are 9000°K (NGC 1068) and ~18000°K (NGC 4151). From the intensity ratios of the [O II] doublet the electron density $N_e > 4 \times 10^3$ cm^{-3}, from the [S II] doublet $N_e > 4 \times 10^4$ cm^{-3}, and from [Ar IV] $N_e \approx 2 \times 10^5$ cm^{-3}, in NGC 1068. Similar methods give $N_e \sim 5 \times 10^3$ cm^{-3} in NGC 4151. In both these objects there is evidence for large density fluctuations with small condensations contained within a much larger lower-density volume. This result is derived from the absolute strengths of the Balmer emission lines, together with the high densities and the observed sizes of the nuclei.

The most plausible source of energy for the emission-line spectrum is photoionization from a source with a continuous spectrum with a power-law spectrum $P(\nu) \propto \nu^{-\alpha}$ (290). In the case of NGC 4151 the nonthermal continuum is well represented by a power law of this type with $\alpha \sim 1.2$, and the observed line intensities of all the ions can be reasonably well explained if this is the source of ionization. Even the lines due to [Fe X] λ6374 and [Fe XIV] λ5303 (298), earlier (195) attributed to a hot low-density collisionally ionized corona, can be explained by this ionization (298). Shklovsky (293) emphasized that another component of continuous radiation with a maximum in the X-ray region may also be important, both in Seyfert galaxies and in radio galaxies. In Cygnus A, for example, the great strength of [O I]λ6300 can only be understood if very short-wavelength photons are involved. On the other hand, Nussbaumer & Osterbrock (437) have supported the earlier suggestion (195) that the [Fe X] and [Fe XIV] lines can arise from collisional ionization in a hot gas with temperatures up to $\geq 2 \times 10^6$ degrees. They point out that if this hot gas exists, it radiates more energy in the ultraviolet than all of the other observed radiation, and it is conceivable therefore that it could be responsible for all of the ionization. However, since there is independent evidence for optical nonthermal radiation in Seyfert nuclei, it would be strange if such radiation was not also powerful in the ultraviolet.

Two major problems stem from the observations of the hydrogen recombination lines and He II λ4686: the Balmer decrement and the line profiles.

The measured Balmer decrements in NGC 1068, NGC 3227, and probably also in NGC 4151 are so steep that they cannot be explained by any existing theoretical calculation, However, it was conjectured that they might be due to reddening. If this were the case the continuum will also be reddened so that the true continuum luminosity is much greater than the observations indicate. Wampler (299) compared the observed intensity ratios of the infrared and violet [S II] lines λλ10287, 10321, and 4069 which arise from transitions from $3p^3$ $^2P_{3/2}$, and λλ10337, 10373, 4076, which arise from transitions from $3p^3$ $^2P_{1/2}$, with the theoretical intensity ratios. He also compared the observed and theoretical ratios of corresponding Balmer and Paschen lines in some cases, and found that substantial reddening is indeed present. He also pointed out that, since the line emission comes from a larger volume than the continuum radiation, the reddening implied by the line ratios does not *necessarily* imply that the continuum is reddened. It all depends on the distribution of the dust. If the continuum is reddened, the absolute violet magnitudes of Seyfert nuclei should be increased by several magnitudes. If it is not, the absolute strengths of the emission lines should be increased and further sources of ionization, perhaps of the kind suggested by Shklovsky, are required. Since Wampler has established that dust is present in Seyfert nuclei, more detailed calculations are required.

An important characteristic of the line spectra is that the hydrogen and helium lines are very broad with extensive wings, and the forbidden lines are much sharper. This characteristic was first found by Seyfert (154). It is also present in the spectra of N galaxies and QSOs. A range of widths of the hydrogen lines is found in the Seyfert galaxies (see Section 5 for the individual objects). While there is little quantitative evidence in the literature, the average widths of the Balmer lines in Seyfert galaxies appear to be somewhat greater than those of N galaxies, and the observed widths of the hydrogen lines in QSOs are comparable with those in Seyfert nuclei (300). However, for the QSOs the redshifts are important, and to obtain a proper comparison, the observed widths must be compared after being divided by $(1+z)$. Since the hydrogen lines have extensive shallow wings, it is clear that the differences between linewidths in different systems may be due largely to the difference in the relative contributions of the (thermally) emitting gas and the (nonthermal) continuum in a given source.

In the early work (154, 163) it was supposed that the large widths were due to Doppler broadening so that very large turbulent motions of the order of several thousand km/sec are present. Since these velocities are much larger than the escape velocity, unless there is an exceedingly large mass concentration [which can be ruled out at least in the case of NGC 1068 (162)], the large widths were thought to give fairly direct evidence for large-scale mass loss from Seyfert nuclei (162, 110). To explain the different widths of permitted and forbidden lines it must be argued that there are both high- and low-velocity gas components, and that the high-velocity component is either cool enough or dense enough so that forbidden-line emission cannot arise in it. The high-resolution studies of some nuclei by Walker and his colleagues show that there are discrete clouds moving at several hundred km/sec in such nuclei, but there is no evidence of discrete clouds with random velocities of thousands of km/sec. Anderson & Kraft (197) have also shown that shells of gas are expanding from the nucleus of NGC 4151 at velocities of 200 to 800 km/sec. In the case of NGC 1275, it was shown that gas outside the Seyfert nucleus is moving at speeds of up to 3000 km/sec with respect to it. Thus there is good evidence that high-velocity clouds are present, so that some of the line broadening is due to the Doppler effect.

However, there have been some suggestions by Field, Oke & Sargent (195), Kaneko & Ohtani (294), and Weymann (438) that the broad wings are due in part to electron scattering. If this were the case one would expect the wings of the lines to be smooth and featureless; it has not been established whether this is the case or not. To explain the difference between the widths of permitted and forbidden lines it is necessary to argue that the hydrogen lines suffer more electron scattering and consequently that Balmer series photons have a longer pathlength for escape from the nucleus. However, as Osterbrock has stressed, this interpretation is unconvincing because probably only the Lyman-line photons suffer resonance line scattering, and if, as a result of their longer pathlength, they undergo electron scattering and are shifted in frequency, then they have only a very small probability of being converted to Balmer-line photons by further resonance fluorescence. It seems unlikely that the gas is optically thick to the Balmer lines unless very small dense regions are involved. However, Weymann (438) has recently pointed out that the observations of Anderson & Kraft (197) and Andrillat & Souffrin (194) may indicate that in some cases electron scattering may be able to explain the observations. In the case of QSOs, electron scattering as a broadening mechanism looks in general more promising since the prominent resonance features are Mg II $\lambda 2798$, C IV $\lambda 1549$, and Ly-α (300).

Thus it must be concluded that in galactic nuclei the broadening is most likely to be due to very large random motions and that it also indicates that extensive mass loss is taking place.

From the analyses of the line spectra of Seyfert nuclei there is no evidence of abnormal chemical composition of the gas as compared with the composition of Population I objects in our own Galaxy. However, this does not mean that abnormalities could not be present. It must always be remembered that one is studying integrated spectra of many regions and clouds with very different physical conditions, and the situation is not as as simple as that encountered, for example, in analysis of the spectra of planetary nebulae.

The Origin of the Continuous Radiation

In the optical part of the spectra of Seyfert nuclei and N galaxies, two components of continuum radiation have been identified. The thermal component simply arises from the hot gas clouds that produce the line spectrum, and it will not be discussed further. The second component appears to be nonthermal in origin, and in most cases it arises in a very small volume compared with the region giving rise to the line spectrum. The nonthermal character has been deduced from a number of properties in different objects: in NGC 1068, from the flat energy distribution and the steep increase into the infrared; in NGC 4151, from the form of the energy curve and the rapid variability; in 3C 120, from the form of the energy curve; in NGC 1275, from the optical variability; in 3C 371 and 3C 390.3, from the form of the energy curve and the variability; and in 3C 109, from the variability alone. In addition to this, significant amounts of linear polarization have been measured in some systems.

It has also been found that a number of these objects, in particular NGC 1068, 1275, 4151, and 3C 120, are powerful infrared sources. It is possible that the infrared flux is thermal radiation from dust heated by a nonthermal ultraviolet source, but it is more likely to have a direct nonthermal origin.

These objects are also known to be nonthermal radio sources with very small nuclear components, and a number of them, particularly NGC 1275 and 3C 120, are rapidly variable at high frequencies. The Seyfert and N galaxies share these properties with the nuclei of galaxies such as M82 and M87, described earlier.

How does one explain this complex continuous spectrum? This question has been considered by Demoulin & Burbidge (236) who concentrated on the optical spectra, by Rees, Silk, Werner & Wickramasinghe (175), by Burbidge & Stein (92) who were concerned with the infrared emission, and by Pauliny-Toth & Kellermann (243, 301) who considered the radio emission and its variability.

It is generally accepted that the radiation mechanism, at least for the optical and radio flux, and perhaps the infrared flux also, either is due to the synchrotron process or, less likely, is generated by Compton scattering, though Colgate (302, 303) has considered that electrostatic bremsstrahlung scattered from plasma waves might be important in these objects as well as in QSOs.

The luminosities range from about 10^{40} ergs/sec for the radio sources in Seyfert nuclei, about $10^{43}-10^{44}$ ergs/sec for the optical flux from Seyfert nuclei, to about 10^{46} ergs/sec for the infrared flux (between $\sim 2-20$ μ) in some objects. These very large nonthermal fluxes are to be compared with the luminosities of the hot gas clouds, which rarely amount to more than 10^{43} ergs/sec, and the total luminosities of bright galaxies, which are about 10^{44} ergs/sec.

If the emission mechanism is the incoherent synchrotron process, calculations of the energetics of the source can be made, if its luminosity, energy spectrum, and size are known. While we know that the optical, infrared, and high-frequency radio flux each come from very small regions, we do not know whether these small regions coexist, or whether different parts of the spectrum arise in different regions.

For the optical flux the energetics have been calculated on the assumption that it is electron synchrotron radiation, also on the less plausible idea that it is generated by Compton scattering from lower-frequency photons, or that it is proton synchrotron radiation. The results of Demoulin & Burbidge (236), for some of the sources, are shown in Table 6.

If the electron synchrotron process operates in the small sources, the minimum magnetic fields are ~ 0.1 G, greater than the equipartition value; total energies in particles and field are given in Table 6 for a variety of assumptions. If the proton synchrotron process operates, the magnetic fields must be much stronger, $\sim 10^3$ G, the proton energies are $\sim 10^{14}$ eV, and the minimum total energies are 10^3-10^4 times those for the electron synchrotron process, while the Compton effect is negligible; however, the ambient gas density must be $\leq 10^4$ particles cm^{-3}, i.e. the continuum source makes a hole in gas clouds. If the optical flux is Compton radiation generated by the interaction of high-energy electrons with an intense radiation field of lower frequency (some part of the high-frequency radio flux), the process is highly

TABLE 6. Energetic properties of optical synchrotron sources in galaxies and typical QSOs according to Demoulin & Burbidge (236)

Object	B_c (G)[a]	Total energy if $B=B_c$ [b] (ergs)	B_{eq} (G)[c]	Total energy if $B=B_{eq}$ [d] (ergs)	Typical electron energy (GeV)	Distance traveled by electron (cm)
M82	2.1 (−6)	1.9 (55) 9.3 (56)	6.2 (−6)	5.0 (54) 1.9 (56) 4.5 (55)	2500 ($B=B_{eq}$)	8 (22)
M87	2.1 (−5)	5.2 (53) 2.6 (55)	5.8 (−5)	1.6 (53) 6.0 (54) 1.5 (54)	810 ($B=B_{eq}$)	3 (21)
NGC 1068	5.4 (−4)	1.4 (54) 6.1 (55)	5.0 (−4)	5.0 (53) 1.9 (55) 4.6 (54)	270 ($B=B_{eq}$)	9 (19)
NGC 4151	2.6 (−1)	2.8 (49) 1.7 (50)	1.1 (−1)	1.6 (49) 5.9 (50) 1.4 (50)	12 ($B=B_c$)	9 (15)
3C 120	3.2 (−1)	1.1 (51) 2.7 (51)	8.8 (−2)	3.0 (50) 1.1 (52) 2.7 (51)	11 ($B=B_c$)	7 (15)
3C 371 (main region)	8.0 (−2)	9.3 (51) 3.0 (52)	2.6 (−2)	3.2 (51) 1.1 (53) 2.9 (52)	22 ($B=B_c$)	6 (16)
3C 371 (small region)	5.1	4.5 (48) 8.5 (48)	1.2	9.0 (47) 3.4 (49) 8.4 (48)	3 ($B=B_c$)	1 (14)
3C 390.3	1.6 (−1)	1.0 (52) 2.3 (52)	4.3 (−2)	2.5 (51) 9.3 (52) 2.2 (52)	15 ($B=B_c$)	2 (16)
3C 109	4.6 (−2)	5.2 (51) 2.6 (52)	1.7 (−2)	2.5 (51) 9.3 (52) 2.2 (52)	30 ($B=B_c$)	1 (17)

[a] Minimum magnetic field if Compton effect is unimportant.
[b] Upper line corresponds to situation in which it is assumed that the energy in protons is equal to that in electrons, and lower line to situation in which energy in protons is 100 times that in electrons.
[c] Equipartition magnetic field.
[d] Total energy calculated under different assumptions concerning equipartition. For details see (236).

TABLE 6. (Continued)

Object	B_c (G)[a]	Total energy if $B=B_c$ [b] (ergs)	B_{eq} (G)[c]	Total energy if $B=B_{eq}$ [d] (ergs)	Typical electron energy (GeV)	Distance traveled by electron (cm)
QSO (cosmological)	1.6 (+1)	4.4 (52) 4.5 (52)	1.2	9.0 (50) 3.5 (52) 8.4 (51)	1.5 ($B=B_c$)	2 (13)
QSO (local)	8.0 (−1)	1.1 (50) 2.7 (50)	2.2 (−1)	3.0 (49) 1.1 (51) 2.7 (50)	7 ($B=B_c$)	2 (15)
QSO (very local)	1.6 (−2)	4.9 (47) 2.3 (49)	2.4 (−2)	3.4 (47) 1.3 (49) 3.1 (48)	40 ($B=B_{eq}$)	3 (17)

inefficient, since a large flux of electrons is required, and only a small fraction of their energy will be transferred in the time taken to cross the source. This process is open to observational test, since in most cases a much larger flux of harder radiation (X rays or γ rays) will be generated.

Demoulin & Burbidge concluded that the electron synchrotron process is likely to be the most efficient for the generation of the nonthermal optical flux. For the infrared flux two possible origins have been discussed. The first is that it is due to ultraviolet radiation, presumably of nonthermal origin, absorbed and reradiated at infrared wavelengths by a surrounding dust cloud; the second that it is synchrotron radiation. The observations of Wampler (299), previously mentioned, show that dust is present in such nuclei, and there is considerable evidence from photographs that fine dust structures are often present in galaxies and can be seen right into the nuclear regions (58, 110).

If the infrared flux is due to dust, one can calculate the amount of material which must be condensed into dust grains using the simplest assumptions; for example, for NGC 1068 and NGC 4151 Burbidge & Stein (92) found that these masses were $\gtrsim 10^3 M_\odot$ and $\gtrsim 2 \times 10^3 M_\odot$, respectively. Now the ratio of the mass of the gas to the mass of the dust that is allowed is given by $\rho_{gas}/\rho_{dust} > n_{gas}/(\mu n_{dust})$ where n is the number density of the least abundant element in each grain molecule and μ is the molecular mass associated with each of the least abundant molecules. For solids likely to exist in the form of grains (graphite and silicates), $\rho_{gas}/\rho_{dust} \gtrsim 300$, provided that the abundances of carbon and heavier elements relative to hydrogen are the same as those in Population I objects in our Galaxy. This means that $\sim 10^6 M_\odot$ of gas should coexist with the dust. However, as can be seen from Table 7, this is larger than the amount of ionized gas deduced to lie within the same

TABLE 7. Physical parameters of dust clouds needed to explain infrared flux from Seyfert nuclei. taken from Burbidge & Stein (92)

Source	D	L (erg sec^{-1})	R (cm)	M_{grains} ($M_\odot$)	M_{gas} ($M_\odot$)	$M_{ionized\ gas}$ (optical observations) ($M_\odot$)	R (optical observations) (cm)
NGC 1068	10 Mpc	$>10^{44}$	$>3\times10^{19}$	$>10^3$	$>3\times10^5$	1.5×10^5	1.5×10^{20}
NGC 4151	10 Mpc	$>10^{44}$	$>3\times10^{19}$	$>2\times10^3$	$>6\times10^5$	3×10^4	1.5×10^{20}
	10 Mpc	$>4\times10^{43}$	$>2\times10^{19}$	$>6\times10^2$	$>2\times10^5$	2.7×10^2	3×10^{17}

volumes from optical observations (166, 195). This is an objection to the dust hypothesis unless either a large amount of undetected gas is present, for example in the form of dense H I regions (which might give rise to the Fe II line spectrum sometimes seen) or the composition of the gas is different in the sense that the heavy elements are overabundant relative to hydrogen. A second and more serious objection to the dust hypothesis is that the size of the dust cloud must be comparatively large if the grains are radiating like blackbodies, and the radiation is emitted close to the peaks of blackbody curves which must lie in the 10 μ range. The sizes come out to be about 150 light years for both NGC 1068 and NGC 4151. However, there appears to be evidence for fairly rapid variations near 2 μ in NGC 1068, and changes appear to have occurred at 10 μ in NGC 4151 over a time of about a year (201). There is still some uncertainty about this because comparison has been made between the measurements of two different groups (174, 201, 304). If such variations are confirmed, they strongly suggest that the region from which the infrared flux arises has dimensions of light years or less, much smaller than the sizes calculated on the blackbody dustgrain hypothesis.

The problem of maintaining dust in the active nucleus of a galaxy is under study. A number of factors, including the effect of an intense ultraviolet radiation field, sputtering by charged particles, and containment by a magnetic field of a charged dustgrain, are not properly understood. If the infrared flux is directly synchrotron radiation, calculations of the energetic conditions in the source can be made in the same way as was done for the optical flux. Again, one can assume either that the total energy of particles and field is a minimum, or that the Compton effect is not important. Because the flux of radiation in the infrared is considerably higher than that in the optical part of the spectrum, if it arises in the same region in the nucleus as the optical flux, or if it arises in a volume of similar size, the radiation density is higher than that of the optical flux, so the magnetic field must be stronger if the synchrotron process is to dominate over Compton scattering.

TABLE 8. Energetic properties of infrared sources in NGC 4151 and NGC 1068 for simple synchrotron models, taken from Burbidge & Stein (92)

	NGC 4151				NGC 1068		
R (assumed size)	$E_{\min}a^{-1}$ [a] (ergs)	$B_{eq}b^{-1}$ [a] (G)	B_c [b] (G)	R (assumed size)	$E_{\min}a^{-1}$ [a] (ergs)	$B_{eq}b^{-1}$ [a] (G)	B_c [b] (G)
10^{18} cm	1.4×10^{52}	0.12	0.36	10^{18} cm	9.7×10^{51}	0.10	0.29
10^{17} cm	7.2×10^{50}	0.85	3.60	10^{17} cm	5.0×10^{50}	0.71	2.90
10^{16} cm	3.7×10^{49}	6.10	36.00	10^{16} cm	2.6×10^{49}	5.10	29.00
10^{15} cm	1.9×10^{48}	44.00	360.00	10^{15} cm	1.3×10^{48}	37.00	290.00

[a] $E_{\min}$ is the minimum total energy required to generate the observed luminosity and is given by $E_{\min} = akR^{9/7}L^{4/7}$ where k is a constant depending on the low-frequency cutoff, and L is the luminosity; B_{eq} is the magnetic-field strength corresponding to $E_{\min}$; a and b are constants determined by the proton/electron energy ratio in the source. If the energy in the protons is equal to that in the electrons, $a = 0.6$ and $b = 1$. If the energy in the protons is 100 times that in the electrons, $a = 22.6$ if the energy residing in electrons and magnetic field is a minimum, and $a = 5.4$ if the energy residing in protons, electrons, and magnetic field is a minimum. In both cases $b = 3.8$.

[b] B_c is the minimum value of the magnetic field for the Compton effect not to dominate; i.e. $B_c = (8L/R^2c)$.

Results of calculations for NGC 1068 and NGC 4151 are given in Table 8. The numbers obtained here are very similar to those which would be obtained for the central exciting source even if the infrared radiation were due to dust, heated by that source. The only differences would come because in that case the initial flux would be made up of ultraviolet photons rather than infrared photons.

The conclusions are that the magnetic field required is rather large, and it may very well be larger than that required if the minimum energy condition (approximate equipartition between particle energy and field energy) is fulfilled. This in turn means that for a single source the Compton problem, first encountered for cosmologically distant QSOs (305), may be present, though not to the extent that it occurs for QSOs at great distances. However, the apparent difficulties can be overcome if it is supposed that the particles are generated in a number of small, highly condensed objects spread throughout the volume of the source. This type of model has already been mentioned in connection with the source at the center of our Galaxy and in M87 (Sections 4 and 5).

The steep rise of the spectra into the infrared means that the particle spectra also rise very steeply. In the infrared $F(\nu) \propto \nu^{-2}$ so that the particle spectrum must have the form $N(E) \propto E^{-5}$. How is variability of the optical and infrared flux to be explained in these models? Certainly the particles will die in the source regions in times which are short compared with R/c, and thus the variability must be determined by variations in the particle-injection rate.

We now turn to the continuum radio sources in these nuclei. There is no doubt that these are synchrotron sources, and there are several points to discuss. They are the overall energetics of the radio components, the question of whether the sources are optically thick at some radio frequencies, and the variability in the centimeter and millimeter wavelength range. The nuclear sources in such galaxies as NGC 1275 and 3C 120 have rather flat spectra as compared with the more extended sources, and they are very small, as was discussed earlier. The spectral-energy distribution is not simply that given by a power law, as is the case for the more extended sources which have been ejected from nuclei or QSOs. In such small variable sources the spectra often show regions of positive curvature and several maxima and minima. This suggests that the sources are composite, and positive curvature combined with very small size indicates that some components are optically thick at short wavelengths (243, 306). Calculations of overall energetics can be made in the usual way, assuming approximate equipartition in energy between particles and magnetic fields, and in general the energies residing in particles and fields are small compared with those required to explain the optical and infrared flux. This is because the radio luminosities are very much smaller. However, for the very tiny nuclear sources, the source is often self-absorbed. In this case the upper limit to the magnetic field set by the self-absorption and the size of the source is less than the equipartition magnetic-field value. Taken at its face value this means that the total energy in the particles must be very much greater than that in the field, and the total energy required to explain the source increases accordingly. A systematic discussion of these problems for radio sources in general has been given by Scheuer & Williams (306).

Much interest has centered on the problem of explaining the variability of such nuclear radio sources. Kellermann & Pauliny-Toth (243) have been largely responsible for model construction. They pointed out that the spectra indicate that the variable components are optically thick at wavelengths where the flux density is increasing, and optically thin at wavelengths where it is decreasing, and that this is just the behavior expected from an expanding cloud of relativistic electrons which is initially optically thick at some wavelengths. As the source expands, both the magnetic field and electron energy decrease and the cloud becomes optically thin at longer and longer wavelengths. At any frequency where the expanding component is optically thick, the flux density will increase when a significant part of the total flux is coming from the expanding component. In the optically thin part the flux will decrease with time in the way described by Shklovsky (307) for expanding supernova remnants.

This simple model can be fitted reasonably well to the observations of radio variations of sources like 3C 120. In that object there appear to have been two successive outbursts since 1965, separated in time by <2 years. In the case of NGC 1275 (3C 84) the nuclear source is more complex. Components from several different outbursts are apparently present.

It is remarkable that such simple models appear to be adequate to explain some sources. However, complications involving the continuous injection of particles, special geometries, etc are likely to be present in most sources. Only a few variable sources have been studied, and the total time of observation so far is <5 years. Already very rapidly varying objects such as BL Lacertae have been found, and if this is a galactic nucleus as appears to be the case, it does not appear that a simple model will be adequate to explain its variations.

To summarize, studies of the continuous spectra of galactic nuclei in optical, infrared, microwave, and radio frequencies suggest that the basic energy-generating mechanism is able to generate a wide energy spectrum of charged particles. The total energy range is probably 10 MeV to 1000 GeV. These sources may be single coherent objects, but in many cases they may be made up of many smaller pieces. The sources are undoubtedly small, almost certainly less than 10^{18} cm in most cases, and conceivably no bigger than 10^{15} cm. The dominant radiation process is most likely to be the electron synchrotron process, and part of the radiation could be coherent, though there is no certainty about this. The radiating process is not steady, and very rapid changes are sometimes seen. There is indirect evidence that magnetic fields stronger than those found anywhere else, except on the surfaces of stars, are present in plasma clouds with sizes much greater than those of ordinary stars. The radiation fields are intense, and the Compton effect may be important in some cases.

10. OBJECTS OF GALACTIC MASS EJECTED FROM GALACTIC NUCLEI

Ambartsumian (308, 309) proposed not only that large energies are released in the active nuclei of galaxies, but that galactic nuclei can split and that large masses can be ejected, either in the form of condensed matter or in the form of clouds which condense into galaxies. In Ambartsumian's view all stars and galaxies have been formed out of very dense matter. Since he proposed that galactic nuclei are the sources of nearly all of the activity seen in and around galaxies, a large body of evidence has grown up which strongly supports his thesis as far as the generation of large nonthermal energies, relativistic particles, and the ejection of gas clouds is concerned. This evidence has been summarized in the previous discussion. However, the even more radical proposal which he espoused that, in effect, galaxies or protogalaxies are ejected from galactic nuclei has not been generally accepted up to the present. This is partly because of the ambiguity involved in the interpretation of the evidence in support of the hypothesis, partly because of the difficulties raised by the proposal, and partly because it departs very strongly from the general ideas concerning galaxy formation. We outline next the evidence and its possible interpretations.

Groups and Clusters of Galaxies as Systems of Positive Energy

Ambartsumian first pointed out that a number of groups and clusters of galaxies appear to have kinetic energies much greater than their potential energies. Not only does it appear that $2E_k+\Omega>0$ so that the systems are expanding, but $E_k+\Omega>0$ so that they have positive total energy. The evidence showing the correctness of this result for nearly all of the systems for which adequate data was then available was presented at the Santa Barbara conference in 1961 (310). At that meeting two general arguments were made which attempted to avoid this conclusion and to suggest instead that in reality the systems are bound and stationary. First, it was argued that for the small groups of galaxies the uncertainties in making estimates of the kinetic energy and potential energy from radial-velocity measurements and the spatial distribution of the galaxies were large enough so that it was not certain that $2E_k>\Omega$. Second, in cases where without any ambiguity $2E_k\gg\Omega$, it was argued that there is a large amount of hidden mass in the groups and clusters, so that they are really bound systems. Markarian (328) showed that there was little chance that the small groups contained galaxies not physically related. The first argument could not be taken seriously for the majority of the systems, since the apparent discrepancies are frequently very large, and in some cases very special situations had to be contrived (cf 311). The second argument was not resolved, and indeed many theoretical investigations have been made since, in which it is tacitly assumed that an amount of unseen mass sufficient to bind the cluster is present. However, the evidence that the majority of groups and clusters appear to have positive total energy has continued to pile up (312, 313), and the indications are that the hidden-mass argument is probably not correct in most of the cases for which data are available. This is particularly true for small groups of galaxies in which one system has a highly discrepant velocity. They include Stephan's Quintet (314), Zwicky's triple system around IC 3481 (315), the chain VV 172 (316), Seyfert's Sextet, and VV 159 (317). Also, small groups involving chains of galaxies that are basically unstable configurations, such as the nine galaxies centered on NGC 383 (318), or No. 330 in Arp's Atlas (317), are very hard to explain in this way. As far as the larger clusters are concerned it is not unreasonable to argue that centrally condensed symmetrical clusters like the Coma cluster are bound (and only a comparatively modest amount of unseen mass is required in this case), but widely dispersed irregular systems like the Hercules cluster and the Virgo cluster are likely to be unstable and expanding.

The time scales for expansion of such systems or for ejection of a single galaxy are $\sim 10^7$–10^9 years and are therefore short when compared with the Hubble time. It follows that the galaxies in such systems must be comparatively young, and this is against the present convention, though it is not in direct contradiction to observation. Ambartsumian has argued that such young systems arise from ejection from some parent galaxies.

The Evidence of Arp

Arp (319–324) has found evidence which he suggests indicates that objects of galactic mass have been ejected from many comparatively nearby systems. He has been led to this view by suggestions of Sérsic (325) and studies of many of the galaxies given in his *Atlas of Peculiar Galaxies* (271). He has suggested that many of the strong nonthermal radio sources, including some identified with faint peculiar elliptical galaxies and QSOs, have been shot out of comparatively nearby galaxies lying at distances no greater than about 20 Mpc. He has also suggested that some giant ellipticals such as NGC 5128, and the brightest galaxies in the Virgo cluster, have given rise to many of the fainter galaxies surrounding them. He has pointed out the existence of remarkable configurations of objects, for example, the peculiar galaxy NGC 520 and four QSOs lying almost in a straight line, which suggests that NGC 520 is the parent galaxy that has ejected the QSOs. Most recently he has suggested that the companion galaxies on the ends of spiral arms of brighter galaxies, the best-known system of this type being NGC 5195 at the end of an arm of NGC 5194 (M51), have been fired out of these systems. In addition, he has speculated that spiral arms are themselves a result of explosive ejection. This latter proposal has earlier been made by Pismis (326) and others.

Arp's work is difficult to evaluate. Quite fundamental problems are raised if his conclusions are accepted. In particular we must then argue that QSOs and many faint radio sources have redshifts of noncosmological origin—an idea that is still quite unpalatable to many. Some of his results have been criticized, correctly so in the reviewer's opinion, by van der Laan & Bash (327) on statistical grounds. At the same time there is no question but that he has discovered a number of very intriguing systems that are very difficult to explain as being chance configurations. In addition to this, other evidence suggests that some objects have been ejected from active nuclei. The radio sources in the Perseus cluster have apparently been ejected from NGC 1275, as small sources are found at great distances from that galaxy (185). Also, it was pointed out many years ago that the direction of the jet in M87 is the same (to within 1°) as the direction of the line joining M87 with the radio source centered on NGC 4374 (143).

It would be rash to disregard all of the evidence presented by Arp, despite the difficulties its acceptance raises, and the lack of satisfactory statistical arguments. If the reviewer may inject a personal opinion, the history of astronomy over the last 20 years, particularly in the extragalactic field, does not suggest that a highly conservative approach is likely to be very fruitful.

Other Evidence

Other evidence could be interpreted as indicating that massive objects are ejected from galactic nuclei. There is the general result previously referred to that the majority of strong radio sources must have been ejected from galactic nuclei. Also, de Jong (255) suggested that there may be a tendency for faint radio sources to cluster about bright galaxies that are normal or weak radio sources.

Holmberg (329) studied a large number of physical groups of galaxies;

one of his conclusions is that in the case of spiral galaxies with an edgewise orientation, the physical satellites have a peculiar distribution. Most of them are found along the elongation of the minor axis, and thus they seem to favor (local) high galactic latitudes. The number of satellites seems to be larger for spirals with exceptionally blue nuclei and for spirals containing much uncondensed hydrogen. Holmberg therefore states that, while the statistical evidence is not conclusive, the results favor the hypothesis that satellite galaxies are produced by matter ejected from the nuclear regions of the spirals. Earlier, Burbidge, Burbidge & Sandage (110) speculated on the possibility that the globular clusters surrounding M87 may have been ejected from it, and recently Unsöld (330) made a similar proposal to explain the halo globular clusters in our own Galaxy.

Morgan & Lesh (331) and more recently Sastry (332) suggested that the major axes of the distributions of galaxies, in clusters containing a giant D galaxy, tend to be oriented in the same direction as the major axes of the D galaxies themselves. Of course this does not necessarily imply that ejection from the D system was responsible for the appearance of fainter systems. The fact that the axes are roughly parallel might be attributable to the direction of the total angular momentum in primeval protogalaxies, if one adheres to a conventional theory of galaxy formation. However, the subclustering of very faint objects around D systems which is sometimes found could be interpreted as evidence for the ejection of satellites.

In summarizing the evidence discussed in this section, it must be admitted that no certain conclusions can be drawn. However, the evidence for Ambartsumian's original hypothesis has grown in the last decade, or put another way, many strange phenomena now known are very hard to understand in the framework of a theory in which it is supposed that galaxies were formed only by condensation at a very early epoch, and that all groups and clusters of galaxies are bound and have remained so since they were formed.

11. ULTRAVIOLET RADIATION FROM GALACTIC NUCLEI

The fact that many galactic nuclei are powerful sources of nonthermal optical, infrared, and radio photons makes it appear likely that they are equally powerful sources of harder radiation. Thus, there have been a number of theoretical speculations—for example, that the diffuse X-ray background radiation is generated by the integrated effect of such sources. But so far only one galaxy has been identified as a powerful X-ray emitter. This is M87, discussed earlier, and it was pointed out that it is not yet known whether or not the X-ray source is confined to the nucleus.

Until very recently, nothing was known about the ultraviolet flux from galaxies. However, while no scientific publication is yet available, preliminary results from the OAO II (1968 110 Å) show that in most, but not all bright galaxies studied, the ultraviolet radiation is considerably in excess of the extrapolated curve from ground-based measurements, in some cases showing a significant excess at 2000 Å. In the case of M31 the bluest region is within 2′ (~300 pc) of the center so that the ultraviolet flux may be related to nuclear activity.

12. THEORY OF GALACTIC NUCLEI

INTRODUCTION

We have described the properties of galactic nuclei as far as their stellar and gas content is concerned, and have summarized the evidence for activity involving the generation of nonthermal flux of various kinds, the generation of relativistic particles, and the ejection of matter.

It is now necessary to discuss the state of our theoretical knowledge concerning such nuclei. Ideally this section should therefore contain a description of the formation of galactic nuclei, and their evolution, with a theory which naturally explains their activity. Unfortunately, the theory is in a rudimentary state, and many fundamental questions remain unanswered.

It is impossible to separate the problems of galactic nuclei from the problem of the formation of galaxies, and this is where the trouble begins. A comprehensive review of the current status of ideas on galaxy formation has recently been given by Field (333), and we shall not discuss them in any detail here.

Most attempts at understanding the formation of galaxies have been made within the framework of an evolving universe, the idea being that the galaxies were largely or completely formed in the distant past when the universe was in a much more condensed state. Added impetus to these ideas has been given by the discovery of the microwave background radiation, which many have taken as proof of the existence of a primeval fireball in a big bang universe. Investigations have been made in which the effect of the radiation on the condensation process is taken into account (334, 335), but the net result has been that no satisfactory theory giving results in accord with observations on the gross properties of galaxies, their masses, angular momenta, and their clustering tendencies has been found.

A basic problem recently stressed again by Harrison (336) is that in all of the theories it has been necessary to *postulate* the existence of initial density fluctuations large enough so that, in an expanding medium, condensation can occur. These fluctuations in an initial fluid have not been justified on the basis of known theory (though they might be a result of highly nonlinear phenomena), and Harrison has pointed out that they also have to be treated as cosmological in origin. Hoyle (337) has stressed that in the big bang cosmologies we must suppose that initial conditions were responsible for the state of the Universe as we now find it. This means that all of the phenomena that we have described so far must be assumed to have been initiated by an early formation process, with evolution of the systems leading to conditions as they are seen at the present epoch.

Not all astronomers have taken this view. Forty years ago, baffled by the problem of the origin of spiral structure, Jeans (338) stated: "The type of conjecture which presents itself, somewhat insistently, is that the centers of the nebulae are of the nature of 'singular points' at which matter is poured into our universe from some other, and entirely extraneous dimension, so that, to a denizen of our universe, they appear as points at which matter is being continuously created."

This is essentially the view of Ambartsumian (308, 309) who arrived at it following his early appreciation of the very active role that nuclei of galaxies appear to play. However, he does not appear to espouse the steady-state cosmology.

In the case of spiral galaxies, studies of the light distribution and the rotation show that the equilibrium configuration is that of a very thin disk with a central more-spheroidal mass concentration. However, most attention has been paid to theories that attempt to explain the smooth light distribution found in ellipticals, and the large velocity dispersions of the stars in the central region. The masses of the bright ellipticals, derived either from the virial theorem, using the velocity dispersion of the stars in the central region, or from studies of double galaxies (45), are hard to explain in terms of stars of known types (15), so that a case can be made for supposing that a significant fraction of the mass is present in the form of collapsed objects (black holes) (345, 346).

The smooth light distribution seen in ellipticals, and their regular shapes led Jeans initially to the conclusion that they must be made up of gas, a conclusion now known to be incorrect. As was first shown by Hubble, the well-studied bright ellipticals have a brightness distribution that is well reproduced by projecting a three-dimensional isothermal distribution into a two-dimensional image. The problem is to understand how they have come to this state and, indeed, to determine whether they are totally relaxed systems and to describe their evolutionary history, both past and future. Extensive theoretical work on the structure of such stellar systems, both for ellipticals and globular clusters, has been carried out by Belzer, Gamow & Keller (347), Spitzer (348), Hénon (344, 350), King (351), Michie (349–355), Michie & Bodenheimer (356), Lynden-Bell (357, 358), and Lynden-Bell & Wood (359). Earlier reviews covering some parts of the problem have been given by King (360) and Michie (361).

In a steady-state universe the problem of galaxy formation is an exceedingly difficult one. Originally attempts were made to argue that galaxies must be formed out of diffuse clouds of newly created matter by the development of gravitational or thermal instabilities, perhaps in the wakes of other galaxies (cf 339). However, this point of view has now been abandoned. More recently it has been proposed that matter is created in regions where the density is high, and that consequently creation and the formation of new galaxies must be closely associated with galactic nuclei (340, 341). These ideas are again closely akin to the original suggestion of Jeans.

We shall return to these more radical ideas. We now attempt to summarize the theoretical ideas based on the more conventional, though not necessarily correct, view that galaxies condensed at an early epoch from a diffuse medium.

FORMATION OF NUCLEI IN AN EVOLVING UNIVERSE

Without an understanding of the details, it is argued that the protogalaxy forms early in the history of the Universe, fragments, and star formation begins. Condensation early in the collapse of a rotating protogalaxy can lead to the formation of a halo population. Attempts to relate this type of evolution to the situation for the halo population in our own Galaxy were made by Eggen, Lynden-Bell & Sandage (342). It has also been demonstrated (343, 344) that the observed angular momentum distribution of spiral galaxies is compatible with the view that they condensed from uniformly rotating protogalaxies.

The classical calculations of relaxation using star-star interactions have shown that the relaxation times are in general very long when compared with the maximum ages of the galaxies. Also, if the stars have reached energy equipartition there will be a tendency toward mass segregation with the more massive stars at the center, and it has been argued that this is in conflict with observation, as far as color differences and stellar populations are concerned. In passing, it should be noted that there is probably not enough good observational evidence available at present to completely settle this question. The best-studied elliptical from this standpoint is NGC 3379 (362) where a small color difference between the nuclear region and the outer parts is observed, and this could not be due to segregation of stars. On the other hand, the less accurate work of Tifft (Section 2) suggests that considerable color differences, and hence possible population differences, are present between the inner and outer parts of many systems.

To understand how a completely smooth distribution and mixing has occurred, it is necessary to argue either that the protosystem condensed to an equilibrium configuration in gaseous form and only then condensed into stars which have not yet reached energy equipartition or, more plausibly, that relaxation has taken place, not by star-star interactions, but through violent changes in the effective gravitational field of the galaxy at some stage. This type of relaxation has been suggested by Hénon (349) and King (351) and has been enthusiastically advocated and elegantly analyzed by Lynden-Bell (358, 359). He suggested that this encounterless relaxation is likely to have taken place early in the condensation of the galaxy. However, it could take place at any period during the life of the galaxy if the effective gravitational field in which the stars move changes in a time short compared with a typical orbit period. Such changes could occur if the gravitational field is weakened by large-scale mass loss. Thus, if it turns out that radiation of gravitational waves from a galactic nucleus leads to a rapid decrease in mass, or if mass ejection of the type seen in Seyfert nuclei is very large, this type of relaxation could be important at stages in the life of a galaxy other than the beginning.

In the observational discussion it has been shown that the structure of galaxies as determined by the stellar distribution is not entirely smooth, and that often at the very center there appears to be some kind of a discontinuity. Examples are the nuclei of M31 and M32 (Section 3). Moreover, the activity that we have described indicates that a large and rapidly evolving mass distribution has been formed.

The classical treatment by Ambartsumian (363) and Spitzer (348; see also 361 and 364) shows that in an isolated cluster, through star-star interactions, stars will occasionally escape from the cluster, which will tend to shrink, attaining a steadily increasing temperature until the density becomes so high that direct star-star collisions become important. Spitzer & Saslaw (365) and Spitzer & Stone (366) have analyzed the evolution of a very dense galactic nucleus along these lines. But a basic problem has been to understand how the star density in a central nucleus could build up to the stage at which rapid evolution occurs. The central densities found in the nuclei of M31 and M32 are only $\sim 10^3$–10^4 stars/pc^3, far below the values required for rapid evolution, and there is no *direct* evidence that higher star densities are present in the nuclei of any other galaxies. They may very well be present, particularly in systems with active nuclei, but for galaxies further away than the local group it has not been possible to investigate the density distribution within a few parsecs of the center. Such studies must await observations from above the Earth's atmosphere, where resolutions $\lesssim 0\rlap{.}{''}1$ ($\lesssim 10$ pc at the distance of the Virgo cluster) will be possible.

A possible solution to the problem of understanding the evolution of a galaxy toward a system with a dense stellar core and a much lower-density smooth envelope has come first from numerical integrations of the n-body problem (367–371), and from theoretical work by Hénon (372). The experiments have shown that a central cusp of density and an extended halo are formed during the first ten or so relaxation times. This led Lynden-Bell & Wood (359) to analyze the thermodynamics of self-gravitating systems [see also an excellent discussion by Thirring (373)]. Since we are dealing with systems with negative specific heat, evolution is always away from equilibrium, since, if heat flows from a hot system to a colder one, the hot one gets hotter and the cold one gets colder. This means that in a star at a stage where no more nuclear fuel is available, the core will contract and become hotter, while the envelope expands and becomes cooler. This is precisely what we know, from detailed calculations, occurs as a star evolves into a red giant. A similar situation can be expected to occur in the evolution of a galaxy where there are no energy sources other than gravity, provided that the heat flow occurs in a time short compared with the age of the galaxy. Lynden-Bell & Wood argue that, provided that the energy exchange takes place through rapid changes in the gravitational potential—the violent relaxation described earlier—this condition will be satisfied, and a core-halo configuration will form. Moreover, the core can evolve again into an inner core-halo distribution, and as the density in the inner core increases, the star-star interactions become increasingly important.

While so far there are no numerical calculations of such an evolution

TABLE 9. Properties of theoretical galactic nuclei from Spitzer & Saslaw (365)

No. of stars	Radius, R (pc)	0.1	1	10	100
$N=10^6$	Stellar, velocity, v_s (km/sec)	147	47	14.7	4.7
	Relaxation time, T_R (years)	4.6×10^6	1.46×10^8	4.6×10^9	1.46×10^{11}
	Collision time, t_c (years)	3.2×10^8	1.09×10^{11}	3.5×10^{13}	1.09×10^{16}
$N=10^8$	Stellar velocity, v_s	1470	470	147	47
	Relaxation time, T_R	3.4×10^7	1.08×10^9	3.4×10^{10}	1.08×10^{12}
	Collision time, t_c	2.8×10^6	5.2×10^9	3.1×10^{12}	1.09×10^{15}
$N=10^{10}$	Stellar velocity, v_s	14700	4700	1470	470
	Relaxation time, T_R	2.7×10^8	8.6×10^9	2.7×10^{11}	8.6×10^{12}
	Collision time, t_c	3.1×10^8	9.7×10^6	2.8×10^{10}	8.5×10^{13}

which can be compared with the observations, it does appear that in this way the nucleus of a galaxy can evolve to a stage at which inelastic star collisions begin to be of importance.

Spitzer (374) has recently shown that there is another mechanism by which a core-halo distribution that appears in the numerical experiments can be generated. He has shown that, if stars of two different masses m_1 and m_2 are present (actually many of the numerical experiments have been made assuming a distribution of masses), rather than a single mass as assumed by Lynden-Bell & Wood, the evolution will proceed as follows. The heavier stars will lose kinetic energy with a time constant about twice the equipartition time, and will gravitate toward the center. If $m_2/m_1 \gg 1$, and if the total masses of the heavy stars and light stars are M_2 and M_1, when

$$M_2 > 0.16 \left(\frac{m_1}{m_2}\right)^{3/2} M_1$$

the self-attraction of the heavy stars requires such a high velocity dispersion that equipartition with the light stars becomes impossible. Under these conditions the heavy stars will contract essentially independently of the light stars with a time constant $\simeq t_{eq}$, forming a dense nucleus comparatively rapidly, while the light stars will remain in an extended distribution.

In Table 9, taken from the paper of Spitzer & Saslaw (365), are numerical values which give some idea of the condition under which very rapid evolution of a dense nucleus will begin. Spitzer and his colleagures have traced the following sequence of events.

When stars collide with relative velocities of some thousands of kilometers per second, a few percent of the mass will be released in the form of hot gas (365, 375). This gas cools and falls to the center and it is assumed that it can then form new stars. The equipartition of energy between the newly formed stars and the older stars leads to a further contraction of the system of old stars, the collision rate increases, and evolution speeds up again. The rate of evolution and hence the rate of energy release is controlled by the initial angular momentum of the configuration.

The very violent processes that will begin at this stage have been suggested by a number of people to be responsible for the very large energy releases associated with QSOs and strong radio galaxies. These proposals were reviewed in detail elsewhere (110, 238, 229) and we shall be brief. It has been proposed by some, including Spitzer and his collaborators, Woltjer (376) and Gold, Axford & Ray (377), that the energy released in collisions between stars is responsible for very high optical luminosities and even radio properties of these objects. Shklovsky (378), Burbidge (379), and Colgate (302, 303) all argued that very frequent supernovae were responsible for the large energy output. Colgate gave the most detailed and plausible discussion of this process. He argued that when collisions are just becoming important, but the mean kinetic energy is smaller than the binding energy of the stars, collisions between stars of solar mass will lead to the formation of more massive stars. These then evolve rapidly to the stage where they explode, and their envelopes are ejected. Colgate believes that it is in this phase that a very large flux of nonthermal radiation and high-energy particles is generated. The bulk of the energy released in his scheme comes from gravitational collapse, in the supernova phase, and the question of whether an implosion can give rise to an explosion in the envelope due to neutrino opacity, etc depends on the mass of the star and other properties. The effects of rotation on such an evolution have not been taken into account.

It is possible also that in a very late stage of evolution the stars that have not experienced collisions leading to coalescence or disruption will be moving so fast in such a small volume that the system becomes a relativistic star cluster. The properties and stability of such star clusters will be discussed later in this section. Greenstein (380) has calculated the gravitational radiation which will be emitted from such a highly evolved cluster, both in the nonrelativistic and in the relativistic situation. Gravitational waves will be emitted both in star-star interaction (a pulsed mode) and continuously, because of the motion of the stars in the field of the cluster. If the stars have normal sizes, the major energy-loss mechanism governing the evolution is star collisions, in which kinetic energy is converted into heat, electromagnetic radiation, and possibly high-energy particles. Only if the stars themselves have evolved to neutron configurations, or black holes, will the major energy-loss mechanism be gravitational radiation. In any case, at this stage the cluster will evolve catastrophically; this is due either to these effects or to relativistic instabilities, in times exceedingly short compared with the time scales associated with the activity observed in galactic nuclei.

The angular momentum originally present in the nucleus will slow down the rate of shrinkage described here, and little has so far been done to allow for this effect. However, it is generally agreed that the endpoint of the evolution will be the formation of a single coherent object which may be treated as a very massive star. It may be a flattened rotating object. Hoyle & Fowler (381, 382), without discussing its origin, orginally proposed that such a massive object is responsible for the explosive events seen in QSOs and strong radio galaxies. It was argued that the energy is released as this object collapses. In order to obtain energy release comparable to that observed in the strong radio sources (10^{58}–10^{62} ergs = 10^4–$10^8 M_\odot c^2$), it is necessary to argue that the energy is released with high efficiency at a stage when the massive superstar has collapsed close to its Schwarzschild radius, $R_s = 3 \times 10^5 M/M_\odot$ cm. If gravitational energy is converted to what is seen at an efficiency $\sim 1\%$, this means that masses $\sim 10^6$–10^{10} $M_\odot$ must be collapsing. Fowler (383) argued that the massive object, before collapsing, could behave as a superstar that would burn in its thermonuclear phase at a luminosity $\sim 10^{46}$ ergs/sec for $\sim 10^6$ years provided that it could be stabilized against gravitational collapse for such a period by rotation or macroscopic turbulence. It has also been suggested that a collapsing rotating massive object of this kind might break up into many small fragments, many of which would be neutron stars (345).

The final point of evolution according to general relativity will be a situation in which the nuclear object collapses and becomes a black hole. There is no escape from this situation, unless it can be shown that the final masses of all of the fragments are less than the critical masses for support by degenerate electron pressure (white dwarfs) or support by degenerate neutron pressure (neutron stars); or unless a field of negative energy is invoked (384). Thus there is good reason to believe that much matter may be present in galaxies in the form of black holes.

Recently Wolfe & Burbidge (346) examined quantitatively the question of how much mass can be present in elliptical galaxies in the form of black holes. They started from the idea that the high mass-to-light ratios might be explained by the presence of a large amount of collapsed matter (cf 345). Their analysis showed that not more than about $10^{10} M_\odot$ can be present in the form of single black holes in the nuclei of the well-studied bright ellipticals in the Virgo cluster, and this is not enough to explain the high mass-to-light ratios. However, the existence of black holes with masses $\sim 10^9$–$10^8 M_\odot$ in the nuclei would be expected following the evolution just described, and the existence of such objects is compatible with the observations. Much more mass can exist in this form if it is distributed throughout the galaxies, and the most plausible idea is that if enough is present to explain the high mass-to-light ratio in bright ellipticals, perhaps $\lesssim 75\%$ of the mass might then be in this form; this may have come from the evolution of stars which condensed with masses large enough so that their thermonuclear evolution times were short compared with the age of the galaxy. Relaxation of the type analyzed by Lynden-Bell is required if models of this type are correct.

Whatever its origin, the existence of a massive black hole at the center of a galaxy will speed up the evolution of the nucleus, because the gravitational field will be much stronger. The processes that we described earlier, evaporation of stars, stellar collisions, etc will be speeded up. At the same time new phenomena can be expected. Stars and gas will be captured and swallowed up by the black hole. However, the rate of star swallowing will be very small except in the case of an exceedingly dense star cluster surrounding the object. Numerical values have been given by Wolfe & Burbidge. The rate at which the stars are swallowed depends ultimately on the relaxation times for dense star clusters to move in and form around the central hole.

Lynden-Bell (385) recently proposed that such black holes at the centers of galaxies are the remnants of dead QSOs and that gas present in the nuclear regions of a galaxy because it has been ejected from stars in the normal course of stellar evolution (it may be present for other reasons also) will be swallowed up by a central black hole. The essential physics of the problem was earlier worked out by Salpeter (386) who considered the general problem of accretion of gas by a collapsed massive object.

If the gas is distributed in the form of a differentially rotating disk about the black hole, atoms will move inward because of a slow drain of angular momentum and energy by turbulent or magnetic dissipation, until they reach the orbit at a radius $6\ GM/c^2$ where, because of general relativistic effects, they become unstable to spontaneous spiraling into the black hole. By this time a fraction $f = 0.057$ of the rest-mass energy has escaped to infinity. Salpeter pointed out that if, at this radius, the atom could be brought to rest by a direct collision with an atom with opposite momentum—a very unlikely situation to arise in the context of a rotating disk—a further fraction, 0.126, of the rest-mass energy could be released. Free fall to even smaller radii could lead to even larger energy release but general relativistic effects would lead to a smaller and smaller fraction of this energy finally escaping. However, at an efficiency of only 5.7%, accretion of matter at a rate of only $10^{-3} M_\odot$/year would lead to an energy output of $\sim 3 \times 10^{42}$ ergs/sec, and Lynden-Bell has suggested that this may account for the energy output from the nuclei of Seyfert galaxies and the like. The problem is then to understand why the energy appears in the form described in the previous sections and how much of it will remain in forms so far undetected. Lynden-Bell argues that the energy is acquired by a magnetic field in the rotating disk, and that the field energy is used in accelerating particles to relativistic energies by a mechanism of the neutral-sheet type. Thus, it is concluded that we have the two ingredients (relativistic particles and magnetic field) required for synchrotron radiation. The scheme is ingenious, but much more work needs to be done to show whether or not the form of the observed nonthermal spectra of nuclei and the ejection of gas clouds, etc can be explained.

Another aspect of the final evolution of a nucleus must be discussed. We mentioned earlier that the final stages may involve a relativistic star cluster, and the question of its stability and the redshift associated with it requires study. Recent interest in this problem is due to the strange effects present in the redshift distribution in QSOs (387–389, 232), particularly a very sharp peak at $z=1.955$ and multiple redshifts found in a number of objects. This has suggested to some that a large part of the redshifts measured in these objects might be intrinsic and not due to the expansion of the universe. If this is correct, the most plausible idea is that the redshifts have a gravitational origin. But some years ago Greenstein & Schmidt (390) showed that the sharpness of the spectral lines could not be explained if they were produced in a region with a strong gradient in the gravitational field, so that the idea that a large gravitational redshift could be explained as coming from a gas cloud surrounding a condensed massive object could not be correct. However, Hoyle & Fowler (391) pointed out that in principle this difficulty is overcome when a gas cloud responsible for the line spectrum is situated at the bottom of a deep potential well, i.e. in the center of a massive object. Since the light must be able to escape, the central object must be made up of a large number of small evolved stars, either neutron stars or white dwarfs. Thus an intrinsic redshift might come from a gas cloud at the center of a highly evolved relativistic star cluster. Two fundamental questions arise then: how large can the redshifts be? and is the cluster stable?

Extensive work on this problem has been carried out by Thorne, Ipser & Fackerell (392–394), Zeldovich & Podurets (395), Bisnovatyi-Kogan & Zeldovich (396), Bisnovatyi-Kogan, Zeldovich & Friedman (397), Bisnovatyi-Kogan & Thorne (398), and Zapolsky (399).

It is to be expected that relativistic star clusters will exhibit properties very similar to those of relativistic stars, as in the Newtonian situation already mentioned. This means that they may show instability against collapse when they become too compact, emission of gravitational waves if they are not stationary, deformation and dragging of inertial frames when they rotate. The importance of relativistic stars as possible sources of energy in QSOs and strong radio galaxies has already been mentioned. Extensive work has been done on their stability, starting with the work of Fowler (400) and Chandrasekhar (401), who showed that for a ratio of the specific heats only slightly in excess of 4/3, dynamical instability will set in long before the mass contracts to anywhere near the Schwarzschild radius. Further studies of nonradial pulsations and of the effect of rotation have been carried out, and it is thought that the instability is understood in general relativity when relativistic effects are small (post-Newtonian approximation) or when the departures from spherical symmetry are small. Study of the analogous problem of relativistic star clusters led Ipser, Fackerell & Thorne to conclude that for polytropic or isothermal clusters it is difficult, and probably impossible, to find a model for a spherical star cluster that is stable, and that has a redshift as large as the values seen in QSOs, which are somewhat greater than 2. In essence what happens is that as one proceeds along a sequence of models with the same velocity distribution function, the ratio of binding energy to total rest-mass energy increases with redshift to a maximum of a few percent when the redshift from the center $z_c \sim 0.5$–1. For higher values of z_c the binding energy becomes negative. Also the models are unstable to gravitational collapse near the maximum in the binding energy.

However, for a rapidly rotating model, a uniformly rotating disk, Bardeen & Wagoner (402) showed that the fractional binding energy appears to increase monotonically with redshift to a value of about 40% as $z_c \to \infty$. The fact that there is no maximum to the binding energy at a finite value of z_c suggests that the disks, although they are unstable to fragmentation, may be stable against overall gravitational collapse up to arbitrarily large redshifts.

Even in the nonrotating case, the belief that the calculations of Ipser, Thorne & Fackerell suggest that no stable clusters can exist with redshifts as large as the values observed has been challenged by Bisnovatyi-Kogan & Zeldovich (396). They constructed models of isotropic star clusters and gas spheres with arbitrarily large values of z_c, which they believed, on intuitive grounds, were stable. Bisnovatyi-Kogan & Thorne (398) have proved that gas spheres with finite central densities and finite radii are stable against all small adiabatic radial perturbations. They believe that clusters are probably stable also, although they were not able to prove stability in this case. Why do these results differ from the early ones? Bisnovatyi-Kogan & Thorne argue that although globally the configurations are stable, locally they are not. Presumably the configurations are not truly isothermal spheres. They argue that the quantities characterizing the local relativistic corrections to Newtonian theory add up to produce a large global redshift, but they do not add up cumulatively to cause a departure from normal Newtonian stability. The objects are unphysical in that they have small velocities, but large gravitational potentials.

What bearing do these results have on the observations? While there is still a considerable amount of confusion, there may be formal solutions which suggest that in principle relativistic star clusters, with or without large net angular momenta, can give rise to large gravitational redshifts over times that would make them observable. While none of the models is the least bit realistic as far as physical clusters in the centers of galaxies are concerned, possibly one psychological barrier against the contemplation of large intrinsic redshifts in astronomical objects has been removed.

There is one final point of some interest. It has recently been proposed by Morrison (403) that QSOs are massive rotating stars—giant pulsars. Such objects might bear some resemblance to the rotating disks studied by Bardeen & Wagoner. Certainly such objects must be proved to be stable over a reasonably long time scale if Morrison's suggestion is correct. Also, arguments based on the form of the nonthermal continuum radiation in the nucleus of our Galaxy and M87, discussed in Sections 4 and 5, have led to the proposal (92, 140) that the basic energy source is composite and that it consists of a cloud of small dense stars with magnetospheres that generate relativistic particles directly. The prototypes of these objects are rotating neutron stars, though they are not necessarily identical with pulsars. If this picture is correct, rotation is again of great importance, and in a conventional evolutionary scheme such a cloud of objects can be thought to have originated from a rapidly rotating supermassive star or a relativistic star cluster. In the breakup of the massive star or cluster, some objects may have been thrown out to give rise to secondary centers of particle emission and synchrotron radiation far from the nucleus (140).

Formation and Evolution of Galactic Nuclei as Delayed Cores

We have so far discussed the evolution of galaxies and their nuclei on the assumption that they all condensed $\sim 10^{10}$ years ago. However, in Ambartsumian's view and in a steady-state universe a different approach is required.

In connection with the Ambartsumian approach and also to explain QSOs, Novikov (404), Ne'eman (405), and Ne'eman & Tauber (406) have proposed that some of the mass energy in the universe resides in part of a Friedman universe whose expansion has been delayed. It is argued that delayed cores expand their Schwarzschild radii at different times for different cores. In the case of expansion beyond the Schwarzschild radius the kinematics of the matter are directly opposite to that of collapse, but that does not mean that there is a simple time reversal of the collapse seen by an external observer. When an object collapses, the observer sees the light emitted outward from the surface, which is collapsing inward. When the collapse is reversed, the light rays corresponding to those emitted outward in collapse are now emitted inward. In an expansion, therefore, we see the object by the light which is emitted inward in collapse, but is now moving outward. These quanta emerge from inside the Schwarzschild radius to the external observer, and consequently an external observer can see the entire expansion from the start even though the object appears to be emerging from an infinite past, as long as the photons are produced from inside the Schwarzschild radius. This implies the existence of a certain amount of matter and a finite number of photons already on the way to the observer at the beginning of our time scale. The reviewer has great difficulty with this concept. Ne'eman & Glauber, to explain the variations in QSOs, attempted to have such an expanding core reach a maximum radius and oscillate, and this led them to complications, involving local nonconservation of matter, i.e. a situation similar to that discussed by Hoyle & Narlikar (341) involving the introduction of a field of negative energy (C field).

Harrison (407) has recently extended this idea, suggesting that such delayed cores with high rotations originate from fluctuations in the metric and give rise to all galaxies. By this means he attempts to avoid the difficulty of understanding how initial condensations form in an expanding, comparatively low-density, gas. In these pictures galaxies behave as miniature universes, each expanding from a very dense core. While this naturally leads to the existence of a dense nucleus in every galaxy, it is not at all clear in what form matter will emerge, nor is the evolution outside the nucleus understood. It is tacitly assumed that it is in the early phases of the expansion that violent activity is to be expected, and that as the objects grow older the activity slows down.

Formation of Galactic Nuclei in the Steady-State Cosmology

Hoyle & Narlikar (341) and McCrea (340) have attempted to explain the formation of elliptical galaxies in the steady-state cosmology by the creation of matter around a central mass concentration. Hoyle & Narlikar argue that such galaxies are formed in an expansion from a steady-state situation in which the mean density is $\sim 10^{-8}$ cm^{-3}. Such an expansion is possible because the inhomogeneous steady-state theory has an instability in which the creation process is essentially cut off, and in which expansion proceeds according to the Einstein-de Sitter law. The characteristic mass of the "observable universe" at the outset of such an instability is $\sim 10^{13} M_\odot$, and they take this to set an upper limit to the masses of galaxies. They argue that because the Einstein-de Sitter expansion law is the limiting case between expansion to infinity at a finite velocity and a fallback situation, in which expansion stops at some minimum but finite density, a central condensation with mass less than that of the associated galaxies suffices to prevent continuing expansion. Thus a mass of $\sim 10^9 M_\odot$ will restrain a mass $\sim 10^{12} M_\odot$ from expanding beyond galactic dimensions.

Comparison of the Theoretical Ideas of the Formation of Galactic Nuclei with the Observations

The observations indicate that activity in galactic nuclei is a widespread and probably quite general phenomenon taking place in many, if not all, types of galaxy. There is a wide range of power levels, and the activity seen is always in the form of generation of nonthermal radiation, relativistic particles, hot gas, outward-moving gas clouds, possibly gravitational waves and the ejection of dense objects with large masses. The activity is going on over the full range of times spanned by observations of nearby nuclei like that of our Galaxy, to the most distant radio galaxies. If the redshifts of the QSOs are cosmological, then the activity extends back to epochs many billions of years ago. In all cases what is observed is the outflow of mass and energy. Inflow, if it is taking place, is never seen.

The theories are of two different types. In one case it is argued that following the condensation of galaxies from low-density matter, with inhomogeneities of cosmological origin, the nuclei gradually evolve by shrinking to higher densities, until they reach such large densities that there is large and violent release of gravitational energy leading to outflow and ejection of matter and energy.

In the other picture it is argued, both in evolutionary cosmology and in steady-state cosmology, that galaxy formation and the activity in nuclei are a direct result of outflow from very dense initial states.

In neither approach has there yet been any satisfactory explanation of why the energy is released in the form in which we see it. There have been no convincing explanations why high-energy particles are generated in such profusion, why infrared and radio fluxes can be so large, how coherent objects can be ejected, if indeed they are, and so on.

In a sense the second approach appears to be the more natural one. The delayed-core model is attractive since it can in principle explain the wide range of evolutionary times involved, continuous activity, the possible existence of young galaxies, systems of positive energy, etc.

Repeated activity, or even continuous activity, may be understood by repeated violent relaxation phenomena in the first scheme. The steady-state model has also the attraction that one would expect continuous activity, formation of new galaxies, etc. The objections to it are largely based on the well-known evidence from observational cosmology consisting of the value of the deceleration parameter q_0, the attempts to explain the log N-log S curve for radio sources and the existence of the microwave background radiation. However, none of this evidence or its interpretation is entirely secure (337).

13. ELEMENT SYNTHESIS IN GALACTIC NUCLEI

In Section 2 we described the evidence on the stellar population and gas in the nuclei of galaxies. There is some indication from this work that there may be composition differences between the stars in the nuclear regions and those further out in M31, in NGC 4472, and perhaps more generally. Not only this, but there are some indications that the CN feature is weaker in the light of dwarf ellipticals than in giant ellipticals (16), suggesting that perhaps dwarf ellipticals have a heavy element-to-hydrogen ratio lower than that found in giant systems.

As far as the gas is concerned, the most striking feature is the change in [N II]/Hα intensity ratio as a function of position in galaxies with old stellar populations, and this may also be attributed to abundance differences between the nuclear regions and outer parts.

This evidence all points to the idea that element synthesis has taken place in the nuclear regions of galaxies. What do we understand of the details of nucleosynthesis in galaxies? The original proposal that synthesis of all of the elements with the possible exception of D, He, and ^{7}Li must have taken place in stars of different types at different stages in their evolution (408) has been amply supported by the evidence from abundance determinations in individual stars. However, studies of the composition of disk and halo star clusters in our own Galaxy show that there must have been a comparatively early and short ($<10^9$ years) epoch of nucleosynthesis in which some of the heavy elements were built. This could not have occurred in an early phase of an evolving universe as Gamow and his colleagues first proposed. The most recent detailed calculations of element synthesis in a big bang (409) show that only D, ^{3}He, ^{4}He, and ^{7}Li could have been built in this way. It is not even clear whether the bulk of the ^{4}He in the Galaxy was built in this way. In a review of the data on the He/H ratio (410) it has been suggested that, if there are stars with very low He/H ratios, as now appears to be the case, the most plausible interpretation is that the bulk of the He also was in the Galaxy in an early evolutionary phase, and that mixing was not complete. Galactic clusters with similar ages have quite different ratios of metals to hydrogen, which suggests also that the mixing was not complete, and there has not been a discernible steady enrichment of heavy elements with time.

These conclusions, together with the very preliminary results suggesting that the metals are more abundant in the central parts of galaxies than in the outer parts, suggest that much nucleosynthesis occurs in the nuclear regions of galaxies, either in the formation period, or intermittently in the life of a galaxy. The products of this nucleosynthesis then naturally enrich the matter out of which the stars in the central region condense, though some matter is thrown out. How can this type of nucleosynthesis take place in an active nucleus, either in an early phase of evolution, or in active phases through the life of a galaxy? There are two possibilities to consider: (a) nucleosynthesis in a massive star of the type discussed in the last section, and (b) nucleosynthesis in normal stars of much smaller mass.

Nucleosynthesis in a massive object has been studied by Wagoner, Fowler & Hoyle (409), by Wagoner (411, 412), and by Bisnovatyi-Kogan (413). [See also Wagoner (414).] To build ^{4}He, very high temperatures $\geq 2 \times 10^{10}$ degrees K are required and little else. Given a mixture containing ^{3}He and ^{4}He in the amounts found in a big bang synthesis, and an implosion followed by an explosion at about 10^9 degrees, K, C, N, and O can be produced. However, the abundances of ^{12}C and ^{13}C will be approximately equal in this type of synthesis. When nucleosynthesis occurs in an object expanding very much faster than the rate determined by gravitation, i.e. an explosion in a flattened object, and initial neutron-proton equality, Wagoner found that, provided the initial object has converted most of its hydrogen to ^{4}He, ^{12}C, ^{14}N, ^{16}O, etc, he could produce heavy elements in the abundance seen in extreme Population II objects, where there are deficits by factors ~ 100 or more.

Thus, massive objects could, in "little big bangs," generate the bulk of the helium seen, they could be responsible for the high abundance of nitrogen seen in the central parts of galaxies, and they might be responsible for the generation of the extreme Population II metal composition.

Stellar nucleosynthesis may also be of great importance. Clearly a number of situations can be envisaged in which star formation and evolution are likely to go on more effectively in the central parts than in the halo. The classical argument, which has only been applied in our Galaxy for stars near the Sun, that the luminosity function and the rate of star formation are proportional to ρ^α, where ρ is the density of interstellar matter and $\alpha = 1-2$, may be applicable in the nuclear region. Since ρ falls off very steeply as the distance from the center increases, the rate of star formation and evolution may be expected to be greater in the center than further out. Also, the type of relaxation discussed by Spitzer (374) in which the more massive stars tend to relax rapidly and move to the center will lead to a situation in which stellar evolution and its products will be accelerated in the central region. Also, if in the initial phases of galactic evolution the formation of more massive stars is favored, the result will be the same.

While we know that nucleosynthesis goes on in stars of all masses, the rate is a strong function of the initial mass, and it is in the more massive stars, therefore, that the processes of nucleosynthesis can be speeded up. Recent work by Arnett & Truran (415), Arnett (416), Truran, Arnett & Cameron (417), and Bodansky, Clayton & Fowler (418) has shown that, in these higher-temperature sources, rapid and finally explosive nucleosynthesis involving carbon and silicon burning can explain very well the relative abundances of the elements from ^{20}Ne to the Fe peak. Since C and O can be ejected from the outer layers of a massive star without too much processing, supernova outbursts in which the elements from Ne to Fe have been synthesized in the inner parts of stars may be responsible for early enrichment of these elements in the nuclear regions of a galaxy. Similar supernovae that are exploding occasionally in the outer parts of a galaxy throughout its life can then be responsible for more local effects. Finally, it must be emphasized that the data on compositions of the nuclear region of galaxies come from composite spectra, and the observed features are due to CN, Mg, Na, Fe, etc so that we have no data on the abundances of elements which can only be produced by the s and r processes of neutron capture.

14. GALACTIC NUCLEI AS GENERAL SOURCES OF PARTICLES AND RADIATION IN THE UNIVERSE

The bulk of the mass energy in the Universe that we can detect appears to be directly associated with galaxies. On the other hand, there are a number of components, magnetic fields, diffuse radiation, and particles, which we either know, or speculate, are present on the cosmological scale, and it is of interest to ask whether they may have been generated in galactic nuclei.

Magnetic Fields

The origin of stellar and galactic magnetic fields, which we know are present, and of intergalactic magnetic fields, for which there is no direct evidence, is obscure. While there have been a number of suggestions that strong magnetic fields detected in our own Galaxy and elsewhere have been amplified from a very weak seed field by some type of dynamo action, the origin of the seed field is unknown. Some have thought that it is fundamental and is in the same category as the baryons at the "beginning" in an evolving universe. Hoyle (419) has recently suggested that the primeval magnetic fields may have been generated in the nuclei of galaxies if these are centers in which massive objects explode. He has argued that a field that is present in a central massive object can be amplified by dynamo action at the expense of the rotational energy of the object. Thus, following adiabatic expansion, it appears that a galactic magnetic field with a strength $\sim 10^{-6}$ G can be generated from a seed field in a condensed central object of $\sim 10^8 M_\odot$ provided that about 10% of the rest-mass energy is converted by rotation into magnetic-field energy.

Antimatter

Hoyle (441) has pointed out that it is conceivable that particle-antiparticle pair creation occurs in the condensed nucleus of a galaxy, that particles are expelled, and that the antiparticles are initially retained in the nucleus. To avoid the difficulty associated with the fact that the mass in a nucleus is much less than the mass outside, Hoyle suggests that much of the antimatter in the nucleus is later ejected in fragments emitted in explosive events, or else it is radiated as gravitational waves.

Dust

Large quantities of dust are known to be present in many galaxies. In many cases it is found in spiral-arm regions, but it is sometimes found in the nuclear regions, and it is very conspicuous in the central regions of radio galaxies in which we know violent outbursts have occurred. Its presence on the intergalactic scale is controversial, but it is certainly sometimes found outside the main bodies of galaxies. Hoyle & Wickramasinghe (420) have suggested that dust may be formed in explosions of massive objects in galactic nuclei. They have calculated the conditions to be expected in the low-density phase following the explosion of a massive superstar in which nucleosynthesis has taken place (Section 13). They have considered two cases, one in which the expanding gas has a mixture of H, O, N, C, Mg, Si, and Fe with solar abundance ratios, and a second with a solar mixture of H, N, O, Mg, Si, and Fe, but with C three times as abundant as O. They have shown that for the oxygen-rich case, only Fe particles can condense and grow to sizes $\sim 10^{-5}$ cm in the time available in an explosion ($\sim 10^5$ years) before the density of the expanding gas falls too low. For the carbon-rich case, C, Si, and Fe can all condense, but the chief contribution must come from graphite since the abundance of carbon is high.

It is possible, therefore, that galactic nuclei are major sources of dust.

Background Radiation

Four distinct radiation fields have so far been discovered in the Universe: diffuse starlight, the nonthermal radio background radiation, the microwave background radiation, and the X-ray background radiation. Three of these are generally thought to be due, directly or indirectly, to discrete sources, i.e. galaxies. The microwave background radiation may be blackbody radiation generated in a big bang universe, or it may also be due to discrete sources. Thus apart from the diffuse starlight, all of the components of the background radiation may have been generated directly or indirectly by galactic nuclei. We discuss them in turn.

Radio background radiation.—The isotropic component of nonthermal radio emission has a maximum brightness temperature of $65 \pm 5°K$ at 178 MHz (421, 422). Measurements at a number of frequencies give a spectral shape which suggests that it is the result of the integrated effect of discrete sources. These are of two types, the strong radio sources, and the so-called normal galaxies. Bridle (422) has argued that this isotropic component is largely due to the strong radio sources. He believes that the normal galaxies cannot make a significant contribution. They will not unless they have followed the same density evolution that the strong radio sources must have gone through if the Cambridge interpretation of the log N-log S curve is accepted, and then they would have produced far too much background emission. On the other hand, Brecher & Morrison (423) have concluded, from a study of the radio luminosity function, that normal galaxies have contributed at least 90% of the background emission. We do not propose to pursue this thorny topic any further here. We only wish to point out that in any case it appears that this background radiation ultimately comes from galactic nuclei. In the case of the strong radio sources it is generally accepted that the energy is generated in violent events in nuclei, and as was discussed in Section 7, it now appears likely that a large part of the radio emission from normal galaxies is generated in the nuclear regions also.

Microwave background radiation.—Since blackbody radiation at a temperature of a few degrees Kelvin was predicted by Gamow and his colleagues to be present at this epoch if the Universe has expanded from a dense initial state, the discovery of a highly isotropic radiation field has been widely interpreted as confirmation of the theoretical prediction. The observed flux of radiation with wavelengths from about 20 cm to a few millimeters fits quite well on a blackbody curve with a temperature of about 2.7°K. The energy density of the radiation is then about 6×10^{-13} erg cm^{-3}. If the blackbody form is confirmed, this will be the strongest evidence for an evolving universe. However, it is not yet certain whether or not the spectrum is really of blackbody form, since the wavelength range directly observed, and not in dispute, is the Rayleigh-Jeans part of the curve where $I(\nu) \propto \nu^2$.

The observations that have been made either directly, or indirectly, close to and just beyond the peak of the blackbody curve are in conflict unless it is argued that some very strong line radiations are present. The indirect observation of the radiation field in the galactic plane which is exciting the CN, CH, and CH$^+$ interstellar molecules suggests that the temperature is less than $\sim$8°K at 0.36 mm, less than $\sim$5°K at 0.56 mm, and less than $\sim$4°K at 1.3 mm, and it appears to be about 2.8°K at 2.64 mm (428). However, direct observation from rockets above the Earth's atmosphere at wavelengths in the range 0.4–1.3 mm gives a flux corresponding to a blackbody temperature of about 8°K (429). Recent balloon observations (442) have confirmed that there is a very large flux of radiation, either in lines or corresponding to a blackbody temperature compatible with the rocket observations at wavelengths in the range ~0.1–1 mm.

If it turns out that the radiation field departs strongly from the blackbody form, then either an extra component is present, or the radiation has a quite different origin. Also, if we live in a steady-state universe the radiation must have another origin. In both cases the answer must be that the excess radiation, or the total flux, ultimately arises in discrete sources. These possibilities have been explored in a number of investigations. The sources are most likely to be the nuclei of galaxies which radiate largely at infrared and microwave frequencies. They cannot be the ordinary population of radio sources discovered at longer wavelengths. Comparison of the discrete source models (424) with the limits on the small-scale fluctuations of the microwave background (425, 426) have led to the conclusion (426, 427) that in steady-state cosmology the number density of point sources must be very much greater ($\times 10^5$) than that of ordinary galaxies, while in evolutionary cosmologies the number density must be at least comparable with the density of normal galaxies. Even if the sources are extended, i.e. the radiation emitted by nuclei is thermalized in extended dust clouds, the density in the steady-state model is still higher than that of galaxies.

Thus, unless the radiation was generated in a primeval fireball, and this now appears to be the most likely possibility, galactic nuclei of so far undiscovered types must be present in large numbers. Some have clearly felt that the very large numbers and power of the discrete sources required is an argument against this interpretation. However, the history of the last few years has shown that we have continuously underestimated the importance of discrete sources in uninvestigated frequency ranges before they were discovered. Thus I feel that this is not a good argument. The most powerful argument for the primeval fireball interpretation would be to show unambiguously that the radiation has a blackbody form, and this has not been achieved so far.

X-ray background radiation.—Two broad classes of model have been proposed to explain this high-frequency nonthermal radiation field. Either it is the sum of the radiation from "normal" X-ray galaxies like our own (430), or strong X-ray galaxies like M87, a proposal similar to that used to explain the radio background, or else it is due to the interaction between charged particles and an ambient intergalactic gas, or between charged particles and the microwave background radiation (423, 431). Many variations on these models have been proposed, and it is not possible now to discriminate between them (432).

Possibly, the bulk of the extragalactic X-ray and γ-ray flux ultimately arises in sources in external galaxies like supernova remnants and Sco X-1.

However, if it arises in objects like M87, it is undoubtedly due to nuclear activity. If it is in part due to the Compton effect, at least the high-energy electrons and perhaps the microwave photons have also been generated in galactic nuclei. Similarly, if some part of the flux is produced by interactions between high-energy protons and electrons and an intergalactic gas, the high-energy particles at least are thought to have come from radio sources generated in galactic nuclei.

High-Energy Particles

We know that galactic nuclei generate relativistic particles in great profusion. Are they therefore responsible for the bulk of the cosmic rays in the Universe? In modern theories of the origin of cosmic rays it has been suggested that the major sources are supernovae and their remnants (pulsars) and galactic nuclei. Ginzburg & Syrovatsky (433) are the main proponents of the supernova origin theory, and they have argued that the cosmic rays entering the solar system are part of a flux that is trapped for comparatively long times, $\sim 10^8$ years, in the Galaxy. They have also suggested that perhaps violent events in the nucleus of our Galaxy (434) have been responsible for some part of the flux. Burbidge & Hoyle (435) have suggested that cosmic rays are universal, or are confined within clusters of galaxies, and that they have largely been generated in strong radio sources. A natural extension of this idea is that all active galactic nuclei contribute to the flux.

Since the energy density of the primary cosmic rays is about 10^{-12} erg cm^{-3}, the energy requirements for extragalactic cosmic rays are large, amounting to about 10^{62}–10^{63} ergs per average galaxy in its lifetime. This hypothesis has been attacked on energetic grounds and for lesser reasons (433). The background γ-ray flux sets limits on the density of intergalactic gas if cosmic rays are universal.

However, in view of the remarkably powerful types of galactic nuclei now being discovered, the large energy requirements are not an overwhelming objection to a universal cosmic-ray theory. The data on radio galaxies (Section 7) now suggest that there is little evidence for galactic halos, the large trapping volumes originally invoked by Ginzburg & Syrovatsky in the galactic theory. Thus even in that theory it may be that the escape times for cosmic rays are as short as 10^5–10^6 years. If this is the case it seems probable that the bulk of the cosmic rays, universal or not, are generated in galactic nuclei.

It is generally agreed that the electron component of cosmic rays, comprising in energy about 2% of the total flux, cannot have an extragalactic origin. This is because the electron flux generated in a galaxy will be degraded by Compton collisions with the microwave background over distances which are small on the cosmological scale.

The energy densities of microwave background radiation and cosmic rays assumed to be universal are both about 10^{-12} erg cm^{-3}. The total mass density in the form of visible galaxies is about 5×10^{-31} g cm^{-3}. If

TABLE 10. Infrared luminosities of some extragalactic objects

Object	Type	IR luminosity out to 25 μ (erg/sec)	IR luminosity extrapolated to 300 μ (erg/sec)
NGC 1068	Seyfert	7.5×10^{44}	2.5×10^{46}
NGC 1275	Seyfert	8.1×10^{44}	2.7×10^{46}
NGC 4151	Seyfert	2.5×10^{43}	8.5×10^{44}
NGC 7469	Seyfert	6.9×10^{44}	2.3×10^{46}
3C 120	Seyfert	8.7×10^{44}	2.9×10^{46}
3C 273	QSO	1.9×10^{47}	6.2×10^{48}
		(assumed distance 630 Mpc)	
		5.0×10^{43}	1.5×10^{45}
		(assumed distance 10 Mpc)	
M82	Irr. exploding nucleus	6.6×10^{43}	2.2×10^{45}
NGC 3077	Irr.	1.0×10^{43}	3.4×10^{44}
NGC 2782	Spiral, emission-line nucleus	3.3×10^{44}	1.0×10^{46}
NGC 5236	Spiral, emission-line nucleus	1.9×10^{43}	6.3×10^{44}
NGC 7714	Spiral, emission-line nucleus	7.8×10^{43}	2.6×10^{45}

about 25% of the mass has been converted to helium in massive objects in galactic nuclei (Section 13), the thermonuclear energy released amounts to about 3×10^{-12} erg cm^{-3}. These numbers may all be coincidentally the same. Alternatively this may indicate that both microwaves and cosmic rays have been generated in galactic nuclei.

This completes our survey of the extended components of matter and radiation known to exist in the Universe. From the discussion it can be seen that many components may have been generated in galactic nuclei.

15. CONCLUSION AND ACKNOWLEDGMENTS

In this paper I have attempted to survey the properties of galactic nuclei and demonstrate that an understanding of them is of the greatest importance for most aspects of extragalactic astronomy. Clearly we are only at the beginning as far as both observations and theory are concerned. Further studies of nuclei are essential if we are ever to make real progress in understanding the formation and evolution of galaxies, the overall energetics of the Universe, and cosmology.

I am indebted to many colleagues and friends who continually bombard me with their ideas, and sometimes with their results. I am particularly grateful to Margaret Burbidge for help at all stages of this discussion. I also wish to thank Jean Fox for doing the bulk of the typing, and Del Crowne for organizing and checking the manuscript and references.

Work in extragalactic astronomy at UCSD is supported in part by the National Science Foundation and in part by NASA under grant NGL 05-005-004.

LITERATURE CITED

1. Baade, W. 1963, *Evolution of Stars and Galaxies* (Harvard Press)
2. Hubble, E. 1930, *Ap. J.*, **71**, 231
3. Smith, S. 1935, *Ap. J.*, **82**, 192
4. Morgan, W. W. 1958, *PASP*, **70**, 364
5. Morgan, W. W. 1959, ibid, **71**, 92
6. Morgan, W. W. ibid, 394
7. Morgan, W. W. 1959, *Astron. J.*, **64**, 432
8. Morgan, W. W. 1962, *Ap. J.*, **135**, 1
9. Morgan, W. W., Mayall, N. U. 1957, *PASP*, **69**, 291
10. Morgan, W. W., Osterbrock, D. E. 1969, *Astron. J.*, **74**, 515
11. Spinrad, H., 1961, *PASP*, **73**, 336
12. Spinrad, H. 1962, *Ap. J.*, **135**, 715
13. Spinrad, H 1966, *PASP*, **78**, 367
14. Deutsch, A. J. 1964, *Ap. J.*, **139**, 532
15. Wood, D. B. 1966, *Ap. J.*, **145**, 36
16. McClure, R. D., van den Bergh, S. 1968, *Astron. J.*, **73**, 313
17. Humason, M. L., Mayall, N. U., Sandage, A. R. 1956, *Astron. J.*, **61**, 97
18. Nassau, J. J., Blanco, V. M. 1958, *Ap. J.*, **128**, 46
19. Baum, W. A. 1959, *PASP*, **71**, 106
20. McClure, R. D. 1969, *Astron. J.*, **74**, 50
21. Hodge, P. W. 1963, *Astron. J.*, **68**, 237
22. Tifft, W. G. 1961, *Astron. J.*, **66**, 390
23. Tifft, W. G. 1963, ibid, **68**, 302
24. Stebbins, J., Whitford, A. E. 1948, *Ap. J.*, **108**, 413
25. Stebbins, J., Whitford, A. E. 1952, ibid, **115**, 284
26. Pettit, E. 1954, *Ap. J.*, **120**, 413
27. de Vaucouleurs, G. 1960, *Ap. J. Suppl.*, **5**, No. 48, 233
28. Oke, J. B., Sandage, A. R. 1968, *Ap. J.*, **154**, 21
29. Holmberg, E. 1950, *Medd. Lunds. Astr. Obs.*, Ser. II, No. 128
30. Tifft, W. G. 1968, *Astron. J.*, **73**, 879
31. Tifft, W. G. 1969, ibid, **74**, 354
32. Miller, R. H., Prendergast, K. H. 1962, *Ap. J.*, **136**, 713
33. Miller, R. H. 1963, *Ap. J.*, **137**, 733
34. Mayall, N. U. 1939, *Lick Obs. Bull.*, **19**, 33
35. Mayall, N. U., 1958, *IAU Symp. No. 5*, 23 (Cambridge Univ. Press)
36. Burbidge, E. M., Burbridge, G. R. 1962, *Ap. J.*, **135**, 694
37. Burbidge, E. M., Burbidge, G. R. 1965, ibid, **142**, 634
38. Burbidge, G. R., Gould, R. J., Pottasch, S. R. 1963, *Ap. J.*, **138**, 945
39. Minkowski, R., 1962, *Problems of Extragalactic Research*, 112 (McVittie, G., Ed., New York: Macmillan)
40. Peimbert, M. 1968, *Ap. J.*, **154**, 33
41. Minkowski, R., Osterbrock, D. E. 1959. *Ap. J.*, **129**, 583
42. Osterbrock, D. E. 1960, *Ap. J.*, **132**, 325
43. Wentzel, D. G., van Woerden, H. 1959, *BAN*, **14**, 335
44. Rogstad, D. H., Rougoor, G. W., Whiteoak, J. B. 1967, Preprint
45. Burbidge, G. R., Burbidge, E. M. 1970, *Stars and Stellar Systems*, **9**, *Galaxies and the Universe* (Sandage, A., Sandage, M., Eds., Univ. Chicago Press)
46. Lallemand, A., Duchesne, M., Walker, M. F. 1960, *PASP*, **72**, 76
47. Hubble, E. 1929, *Ap. J.*, **69**, 154
48. Kinman, T. D. 1965, *Ap. J.*, **142**, 1376
49. Sandage, A. R., Becklin, E. E., Neugebauer, G. 1969, *Ap. J.*, **157**, 55
50. Münch, G. 1960, *Ap. J.*, **131**, 250
51. Rubin, V. C., Ford, W. K. 1969, Preprint
52. Walker, M. F. 1962, *Ap. J.*, **136**, 695
53. Walker, M. F. 1964, *Astron. J.*, **69**, 744
54. Mayall, N. U., Aller, L. H. 1942, *Ap. J.*, **95**, 5
55. Sandage, A. R. 1963, *Ap. J.*, **138**, 863
56. Carranza, G., Courtès, G., Georgelin, Y., Monnet, G., Pourcelot, A. 1968, *Ann. Ap.*, **31**, 63
57. Münch, G. 1959, *PASP*, **71**, 101
58. Sandage, A. R. 1961, *The Hubble Atlas of Galaxies* (Washington, D. C.: Carnegie Inst. Wash.)
59. Burbidge, E. M., Burbidge, G. R. 1964, *Ap. J.*, **140**, 1445
60. Carranza, G., Crillon, R., Monnet, G., 1969, *Astron. Ap.*, **1**, 479
61. Burbidge, E. M., Burbidge, G. R., Prendergast, K. H. 1962, *Ap. J.*, **136**, 339
62. Demoulin, M.-H., Burbidge, E. M. 1970, *Ap. J.*, **159**, 799
63. Fomalont, E. B. 1968, *Ap. J. Suppl.*, **15**, 203
64. Gardner, F. F., Morris D., Whiteoak, J. B. 1969, *Aust. J. Phys.*, **22**, 79
65. Ekers, R. D. 1969, *Aust. J. Phys. Ap. Suppl.* No. 6
66. Demoulin, M.-H., Burbidge, E. M. 1969, *Ap. Lett.*, **4**, 89
67. Burbidge, E. M., Burbidge, G. R. 1960, *Ap. J.*, **132**, 30
68. Burbidge, E. M., Burbidge, G. R., Prendergast, K. H. 1962, *Ap. J.*, **136**, 119

69. Burbidge, E. M., Burbidge, G. R., Prendergast, K. H. 1960, ibid, **132**, 661
70. Lampland, C. 1921, *PASP*, **33**, 167
71. Walker, M. F. 1967, *PASP*, **79**, 593
72. Walker, M. F., Chincarini, G. 1967, *Ap. J.*, **147**, 416
73. Piddington, J. H., Minnett, H. C. 1951, *Aust. J. Sci. Res.*, **A4**, 459
74. Haddock, F. T., Mayer, C. H., Sloanaker, R. M. 1954, *Ap. J.*, **119**, 456
75. Drake, F. D. 1959, *Astron. J.*, **64**, 329
76. Pariiskii, Y. N. 1959, *Dokl. Akad. Nauk SSSR*, **129**, 1261
77. Biraud, F., Lequeux, J., LeRoux, E. 1960, *Observatory*, **80**, 116
78. Burke, B. F. 1965, *Ann. Rev. Astron. Ap.*, **3**, 275
79. Downes, D., Maxwell, A. 1966, *Ap. J.*, **146**, 653
80. Maxwell, A., Taylor, J. H. 1968, *Ap. Lett.*, **2**, 191
81. Thompson, A. R., Riddle, A. C., Lang, K. R. 1969, *Ap. Lett.*, **3**, 49
82. Lequeux, J. 1967, *Radio Astronomy and the Galactic System, IAU Symp. No. 31*, 393 (New York: Academic)
83. Mezger, P. G., Höglund, B. 1967, *Ap. J.*, **147**, 490
84. Reifenstein, E. C., Wilson, T. L., Burke, B. F., Mezger, P. G. 1967, *Am. URSI Spring Meet., Ottawa, May*
85. Kerr, F., Sinclair, M. W. 1966, *Nature*, **212**, 166
86. Cameron, M. J. 1969, Preprint
87. Dworetsky, M. M., Epstein, E. E., Fogarty, W. G., Montgomery, J. W. 1969, *Ap. J.*, **158**, L183
88. Becklin, E. E., Neugebauer, G. 1968, *Ap. J.*, **151**, 145
89. Becklin, E. E., Neugebauer, G. 1969, ibid, **157**, L31
90. Low, F. J., Kleinmann, D. E., Forbes, F. F., Aumann, H. H. 1969, *Ap. J.*, **157**, L97; Aumann, H. H., Low, F. J. 1970, *Ap. J.*, **159**, L159
91. Hoffmann, W. F., Frederick, C. L. 1969, *Ap. J.*, **155**, L9
92. Burbidge, G. R., Stein, W. A. 1970, *Ap. J.*, **160**, May
93. Clark, G. W., Garmire, G., Kraushaar, W. L., 1968, *Ap. J.*, **153**, L203
94. van Woerden, H., Rougoor, G. W., Oort, J. H. 1957, *Compt. Rend.*, **224**, 1691
95. Rougoor, G. W., Oort, J. H. 1960, *Proc. Nat. Acad. Sci.*, **46**, 1
96. Rougoor, G. W. 1964, *BAN*, **17**, 381
97. Kerr, F., Vallak, R. 1967, *Aust. J. Phys. Ap. Suppl. No. 3*, 1
98. Cugnon, P. 1968, *BAN*, **19**, 363
99. van der Kruit, P. C. 1969, Preprint
100. Oort, J. H. 1966, *BAN*, **18**, 421
101. Oort, J. H. 1967, *IAU Symp. No. 31*, 279
102. Oort, J. H. 1969, *Nature*, **224**, 1158
103. Hulsbosch, A. N. M. 1968, *BAN*, **20**, 33
104. Robinson, B. J., McGee, R. X. 1967, *Ann. Rev. Astron. Ap.*, **5**, 183
105. Bolton, J. G., Gardner, F. F., McGee, R. X., Robinson, B. J. 1964, *Nature*, **204**, 30
106. Weber, J. 1969, *Phys. Rev. Lett.*, **22**, 1320; see also Beron, B. L., Hofstadter, R. 1969. *Phys. Rev. Lett.*, **23**, 184
107. Weber, J. 1969, Unpublished
108. Field, G. B., Rees, M. J., Sciama, D. W. 1969, *Comments Ap. Space Phys.*, **1**, 187
   Sciama, D. W., Field, G. B., Rees, M. J. 1969, *Phys. Rev. Lett.*, **23**, 1514
109. Sciama, D. W. 1969, *Nature*, **224**, 1263
110. Burbidge, G. R., Burbidge, E. M., Sandage, A. R. 1963, *Rev. Mod. Phys.*, **35**, 947
111. Lynds, C. R., Sandage, A. R. 1963, *Ap. J.*, **137**, 1005
112. Lequeux, J. 1962, *Compt. Rend.*, **255**, 1865
113. Sandage, A. R., Miller, W. C. 1964, *Science*, **144**, 405
114. Sandage, A. R., Visvanathan, N. 1969, *Ap. J.*, **157**, 1065
115. Elvius, A. 1963, *Lowell Obs. Bull.*, **5**, 281
116. Elvius, A. 1969, ibid, **7**, 117
117. Burbidge, E. M., Burbidge, G. R., Rubin, V. C. 1964, *Ap. J.*, **140**, 942
118. Solinger, A. B. 1969, *Ap. J.*, **155**, 403
119. van den Bergh, S. 1969, *Ap. J.*, **156**, L19
120. Bertola, F., D'Odorico, S., Ford, W. K., Rubin, V. C. 1969, *Ap. J.*, **157**, L27
121. Raff, M. I. 1969, *Ap. J.*, **157**, L29
122. Peimbert, M., Spinrad, H. 1969, Preprint
123. Kleinmann, D. E., Low, F. J. 1969, *129th AAS Meet., Abstr.*
124. Curtis, H. D. 1918, *Publ. Lick Obs.*, **13**, 31
125. Bolton, J. G., Stanley, G. J., Slee, O. B. 1949, *Nature*, **164**, 101
126. Baade, W., Minkowski, R. 1954, *Ap. J.*, **119**, 215
127. Shklovsky, I. S. 1955, *Astron. J. USSR*, **32**, 215
128. Baade, W. 1956, *Ap. J.*, **123**, 550
129. Hiltner, W. A. 1959, *Ap. J.*, **130**, 340
130. Burbidge, G. R. 1956, *Ap. J.*, **124**, 416

131. Bless, R. C. 1962, *Ap. J.*, **135**, 187
132. Felten, J. E. 1968, *Ap. J.*, **151**, 861
133. Felten, J. E., Arp. H. C., Lynds, C. R. 1970, *Ap. J.*, **159**, 415
134. de Vaucouleurs, G., Angione, R., Fraser, C. W. 1968, *Ap. Lett.*, **2**, 141
135. Arp, H. C. 1967, *Ap. Lett.*, **1**, 1
136. Friedman, H., Byram, E. T. 1967, *Science*, **158**, 257
137. Bradt, H., Mayer, W., Naranan, S., Rappaport, S., Spada, G. 1967, *Ap. J.*, **150**, L199
138. Hudson, H. S., Peterson, L. E., Schwartz, D. A. 1969, *Solar Phys.*, **6**, 205
139. Haymes, R. C., Ellis, D. V., Fishman, G. J., Glenn, S. W., Kurfess, J. D. 1968, *Ap. J.*, **151**, L131
140. Burbidge, G. R. 1970, *Ap. J.*, **159**, L105
141. Cohen, M. H., Moffet, A. T., Shaffer, D., Clark, B. G., Kellermann, K. I., Jauncey, D. L., Gulkis, S. 1969, *Ap. J.*, **158**, L83
142. Hogg, D. E., McDonald, G. H., Conway, R. G., Wade, C. M. 1969, *Astron. J.*, **74**, 1206
143. Wade, C. M. 1960, *Observatory*, **80**, 235
144. Minkowski, R. 1959, *Radio Astronomy, IAU Symp. No. 9*, 335 (Bracewell, R. N., Ed., Stanford Univ. Press)
145. Baade, W., Minkowski, R. 1954, *Ap. J.*, **119**, 215
146. Shklovsky, I. S. 1962, *Astron. Zh.*, **39**, 591
147. Shklovsky, I. S. 1963, ibid, **40**, 972
148. Walker, M. F., Hayes, S. 1967, *Ap. J.*, **149**, 481
149. Mayall, N. U. 1936, *PASP*, **48**, 14
150. Heeschen, D. S. 1968, *Ap. J.*, **151**, L135
151. Wills, D. 1968, *Ap. Lett.*, **2**, 187
152. Lang, K. R., Terzian, Y. 1968, *Ap. J.*, **152**, L63
153. de Jong, M. L. 1967, *Ap. J.*, **150**, 1
154. Seyfert, C. K. 1943, *Ap. J.*, **97**, 28
155. Fath, E. A. 1908, *Lick Obs. Bull.*, **5**, 71
156. Slipher, V. M. 1917, *Lowell Obs. Bull.*, **3**, 59
157. Campbell, W. W., Moore, J. H. 1918, *Lick Obs. Publ.*, **13**, 88
158. Hubble, E. P. 1926, *Ap. J.*, **64**, 328
159. Mayall, N. U. 1934, *PASP*, **46**, 134
160. Lynds, C. R. 1970, *Ap. J.*, **159**, L151
161. Burbidge, E. M., Burbidge, G. R., Prendergast, K. H. 1963, *Ap. J.*, **137**, 1022
162. Burbidge, E. M., Burbidge, G. R., Prendergast, K. H. 1959, ibid, **130**, 26
163. Woltjer, L. 1959, *Ap. J.*, **130**, 38
164. Pacholczyk, A. G., Weymann, R. 1968, *Proc. Seyfert Galaxies Related Objects Conf.*, *Astron. J.*, **73**, 836
165. Walker, M. F. 1968, *Ap. J.*, **151**, 71
166. Osterbrock, D. E., Parker, R. A. 1965, *Ap. J.*, **141**, 892
167. Dibai, E. A., Pronik, V. I. 1965, *Astrofizika*, **1**, 78
168. Code, A. D. 1959, *PASP*, **71**, 118
169. Visvanathan, N., Oke, J. B. 1968, *Ap. J.*, **152**, L165
170. Dibai, E. A., Shakhovskoi, I. S. 1966, *Astron. Cirk.*, **375**, 1
171. Hagen-Thorn, V. A., Dombrowski, V. A. 1967, *Astron. Cirk. No. 454*
172. Kruszewski, A. 1968, *Astron. J.*, **73**, 852
173. Pacholczyk, A. G., Wisniewski, W. A. 1967, *Ap. J.*, **147**, 394
174. Low, F. J., Kleinmann, D. E. 1968, *Astron. J.*, **73**, 868
175. Rees, M. J., Silk, J. I., Werner, M. W., Wickramasinghe, N. C. 1969, *Nature*, **223**, 788
176. Pacholczyk, A. G., Weymann, R. J. 1968, *Astron. J.*, **73**, 870
177. Bash, F. N. 1967, PhD thesis, Univ. Virginia; 1968, *Ap. J. Suppl.*, **16**, 373
178. Cohen, M. H., Gunderman, E. J., Harris, D. E. 1967, *Ap. J.*, **150**, 767
179. Epstein, E. E., Fogarty, W. G. 1968, *Astron. J.*, **73**, 873
180. Burbidge, E. M., Burbidge, G. R. 1965, *Ap. J.*, **142**, 1351
181. Dibai, E. A., Pronik, V. I. 1966, *Izv. Krymskoi Astrofiz. Obs.*, **35**, 87
182. Minkowski, R. 1968, *Astron. J.*, **73**, 842
183. Minkowski, R. 1957, *IAU Symp. No. 4, Radio Astronomy*, 107 (van de Hulst, H. C., Ed., Univ. Cambridge Press)
184. Oke, J. B. 1968, *Astron. J.*, **73**, 849
185. Ryle, M., Windram, M. D. 1968, *MNRAS*, **138**, 1
186. Kellermann, K. I., Pauliny-Toth, I. I. K. 1968, *Astron. J.*, **73**, 874
187. Clark, B. G., Kellermann, K. I., Bare, C. D., Cohen, M. H., Jauncey, D. L. 1968, *Ap. J.*, **153**, 705
188. Sérsic, J. L. 1957, *Observatory*. **77**, 146
189. de Vaucouleurs, G, de Vaucouleurs, A. 1961, *Mem. RAS*, **68**, 69
190. de Vaucouleurs, G., de Vaucouleurs, A. 1968, *Astron. J.*, **73**, 858
191. Dibai, E. A., Pronik, V. I. 1968, *Sov. Astron. AJ*, **11**, 767
192. Rubin, V. C., Ford, W. K. 1968, *Ap. J.*, **154**, 431
193. Wade, C. M. 1968, *Astron. J.*, **73**, 876

194. Andrillat, Y., Souffrin, S. 1968, *Ap. Lett.*, **1**, 111
195. Oke, J. B., Sargent, W. L. W. 1968, *Ap. J.*, **151**, 807
196. Ford, W. K., Purgathofer, A. T., Rubin, V. C. 1968, *Ap. J.*, **153**, L39
197. Anderson, K. S., Kraft, R. P. 1969, *Ap. J.*, **158**, 859
198. Danielson, R., Savage, B. D., Schwarzschild, M. 1968, *Ap. J.*, **154**, L117
199. Walker, M. F. 1968, *Astron. J.*, **73**, 854
200. Fitch, W. S., Pacholczyk, A. G., Weymann, R. J. 1967, *Ap. J.*, **150**, L67
201. Stein, W. A., Gillett, F. C. 1969, *Nature*, **224**, 675
202. Dibai, E. A., Esipov, B. F., Pronik, V. I. 1968, *Sov. Astron. AJ*, **11**, 553
203. Bardin, C., Chopinet, M., Duflot-Augarde, R. 1967, *Compt. Rend. B*, **265**, 1149
204. Zwicky, F. 1964, *Ap. J.*, **140**, 1467
205. Arp, H. 1968, *Ap. J.*, **152**, 1101
206. Clarke, M. E., Bolton, J. G., Shimmins, A. J. 1966, *Aust. J. Phys.*, **19**, 375
207. Burbidge, E. M. 1967, *Ap. J.*, **149**, L51
208. Sargent, W.L. W. 1967, *PASP*, **79**, 369
209. Kinman, T. D. 1968, *Astron. J.*, **73**, 885
210. Oke, J. B., Sargent, W. L. W., Neugebauer, G., Becklin, E. E. 1967, *Ap. J.*, **150**, L173
211. Zwicky, F., Oke, J. B., Neugebauer, G., Sargent, W. L. W., Fairall, A. P. 1969, Preprint
212. Sargent, W. L. W. 1968, *Ap. J.*, **152**, L31
213. Wampler, E. J., Oke, J. B. 1967, *Ap. J.*, **148**, 695
214. Khachikian, E. Ye. 1968, *Astron. J.*, **73**, 891
215. Weedman, D. W., Khachikian, E. Ye. 1969, *Astrofizika*, **5**, 113
216. Sargent, W. L. W. 1970, Preprint
217. Fairall, A. P., 1968, *PASP*, **80**, 235
218. Sargent, W. L. W. 1968, *Astron J.*, **73**, 894
219. Burbidge, E. M., Burbidge, G. R. 1961, *Astron J.*, **66**, 541
220. Burbidge, E. M., Burbidge, G. R. 1964, *Ap. J.*, **140**, 1307
221. Sargent, W. L. W. 1970, Preprint
222. Matthews, T. A., Morgan, W. W., Schmidt. M. 1964, *Ap. J.*, **140**, 35
223. Schmidt, M. 1965, *Ap.. J.*, **141**, 1
224. Sandage, A. R. 1966, *Ap. J.* **145**, 1
225. Burbidge, E. M. 1969, Private communication
226. Lynds, C. R. 1968, *Astron. J.*, **73**, 888
227. Oke, J. B. 1967, *Ap. J.*, **150**, L5
228. Sandage, A. R. 1967, *Ap. J.*, **150**, L9
229. Burbidge, G. R., Burbidge, E. M. 1967, *Quasi-Stellar Objects* (San Francisco: Freeman)
230. Burbidge, E. M. 1967, *Ann. Rev. Astron. Ap.*, **5**, 399
231. Schmidt, M. 1969, *Ann. Rev. Astron. Ap.*, **7**, 527
232. Burbidge, G. R., Burbidge, E. M. 1969, *Nature*, **224**, 21
233. Braccesi, A., Lynds, R., Sandage, A. 1968, *Ap. J.*, **152**, L105
234. Lynds, C. R. 1969, Private communication
235. Oke, J. B. 1969, *Ap. J.*, **158**, L9
236. Demoulin, M.-H., Burbidge, G. R. 1968, *Ap. J.*, **154**, 3
237. Gubbay, J., Legg, A. J., Robertson, D. S., Moffet, A. T., Ekers, R. D., Seidel, B. 1969, *Nature*, **224**, 1094
238. Burbidge, E. M., Burbidge, G. R. 1965, *The Structure and Evolution of Galaxies*, Proc. 13th Conf. Phys., Univ. Brussels, 137 (Interscience: Wiley)
239. Moffet, A. T. 1966, *Ann. Rev. Astron. Ap.*, **4**, 145
240. Burbidge, G. R. 1962, *Progr. Theor. Phys. Japan*, **27**, 999
241. Schmidt, M. 1966, *Ap. J.*, **146**, 7
242. Rogstad, D. H., Ekers, R. D. 1969, *Ap. J.*, **157**, 481
243. Kellermann, K. I., Pauliny-Toth, I. I. K. 1968, *Ann. Rev. Astron. Ap.*, **6**, 417
244. Bolton, J. G., Clark, B. G. 1960 *PASP*, **72**, 29
245. McDonald, G. H., Kenderdine, S., Neville, A. C. 1968, *MNRAS*, **138**, 259
246. van der Laan, H., Perola, G. C. 1969, *Astron. Ap.*, **3**, 468
247. van der Laan, H. 1963, *MNRAS*, **126**, 532
248. Ryle, M., Longair, M. S. 1967, *MNRAS*, **136**, 123
249. Gold, T. 1965, *Proc. 9th Int. Conf. Cosmic Rays, London*, 132 (Inst. Phys. and Phys. Soc.)
250. De Young, D. S., Axford, W. I. 1967, *Nature*, **216**, 129
251. Mills, D. M., Sturrock, P. A. 1970, *Ap. Lett.*, **5**, 105
252. Burbidge, G. R. 1967, *Nature*, **216**, 1287
253. Grewing, M., Demoulin, M.-H., Burbidge, G. R. 1968, *Ap. J.*, **154**, 447
254. Heeschen, D. S., Wade, C. M. 1964, *Astron. J.*, **69**, 277
255. De Jong, M. L. 1966, *Ap. J.*, **144**, 553

256. Cameron, M. J., Glanfield, J. R. 1968, *MNRAS*, **141**, 145
257. Tovmassian, H. M. 1966, *Aust. J. Phys.*, **19**, 565
258. Tovmassian, H. M. Ibid, 883
259. Lang, K. R., Terzian, Y. 1969, *Ap. Lett.*, **3**, 29
260. Purton, C. R. 1966, *MNRAS*, **133**, 463
261. Penzias, A. A., Wilson, R. W. 1966, *Ap. J.*, **146**, 666
262. Baldwin, J. E., Costain, C. H. 1960, *MNRAS*, **121**, 413
263. Heeschen, D. S. 1970, *Astron. J.*, **75**, June, and preprint, *New York Meet.*, December
264. Sérsic, J. L., Pastoriza, M. 1965, *PASP*, **77**, 287
265. Sérsic, J. L., Pastoriza, M. 1967, *PASP*, **79**, 152
266. Sérsic, J. L. 1968, *Astron. J.*, **73**, 892
267. Vorontsov-Velyaminov, B. A. 1959, *Atlas of Interacting Galaxies* (Moscow)
268. Vorontsov-Velyaminov, B. A., Krasnogorskaja, A. 1962, *Trudy Sternberg State Astron. Inst.*, 32
269. Vorontsov-Velyaminov, B. A., Artipova, V. 1963; 1964, *Trudy Sternberg State Astron. Inst.*, **33**; **34**
270. Vorontsov-Velyaminov, B. A., Artipova, V. 1968, ibid, **38**
271. Arp, H. 1966, *Atlas of Peculiar Galaxies* (California Inst. Tech., Pasadena
272. Zwicky, F., Herzog, E., Wild, P. 1961, *Catalogue of Galaxies and Clusters of Galaxies*, **1** (California Inst. Tech., Pasadena)
273. Zwicky, F., Herzog, E. 1963, ibid, **2**
274. Zwicky, F., Herzog, E. 1966, ibid, **3**
275. Zwicky, F., Herzog, E. 1968, ibid, **4**
276. Zwicky, F., Karpowicz, M., Kowal, C. 1965, *Catalogue of Galaxies and Clusters of Galaxies*, **5** (California Inst. Tech., Pasadena)
277. Zwicky, F., Kowal, C. 1968, *Catalogue of Galaxies and Clusters of Galaxies*, **6**
278. Fairall, A. P. 1968, *Mon. Not. S. Afr. Astron. Soc.*, **7**, 67
279. Haro, G. 1956, *Bol. Obs. Tonantzintla Tacubaya*, No. 14, 8
280. DuPuy, P. L. 1968, *Astron. J.*, **73**, 882
281. Kiang, T. 1967, *Ap. J.*, **150**, L31
282. Markarian, B. E. 1963, *Soobcheniya Byur. Obs.*, **34**, 3
283. Markarian, B. E., Oganesyan, E. Ya., Arakelyan, S. N. 1965, *Astrofizika*, **1**, 38
284. Markarian, B. E., Oganesyan, E. Ya., Arakelyan, S. N. 1966, ibid, **2**, 85
285. Markarian, B. E. 1967, *Astrofizika*, **3**, 55
286. Sandage, A. R. 1965, *Ap. J.*, **141**, 1560
287. Sandage, A. R., Luyten, W. J. 1967, *Ap. J.*, **148**, 767
288. Sandage, A. R., Luyten, W. J. 1969, ibid, **155**, 913
289. Braccesi, A. 1967, *Nuovo Cimento*, **49**, 148
290. Williams, R. E., Weymann, R. J. 1968, *Astron. J.*, **73**, 895
291. Souffrin, S. 1969, *Astron. Ap.*, **1**, 305
292. Souffrin, S. Ibid, 414
293. Shklovsky, I. S. 1966, *Sov. Astron.*, **9**, 683
294. Kaneko, N., Ohtani, H. 1968, *Astron. J.*, **73**, 899
295. Osterbrock, D. E. 1968, *Mem. Soc. Roy. Sci. Liège*, **XVI**, 391
296. Osterbrock, D. E. 1968, *Astron. J.*, **73**, 904
297. Osterbrock, D. E. Ibid, 916
298. Osterbrock, D. E. 1969, Preprint
299. Wampler, E. J. 1968, *Ap. J.*, **154**, L53
300. Burbidge, E. M., Burbidge, G. R., Hoyle, F., Lynds, C. R. 1966, *Nature*, **210**, 774
301. Pauliny-Toth, I. I. K., Kellermann, K. I. 1966, *Ap. J.*, **146**, 634
302. Colgate, S. A. 1967, *Ap. J.*, **150**, 163
303. Colgate, S. A. 1968, *Astron. J.*, **73**, 905
304. Low, F. J. 1969, *Science*, **164**, 501
305. Hoyle, F., Burbidge, G. R., Sargent, W. L. W. 1966, *Nature*, **209**, 751
306. Scheuer, P. A. G., Williams, P. J. S. 1968, *Ann. Rev. Astron. Ap.*, **6**, 321
307. Shklovsky, I. S. 1960, *Astron. J. USSR*, **37**, 256; 1960, *Sov. Astron*, **4**, 243
308. Ambartsumian, V. A. 1958, *Solvay Conf. Struct. Evolution Universe*, 241 (Stoops, R., Ed., Brussels)
309. Ambartsumian, V. A. 1965, *The Structure and Evolution of Galaxies, Proc. 13th Conf. Phys., Univ. Brussels*, **1** (Interscience: Wiley)
310. Neyman, J., Page, T., Scott, E. 1961, *Astron. J.*, **66**, 533
311. Limber, D. N., Mathews, W. G. 1960, *Ap. J.*, **132**, 286
312. Karachentsev, J. D. 1966, *Astrofizica*, **2**, 81
313. Burbidge, G. R., Sargent, W. L. W. 1969, *Comments Ap. Space Phys.*, **1**, 220
314. Burbidge, E. M., Burbidge, G. R. 1961, *Ap. J.*, **134**, 244
315. Zwicky, F. 1956, *Ergeb. Exakt. Naturw.*, **29**, 544
316. Sargent, W. L. W. 1968, *Ap. J.*, **153**, L135
317. Sargent, W. L. W. 1970, ibid In press

318. Burbidge, E. M., Burbidge, G. R. 1961, *PASP*, **73**, 191
319. Arp, H. 1967, *Ap. J.* **148**, 321
320. Arp, H. 1968, *PASP*, **80**, 129
321. Arp. H. 1968, *Astrofizica*, **4**, 59
322. Arp, H. 1969, *Sky Telescope*, **38**, 385
323. Arp, H. 1969, *Astron. Ap.*, **3**, 418
324. Arp, H. 1970, *Nature*, **225**, 1033
325. Sérsic, J. L. 1960, *Zs. Ap.*, **50**, 168 and **51**, 64; 1968, *Astrofizica*, **4**, 105; 1967, *ASP Leaflet No. 453*, March
326. Pismis, P. 1963, *Bol. Tonantzintla Tacubaya*, No. 23
327. van der Laan, H., Bash, F. N. 1968, *Ap. J.*, **152**, 621
328. Markarian, B. E. 1961, *Astron. J.*, **66**, 555
329. Holmberg, E. 1969, *Arkiv Astron.*, **5**, No. 20, 305
330. Unsöld, A. 1969, *Science*, **163**, 1015
331. Morgan, W. W., Lesh, J. R. 1965, *Ap. J.*, **142**, 1364
332. Sastry, G. N. 1968, *PASP*, **80**, 252
333. Field, G. B. 1970, Preprint
334. Peebles, P. J. E. 1965, *Ap. J.*, **142**, 1317; 1967, **147**, 859. Peebles, P. J. E., Dicke, R. H. 1968, *Ap. J.* **154**, 891
335. Doroshkevich, A. G., Zeldovich, Ya. B., Novikov, I. D. 1967, *Astron. Zh.*, **44**, 295; 1967, *Sov. Astron. A J*, **11**, 233
336. Harrison, E. R. 1968, *MNRAS*, **141**, 397
337. Hoyle, F. 1968, *Proc. Roy. Soc. A*, **308**, 1
338. Jeans, J. H. 1929, *Astronomy and Cosmogony*, 352 (Cambridge Univ. Press)
339. Sciama, D. W. 1955, *MNRAS*, **115**, 3
340. McCrea, W. H. 1964, *MNRAS*, **128**, 335
341. Hoyle, F., Narlikar, J. V. 1966, *Proc. Roy. Soc. A*, **290**, 177
342. Eggen, O. J., Lynden-Bell, D., Sandage, A. R. 1962, *Ap. J.*, **136**, 748
343. Crampin, J., Hoyle, F. 1964, *Ap. J.*, **140**, 99
344. Saslaw, W. C. 1969, Preprint
345. Hoyle, F., Fowler, W. A., Burbidge, E. M., Burbidge, G. R. 1964, *Ap. J.*, **139**, 909
346. Wolfe, A., Burbidge, G. R. 1970, *Ap. J.*, **161**, August
347. Belzer, J., Gamow, G., Keller, G. 1950, *Ap. J.*, **113**, 166
348. Spitzer, L. 1940, *MNRAS*, **100**, 396
349. Hénon, M. 1964, *Ann. Ap.*, **27**, 83
350. Hénon, M. 1965, ibid, **28**, 62
351. King, I. R. 1966, *Astron. J.*, **71**, 64
352. Michie, R. W. 1961, *Ap. J.*, **133**, 781
353. Michie, R. W. 1963, *MNRAS*, **125**, 127
354. Michie, R. W. Ibid, **126**, 331
355. Michie, R. W. Ibid, **449**
356. Michie, R. W., Bodenheimer, P. 1963, *MNRAS*, **126**, 269
357. Lynden-Bell, D. 1962, *MNRAS*, **124**, 279
358. Lynden-Bell, D. 1967, Ibid, **136**, 101
359. Lynden-Bell, D, Wood, D. B. 1968, *MNRAS*, **138**, 495
360. King, I. R. 1963, *Ann. Rev. Astron. Ap.*, **1**, 179
361. Michie, R. W. 1964, *Ann. Rev. Astron. Ap.*, **2**, 49
362. Miller, R. H., Prendergast, K. H. 1962, *Ap. J.*, **136**, 713
363. Ambartsumian, V. A. 1938, *Ann. Leningrad State Univ. No. 22*
364. Spitzer, L., Härm, R. 1958, *Ap. J.*, **127**, 544
365. Spitzer, L., Saslaw, W. C. 1966, *Ap. J.*, **143**, 400
366. Spitzer, L., Stone, M. E. 1967, *Ap. J.*, **147**, 519
367. von Hoerner, S. 1960, *Z. Ap.*, **50**, 184
368. von Hoerner, S. 1963, ibid, **57**, 47
369. von Hoerner, S. 1967, *Symp. n-Body Problem, Paris*
370. Aarseth, S. J. 1963, *MNRAS*, **126**, 233
371. Wielen, R. 1966, *Mitt. Astron. Rech. Inst. Heidelberg Ser. B*, No. 14
372. Hénon, M. 1960, *Ann. Ap.*, **23**, 668
373. Thirring, W. 1969, Preprint
374. Spitzer, L. 1969, *Ap. J.*, **158**, L139
375. De Young, D. S. 1968, *Ap. J.*, **153**, 633
376. Woltjer, L. 1964, *Nature*, **201**, 807
377. Gold, T., Axford, W. I., Ray, E. C. 1965, *Quasi-Stellar Sources and Gravitational Collapse*, 93 (Univ. Chicago Press)
378. Shklovsky, I. S. 1960, *Astron. Zh.*, **37**, 945; 1960, **4**, 885
379. Burbidge, G. R. 1961, *Nature*, **190**, 1053
380. Greenstein, G. 1969, *Ap. J.*, **158**, L145
381. Hoyle, F., Fowler, W. A. 1963, *MNRAS*, **125**, 169
382. Hoyle, F., Fowler, W. A. 1963, *Nature*, **197**, 533
383. Fowler, W. A. 1964, *Rev. Mod. Phys.*, **36**, 545
384. Hoyle, F., Narlikar, J. V. 1964, *Proc. Roy. Soc. A*, **278**, 465
385. Lynden-Bell, D. 1969, *Nature*, **223**, 690
386. Salpeter, E. E. 1964, *Ap. J.*, **140**, 796
387. Burbidge, G. R. 1967, *Ap. J.*, **147**, 851
388. Burbidge, E. M., Burbidge, G. R. 1967, *Ap. J.*, **148**, L107

389. Burbidge, G. R., Burbidge, E. M. 1969, *Nature*, **222**, 735
390. Greenstein, J. L., Schmidt, M. 1964, *Ap. J.*, **140**, 1
391. Hoyle, F., Fowler, W. A. 1967, *Nature*, **213**, 373
392. Ipser, J. R., Thorne, K. S. 1968, *Ap. J.*, **154**, 251
393. Ipser, J. R. 1969, *Ap. J.*, **156**, 509 and **158**, 17
394. Fackerell, E. D. 1968, *Ap. J.*, **153**, 643; 1970, Preprint
395. Zeldovich, Ya. B., Podurets, M. 1965, *Astron. Zh.*, **42**, 963; 1966, *Sov. Astron.*, **9**, 742
396. Bisnovatyi-Kogan, G. S., Zeldovich, Ya. B. 1969, *Astrofizika*, **5**, 223
397. Bisnovatyi-Kogan, G. S., Zeldovich, Ya. B., Friedman, A. M. 1968, *Dokl. Akad. Nauk SSSR*, **182**, 794; *Sov. Phys. Dokl.*, **13**, 969
398. Bisnovatyi-Kogan, G. S., Thorne, K. S. 1970, Preprint
399. Zapolsky, H. S. 1968, *Ap. J.*, **153**, L163
400. Fowler, W. A. 1964, *Rev. Mod. Phys.*, **36**, 545
401. Chandrasekhar, S. 1964, *Phys. Rev. Lett.*, **12**, 114, 437
402. Bardeen, J. M., Wagoner, R. V. 1969, *Ap. J.*, **158**, L65
403. Morrison, P. 1969, *Ap. J.*, **157**, L73
404. Novikov, I. D. 1964, *Astron. Zh.*, **41**, 1075; 1965, *Sov. Astron.*, **875**; see also Novikov, I. D., Zeldovich, Ya. B. 1966, *Suppl. Nuovo Cimento*, **4**, 810
405. Ne'eman, Y. 1965, *Ap. J.*, **141**, 1303
406. Ne'eman, Y., Tauber, G. 1967, *Ap. J.*, **150**, 755
407. Harrison, E. R. 1970, *MNRAS*, In press
408. Burbidge, E. M., Burbidge, G. R., Fowler, W. A., Hoyle, F. 1957, *Rev. Mod. Phys.*, **29**, 547
409. Wagoner, R. V., Fowler, W. A., Hoyle, F. 1967, *Ap. J.*, **148**, 3
410. Burbidge, G. R. 1969, *Comments Ap. Space Phys.*, **1**, 101
411. Wagoner, R. V. 1968, *Ap. J.*, **151**, L103
412. Wagoner, R. V. 1969, *Ap. J. Suppl.*, **18**, 247
413. Bisnovatyi-Kogan, G. S. 1968, *Astron. Zh.*, **45**, 74; *Sov. Astron. AJ*, **12**, 58
414. Wagoner, R. V. 1969, *Ann. Rev. Astron. Ap.*, **7**, 553
415. Arnett, W. D., Truran, J. W. 1969, *Ap. J.*, **157**, 339; 1970, *Ap. J.*, **160**, 181
416. Arnett, W. D. 1969, *Ap. J.*, **157**, 1381
417. Truran, J. W., Arnett, W. D., Cameron, A. G. W. 1967, *Can. J. Phys.*, **45**, 2315
418. Bodansky, D., Clayton, D. D., Fowler, W. A. 1968, *Ap. J. Suppl.*, **16**, 299
419. Hoyle, F. 1969, *Nature*, **223**, 936
420. Hoyle, F., Wickramasinghe, N. C. 1968, *Nature*, **218**, 1126
421. Turtle, A. J., Baldwin, J. E. 1962, *MNRAS*, **130**, 379
422. Bridle, A. H. 1967, *MNRAS*, **136**, 219
423. Brecher, K., Morrison, P. 1969, *Phys. Rev. Lett.*, **23**, 802
424. Wolfe, A. M., Burbidge, G. R. 1969, *Ap. J.*, **156**, 345
425. Conklin, E. K., Bracewell, R. N. 1967, *Nature*, **216**, 777
426. Penzias, A. A., Schraml, J., Wilson, R. W. 1969, *Ap. J.*, **157**, L49
427. Hazard, C., Salpeter, E. E. 1969, *Ap. J.*, **157**, L89
428. Bortolot, V. J., Clauser, J. F., Thaddeus, P. 1969, *Phys. Rev. Lett.*, **22**, 307
429. Shivanandan, K., Houck, J. R., Harwit, M. 1968, *Phys. Rev. Lett.*, **21**, 1460. Houck, J. R., Harwit, M. 1969, *Ap. J.*, **157**, L45
430. Gould, R. J., Burbidge, G. R. 1963, *Ap. J.*, **138**, 969
431. Felten, J. E., Morrison, P. 1966, *Ap. J.*, **146**, 686
432. Brecher, K., Burbidge, G. R. 1970, *Comments Ap. Space Phys.*, **2**, 75
433. Ginzburg, V. L., Syrovatsky, S. I. 1964, *Origin of Cosmic Rays* (Oxford: Pergamon); 1967, *Radio Astronomy and the Galactic System*, IAU Symp. No. 31, 411 (Academic)
434. Burbidge, G. R., Hoyle, F. 1963, *Ap. J.*, **138**, 57
435. Burbidge, G. R. 1962, *Progr. Theor. Phys.*, **27**, 999. Burbidge, G. R., Hoyle, F. 1964, *Proc. Phys. Soc.*, **84**, 141
436. de Veny, J. B., Lynds, C. R. 1969, *PASP*, **81**, 535
437. Nussbaumer, H., Osterbrock, D. E. 1970, Preprint
438. Weymann, R. J. 1970, *Ap. J.*, **160**, 31
439. Cromwell, R., Weymann, R. J. 1970, *Ap. J. Lett.*, **159**, L147
440. Kleinmann, D. E., Low, F. J. 1970, *Ap. J. Lett.*, **159**, L165
441. Hoyle, F. 1969, *Nature*, **224**, 477
442. Muehlner, D., Weiss, R. 1970, *Phys. Rev. Lett.*, **24**, 742

SOME RELATED ARTICLES APPEARING IN OTHER *ANNUAL REVIEWS*

Articles of direct interest to our readers
From the *Annual Review of Fluid Mechanics*, Volume 2 (1970):

Lick, Wilbert. Nonlinear Wave Propagation in Fluids
Monin, A. S. The Atmospheric Boundary Layer
Phillips, Norman A. Models for Weather Prediction
Spreiter, John R., and Alksne, Alberta Y. Solar-Wind Flow Past Objects in the Solar System

From the *Annual Review of Nuclear Science*, Volume 20 (1970):

Firk, F. W. K. Low-Energy Photonuclear Reactions
Hollander, Jack M., and Shirley, D. A. Chemical Information from Photoelectron and Conversion Electron Spectroscopy
Mandula, Jeffrey, Weyers, Jacques, and Zweig, George. Patterns of Exchange Degeneracies
Morpurgo, G. A Short Guide to the Quark Model
Shapiro, Maurice M., and Silberberg, M. Heavy Cosmic Ray Nuclei

Articles of peripheral interest
From the *Annual Review of Physical Chemistry*, Volume 21 (1970):

Chu, Ben. Laser Light Scattering
Hastie, J. W., Hauge, R. H., and Margrave, J. L. High Temperature Chemistry: Stabilities and Structures of High Temperature Species
Overend, John. Vibrational Spectroscopy
Robinson, G. Wilse. Electronic and Vibrational Excitons in Molecular Crystals
Rudolph, H. D. Microwave Spectroscopy
Turner, D. W. Photoelectron Spectroscopy

AUTHOR INDEX

A

Aarseth, S. J., 437
Abbe, C., 68
Ables, J., 149, 150
Abouaf, R., 361
Abrines, R., 352, 353
Adams, W. S., 17
Aizenman, M. L., 167
Akasofu, S.-I., 67, 69, 70, 71, 73, 74, 75, 76, 77, 80, 81, 82
Alfven, H., 5, 43, 79
Alder, K., 341
Alexander, J. B., 165
Alfonso-Faus, A., 298, 317
Allen, C. W., 318
Allen, J., 362
Aller, L. H., 168, 169, 170, 256, 383
Alloucherie, Y., 54
Alpher, R. A., 175
Altenhoff, W., 170, 231, 244, 253, 255, 261, 262, 263
Altschuler, M. C., 2, 5
Ambartsumian, V. A., 179, 430, 435, 437
Ambronn, L., 304
Anderson, K. A., 78, 79
Anderson, K. S., 399, 422, 423
Andrew, B., 149, 150
Andrews, M. H., 239, 246
Andrillat, Y., 399, 423
Angione, R., 392, 393
Ångström, A. J., 71
Anzer, U., 3
Appenzeller, I., 94
Appleton, E. V., 80
Arakelyan, S. N., 419
Argelander, F. W. A., 115
Arm, M., 134
Arnett, W. D., 448
Arnoldy, R., 78
Arp, H., 116, 123, 392, 393, 402, 403, 418, 431, 432
Artipova, V., 418
Asano, N., 168
Asbridge, J. R., 167
Ashbrook, J., 299
Audretsch, J., 299
Auman, J. R., 100, 101, 102, 103, 104, 105, 106, 107, 108, 109
Aumann, H. H., 388, 389
AXFORD, W. I., 31-60; 31, 32, 35, 40, 41, 48, 49, 55, 82, 414, 439

B

Baade, W., 143, 371, 392, 393, 396
Babcock, H. D., 2
Babcock, H. W., 2, 17, 21, 92, 122
Backus, G. E., 6, 13
Bacon, F. M., 351, 360, 361
Bader, M., 227
Bagariazky, B. A., 71
Bahcall, J. N., 166, 171, 184, 200
Bahcall, N. A., 166
Baker, J. G., 239
Baldwin, J. E., 416, 450
Balick, B., 231
Ball, J. A., 259
Bame, S. J., 167
Banks, P. M., 32, 55, 56, 57, 58
Bappu, M. K. V., 2, 3
Barbier, D., 81
Bardeen, J. M., 443
Bardet, M., 226
Bardin, C., 402
Bare, C. D., 398
Barker, B. M., 181
Barnes, A., 43
Barth, C. A., 77
Bartlett, J. F., 133
Baschek, B., 173
Bash, F. N., 396, 432
Batchelor, G. K., 5
Batchelor, R. A., 170
Bates, D. R., 71
Bates, R. H. T., 135
Bauer, E., 102
Bauer, W., 68
Baum, W. A., 374
Baym, G., 199, 200, 204, 205, 206, 291, 294
Beard, M., 123
Beckers, J. M., 2, 3
Becklin, E. E., 143, 147, 150, 282, 382, 388, 403
Bederson, B., 350
Beer, R., 226
Beigman, I. L., 350, 360
Belian, R., 143, 156
Bell, K., 151
Bell, S. J., 135, 265, 281, 290
Belon, A. E., 67, 77, 78
BELY, O., 329-68; 333, 337, 339, 340, 342, 344, 347, 348, 350, 351, 357, 359, 360, 361, 362, 363
Belzer, J., 435
Benedict, W. S., 101, 215, 226
Berg, R. A., 104, 105
Berge, G. L., 135
Bergmann, P., 297
Beron, B. L., 390
Berry, R. S., 98, 99
Bertola, F., 391
Bertotti, B., 290
Bethe, H. A., 356
Bhatia, M. S., 181, 193
Biermann, L., 5, 43
Billings, D. E., 3
Biraud, F., 387
Biraud, Y., 135, 226
Birkeland, K., 73
Bisnovatyi-Kogan, G. S., 442, 443, 447
Biswas, S., 166
Blaauw, A., 93
Black, D. C., 290
Blaha, M., 340, 348
Blanco, V., 140, 141, 155, 156
Blanco, V. M., 89, 373
Bless, R. C., 392
Bodansky, D., 448
Bodenheimer, P., 103, 435
Boesgaard, A. M., 91
Böhm, K. H., 102
Böhm-Vitense, E., 103, 105, 293
Bohr, A., 341
Boland, B. C., 351
Bolton, J. G., 390, 392, 393, 403, 414
Bonazzola, S., 193
Bond, F. R., 65
Bondi, H., 33, 35, 38, 39
Booker, H. G., 81
Bopp, B. W., 215, 226
Bortolot, V. J., 450
Bouigue, R., 215, 226
Bowers, H. C., 221
Bowyer, S., 140, 142, 151
Boyce, P. B., 225
Boynton, P. E., 272, 273
Braccesi, A., 410, 420
Bracewell, R. N., 132, 134, 135, 451
Brackmann, R. T., 350, 353, 360
Bradt, H., 140, 141, 142, 143, 144, 152, 155, 156, 157, 276, 392

AUTHOR INDEX

Braes, L. L. E., 122
Braginskii, S. I., 13, 14, 17
Brandt, J. C., 4
Brans, C., 297, 298
Branscomb, L. M., 94, 99, 330, 360, 362
Bransden, B. H., 334
Bratsev, V. F., 350
Bray, R. J., 2
Brecher, K., 450, 451
Brice, N. M., 82
Bridle, A. H., 450
Brink, G., 357
Brocklehurst, M., 239
Brooker, A. A., 90
Brooks, J. W., 170
Brown, R. R., 78
Brownell, D. H., 199
Brush, S. G., 96
Buijs, H. L., 226
Bullard, E., 8
Burbidge, E. M., 154, 171, 374, 378, 379, 381, 384, 385, 386, 390, 391, 395, 396, 397, 402, 403, 406, 407, 411, 412, 415, 418, 420, 422, 423, 426, 431, 433, 435, 439, 440, 442, 446
BURBIDGE, G. R., 369-460; 161, 171, 176, 290, 374, 378, 379, 381, 384, 385, 386, 388, 389, 390, 391, 392, 395, 396, 397, 402, 406, 407, 411, 412, 413, 414, 415, 418, 420, 422, 423, 424, 425, 426, 427, 428, 431, 433, 435, 439, 440, 442, 443, 444, 446, 447, 451, 452
Burgess, A., 239, 339, 340, 341, 342, 343, 350, 352, 353, 363
Burke, B. F., 170, 231, 244, 253, 255, 261, 262, 263, 387, 389
Burke, J. A., 43
Burke, P. G., 329, 334, 335, 338, 339, 346, 348, 350, 351, 357
Burns, W. R., 130, 135
Burroughs, W. J., 226, 227
Buselli, G., 150
Byram, E. T., 140, 142, 151, 155, 157, 392

C

Cade, P. E., 99
Callaway, J., 98, 199
CAMERON, A. G. W., 179-208; 161, 166, 174, 176, 179, 180, 181, 182, 185, 186, 187, 194, 195, 200, 201, 203, 205, 206,
294, 448
Cameron, M. J., 387, 416
Cameron, R. M., 227
Campbell, D. B., 273, 275, 276, 283, 287
Campbell, W. W., 395
Canuto, L. F., 290, 294
Canuto, V., 198, 204, 290, 294
Carbon, D., 104, 105, 106
Carleton, N. P., 351
Carranza, G., 384
Carrigan, A., 67, 71
Carson, T. W., 96
Cartwright, D. C., 350
Castellani, V., 168
Cavaliere, A., 290
Cayrel, R., 163
Cazzola, P., 181
Celsius, A., 66
Cesarsky, D., 249
Chamberlain, G., 350
Chamberlain, J., 226, 227
Chamberlain, J. W., 48, 50, 71, 77, 78
Chandrasekhar, S., 5, 6, 442
Chang, C. C., 48, 50
Chao, N. C., 198
CHAPMAN, S., 61-86; 66, 69, 71, 77, 79, 82
Chau, W. Y., 203
Chauville, J., 215, 226
Chazy, J., 297
Chedin, A., 226
Chibisov, M. I., 362
Childress, S., 14
Chincarini, G., 386
Chitnis, E., 156
Chitre, S. M., 2, 103
Chiu, H. Y., 180, 195, 204, 290, 294
Chiuderi, C., 204
Chodil, G., 143, 153, 156
Chopinet, M., 402
Christy, R. F., 168
Chubb, T. A., 140, 142, 155, 157, 282
Chudakov, A. E., 79
Churchwell, E., 170, 232, 246, 248, 251, 253, 255, 257, 258, 259, 261
Chvojkova, E., 23
Cihla, Z., 226
Clancy, M., 150
Clark, A., 3, 299, 315
CLARK, B. G., 115-38; 133, 135, 257, 393, 398, 414
Clark, G. W., 140, 141, 145, 155, 156, 389
Clark, J. W., 198, 199
Clark, P. A., 299, 315
Clark, R. R., 278, 279,
292
Clarke, E. M., 355
Clarke, M. E., 403
Clauser, J. F., 450
Clauser, T., 38
Clayton, D. D., 448
Clemence, G. M., 297, 299
Climenhaga, J., 92
Cocke, W. J., 4, 143, 265, 299, 312, 313
Code, A. D., 396
Codling, K., 340
Coffeen, M. F., 256
Cohen, J. G., 163
Cohen, J. M., 181, 182, 185, 187, 206
Cohen, M. H., 393, 396, 398
Cole, K. D., 80, 81
Cole, T., 130
Cole, T. W., 272
Coleman, I., 222, 225, 227, 228
Coleman, P. J., Jr., 318
Colgate, S. A., 299, 313, 316, 424, 439
Collins, R. A., 135, 265, 281, 290
Comella, J. M., 265, 272, 273, 275, 276, 278, 283, 285, 288
Condon, E. U., 333
Conklin, E. K., 285, 288, 451
Conner, J., 143, 156
Connes, J., 127, 210, 211, 214, 215, 216, 218, 219, 220, 221, 222, 223, 224, 225, 226
CONNES, P., 209-30; 127, 209, 210, 214, 215, 216, 218, 219, 220, 221, 222, 223, 224, 225, 226, 230
Conrath, B., 228
Conti, P. S., 91, 301
Conway, R. G., 133, 393
Cook, J., 61
Cooke, D. J., 272, 276, 282, 285, 293
Cooley, J. W., 136
Cooper, B. R. C., 130
Cooper, J. W., 99, 350
Coppi, B., 3, 82
Costain, C. H., 416
Costero, R., 169, 256
Counselman, C. C. III, 272, 273, 275, 276, 283, 287
Courtes, G., 384
Cowley, C. R., 91
Cowling, T. G., 2, 5, 6,

AUTHOR INDEX

22, 298, 302
Cox, J. P., 103, 105, 107
Craft, H. D., 265, 273, 275, 276, 278, 279, 282, 283, 285, 287, 288
Crampin, J., 436
Crillon, R., 384
Cromwell, R., 401
Crothers, D., 337
Cugnon, P., 389
Curott, D., 164
Currie, B. W., 68
Curtis, H. D., 392
Czyzak, S. J., 170, 346, 348, 351

D

Dahlberg, E., 33
Dalgarno, A., 98, 100, 306
Dalton, J., 61
Damburg, R. J., 335
Dance, D. F., 339, 350, 360, 362
Danielson, R., 401
DANZIGER, I. J., 161-78; 91, 174, 256, 302, 325
David, C. W., 99
Davidson, K., 280
Davies, J. G., 272
Davies, L. G., 252
Davies, R. D., 130, 246, 247, 248, 249, 251, 252
Davis, D. N., 91
Davis, L., Jr., 4, 298, 317
Davis, R., Jr., 165
Davis, T. N., 66, 67, 73, 74, 75
Davison, P., 150
De Heer, F. J., 360, 361
Dehnen, H., 299
de Jager, G., 130, 281
De Jong, M. L., 394, 416, 432
Delache, P., 55
Delbouille, L., 225
Delhaye, J., 93
Delouis, H., 214, 215, 216, 218, 219, 223, 225
de Mairan, J. J. d'O., 61
Demarque, P., 162, 165, 167
Demoulin, M., 154
Demoulin, M.-H., 384, 385, 412, 415, 424, 425
Dennis, T. R., 163
Dessler, A. J., 31
Deutsch, A. J., 88, 91, 297, 372, 374
de Vaucouleurs, A., 398
de Vaucouleurs, G., 16,

377, 382, 392, 393, 398
de Veny, J. B., 405
Devlin, J. J., 67, 71
De Young, D. S., 414, 438
Dibai, E. A., 396, 397, 398, 399, 402
DICKE, R. H., 297-328; 17, 24, 93, 103, 175, 297, 298, 299, 302, 303, 306, 307, 308, 309, 310, 312, 314, 315, 317, 318, 325, 434
Dieter, N. H., 122, 231, 244, 245, 247
Disney, M. J., 143, 265
Ditchburn, R. W., 100
Dixon, R. S., 123, 134
D'Odorico, S., 391
Dolder, K. T., 357, 360, 361, 362
Dombrowski, V. A., 396, 397, 398, 402
Donn, B., 94, 97
Doremus, C., 168
Doroshkevich, A. G., 434
Doughty, N. A., 98
Downes, D., 387
Downs, G. S., 273
Doyle, R. O., 100
Drake, F. D., 272, 276, 278, 279, 282, 285, 387
Dravskikh, A. F., 231
Dravskikh, Z. V., 231
Drawin, H. W., 363
Dubau, J., 361, 362
Duchesne, M., 381, 383
Duflot-Augarde, R., 402
Duncombe, R. L., 297, 298
Dungey, J. W., 3, 82
Dunn, G. H., 329, 353
DUPREE, A. K., 231-64; 237, 238, 239, 240, 242, 257, 259, 362
DuPuy, P. L., 419
Durgopal, M. C., 181
Durney, B. R., 289, 294, 299
Dworetsky, M. M., 388
Dyck, H. M., 94
Dyson, F. J., 198, 290
Dyson, J. E., 245

E

Eastlund, B. J., 290, 292
Eddington, A. S., 107
Eddy, J. A., 227
Edmonds, A. R., 347
Edmonds, F. N., 102, 215, 226
Edrich, J., 259, 261
Edwards, P., 150

Eggen, O. J., 155, 156, 163, 164, 165, 436
Ehlers, J. G., 97
Ehrhardt, H., 350
Eichhorn, H. K., 125, 151, 152
Eissner, W., 224, 333, 346, 348, 349
Ekers, R. D., 277, 278, 283, 288, 292, 385, 412, 414, 416
Ekre, H., 130
Ellder, J., 131
Ellis, D. G., 200
Ellis, D. V., 392
Ellis, S. A., 246, 252, 260, 261
Elsasser, K., 4
Elsasser, W. M., 6, 8, 22
Elsmore, B., 131
Elton, L. R. B., 350
Elvius, A., 391
Elwert, G., 363
Englander, P., 350
Epstein, E. E., 388, 396
Esipov, B. F., 402
Evans, W., 143, 156
Ezer, D., 166

F

Fackerell, E. D., 442
Fahlman, G. G., 294
Fairall, A. P., 403, 405, 418
Fassio-Canuto, L., 204
Fastie, W. G., 71
Fath, E. A., 395
Faulkner, D. J., 169, 170
Faulkner, J., 164, 167, 173, 289, 290
Feast, M., 155, 156
Feast, M. W., 91
Feautrier, N., 341, 342
Fedorova, N. I., 71
Fejer, J. A., 83
Feldman, B. A., 294
Feldman, P. D., 212
Feldstein, Y. I., 65, 66, 76
Fellgett, P. B., 209, 221, 225
Felten, J., 151
Felten, J. E., 392, 393, 451
Feltsan, P. V., 351
Fernie, J. D., 90
Ferraro, V. C. A., 79
Fichtel, C. E., 3, 166
Field, G., 151
Field, G. B., 257, 390, 434
Finzi, A., 200, 203
Fishman, G. J., 282, 392
Fitch, W. S., 402
Fite, W. L., 329, 335, 350,

AUTHOR INDEX

353, 360
Flather, E. M., 252, 256
Flower, D. R., 244
Fogarty, W. G., 388, 396
Follen, J. W., 175
Fomalont, E. B., 133, 385
Forbes, F. F., 222, 227, 388, 389
Ford, W. K., 383, 391, 398, 399
Forman, M., 219, 223, 226
Fowler, W. A., 175, 435, 440, 442, 446, 447, 448
Fox, R. E., 350, 357
Frank, L. A., 79
Fraser, C. W., 392, 393
Fraser, P. A., 98
Frederick, C. L., 388
Freeman, K., 155, 156
Fricke, K., 313, 315, 316
Friedman, A. M., 442
Friedman, H., 140, 142, 151, 155, 157, 282, 392
Fritz, G., 151, 282
Fritz, H., 63
Fujita, S., 225
Fujita, Y., 88, 90, 91, 92

G

Gabriel, A. H., 350
Gailitis, M., 344
Galperin, Yu. I., 71
Gamow, G., 435
Gardner, F. F., 170, 231, 244, 245, 247, 252, 260, 261, 262, 263, 276, 285, 390
Garmire, G., 140, 142, 143, 144, 147, 150, 152, 389
Gartlein, C. W., 63, 67, 72
Gartlein, H. E., 67
Garz, T., 166
Gassendi, P., 61
Gatewood, G., 125, 151, 152
Gaustad, J. E., 102, 166, 196
Gay, J., 226
Gaydon, A. G., 97
Gayet, R., 242
Gebbie, H. A., 225, 226, 227
Geddes, M., 65
Gehlot, G. L., 181
Gehrels, T., 94, 143
Geiss, J., 54
Gellman, H., 8
Geltman, S., 98, 99, 335, 355, 357, 362
Georgelin, Y., 384
Gerlach, U. H., 181

Giacconi, R., 140, 142, 143, 144, 152, 154
Giannone, P., 96, 168
Gilbert, H. E., 167
Gillett, F. C., 402, 427
Gilman, R. C., 97
Gilvarry, J. J., 299, 312
Gingerich, O. J., 104, 105, 106
Ginzburg, V. L., 198, 290, 292, 452
Giovanelli, R. G., 2
Glanfield, J. R., 416
Glass, N. W., 67
Gleissberg, W., 2
Glenn, S. W., 392
Gliese, W., 93
Gold, T., 79, 195, 289, 290, 292, 293, 414, 439
GOLDBERG, L., 231-64; 233, 237, 238, 242, 244, 249, 257, 259, 362
Golden, S. A., 102
Goldenberg, H. M., 299, 312
Goldreich, P., 4, 291, 294, 299, 302, 303, 313, 315, 316, 319, 325
Goldsmith, D. W., 257
Goldstein, S. J., 272, 276
Goldwire, H. C., 170
Goldwire, H. C., Jr., 236
Golson, J., 144
Googe, W. D., 125, 126
Goon, G., 102
Gordon, C. P., 90, 92, 130, 281
Gordon, K. J., 130, 281
Gordon, M. A., 169, 245, 247, 260, 261
Gordon, S., 97
Gorenstein, P., 140, 142, 143, 144, 150, 152, 154
Goss, W. M., 169, 170
Gottlieb, M. B., 79
Götz, F. W. P., 69
Gould, R., 151
Gould, R. J., 379, 451
Grader, R., 151
Graham, D., 152
Graham, G., 66
Graham, J. A., 168
Gratton, L., 181
Green, A. E. S., 77, 181
Green, R. M., 3
Greenberg, D. W., 237
Greene, T. F., 91, 226
Greenstein, G., 439

Greenstein, G. S., 176, 294
Greenstein, J. L., 90, 91, 173, 174, 442
Greenstein, S., 206
Grewing, M., 281, 451
Gribbin, J. R., 289
Griem, H. R., 261, 350
Griffin, R., 91
Grigorjev, V. M., 2, 3
Grossman, A. S., 96
Groth, E. J., 272, 273
Gryzinski, M., 342, 343, 352
Gubbay, J., 412
Guelachvili, G., 214, 215, 216, 218, 219, 223, 225
Guelin, M., 281
Guibert, J., 281
Gulkis, S., 393, 398
Gunderman, E. J., 396
Gunn, J. E., 290, 291
Gursky, H., 140, 142, 143, 144, 150, 152, 154
Gush, H. P., 226

H

Habing, H. J., 257, 281
Haddock, F. T., 387
Hagen-Thorn, V, A., 396, 397, 398, 402
Hale, G. E., 1
Halley, E., 61
Hamilton, P. A., 276, 282
Hanel, R., 219, 223, 224, 226, 228
Hansen, C. J., 161, 174, 203
Harang, L., 68, 77
Hardie, R., 124
Hargreaves, K. K., 67
Härm, R., 437
Harmer, D. S., 165
Harmon, L. D., 118
Harnden, F. R., 282
Haro, G., 419
Harrington, J. P., 94
Harris, D. E., 396
Harris, D. L., 90, 162
Harris, J. E., 227
Harrison, B. K., 179, 180, 187, 188
Harrison, E. R., 175, 434, 444
Harrison, M. F. A., 339, 350, 360, 362
Hartle, R. E., 32, 51, 52, 55, 58, 59
Harvey, J. W., 3
Harwit, M., 451
Haurwitz, M., 17
Hawking, S. W., 175

AUTHOR INDEX 467

Hayashi, C., 175
Hayes, S., 393
Haymes, R. C., 282, 290, 392
Hazard, C., 451
Heddle, D. W. O., 329, 349, 350, 351
Hedges, P., 123
Heeschen, D. S., 116, 124, 394, 416
Hegyi, D., 164
Heiles, C., 122, 170, 276
Heimel, S., 97
Helfer, H. L., 91
Henderson, A. P., 236
Henon, M., 435, 436, 437
Henry, R. C., 151, 282
Henry, R. J. W., 99, 334, 335, 344, 346, 348
Henyey, L. G., 103
Herbig, G. H., 301
Herman, R. C., 175
Hershey, J. L., 104, 105
Herzberg, G., 97
Herzenberg, A., 13
Herzog, E., 418
Hesser, J., 147, 156
HEWISH, A., 265-96; 132, 135, 265, 281, 290
Hickam, W., 357
Hill, J. E., 68
Hill, R., 151
Hils, D., 350
HILTNER, W. A., 139-60; 140, 141, 145, 146, 147, 151, 152, 153, 155, 156, 392
Hines, C. O., 82
Hiorter, O. P., 66
Hirt, P., 54
Hjellming, R. M., 239, 240, 246, 247, 248, 249, 250, 251, 252
Hoang Binha, D., 242
Hodge, P. W., 165, 376
Hoffman, K. C., 165
Hoffman, R., 78
Hoffmann, W. F., 388
Hofstadter, R., 390
Högbom, J., 135
Hogg, D. E., 133, 393
Höglund, B., 231, 244, 245, 387
Holmberg, E., 377, 432
Holt, H. K., 350
Holton, J. R., 315
Holweger, H., 166
HOLZER, T. E., 31-60; 32, 55, 56, 57, 58
Hones, E. W., Jr., 83
Hönl, H., 299

Hooper, J. W., 351, 360, 361
Houck, J. R., 451
Hovis, W. A., 228
Howard, H. T., 285, 288
Howard, L. N., 299, 312
Howard, R., 2, 3
Howe, M. S., 2
Hoyle, F., 94, 171, 175, 290, 302, 303, 422, 423, 428, 434, 435, 436, 440, 442, 444, 445, 446, 447, 448, 449, 452
Hubble, E. P., 371, 382, 395
Huchtmeier, W., 281
Hudson, H. S., 145, 153, 392
Hudson, J. P., 94, 97
Hughes, G. F., 67
Huguenin, G. R., 284, 286, 292
Hulburt, E. O., 66, 79
Hulsbosch, A. N. M., 389
Hultquist, B., 65, 66
Humason, M. L., 372, 378
Hummer, D. G., 339, 350
Hundhausen, A. J., 4, 31, 52, 53, 58, 82, 167
Hunger, K., 91
Hunt, G. C., 271, 272
Hunten, D. M., 127
Huus, T., 341

I

Iben, I., Jr., 90, 165, 166, 167, 168, 173
Ingram, L. J., 80
Inokuti, M., 356
Inoue, M., 225
Ipser, J. R., 289, 442
Isler, R. C., 71
Israel, W., 290
Itoh, N., 198

J

Jacka, F., 65
Jacquinot, P., 209, 210, 213
Jaggi, R. K., 3
Jahoda, F. C., 351
James, J. F., 221
Jauncey, D. L., 393, 398
Jayanthan, R., 6
Jayanthi, U., 147, 156
Jeans, J. H., 434
John, T. L., 98
Johnson, H., 143, 144,
152
Johnson, H. E., 43
Johnson, H. L., 89, 90, 94, 209, 222, 225, 227, 228
Johnson, H. M., 170
Jokipii, J. R., 55
Joly, F., 242
Jones, P. B., 203
Jones, R. G., 227
Jones, T. J. L., 351
Jordan, P., 297
Joy, A. H., 87, 92, 93
Jugaku, J., 143, 144, 152
Julian, W. H., 291, 294
Jura, M., 284, 286, 292
Jura, M. A., 174

K

Kahn, F. D., 212
Kaler, J. B., 256
Kamijo, F., 88, 90, 97
Kandel, R., 96
Kaneko, N., 420, 422
Kaneko, Y., 357
Kaper, H. G., 134
Kaplan, J., 77
Kaplan, L. D., 101, 215, 226
Karachentsev, J. D., 431
Kardashev, N. S., 237, 253
Karpowicz, M., 418
Karule, E. M., 335, 350
Kasper, J. E., 79
Katem, B., 167
Kaupilla, W. E., 335
Keenan, P. C., 87, 88, 89, 91
Keesing, R. G. W., 329, 349, 351
Keller, G., 414, 424, 429, 435
Kellermann, K. I., 393, 397, 398, 403, 424
Kellman, E., 87
Kelsall, T., 162
Kenderdine, S., 132, 414
Kern, J. W., 83
Kerr, F., 263, 387, 389
Keys, J. G., 67
Khachikian, E. Ye., 405
Khorosheva, O. V., 76
Kiang, T., 419
Kieffer, L. J., 329, 353, 356
Kimball, D. S., 67
King, I. R., 435, 436
Kingston, A., 151
Kingston, A. E., 100, 343,

AUTHOR INDEX

352
Kinman, T. D., 168, 256, 382, 403
Kippenhahn, R., 3, 4, 314
Kirzhnits, D. A., 198
Kistemaker, J., 360, 361
Kjeldaas, T., 357
Kleinmann, D. E., 388, 389, 391, 396, 397, 402, 403, 406, 427
Kleinpoppen, H., 350
Klinglesmith, D., 91
Kneubuhl, F., 226
Knowlton, K. C., 118
Kock, M., 166
Kollberg, E., 131
Komesaroff, M. M., 272, 276, 282, 288, 293
Konyukov, M. V., 44, 50
Kopecky, M., 22, 23
Kopp, R. A., 3
Koppendorfer, W., 350
Korchevoi, Y., 361
Koschmieder, H., 350
Kovetz, A., 103
Kowal, C., 418
Kozlovsky, B., 171
Kraft, R., 144, 145, 154, 155, 298, 301, 319
Kraft, R. P., 399, 422, 423
Kraichnan, R. H., 5
Krascella, N. L., 102
Krasnogorskaja, A., 418
Krassovsky, V. I., 71, 80, 81, 82
Kraus, J. D., 123, 134
Krause, F., 12, 15, 17, 23
Kraushaar, W. L., 389
Krishna Swamy, K. S., 107
Krisitan, J., 143, 144, 145, 150, 151, 152, 154, 167, 276, 282, 289
Krogdahl, W. S., 98
Kronberg, P. P., 133
Krook, M., 9
Krotkov, R., 350
Krueger, T. K., 346, 348, 351
Kruszewski, A., 94, 396, 402
Kuhi, L. V., 87, 94
Kuiper, G. P., 165, 216, 220, 222, 227
Kuklin, G. V., 22
Kulsrud, R. M., 3
Kumar, S. S., 104, 105, 106
Kunde, V. G., 101, 224, 228
Kunkel, W., 140, 141, 155, 156
Kunze, H. J., 350

Kuperus, M., 3
Kupperman, A., 350
Kurfess, J. D., 392
Kurucz, R. L., 102
Kyle, T. G., 102

L

LaBahn, R. W., 98
Lallemand, A., 381, 383
Lambert, D. L., 90, 166
Lambert, L. B., 134
Lampland, C., 386
Lan, Vo Ky, 333
Landau, L., 179
Landstreet, J. D., 201
Lane, N. F., 98
Lang, K. R., 288, 387, 394, 416
Langer, W. D., 181, 182, 184, 185, 187, 203
Langhans, L., 350
Large, M. I., 265, 267, 268, 280
Lasker, B., 147, 156
Latham, D. W., 104, 105, 106
Lauer, I. E., 343
Laval, G., 82
Layzer, D., 290
Ledoux, P., 28, 92, 93
Lee, A. R., 351
Lee, H. J., 204
Lee, R. H., 227
Legg, A. J., 412
Leighton, R. B., 2, 3, 17, 20, 24, 25, 143, 282
Lena, P. J., 227
Lequeux, J., 226, 387, 391
LeRoux, E., 387
Lesh, J. R., 433
Leutwyler, H., 54
Leverrier, U. J., 297
Levin, G. V., 228
Levinger, J. S., 184, 185, 187
Lewin, W., 145, 156
Lewis, B. A., 351
Lewis, J. S., 215, 226
Libby, L. M., 186
Lichten, W., 350
Lighthill, M. J., 43
Liller, W., 168, 256
Lilley, A. E., 169, 237, 240, 244, 245, 246, 247, 253, 254, 255, 256, 259
Lilliequist, C. G., 2
Limber, D. N., 431
Limber, N., 90
Lindblad, B., 99
Lindemann, F. A., 73
Linder, F., 350
Lineberger, W. C., 360,

361
Linsky, J. L., 101, 104, 105, 106
Lipsky, L., 350
Liska, J. N., 361
Little, C. G., 80
Little, L. T., 286
Liu, C. K., 50
Livingston, W. C., 2, 17, 312
Loewenstein, E., 209
Logachev, Yu. I., 79
Longair, M. S., 414
Loomis, E., 63
Lord, H. C., 97
Lötz, W., 363
Loughhead, R. E., 2
Lovelace, R. V. E., 265, 283, 285
Low, F. J., 388, 389, 391, 396, 397, 402, 403, 406, 427
Lowman, P. D., 228
Lucaroni, L., 181
Ludden, D., 152
Luyten, W. J., 420
Lykoudis, P. S., 3
Lynden-Bell, D., 435, 436, 437, 441
Lynds, C. R., 154, 171, 391, 392, 393, 397, 405, 407, 410, 420, 422, 423
Lynds, R., 265
Lyne, A. G., 266, 273, 275, 276, 281, 283, 284, 286
Lynga, G., 140, 141, 155, 156

M

Macdonald, G. H., 133
MacFarlane, M., 143, 276
Mackie, J. C., 99
MacQueen, R. M., 227
Madden, R. P., 340
Maehara, H., 92
Maggs, J. E., 66, 67, 74, 75
Mahoney, M. J., 281
Maillard, J. P., 214, 215, 216, 218, 219, 220, 221, 223, 225, 226
Main, R. P., 102
Malik, F. B., 360
Malville, J. M., 3
Manchester, R. N., 273
Mangus, J., 228
Mannery, E., 153
Mapleton, R. A., 352
Maran, S. P., 180, 265
Marino, L. L., 353, 360
Mark, H., 143, 153, 156, 157

AUTHOR INDEX

Markarian, B. E., 419, 431
Marovich, E., 81
Marriott, R., 350
Martin, J. B., 99
Martin, S. O., 361
Martins, P. de A. P., 244
Martz, D. E., 223
Marx, G., 198
Massey, H. S. W., 329, 339
Mather, K. B., 67
Mathews, W. G., 431
Mathis, J. S., 168, 169, 170, 256
Matthews, T. A., 406
Maunder, A. S. D., 17
Max, J., 135
Maxwell, A., 387
Mayall, N. U., 372, 378, 383, 394, 395
Mayer, C. H., 387
Mayer, W., 155, 157, 276, 392
McBridge, B. J., 97
McCarroll, R., 242, 337, 350
McCarthy, M. F., 88
McClure, R. D., 372, 375, 446
McCracken, K., 150
McCrea, H. C., 38
McCrea, W. H., 435, 445
McCullough, P. M., 276, 282, 288
McDaniel, E. W., 360
McDonald, D. E., 299, 315
McDonald, F. B., 3
McDonald, G. H., 393, 414
McDowell, M. R. C., 98, 99, 342, 360, 362
McEachran, R. P., 98
McFarland, R. H., 360, 361
McGee, R. X., 170, 244, 245, 247, 252, 260, 390
McGowan, J. W., 355
McIlraith, A. H., 290
McInnes, B., 67
McKee, J. S. C., 334
McKinney, J. D., 360, 361
McLean, D. J., 134
McLellan, A., 3
McLennan, J. C., 72
McNutt, D. P., 212
McVicar, D., 339, 350
McWhirter, R. W. P., 351
Mebold, U., 281
Meekins, J. F., 151, 282
Meeks, M. L., 169, 245, 247, 260, 261
Meilleur, T., 219, 223, 224, 226
Meinel, A. B., 71, 72

Meisel, D. D., 272, 276
Mendoza, V. E. E., 89, 90
Meng, C.-I., 67
Menon, T. K., 246, 252, 261
Menzel, D. H., 236, 239, 246
Meredith, L. H., 79
Merrill, P. W., 87, 91, 92, 93, 97
Mertz, L., 209, 222, 225
Mestel, L., 4, 5, 40, 41
Meyer, F., 50
Meyer, P., 176
Mezger, P. G., 170, 231, 232, 236, 237, 240, 244, 245, 246, 252, 253, 254, 255, 256, 257, 258, 259, 260, 261, 262, 263, 387
Michel, F. C., 290
Michel, G., 214, 215, 216, 218, 219, 223, 225
Michie, R. W., 435, 437
Middleton, D., 128
Mielczarec, S. R., 350
Migdal, A. B., 198
Mihalas, D., 102, 103, 171
Miller, J. E., 98
Miller, J. S., 143, 155, 168, 276, 293
Miller, R. E., 71
Miller, R. H., 167, 378, 436
Miller, W. C., 391
Mills, B. Y., 268
Mills, D. M., 414
Milne, D. K., 170, 231, 261, 262, 263
Minami, S., 225
Minkowski, R., 380, 381, 392, 393, 394, 396, 397
Minnaert, M., 17
Minnett, H. C., 387
Mishin, V. M., 77
Misner, C. W., 175
Mitchell, R. I., 222, 225, 227, 228
Mock, M., 203
Modiesette, J. L., 298, 317
Moffat, K., 5
Moffet, A. T., 277, 278, 283, 288, 292, 393, 398, 412, 413
Moiseiwitsch, B. L., 329, 351
Monnet, G., 384
Montgomery, E. F., 215, 226
Montgomery, J. W., 388
MOOK, D. E., 139-60; 145, 146, 147, 149, 152
Moore, C. E., 97

Moore, D. W., 3, 299, 312
Moore, J. H., 395
Moores, D., 338, 351, 360, 361, 362
Moreton, G. E., 3
Morgan, W. W., 87, 88, 372, 379, 385, 406, 417, 433
Morimoto, M., 123
Moroz, W. I., 225
Morris, D., 133, 272, 276, 293, 385
Morris, S., 90
Morrison, P., 139, 443, 450, 451
Morton, D. C., 162, 180, 194, 200, 257
Mottelson, B., 341
Muehlner, D., 451
Müller, C. A., 122
Muller, E. A., 226
Mulyarchik, T. M., 80
Münch, G., 150, 173, 244, 247, 256, 382, 384
Muncke, G. W., 63
Murray, B. C., 223
Myerscough, V. P., 98, 99, 100

N

Nagarajan, S., 5
Naismith, R., 80
Nakada, M. P., 55
Nakagawa, Y., 2, 4
Nakano, K., 225
Naranan, S., 140, 141, 155, 156, 157, 392
Narlikar, J. V., 290, 435, 440, 444, 445
Nassau, J. J., 373
Nather, R. E., 143, 276
Ne'eman, Y., 444
Negaard, J. B., 71
Nemeth, J., 184
Ness, N. F., 2, 3, 4, 16, 83
Neugebauer, G., 143, 147, 150, 282, 382, 388, 403
Neugebauer, M., 318
Neville, A. C., 131, 414
Newcomb, J. S., 126
Newcomb, S., 297
Newell, E. B., 168, 173
Newman, R. C., 35, 40, 48, 49
Neyman, J., 431
Neynaber, R. H., 353, 360
Nichols, J., 5
Nicolet, W. E., 100
Nikolaev, A. G., 79
Nishida, A., 54, 82
Noble, L. M., 48, 50
Norris, J., 173
Novikov, I. D., 434, 444

AUTHOR INDEX

Noyes, R. W., 3, 17
Nussbaumer, H., 244, 333, 346, 348, 349, 360, 361, 362, 420, 421

O

Obridko, V., 22
O'Brien, B. J., 79
Occhionero, F., 203
Ochkur, V. I., 336, 339, 350, 352, 357
O'Connell, R. F., 299
Oda, M., 139, 140, 142, 143, 144, 151, 152
O'Dell, C. R., 94, 168, 256
Oganesyan, E. Ya., 419
Öhman, Y., 88
Ohtani, H., 420, 422
Oinas, V., 90
Oke, J. B., 123, 124, 143, 147, 150, 282, 377, 396, 397, 399, 400, 403, 404, 407, 410, 421, 422, 427
Oldenberg, O., 77
Oliver, N. J., 67, 71
Olson, C. A., 3
Omidvar, K., 337, 350, 356, 360
Onyejuba, P. E., 196
Oort, J. H., 389
Öpik, E. J., 299
Öpik, U., 100
Oppenheimer, J. R., 179, 188
Ormes, J. F., 91
Ormonde, S., 335, 350
Orszag, S. A., 196
Osawa, K., 143, 144, 152
Osmer, P., 143, 144, 152, 154
Oster, L., 236
Osterbrock, D. E., 3, 43, 171, 252, 348, 372, 379, 380, 393, 394, 396, 420, 421, 427
Ostriker, J. P., 289, 290, 291
Osvalds, V., 89
Ott, W. R., 335

P

Pacholczyk, A. G., 395, 396, 399, 402
Pacini, F., 181, 290, 291
Page, T., 431
Pagel, B. E. J., 91, 330, 362
Palmer, P., 169, 236, 237, 240, 244, 245, 246, 247, 253, 254, 255, 256, 257, 258, 259, 261
Pande, M. C., 108

Pao, Y. H., 5
Paolini, F., 143
Papagiannis, M. D., 292
Pariiskii, Y. N., 387
PARKER, E. N., 1-30; 2, 3, 4, 5, 8, 9, 10, 11, 12, 13, 17, 19, 20, 21, 31, 38, 48, 55, 73, 82
Parker, R. A., 171, 396, 427
Partridge, R. B., 175, 272, 273
Pastoriza, M., 417
Pauliny-Toth, I. I. K., 397, 414, 424, 429
Payne, J., 246, 252, 261
Peach, G., 100, 360, 361
Peach, J., 154
Pearson, J. M., 199
Peart, B., 357, 360, 361, 362
Peat, D. W., 91
Pecker, J. C., 95, 103
Pedlosky, J., 315
Peebles, P. J. E., 175, 434
Peimbert, M., 168, 169, 170, 171, 244, 256, 379, 384, 391
Pellat, R., 82
Pemberton, A. C., 91
Penfield, H., 169, 231, 237, 240, 244, 245, 246, 247, 253, 254, 255, 256, 259, 261
Pengelly, R. M., 342
Penzias, A. A., 175, 416, 451
Percival, I. C., 333, 343, 350, 352, 353, 360, 363
Percy, J. R., 162
Perel, J., 350
Perola, G. C., 414
Peterkop, R., 329, 350, 353
Peterson, D. M., 103, 104
Peterson, J. R., 360, 361
Peterson, L. E., 78, 145, 153, 392
Pethick, C., 199, 200, 204, 205, 206, 291, 294
Petrie, W., 71
Petrini, D., 338, 344, 348, 351
Petrunkin, A. M., 352
Petschek, H. E., 3, 82
Pettengill, G. H., 272, 273
Pettit, E., 377
Phelps, A. V., 350
Pichanick, R. M. T., 350
Piddington, J. H., 82, 290, 387
Pilkington, J. D. H., 135, 265, 281, 290

Pinard, J., 214, 223, 225
Pines, D., 199, 200, 204, 205, 206, 291, 294
Pismis, P., 432
Plaskett, H. H., 298
Podurets, M., 442
Pointon, L., 130, 281
Ponsonby, J. E. B., 130, 281
Pooley, G. G., 132
Popov, G. V., 77
Pottasch, S. R., 169, 281, 379
Pourcelot, A., 384
Powell, A. L. T., 91
Prabhakara, C., 228
Praderie, F., 342
Prakasarao, A., 147, 156
Prasad, S. S., 357, 360
Prendergast, K. H., 378, 384, 385, 386, 395, 402, 422, 436
Prentice, A. J. R., 280, 281
Presnyakov, L., 337
Price, R., 153, 157
Pritchard, J., 219, 223
Pronik, V. I., 396, 398, 399, 402
Pronski, A., 361
Przybylski, A., 174
Purgathofer, A. T., 399
Purton, C., 149, 150
Purton, C. R., 416
Puzanov, V. A., 231, 247

Q

Querci, M., 215, 226

R

Rabik, L. L., 350
Radhakrishnan, V., 272, 273, 276, 293
Radler, K. H., 12, 15, 17, 23
Raff, M. I., 391
Ramaty, R., 150
Rankin, J. M., 272, 273, 275, 276, 283, 287
Rao, U., 147, 156
Rappaport, S., 140, 141, 155, 156, 157, 276, 392
Ray, E. C., 439
Rea, D. G., 215, 226
Rees, M. H., 77
Rees, M. J., 390, 396, 424
Reichely, P., 273
Reifenstein, E. C. III, 170, 231, 244, 253, 255, 261, 262, 263, 265, 285, 387
Reimann, C. W., 99

AUTHOR INDEX

Renson, P., 28, 92
Renzini, A., 168
Rice, M. J., 199
Richards, D., 343
Richards, D. W., 272, 273, 275, 276, 283, 287
Richter, J., 166
Rickett, B. J., 266, 273, 275, 283, 284, 286, 288
Riddle, A. C., 387
Riegler, G., 150
Rikitake, T., 8
Ring, J., 213, 252
Risley, A. M., 89
Roach, F. E., 81
Roberts, G. O., 15
Roberts, J. A., 132, 134, 272, 273, 294
Roberts, M. S., 122
Roberts, P. H., 14, 15
Robertson, D. S., 412
Robinson, B. J., 390
Robinson, E. J., 99
Rodgers, A., 155, 156
Rodgers, A. W., 168, 173
Rodrigues, R., 143, 153, 156, 157
Rogalski, M., 362
Rogstad, D. H., 381, 414, 416
Rohlfs, K., 281
Roizen-Dossier, B., 210
Roland, G., 225
Romick, G. J., 77, 78
Rönnäng, B., 131
Rood, R., 167, 168
Rose, W. K., 227, 228, 289
Rosen, L. C., 181, 182, 184, 185, 187, 195
Rossi, B., 143
Rothe, E. W., 353, 360
Roucher, J., 226
Rougoor, G. W., 381, 389
Roxburgh, I. W., 289, 298, 299, 312
Rubin, R. H., 231, 253
Rubin, V. C., 383, 391, 398, 399
Ruderman, M., 197, 198, 199, 203, 204, 205, 294
Rudge, M. R. H., 329, 336, 353, 355, 356, 357, 360, 361, 362
Runcorn, S. K., 4
Rundel, R. D., 360, 362
Rust, D. M., 3
Rydbeck, O. E. H., 131
Ryle, M., 131, 132, 133, 397, 414, 432

S

Saakyan, G. S., 179
Saffman, P. G., 5
Sahal-Brechot, S., 341

Sakai, H., 209
Sakurai, T., 315
Salmona, A., 193
Salomonovich, A. E., 231, 247
Salpeter, E. E., 195, 196, 286, 287, 441, 451
Sandage, A. R., 143, 144, 145, 150, 151, 152, 154, 155, 156, 163, 167, 168, 372, 377, 378, 382, 384, 390, 391, 407, 410, 411, 412, 415, 420, 422, 426, 433, 436, 439
Sanford, B. P., 76
Sanford, R. F., 92
Saraph, H. E., 244, 342, 343, 346, 348, 349, 350
Sargent, W. L. W., 172, 173, 399, 400, 403, 405, 406, 421, 422, 427, 428, 431
Saslaw, W. C., 290, 436, 437, 438
Sastry, G. N., 433
Saunier, G., 199
Savage, B. D., 401
Sawyer, C., 2, 3
Scarf, F. L., 48, 50
Scargle, J. D., 143, 276, 293
Scarini, C., 181
Schadee, A., 91
Schatten, K. H., 3, 16
Schatzman, E., 298
Schaub, W., 303
Scheuer, P. A. G., 286, 429
Schiff, L. I., 357
Schlachman, B., 228
Schlenker, S. L., 289
Schlesinger, B., 167
Schlüter, A., 3, 5
Schmidt, H. U., 50
Schmidt, M., 93, 170, 171, 406, 407, 414, 415, 442
Schnopper, H. W., 140, 225
Schofield, J. W., 213
Schram, B. L., 360, 361
Schraml, J., 451
Schroeder, M. R., 118
Schröter, E. H., 2, 22
Schubert, G., 4, 299, 302, 303, 313, 315, 316, 319, 325
Schulte, D. H., 116
Schultz, S., 350
Schulz, G. J., 350
Schur, W., 304
Schuster, A., 82
Schwartz, D. A., 145, 153, 392
Schwartz, S. B., 356, 357, 360, 361, 362, 363

Schwarz, R. A., 294
Schwarz, U., 134
Schwarzschild, M., 227, 228, 315, 316, 326, 401
Sciama, D. W., 390, 435
Scott, E., 431
Scott, P. F., 135, 265, 281, 290
Searle, L., 168, 172, 173
Seaton, M. J., 80, 238, 239, 244, 329, 333, 338, 340, 341, 342, 345, 346, 348, 349, 353, 355, 356, 357, 360, 363
Seidel, B., 412
Seielstad, G. A., 133
Sejnowski, T. J., 239, 240, 250
Selby, M. J., 252
Seling, T. V., 170
Seman, M. L., 94, 99
Serkowski, K., 94
Sersic, J. L., 398, 417, 431
Severny, A. B., 3
Seward, F., 151, 153, 156, 157
Seyfert, C. K., 395, 396, 398, 422
Shaffer, D., 393, 398
Shakhovskoi, I. S., 396, 397
Shalloway, A. M., 130, 281
Shapiro, I. I., 299
Sharpless, S., 87
Shaviv, G., 103, 166
Shcheglov, P. V., 80
Shefov, N. N., 72
Sheglov, P. V., 256
Shemming, J., 346, 348, 349
Shiba, K., 225
Shimmins, A. J., 403
Shipman, H. S., 171
Shivanandan, K., 212, 451
Shklovsky, I. S., 392, 393, 420, 421, 429, 439
Shortley, G. H., 333
Shpenik, O. B., 351
Shrum, G. R., 72
Shteinshleger, V. B., 231, 247
Shuter, W. L. H., 281
Sil, N. C., 350
Silk, J. I., 237, 294, 396, 424
Silpher, V. M., 395
Silverstein, S. D., 199
Simmons, L. M., 184, 185, 187

AUTHOR INDEX

Simon, G. W., 2, 3, 17
Simpson, E., 166
Simpson, J. A., 350
Sinclair, M. W., 170, 387
Sinfailam, A. L., 334, 346, 348
Singer, S. F., 79
Sinton, W. M., 223, 225
Slee, O. B., 288, 392, 393
Sloan, I. H., 360
Sloanaker, R. M., 387
Smak, J., 89
Small, R., 71
Smirnov, B. M., 362
Smirnov, Yu. N., 175
Smith, A. C. H., 339, 350
Smith, E. P., 3
Smith, F. G., 272, 276, 278, 279, 292
Smith, H. J., 3
Smith, K., 334, 335, 344, 346, 348
Smith, S., 371
Smith, S. J., 329, 350, 351
Smith, S. M., 3
Smith, W., 145, 156
Smits, D. W., 134
Snellen, G., 143, 282
Snyder, C. W., 318
Snyder, E. J., 65
Sobelman, I. I., 337
Sofia, S., 151, 152
Solinger, A. B., 3, 391
Solomatina, E. K., 66
Solomon, P. M., 100
Somerville, W. B., 98, 338
Sorochenko, R. L., 231, 247
Soshnikov, V. N., 100
Souffrin, S., 399, 420, 423
Spada, G., 140, 141, 142, 155, 156, 157, 392
Speiser, T. W., 83
Spiegel, E. A., 3, 103, 299, 312
Spinrad, H., 87, 88, 90, 94, 98, 101, 109, 170, 171, 226, 372, 373, 374, 376, 391
Spitzer, L., 43, 435, 437, 438, 448
Spitzer, L., Jr., 237
Sprung, D. W. L., 184
Sreekantan, B., 140, 142, 143, 144, 152
Stabler, R. C., 342, 352
Staelin, D. H., 136, 265, 285
Stambach, G., 223, 224, 226

Stanley, G. J., 392, 393
Starkov, G. V., 76
Stauffer, A., 342
Stebbings, R. F., 350
Stebbins, J., 377
Stecher, T. P., 94, 97
Steel, W. H., 213, 215
Steenbeck, M., 12, 15, 17, 23
Stein, W. A., 388, 389, 396, 402, 424, 426, 427, 428, 443
Steinmetz, D. L., 222, 225, 227, 228
Stepanov, V. E., 2, 3
Stephenson, C., 143, 152
Stettler, P., 226
Stewart, J. M., 175
Stockton, A., 154
Stoeckly, R., 174
Stoffregen, W., 67
Stone, M. E., 257, 437
Stone, P. H., 315
Störmer, C., 68, 73
Stothers, R., 162, 289, 290
Strand, K. A., 90, 116, 162
Strittmatter, P. A., 172, 290
Strockhausen, R., 228
Strom, K. M., 163, 174
Strom, S. E., 102, 163, 171, 174
Stromgren, B., 162
Strong, I. B., 167
Strong, J., 223, 225
Sturrock, P. A., 3, 32, 51, 52, 55, 58, 59, 299, 312, 414
Suemoto, Y., 225
Sugimori, H., 225
Sugimoto, D., 168
Summers, H. P., 239
Sural, D. P., 350
Sutherland, P., 206
Sutton, J. M., 265, 283
Swarztrauber, P., 4
Sweet, P. A., 3, 82
Swenson, G. W., 131
Swift, C., 143, 153, 156, 157
Swift, D. W., 83
Swings, P., 215, 226
Syrovatskii, S. I., 292, 452
Szamosi, G., 181, 193

T

Tait, J. H., 351
Takakubo, K., 134

Takeuchi, H., 8
Tandberg-Hanssen, E., 3
Tarafdar, S. P., 98, 100
Tassoul, J. L., 289
Tatum, J. B., 97
Tauber, G., 444
Tayler, R. J., 161, 175
Taylor, A. J., 335, 339, 350, 357
Taylor, D. J., 265
Taylor, H. E., 83
Taylor, H. S., 98
Taylor, J. H., 284, 286, 292, 387
ter Haar, D., 280, 281
Terzian, Y., 231, 280, 394, 416
Thaddeus, P., 450
Thirring, W., 437
Thirry, Y. R., 297
Thomas, F. J., 186
Thomas, J. H., 299, 315
Thomas, R., 150
Thompson, A. R., 387
Thompson, R. I., 140, 225
Thompson, W. I., 3
Thomson, J. J., 342, 351, 353
Thonemann, P. C., 360
Thorne, K. S., 175, 180, 187, 188, 202, 289, 442, 443
Thorne, R. M., 3
Thruhlar, D. G., 350
Tifft, W. G., 376, 377
Tisone, G., 360, 362
Tomasko, M. G., 237
Torres-Peimbert, S., 92, 166
Tovmassian, H. M., 416, 417
Traving, G., 174
Treanor, P. J., 88, 94
Trefftz, E., 360
Trehan, S., 4
Trujillo, S. M., 353, 360
Trumbo, D. E., 265
Truran, J. W., 161, 174, 448
Tsuji, T., 88, 90, 91, 92, 97, 101, 104, 105, 106, 108
Tsuruta, S., 180, 181, 194, 195, 200, 203, 205
Tucker, W. H., 290
Tukey, J. W., 136
Tully, J., 333, 339, 340, 342, 350, 351
Tuominen, J., 23
Turiel, I., 153
Turner, B. E., 231

AUTHOR INDEX

Turon-Lacarrieu, P., 226
Turtle, A. J., 450
Tyler, G. L., 265, 283

U

Ulrich, R. K., 103, 106, 166
Underhill, A. B., 171, 172
Ungstrup, E., 80
Unno, W., 105
Unsöld, A., 433
Unsöld, A. O. J., 167
Usher, P., 154
Utsumi, K., 88, 90, 92

V

Vainshtein, L. A., 337, 350, 360
Vaisberg, O. L., 71
Valentine, N. A., 352, 353
Vallak, R., 389
Vallance Jones, A., 71
Valukov, P. V., 79
Van Allen, J. A., 79
Vanasse, G. A., 209
Van de Kamp, P., 90
van den Bergh, S., 372, 375, 391, 446
Van den Bos, W. H., 115
Van den Heuvel, E. P. J., 102, 113
van der Kruit, P. C., 389
van der Laan, 414, 432
Vandervort, G. L., 90
Van Der Wiel, M. J., 360, 361
van Genderen, A., 147
Van Horn, H. M., 196
VAN REGEMORTER, H., 329-68; 333, 338, 340, 342, 351
Van Speybroeck, L., 140
Van Vleck, J. H., 128
van Woerden, H., 134, 381, 389
VARDYA, M. S., 87-114; 90, 96, 97, 98, 99, 100, 102, 103, 104, 105, 106, 109, 110
Varsavsky, C. M., 100
Vasilevskis, S., 116
Vaughan, A. E., 265, 267, 268, 280
Vaughan, A. H., Jr., 209
Vegard, L., 72, 77
Veldre, V., 329, 350, 353, 357
Venugopal, V. R., 281
Verdet, J. P., 226
Vernov, S. N., 79

Vestine, E. H., 64, 65
Vila, S. C., 286
Vinkalns, I., 357
Virgopia, N., 96
Visvanathan, N., 391, 396
Volkoff, G. M., 179, 188
von Hoerner, S., 134, 437
von Rosenvinge, T. T., 91
von Zeipel, H., 307, 308
Vorontsov-Velyaminov, B. A., 418
Vostry, J., 23
Vriens, L., 350, 352, 363

W

Wade, C. M., 133, 135, 393, 398, 399, 402, 416, 417, 432
Wagoner, R. V., 175, 443, 446, 447
Wakano, M., 179, 180, 187, 188, 203
Waldmeier, M., 2
Walker, M. F., 170, 381, 383, 386, 393, 395, 396, 397, 402
Wallerstein, G., 91, 92, 150, 151, 152, 153, 165, 301
Walraven, T., 93
Wampler, E. J., 143, 171, 276, 293, 404, 421, 426
Wang, C. G., 289
Wannier, G. M., 355
Warner, B., 91, 143, 276
Waters, J., 140, 142, 143, 144, 152, 154
Watt, S., 140
Wayman, P., 165
Wayman, P. A., 297
Webber, W. R., 91
Weber, E. J., 4, 298, 317
Weber, J., 390
Webster, H. F., 82
Webster, W. J., Jr., 247
Weedman, D. W., 244, 405
Weinberg, M., 98
Weinreb, S., 128
Weiss, R., 451
Weiss, R. A., 186
Welch, W. J., 215, 226
Weliachew, L., 281
Wentzel, D. G., 3, 381
Werner, M. W., 237, 396, 424
Westcott, R., 219, 223
Westerhout, G., 122
Westphal, J. A., 143, 144, 145, 150, 151, 152, 154, 223, 282

Weymann, R. J., 314, 317, 319, 323, 326, 395, 396, 399, 401, 402, 420, 421, 422, 423
Whang, Y. C., 48, 50
Whitaker, W., 335, 350
Whitaker, W. A., 43
White, F. G. W., 65
Whiteoak, J. B., 276, 381, 385
Whitford, A. E., 377
Wickramasinghe, N. C., 94, 97, 396, 424, 449
Wielebinski, R., 265, 267, 268, 280
Wielen, R., 437
Wilcke, J. C., 66
Wilcox, J. M., 3, 4, 16
Wild, J. P., 123, 133
Wild, P., 418
Wildey, R., 167
Wildt, R., 98, 100
Wilkinson, D. T., 175, 272, 273
Williams, D. A., 100
Williams, D. R. W., 245, 247
Williams, P. J. S., 429
Williams, R. E., 420, 421
Williamson, J. H., 98, 362
Wills, D., 394
Wilson, C. R., 78
Wilson, D. C., 318, 323
Wilson, K. H., 100
Wilson, O. C., 88, 92, 93, 165, 247, 256
Wilson, P. R., 2
Wilson, R. L., 11
Wilson, R. W., 133, 175, 416, 451
Wilson, T. L., 170, 231, 244, 253, 255, 260, 261, 262, 263, 387
Winckler, J. R., 78
Windram, M. D., 397, 432
Wing, R. F., 87, 88, 94, 98, 226
Winnberg, A., 131
Winterberg, F., 3
Winther, A., 341
Wisniewski, W. A., 396
Wolf, R. A., 184, 198, 200, 203
Wolfe, A., 435, 440, 451
Woltjer, L., 169, 395, 420, 422, 439
Wood, D. B., 88, 372, 374, 375, 435, 436, 437
Woolf, N. J., 221, 227, 228
Worley, C. E., 90, 162
Wyller, A. A., 90
Wyndham, J. D., 133

AUTHOR INDEX

Y

Yamashita, Y., 88, 90
Yao, S. S., 130
Yeh, T., 54
Yong-Ki-Kim, 356
Yoshida, S., 225
Yoshinaga, H., 225
Young, L. D. J., 215, 226

Z

Zaitsev, V. V., 290, 292
Zapesochyi, I. P., 351
Zapolsky, H. S., 442
Zappala, R. R., 94
Zeau, Y., 226
Zeldovich, Ya. B., 434, 442, 443, 444
Ziesel, J. P., 361
Zisk, S., 135
Zuckerman, B., 169, 231, 240, 244, 245, 246, 247, 253, 254, 255, 256, 257, 258, 259, 261
Zwaan, C., 2
Zwicky, F., 116, 402, 403, 418, 431

SUBJECT INDEX

A

Absorption coefficient, 233, 235, 250
Accretion
  of matter, 441
  of stellar, 35-40
  see also Stellar winds
Adiabatic exponent, 185-86
Aerodynamic drag, 103
Alfven velocity, 318
Angular momentum
  diffusion of
    from core to shell, 298, 302, 323
  distribution, 436, 439
  orbital
    degeneracy, 340
    of solar system, 302
    of solar-type stars, 302
    transfer of, 302, 315-17, 319-20, 325
  see also Core, Rotation, Sun
Antenna
  multielement array, 133
Antimatter, 449
Antisymmetrizing operator, 332
AP 0823, 270, 286
AP 2015, 270, 278
Aperture synthesis, 131, 133
Apodization, 209-10
Arp 330, 431
Ascafilms, 73-74
Astigmatism
  in mirror, 305
Astrolabes, 124
Astrometry
  plate-overlap, 124-25
Atmospheric
  refraction, 304-5
  thermal emission, 223
  transmission difficulties, 219-21
    achromatic fluctuations, 219
    chromatic fluctuations, 219-20
    turbulence difficulties, 221-24
    in intermediate infrared, 223-24
    in near infrared, 223-24
  see also Stellar atmospheres
Auroras
  alignment with field lines, 74
  arcs, 67-68, 74-76, 78, 80-82
    multiple, 75
    artificial, 67
    australis, 62, 65
    borealis, 61
    breakup, 74, 82
    conjagacy of, 66-67, 74, 81
  corona, 62, 66
  Doppler broadening in hydrogen lines, 73
  Doppler displacement of hydrogen lines, 73, 79
  echoes, 81
  electron temperature of, 80
  emission from H, He I, 72
    from N I, 69
    from N_2, O_2, O, 72
    of radio waves, 78
  energetic, 80
  forbidden lines, 72
    oxygen, 72
  height of, 68
  height distribution
    of emission in, 77-78
  latitude, 71
  low-altitude, 71
  morphology, 73-79
  observations of, 61, 67-68, 73-74
    IGY, 67, 73-74
  oval, 76-77, 81
  particles, 81-82
  physics, 61-83
  polaris, 62
  polar
    type A, 72
    type B, 72
  polar-glow, 76
  production of
    by charged particles, 73, 79-81
    proton, 73, 81
  radiant point of, 74
  rays, 66, 74, 76
  relations with geomagnetism and sunspots, 66-71
  satellite observations of, 68-69
  sheets, 80-81
  spectrum of, 71-73
  subauroral displays of, 71
  substorms, 73-74, 76
  synoptic patterns of, 76-77
    thermal, 80-81
  VLF emission, 73
  X-ray events in, 73, 78-79
  zone of, 63-65, 77
    southern, 65
Autocorrelation function, 127-28, 131
Autoionization, 344, 357-62
  vs ionization and excitation, 357-62
  levels, 359
  rates, 359

B

B264, 407, 420
Background radiation
  energy density of, 450
  microwave, 175-76, 434, 446, 449-52
  nonthermal radio, 449-50
  X ray, 449, 451-52
  see also Blackbody, X rays
Balmer decrement, 421
  in quasistellar objects, 171
  in Seyfert galaxies, 171
  see also Hydrogen
Balmer emission lines, 421-23
  see also Hydrogen
Bandwidth, 266-67, 284, 286, 288
  instantaneous, 266
Baryons, 181, 184, 187, 448
Baseline drifts
  removal of, 134
Beerfoam process, 302-3
Beryllium
  non-depletion of
    in solar-type stars, 302, 317, 319, 326
  see also Elemental abundances
Bethe approximation, 340-41, 347
Big bang
  cosmology, 434
  synthesis, 447
  universe, 449
Binary
  eclipsing, 162
  encounter, 342
Binding energy, 181, 442-43
  gravitational, 188, 194, 206

SUBJECT INDEX

neutron, 181
proton, 183
Bipolar region, 2-3, 24-25
  see also Magnetic fields, Sun, Sunspots
Blackbody radiation, 379, 449-51
  temperature of, 450-51
Black holes, 435, 439-41, 444
Blanketing
  line, 163-64
  in late-type stars, 102, 105
  statistical models for, 102
BL lacertae, 430
Bolometric corrections
  of late-type stars, 89
  magnitudes, 89
Boltzmann factor, 242, 259
  relation, 330
Boreal pole, 67
  see also Auroras
Born approximation, 335-37, 341-43, 347, 350, 356-57, 360-61
  -exchange, 357, 360-61
  -Oppenheimer, 336, 356-57, 360
  unitarized, 337-38
  see also Quantal approximations
Bosons, 198
Bound-bound transitions, 100-2
  of H_2O, 101, 105-6, 108
  CaH, 101, 105
  CN, 101, 105
  CO, 101, 105
  MgH, 101, 105
  SiH, 101, 105
  TiO, 101, 105-6
Bound-free absorption, 99
  transitions, 194-95
Boussinesq approximation, 103
Branching ratio, 359, 362
Brans-Dicke scalar-tensor theory, 193
  coupling constant of, 298
  see also Gravitation
Bremsstrahlung, 389, 424
Broadening
  collisional, 102
  Doppler, 102, 108, 231, 261, 422
  natural, 102
  pressure, 261
  Stark, 261

C

3C and 4C sources, 403, 404,
408, 409, 410, 415, 423, 425
Ca II H and K emission
  in late-type stars, 92-94
  see also Wilson-Bappu effect
Camera
  all-sky, 68
Carbon stars, 88, 90-92, 94, 99
  high-velocity, 88
  see also Molecules, Stars
Cascading, 239
Cassiopeia A, 155
Catalogue
  Astrographic, 126
  BD, 115
  Shapley-Ames, 417
  star, 124
  magnitude and color corrections to, 125
Centrifugal force, 309, 313
  potential, 309
Cen X-2, 155-57
Cen X-4, 155-56
Cerenkov radiation, 78
Chains
  galaxy, 431
  see also Galactic nuclei, Galaxies
Characteristic time, 270-72
Charge
  neutrality, 181-82, 184
  resonant exchange reaction, 56
Chromosphere, 307, 318
  of late-type stars, 92
  see also Sun
Circumstellar envelope
  of late-type stars, 94, 97
Classification of galaxies
  according to color
    Haro's method, 419
    Markarian's method, 419-20
    Sandage and Luyten's method, 420
  according to form, 417-19
  see also Spectral classification
Clouds
  coherent
    of plasma, 393
  discrete
    in galactic nuclei, 395-96, 398, 422
  gas
    ejection of, 397-98, 401
    in galactic nuclei, 424, 430, 441-42, 446
    high-velocity, 389, 422
    see also Galactic nuclei
Clusters
  of galaxies, 410
  hidden
    mass in, 431
  positive
    total energy of, 430-31
  time scale
    for expansion of, 431
CNO cycle, 162
Coherent objects, 414
Collapse
  energy release from, 440
  gravitational, 439-40, 443-44
Collapsed
  matter, 370
  objects (black holes), 435, 439-41, 444
Collimators
  modulation, 140, 142, 156
  slat, 140
Collision
  Coulomb, 51-52
  cross sections, 239, 244
  proton-antiproton, 392
  proton-proton, 392
  star-star, 437-39, 441
  strength, 330-31, 338, 346
  time, 438
  see also Cross sections, Excitation, Ionization, Relaxation
Color differences
  in galaxies, 436
Coma cluster, 431
Compact galaxy, 402-3, 406, 418-19
  see also Galaxies
Compact radio components, 414, 416, 424, 429
  see also Galactic nuclei
Composition
  differences in galaxies, 446-47
  stellar, 370
  see also Elemental abundances, Hydrogen, Helium
Compton effect, 452
  inverse, 412
Compton scattering, 194-95, 370, 392, 424-25, 427-28
  see also Scattering
Computation
  real time
    of transform, 224-25
    see also Fourier spectroscopy
Computer
  analog, 133
  general-purpose online, 123, 131, 133
  wired, 225
Condensation

SUBJECT INDEX 477

in envelopes of cool stars, 97
see also Graphite
Conjagacy
of auroras, 66-67, 74, 81
see also Auroras
Continuum energy distribution, 397, 403-4, 410, 419-24, 429-30
see also Spectra, Galactic nuclei
Continuum observations of the Galaxy, 387-89
see also Galaxy
Continuum radiation, 410-11, 421-30
origin of
in galaxies, 423-30
see also Galactic nuclei, Galaxies
Contour diagram, 118-19, 131
Convection, 164
effect
of Coriolis forces on, 7
instability criterion for, 103
in late-type stars, 103-7
zone, 298, 302-3, 313-20, 324, 326
in Sun, 16-17, 24
see also Sun
Convective
cells, 8-9, 11-13, 15-18, 20
period of, 16
motions
in Earth's core, 7-8, 10
thermal
forces, 22
Core
delayed, 444, 446
formation of galactic nuclei as, 444-45
Earth's
convective motions in, 7-8, 10
liquid, 16
expanding, 444
-halo configuration, 437-38
radiative, 313-14
rapidly rotating solar, 297-99, 303, 319-23, 326-27
stellar, 437
superfluid
of neutron star, 198-200
see also Galactic nuclei, Neutron stars, Pulsars, Rotation, Sun
Coriolis forces, 10, 15-16, 22, 312

effect of
on convection, 7
Corona, 3, 4, 317-19, 421
auroral, 62, 66
infrared spectrum of, 228
lines, 399
solar, 55, 345, 346, 348
temperature maximum of, 55
see also Auroras, Infrared emission, Sun
Correlation coefficient, 128-29
Cosmic rays
He abundance in, 166, 176
from neutron stars, 195-96
origin of, 452-53
primary
energy density of, 452
see also Helium, Neutron stars
Cosmology
big bang, 434
see also Big bang
Coulomb
collisions, 51-52
crystal, 196-97
exchange approximation, 349-50
field, 336, 343, 353, 355, 362
long-range interaction, 338
potential, 337
wavefunctions, 355
waves, 336, 338
see also Interaction, Quantal approximations
Coulomb-Born approximation, 336-40, 342, 347, 351
II, 337, 339-42, 350, 361
see also Quantal approximations
Coupling
close
approximation, 334-36, 343, 348-51
to excited states, 334
intermediate, 348
jj, 347
LS, 340, 346-48
resonant state, 334
weak
assumption of, 337
cutoff, 341
see also Quantal approximations, Quantum theory
Cowling's theorem, 6, 13
CP 0329, 270, 274, 278, 281-82, 284, 288, 290
CP 0808, 270, 274, 284, 291
CP 0834, 270, 274, 282,

285-86
CP 0950, 270, 274-79, 282-83, 285
CP 1133, 270, 274, 277-78, 285
CP 1919, 265, 271, 274, 276, 282, 284-85
see also Pulsars, Supernova remnants, Crab Nebula
Crab Nebula, 140, 142-43, 180, 203, 205-6, 265, 281, 283, 290, 397
see also NP 0532, Neutron stars, Pulsars, Supernova remnant
Cross sections, 330, 332-34, 340-46
collision, 239, 244
dipole, 361
elastic, 334
excitation, 334, 340, 342, 346, 349
table of, 350-51
forbidden transitions, 330-31, 346, 361
inelastic, 332, 334, 345
ionization, 349, 351-52, 355, 357, 360-64
H^-, 362
reduced, 351, 362-63
see also Collisions, Excitation, Forbidden lines, Ionization, Transitions
Cup of tea, 313
Cyclonic convective cells, 8-10
interaction with poloidal field, 10
Cyclonic motions, 12-14, 17, 19-20, 22, 24, 26, 27
see also Convection
Cygnus A, 409, 414, 421
Cyg X-2, 142, 154-55, 157
binary nature of, 154-55
colors of, 154
He II 4686 emission in, 154-55
optical identification of, 154
optical spectrum of, 154-55
position of, 154
variable radial velocity of, 154-55
see also X rays

D

Data
display, 118, 131
handling systems
for continuum, 134-35
processing, 118
real-time reduction device for, 117

SUBJECT INDEX

see also Computer
Debye-Huckel approximation, 98
see also Quantal approximations
Deceleration parameter, 446
Degenerate
  electron gas, 181, 188, 194
  electron pressure, 440
  neutrinos, 175
  neutron gas, 182
  neutron pressure, 440
  see also Neutrino, Neutron, Neutron stars
Density
  central, 437
  electron, 252, 257, 380, 384, 404, 420
  fluctuations, 421, 434
  number
    of discrete sources, 451
Density-modulated map
Departure coefficients, 233, 237-40, 242-43, 253, 259
Detachment energies, 97
Detailed balancing, 330
Detector
  Ge-Hg, 223
  PbS, 212, 217, 227-28
Dielectronic recombination, 242-43, 255, 257
  see also Ionization, Recombination
Diffusion
  molecular
    in Sun, 317, 319, 321
    in neutron stars, 195
Digital circuits, 123
Digitization
  of signal, 128-30
  one-bit, 128, 130
  two-bit, 130
Dipole
  approximation, 341
  axis, 64
  cross section, 361
  field, 7-9, 12
  radiator
    classical, 291-92
  transition probability, 359
  see also Magnetic field, Sun, Quantal approximation
Discrete sources
  number density of, 451
Disk population, 376
Dispersion, 266, 268, 281, 284, 288
Dispersion measure, 136
  of pulsars, 266-67, 270, 279-81, 284, 286-87
Dispersion relation, 19

Displays
  two-dimensional, 118
  see also Data handling
Dissipation
  magnetic, 441
Dissociation
  of H_2, 104, 107
  pressure, 96
Distance
  estimates of Sco X-1, 149-52
  modulus of Hyades, 165
Distances
  of pulsars, 280-82
  from redshifts, 410-11, 413
Doppler broadening, 231, 261, 422
  see also Broadening
D-type systems, 413, 433
  see also Galaxies, Galactic nuclei
Dust
  in galaxies, 385, 391-92, 394, 421, 423, 426-27, 449
  see also Galactic nuclei, Galaxies
Dwarf population, 373-74
Dynamics of stars and gas
  in galactic nuclei, 381-87
  see also Galactic nuclei, Galaxies
Dynamo
  action, 448-49
  Babcock's model of, 21-23
  equations of, 14, 18
  hydromagnetic, 6, 8, 10-11
  Leighton's model of, 24-27
  migratory solar, 15-27
  stationary, 13, 23, 27
  theory of
    formal, 6, 8, 10-11, 17-21
  two-stage, 13
  waves, 19-20
    phase velocity of, 19
  see also Magnetic field, Sun

E

Earth
  emission spectra of, 228
Eclipsing binaries, 162
Eddington-Sweet circulation currents, 315-16
Effective temperatures, 89, 162, 165, 307
  of late-type stars, 89
  see also Temperature
Eigenfunction, 332, 335

Einstein-de Sitter law, 445
Ejection
  of galactic mass objects
    Arp's evidence for, 431-32
  of gas
    clouds, 441
    from galactic nuclei, 384-85, 387, 389, 391, 394
    see also Galactic nuclei, Galaxies
Ekman pumping, 299, 313, 315
Electrical conductivity
  in neutron stars, 204
Electrojets, 78
Electron
  capture
    on nuclei, 181-82
    on protons, 203
  degenerate
    gas, 181, 188, 194
    pressure, 440
  density, 252, 257, 380, 384, 404, 420
  donors, 96
  Fermi level, 181-82, 184, 203
  impact
    excitation and ionization by, 329-64
  indistinguishability, 336, 354
  -ion energy exchange, 50-51
  -proton-alpha particle plasma, 54
  -proton plasma, 54-55
  relativistic, 417, 429
  scattering, 100, 336, 344, 391, 422, 423
  temperature
    in stellar winds, 50, 52-53, 58-59
  see also Galactic nuclei, Neutron stars, Scattering, Temperature
Elemental abundances, 90-92, 109, 375-76
  differences
    between galactic nuclei and outer parts, 375, 377
  in late-type stars, 90-92, 109
  see also Helium, Hydrogen, Metal/hydrogen ratio, and individual elements
Element synthesis
  in galactic nuclei, 446-48
Elliptical galaxies, 371, 373-74, 377-81, 383, 392, 394, 413-17, 432, 435-36, 440, 446
  light distribution in, 371

SUBJECT INDEX

see also Galactic nuclei,
  Galaxies
Emission
  coefficient, 233
  in galaxies, 378
  frequency of, 378-79
  infrared night-sky, 226
  line
    in late-type stars, 92, 94,
      108
    in nucleus of NGC 4151,
      400-1
  line profiles, 421-22
  line spectra
    high excitation, 395,
      399
  measure, 236-37, 245, 249-
    50, 252, 257
  radio, 413, 416-17, 424
  nonthermal, 412, 432-
    33
  spectrum of Earth, 228
  stimulated, 233, 236, 238,
    243, 249, 251, 260
  see also Galactic nuclei,
    Galaxies, Helium, Hydro-
    gen, Recombination, Syn-
    chrotron radiation
Energy
  density
    of background radiation,
      450
    of primary cosmic rays,
      452
  detachment, 97
  equipartition, 436, 438-
    39
  interaction, 336-37
  ionization, 97, 351
    reduced, 351
  kinetic, 430-31, 438-39
  lattice interaction,
    196
  positive total
    of clusters of galaxies,
      430-31
  potential, 430-31
  radiated
    total, 414
  transfer
    convective, 194
    radiative, 194
  see also Background radia-
    tion, Excitation, Ioniza-
    tion, Galactic nuclei
Equation of state
  for late-type stars, 96-
    97
  for neutron stars, 180-
    89
Equipartition, 424, 425, 428-
  29
  see also Energy
Error beam
  removal of, 131
Evolution
  of galactic nuclei, 437-39,

441, 444-45, 448
  thermonuclear, 440
  see also Galactic nuclei
Exact-resonance approximation, 348
  see also Quantal approxima-
    tions
Exchange
  amplitude, 336
  between two electrons,
    352
  Coulomb approximation,
    349-50
  resonant charge reaction,
    56
  see also Quantal approxima-
    tions
Excitation
  collisional
    rate of, 330-31
    cross sections, 334, 340,
      342, 346, 349
    for 1s-2s in He II, 339
    table of, 350-51
    by electron impact, 330-
      49
    from ground state, 343
  of hydrogen
    1s-2s, 334-35
    inelastic threshold of,
      343
  of interstellar molecules
    by background radiation,
      450
  radiative, 379
  two-step, 344
  see also Collisions, Cross
    sections, Forbidden
    lines, Helium, Hydrogen,
    Ionization
Expanding arm, 389
Expansion
  dynamical time scale of,
    413, 444
  in equatorial plane,
    383
  of galaxies, 445
    clusters of, 431
  speeds, 412, 415
  see also Galaxies

F

Fabry-Perot
  etalon, 213, 217
  interferometer, 127,
    215
  spectroscopy, 209
  see also Fourier spectros-
    copy, Spectrometer
Faraday rotation, 266,
  276
Feige 11, 173
Feige 13, 173
Feige 65, 173
Feige 86, 173
Fermi gas, 179

of neutrons, 188
Fermi level
  electron, 181-82, 184,
    203
  neutron, 182, 184, 203
  proton, 184, 203
Fermi sea, 196, 198-99
  see also Degenerate, Neutron,
    Neutron stars
Ferromagnetism
  in neutron stars, 199, 204-
    5
Filamentary structure, 397
Filaments in M82, 391
Filter
  matched-bandpass,
    130
  optimum linear
    for pulsar detection, 135-
      36
  see also Pulsars
Filtering
  of low-frequency components,
    134
Finesse, 213
Flares
  of Sco X-1, 144-45, 153
  simultaneous optical and
    X ray, 153
  solar, 2
Flare stars, 92-93
  UV Ceti-type, 93
Flash stars, 93
Forbidden lines, 244,
  422
  in auroras, 72
  excitation of, 344, 346
Forbidden transitions, 346,
  361
  cross sections of, 330,
    331
  induced
    in late-type stars, 98
    within ground configuration,
      348-49
  see also Galactic nuclei,
    Galaxies, H II regions,
    N II, O II, Seyfert galaxies
Formation
  of galaxies, 430, 433-
    35
  see also Galaxies, Star
    formation
Form
  classification of galaxies by,
    417-19
  compact galaxies, 418-
    19
  Morgan's method, 417
  Sersic and Pastorzia, 417-
    18
  Vorontsov-Velyaminov,
    418
  see also Classification,
    spectral type, Galaxies
Form of galaxies
  correlation with spectral

SUBJECT INDEX

type, 372-73
Fourier algorithm
　fast, 136
Fourier inversion, 214
　sampling theorem, 130
Fourier spectroscopy, 209-29
　accuracy requirements of, 217-19
　advantages of, 211-17
　from aircraft, balloons, and satellites, 226-28
　computational difficulties of, 224-25
　consequences
　　for astronomy, 215-17
　difficulties of, 217-25
　encircling the earth, 228
　errors
　　intensity, 218-19
　　path difference, 217
　experimental results of, 225-28
　size and weight reductions in, 214-15
　see also Fourier spectrum, Interferometer, Spectrometer
Fourier spectrum, 210
　of Venus, 220
　see also individual objects
Fourier transform, 127, 131, 135, 210-11, 224
Fragmentation, 443
Free-bound absorption by quasi-H_2, 100
Free-free absorption, 98-99
　transitions
　　in H, 236
　　in He, 236
Friction
　dynamical, 55
Friedman universe, 444

G

Galactic nuclei, 434-46
　as delayed cores, 444-45
　dynamics of stars and gas in, 381-87
　ejection
　　of gas from, 384-85, 387, 389, 391, 394
　　of objects of galactic mass, 431-32
　element synthesis in, 446-48
　evolution of, 437-39, 441, 444-45, 448
　flux variations of, 370
　limits on size from, 370

highly active, 390-407
　M82, 390-92
　M87, 392-94
　mass of, 382-83, 395, 398-99, 402
　mass distribution in, 370, 395
　mass of ionized gas in, 380, 384, 387
　nonthermal energy sources in, 370, 372, 379, 387-88, 392, 394, 396, 399
　optical size of, 369, 381, 383-84, 386, 393-94, 396, 402-3
　rotation of, 381-83, 385
　small-scale structure in, 370
　as sources of
　　dust, 449
　　particles and radiation in the Universe, 448-53
　theory of, 434-46
　ultraviolet radiation from, 433-34
　see also Cores, Ejection, Clouds, Gas, H II regions, Rotation, Synchrotron radiation
Galactic structure, 268
Galactic winds, 41-44
　equations of motion for, 41
　see also Stellar wind
Galaxies
　ages of, 413, 431, 436
　chains of, 431
　classification of
　　according to color, 419-20
　　according to form, 417-19
　clusters of, 410, 431
　compact galaxies, 402-3, 406, 418-19
　correlation of form with spectral type, 372-73
　elliptical, 371, 373-74, 377-81, 383, 392, 394, 413-17, 432, 435-36, 440, 446
　expansion of, 445
　infrared luminosities
　　table of, 453
　irregular, 372, 376, 378, 380-81
　line spectrum of, 396-99, 402-3, 405-7, 410-11, 413, 415, 419-23
　N-type, 404, 406-7, 410-11, 413, 415, 417, 419-20, 422-24
　radio, 412-17, 439-40, 442, 445, 449-52
　rotation of, 435

satellites of, 432-33
Shapley-Ames, 377, 395
spiral, 371-74, 376-81, 385, 397, 410, 416-18, 433, 435-36
see also Classification of galaxies, Clusters of galaxies, Compact galaxies, Elliptical galaxies, N-type galaxies, Rotation, Seyfert, Spiral galaxies, Galactic structure, individual objects
Galaxy, 416, 423, 426, 436, 445-48
　continuum observations of, 387-89
　large-scale structure of, 244, 260-63
　nucleus of, 369, 373, 382, 387-90
　see also Galactic nuclei, H II regions, Pulsars
Gamma radiation, 389, 426, 452
Gas
　clouds, 442
　ejection of, 441
　hot, 424, 430
　to dust ratio, 426
　hot, 438, 445
　intergalactic, 451-52
　ionized, in galactic nuclei, 378-80, 383-84, 387
　cloud in nucleus, 370
　in nuclei, 446
　motion of
　　in nuclear region, 382
　neutral
　　in galactic nuclei, 380-81
　see also Clouds, Galactic nuclei, H II regions
Gaunt factor, 237, 240
Gaussian fitting, 134
Geomagnetic pole, 6
Geomagnetism
　relation of auroras to, 66-71
　see also Magnetic fields, Auroras
Globular clusters, 373-74, 433, 435
　He abundance in
　　sensitivity of horizontal branch to, 167
　　variation of, 168
　see also Helium, Horizontal branch, individual clusters
Goldreich-Schubert-Fricke instability, 315-17
Graphite, 449
　in M stars, 94
Graph plotter, 118
Gravitation
　general relativistic theory

SUBJECT INDEX

of, 297-98
scalar-tensor theory of, 297-99
see also Brans-Dicke
Gravitational
distortion, 304
field, 436, 441-42
potential, 308, 314
quadrupole moment of Sun, 297-99, 307-12, 319, 326
radiation
mass loss implied by, 390
from neutron stars, 180, 202
redshift
in neutron stars, 194, 206
waves, 202, 436, 439, 442, 445, 449
excitation of, 315
see also Neutron stars, Rotation, Sun
GX 3 + 1, 140-41, 155-56

H

H I clouds, 237-38, 240
absorption by, 238
emission by, 237
H I regions, 233, 240, 257-60, 263, 427
H II regions, 169-70, 231, 233, 235-38, 244, 246-47, 251-53, 255-57, 259-63, 280-81, 372, 384-85, 387, 417, 420
galactic distribution of, 262
internal structure of, 244, 260-61
kinematics of, 260-63
in M31, 171
physical properties of, 244
temperature fluctuations in, 244
as tracers of spiral arms, 261
see also Balmer emission, Gas, Hydrogen, Helium, Recombination
Halo, 437, 447
B stars, 172-73, 176
population, 436
radio emission, 393
Hamiltonian, 355
Hayashi phase, 320
He I
- $\lambda 3889$, absorption features in, 401
emission from auroras, 72
singlet-triplet ratio, 173
He II emission in Cyg X 2, 154-55

He III abundance in H II regions, 237
Heavy element enrichment, 447
Heliometer, 303-4
Helium
autoionized states of, 340
cosmological significance of, 161
excitation cross sections of, 339
gravitational diffusion of, 176
to hydrogen ratio, 161, 168-72, 176, 244, 253, 256, 447
production site
fireball, 175
supermassive objects as, 175-76
recombination lines, 231, 253-56
variable line strength, 172
Helium abundance, 446-47
cosmic, 161-76
in cosmic rays, 176
solar, 166
in Crab Nebula, 169
extragalactic determinations of, 170-71
in galactic clusters, 167
in globular clusters, 167-68
in interstellar medium, 244
in moving groups, 165, 176
in novae, 169
in Orion Nebula, 255
in planetary nebulae, 168-69
from Population I stars, 171-72
primeval, 174-76
in quasistellar sources, 171
in Seyfert galaxies, 171
from stellar photospheres, 171-74
from Sun, 165-67
Hercules cluster, 431
High-resolution spectra
of planets, 226-27
of Sun, 226
High-velocity clouds, 389, 422
Horizontal branch, 167-68
sensitivity to He abundance, 167
stars, 167-68, 173-74, 380
HP 0904, 271, 274, 282
HP 1506, 270, 275
Hubble

classification, 372
constant, 369
law, 372
Hyades
rotation in, 301, 319-20, 322
standard sequence, 163-64
Hyades-Pleiades mass-luminosity law, 164-65
Hydra A, 409, 415
Hydrogen
emission
in auroras, 72
Balmer, 421-23
Balmer decrement, 171, 421
Hα, 378-79, 385-86, 398, 404, 408, 415, 420
in M stars, 93
excitation of 1s-2s in, 334-35
exhaustion, 162, 167
ionization of, 354
ionization zone
in late-type stars, 104, 107
to metal ratio, 375, 404
neutral
mass of, 389
21-cm line of, 380-81, 389
Paschen lines, 421
photoionization of, 164
recombination lines of, 231, 244-54
see also H I, H II regions, Excitation, Ionization, Recombination
Hydrogen, molecular
dissociation of, 104, 107
dissociation zone
in late-type stars, 107
pressure-induced opacity of, 101, 105
quadrupole lines of, 101
see also Molecules
Hydromagnetic equation, 4, 11
Hydrostatic equilibrium
general relativistic equation of, 187, 193
in late-type stars, 96, 107
Hyperfine splitting
in He^3, 170
Hyperons, 184-86, 189-90, 200, 203
Σ^- 184, 186, 190, 200
Σ^0 184, 186
Δ^- 184, 186
Δ^0 184, 186
Λ^0 184, 186

I

IC 1569, 170

SUBJECT INDEX

IC 3481, 431
IC 5532, 408
Imaging devices
  photoelectric, 116
Impact parameter, 331, 341-42
  method, 350, 352
  for permitted transitions, 341-42
  see also Quantal approximations
Information
  processing
    general principles of, 117-22
    online devices, 122-24, 131
  systems, 115-36
  storage devices, 117
  see also Data handling
Infrared
  emission, 424, 426-27, 429, 433
  night-sky, 226
  luminosites
    of galaxies, 453
  observations, 388
  photometry of Sco X-1, 147, 150
  radiation, 396, 403, 406
    from galactic nuclei, 371
  sources, 423, 428
  spectrum of
    Moon, 223
    pulsars, 282-83
    solar corona, 228
    Venus, 216-20, 223-24, 227
  see also Fourier spectroscopy, Galactic nuclei
Instability
  β Cepheid, 162
  dynamical, 442
  strip, 168
    dependence on He content, 168
Instrumental lineshape, 210, 218
Integrated light, 411
Interaction
  configuration, 340, 348
  Coulomb long-range, 338
  electron-electron, 336-37
  energy, 336-37
  neutron-baryon, 181, 184
  nuclear potential, 185
  p-state, 199
  star-star, 436-39
  see also Collisions
Interferogram, 210, 218-19, 221-22, 224
Interferometer
  Fabry-Perot, 127, 215

fast-scanning, 222, 225, 227
  Fourier, 211, 212, 214
  intercontinental, 370
  interpolation of unevenly spaced data, 135
  lamellar grating, 226
  Michelson, 127, 213, 225-27
  Pepsios, 217
  polarization, 225
  radio, 131
    large-field aperture synthesis, 131
  SISAM, 217
  three-element, 132-33
  two-beam, 210
  see also Fourier spectroscopy
Intergalactic
  magnetic field, 414
  medium, 414
Interiors
  stellar
    He abundance from, 161-64
International Brightness Coefficient, 77
Internal modulation, 222
Interpulse, 275, 283, 294
  see also Pulsars
Interstellar
  absorption, 391
  excitation of CN, CH, CH^+
    by background radiation, 450
  gas, 374
  helium abundances, 169-70, 244
  plasma, 279, 281, 286-87
    irregularities of, 286-87
    mean density of, 281
    scintillation, 276, 285-88
  X ray absorption, 150-51
  see also Clouds, Gas, H I, H II regions, Dispersion, Pulsars
Inverse predissociation
  of AlH, 108
Ionization, 343
  of atomic hydrogen, 354
  classical theories of, 351-53
  modified, 352-53
  collisional, 239, 342, 421
  cross sections, 349, 351-52, 355, 357, 362-63
  empirical formulas for, 362-64
  for H (1s), 353
  for H^-, 362
  table of, 360-61

by electron impact, 349-64
  energies, 97, 351
    reduced, 351
  via excitation and autoionization, 357-62
  limit, 358, 362
    bound states above, 358
  of negative ions, 362
  nocturnal, 80
  pressure, 96
  source of, 421-22
  threshold law, 355-56
  see also Collisions, Cross section, Autoionization
Ionized gas
  in galactic nuclei, 370, 378-80, 382-84, 387
  line emission from, 393-94
  see also Clouds, Gas, H II regions
Ionosphere
  terrestrial, 55-56, 78
  H^+ ions in, 56
  O^+ ions in, 56
  polarization electric field in, 56
Iron abundance
  in Sun, 166
Irregular galaxies, 372, 376, 378, 380-81
  see also Galactic nuclei, individual galaxies
Isoaurores, 66
Isochasms, 63-66, 69
Isodensitometer, 116
Isoelectronic sequence, 340, 345, 359
Isometric projection of 3-D graph, 118, 120-21
Isotopes
  C in late-type stars, 91-92
  He^3, 170, 172
  in 3 Cen A, 172
  Mg in late-type stars, 91
  see also Elemental abundances, individual elements

J

Jacobi ellipsoid, 205
Jets
  associated with nuclei of galaxies, 419, 432

K

K648, 168
Kelvin-Helmholtz instability, 314
Kinematic distance, 261

SUBJECT INDEX

L

Large Magellanic Cloud, 155, 157
Lattice interaction energy, 196
Least-squares
 analysis, 125
 nonlinear
  fitting, 135
Leptons, 184
Level populations
 of C, 240-44
 of H and He, 238-40
 see also Departure coefficients
Light distribution, 435
 in ellipticals, 371
Light-to-mass ratio, 370
Limb darkening
 in late-type stars, 94
 in Sun, 306-8
Line emission
 from ionized gas, 393-94
 see also H II regions
Line intensity, 234-36, 239, 246, 259
 theory of, 233-38
Line radiation
 from the nucleus of the Galaxy, 389-90
Line spectrum
 of galaxies, 396-99, 402-3, 405-7, 410-11, 413, 415, 419, 420-23
 see also Galactic nuclei, Galaxies
Lithium
 abundance of, 299, 301, 302, 317, 324-26, 446
 in late-type stars, 91-92
 burning, 324-26
 depletion of
  in solar-type stars, 302, 317, 319, 323-27
  diffusion of, 323
Lofer, 204
Log N-log S relation, 446
Lorentz forces, 11
LTE, 233, 235-38, 244, 246, 249-51, 257-59
 departures from, 92, 108, 240, 242, 244-45, 248, 255
 see also non-LTE
Luminosity
 effective temperature plane, 162, 165
  subdwarfs in, 162-64
 function, 374, 447
 irregular, 450
 radio, 450
 infrared
  of galaxies, 453
 intrinsic, 411
 of late-type stars, 89

 optical, 411, 415, 419, 421, 424, 439-40
 profile, 382, 417
 radio, 412, 415, 424, 429
 see also Galaxies

M

M3, 168, 173-74
M8, 169
M13, 173-74
M15, 167-68, 173
M17, 169-70, 245-47, 253, 255
M31, 171, 371, 373-74, 381-83, 416, 434, 437, 446
M32, 371, 383, 437
M33, 170, 383
M42, 252, 260
M43, 252, 260
M51, 379, 384, 432
M67, 167
M81, 373-74, 384, 416
M82, 171, 390-92, 408, 424-25, 453
M87, 155, 157, 392-94, 413, 424-25, 428, 432-33, 443, 451-52
M92, 167-68, 173
M101, 371
Mach number, 31, 33, 35, 37-38, 40, 42, 44-46
MacLaurin ellipsoid, 205
Magnetic braking, 303
 see also Solar wind, Stellar wind
Magnetic buoyancy, 21, 26, 28, 303
Magnetic field, 370, 388-89, 392, 398, 424, 427-30, 441, 448-49
 decay time of, 200
 intergalactic, 414
 in late-type stars, 92
 migratory, 20, 26
 in neutron stars, 204-5, 290-92
 poloidal, 316
 production of
  from fluid motions, 13
  solar, 299-300, 308-10, 312, 315-16, 326
 bipolar regions of, 21, 23-27
 diffusivity of, 21
 dipole, decay time of, 12
 dynamo theory of, 6, 8, 10-11, 17-21
 generation of poloidal from toroidal, 12
 irregular, 2, 5
 origin of, 1-28
 polar, 2, 4, 20-23
 poloidal, 12-13, 15, 18-24, 27

 production of, 1
 reversals of, 2, 11
 strength, 2
 toroidal, 2, 4, 12-13, 15, 18-26
 variability of, 2
 vector potential of, 12, 14
 in solar wind, 298
 trapping in, 318
 in stellar winds, 58
 surface, 310-12
 terrestrial
  dipole, 7-9
  generation of, 6, 8
  geomagnetic pole of, 63
  inhomogeneities in, 8
  maintenance of, 9
  poloidal, 8, 10-11
  secular variation of, 66
  toroidal, 8, 10, 16
  transient changes of, 66
 torque, 302
 trapping of internal, 302, 323
 see also Dynamo, Neutron stars, Solar wind, Stellar wind, Sun
Magnetic pressure, 318
Magnetic storms, 69, 73, 76-77
Magnetograph, 122
Magnetosphere, 79, 81-83
 cavity, 82
 electric ring current in, 79, 82
 substorms in, 73, 82
 tail of, 74
 trapping of particles by in solar wind, 82-83
Magnitude
 absolute, 419, 421
Markarian objects, 405, 420
Mass
 of ellipticals, 435
 of galactic nuclei, 382-83, 395, 398-99, 402
 of ionized gas
  in galactic nuclei, 380, 384, 387
 late-type stars, 90
Mass distribution, 437
 in galactic nuclei, 370, 395
Mass energy, 448
Mass-to-light ratio, 382, 383, 440
Mass loss, 167, 189, 205, 401, 422-23, 436
 rate, 401
Mass-luminosity plane, 162, 165
 B stars in, 162

SUBJECT INDEX

Mass-luminosity relation
  for Hyades-Pleiades, 164-65
  for late-type stars, 90
Mass number
  of nuclei, 181-82
Mass segregation, 436, 438
Matrix
  reactance, 333, 338, 348
  scattering, 333
  transmission, 333, 338
Maxwellian velocity distribution, 330
Measuring engines, 116, 125
Meissner effect, 199
Meridian circles, 124
Metal-to-hydrogen ratio, 404, 426-27, 446-47
  see also Hydrogen
Metric
  fluctuations in, 444
Mg isotopes
  in late-type stars, 91
Microdensitometer, 116
Micropulsations
  geomagnetic, 73
Microturbulence
  in late-type stars, 102, 107
  see also Turbulence
Microwave background radiation, 175-76, 434, 446, 449-52
  see also Background radiation, Blackbody
Mixing, 436-37
  in solar core, 316
Mixing-length theory, 103
Model atmospheres
  for late-type stars, 87-110
  construction of, 103-4
  density inversion in, 104-5
  theoretical, 95-109
  T-τ relation in, 105
  validity checks on assumptions of, 107-9
  see also Molecules, Opacity
Molecular bands
  in auroras
    Kaplan-Meinel, of O_2, 72
    nitrogen, 72
  in late-type stars
    C_2, 88
    CaH, 88
    CH, 88, 109
    CN, 88
    CO, 88
    H_2, 101, 104-5, 107
    H_2O, 101, 105-6, 108-9
    LaO, 88

MgH, 88
SiC_2, 88
TiO, 87-88
VO, 88
ZrO, 88, 91
  rotation-vibration, 101
  Sanford, 88
  Swan, 88
  in Venus
    CO_2, 220, 224, 226
  see also Hydrogen
Molecular diffusion, 317, 319, 321
Molecules
  photodissociation of, 100
  photoionization of, 100
  see also Dissociation, Ionization
Moment of inertia
  of Sun, 302
Morgan class f, fg, 372
Morphology, 407, 413, 418
  see also Classification, Form, Galaxies
Moving groups, 164-65
  helium abundance in, 165, 176
  Hyades-Pleiades, 176
    mass-luminosity law for, 164-65
  Sun-Sirius, 164-65, 176
MSH 03-31, 408
MSH 05-43, 404, 409
MSH 12 + 04A, B and C, 409
MSH 13-42, 408
MSH 23-112, 409
Multiplexing
  ability, 211-12, 215-16
  factor, 211
  see also Fourier spectroscopy, Interferometers
Muon, 184-85
  negative
    threshold, 184, 186

N

Na I D lines
  in galaxies, 374
National Geographic Society-Palomar Observatory Sky Survey, 126
Navier-Stokes equations, 32-33, 46
N-body problem, 437
Neutrino
  -antineutrino pairs, 200-1
  degeneracy, 175
  emission from
    bremsstrahlung process, 200
    plasma process, 200
    URCA process, 200

flux
  from the Sun, 165-66, 168
Neutron
  degenerate
    gas, 182
    pressure, 440
  drip line, 181
  Fermi level, 182, 184, 203
  free, 182
  number density of
    in neutron stars, 182, 184
  -proton-electron fluid, 198
  proton ratio
    in neutron stars, 181
  see also Degeneracy, Fermi, Neutron stars
Neutron star, 179-206, 289-91, 370, 440, 442-43
  atmosphere, 193-96
    composition of, 195
  cosmic-ray generation by, 195-96
  effect of magnetic field on, 195
  composition of, 179, 184
  cooling of, 180, 200-1
    by electromagnetic radiation, 200
    by neutrino-antineutrino pairs, 200-1
    by synchrotron radiation, 201
  crust of, 190, 196
    crystalline, solid, 196
    density variation in, 197
    lattice array in, 196
    melting temperature of, 196
    seismic activity in, 198
    volcanic activity in, 198
  damping of vibrations in
    by gravitational waves, 202
    by rotation, 203
    by URCA process, 203
  density distribution
    in interior of, 189, 191-92
  diffusion in, 195
  electrical conductivity in, 204
  electrodynamics
    of spinning, 291
  equation of state for, 180-89
  ferromagnetism in, 199, 204-5
  gravitational binding energy of, 188, 194, 206
  gravitational radiation

SUBJECT INDEX

from, 180, 202
magnetic fields in, 204-5, 290-92
ohmic dissipation of, 204
matter density in, 182
maxium stable mass of, 189
models, 186-94
neutron-proton ratio, 181
number density of neutrons in, 182, 184
radial electric field of, 193
radius of, 190
rotation of, 197, 203, 205-6
 rapid, of core, 206
 slowing down of, 205-6
stable, 188-89
superfluid core of, 198-200, 294
superfluidity of, 198-200, 206
surface gravitational redshift of, 194, 206
surface temperature of, 195, 201
vibrations of, 180, 201-4
 axial motion, 204
 fundamental period of, 202
 nonradial, 202
 torsional, 203-4
see also Degeneracy, Neutrino, Neutron, Gravitational waves, Magnetic fields, Rotation
Neutron threshold, 192-93
N galaxies, 404, 406-7, 410-11, 413, 415, 417, 419-20, 422-24
 classification of, 406
 spectroscopic properties of, 404
see also Galactic nuclei, Galaxies, and individual galaxies
NGC galaxies, 155-56, 167, 168, 169, 170, 171, 173, 231, 237, 245-47, 254-55, 257-59, 261, 371, 372, 373-74, 375, 378, 380, 383, 384, 385, 386, 394, 395-98, 399-402, 406, 408, 412, 416, 417, 420-24, 425-28, 429, 431, 432, 436, 446, 453
Night-sky emission
 infrared, 226
NML Cyg, 94
NML tau, 94
Noise
 acoustic, 318

equivalent power, 212
 receiver, 219
 source, 219
see also Signal-to-noise
Noncircular motions, 381-82, 384-86, 391
see also Galactic nuclei, Galaxies, Rotation
Non-LTE, 233, 236-37, 241, 245-46, 248-49, 251, 257-58
see also LTE, departures from
Nonthermal radiation, 407, 410-11, 415-17, 422-24, 426, 430, 432-33, 439, 443, 445, 449, 451
see also Bremsstrahlung, Emission, Synchrotron
Nonthermal radio background, 449-50
see also Background radiation
Novae
 helium abundance in, 169
NP 0527, 270, 274
NP 0532, 143, 265, 270, 273-74, 276-77, 281-84, 286, 288-91, 293-94
see also Crab Nebula, NGC 1952, Pulsars, Tau X-1
Nuclear
 breakup, 192-93
 forces, 184
 potential, 181, 183-84, 189
 -shell effects, 181
Nuclear (galactic) activity, 390, 392, 412, 414-17, 430, 433, 452
Nucleation, 97
Nuclei of galaxies, 369-453
 blue, 377
 hot-spot, 417
 peculiar
  surveys of, 417-20
 red, 377
 Seyfert, 395-406
 visibility of, 371
see also Galactic nuclei
Nucleosynthesis, 446-49
 in a massive object, 447
Nyquist frequency, 129-30

O

Objective prism, 419
Oblateness of Sun, 299, 303-7, 319, 326
 diagonal component of, 305-6
 excess, 312
 measurement of, 299-300, 303-6
 telescope for, 300, 304-5

value of, 307
see also Sun, Rotation
Ochkur approximation, 336, 350, 357
see also Quantal approximations
OH lines, 390
Opacity, 194
 bound-free
  due to metals, 162-63
 continuous, 98-100
 H^-, 96, 98, 106
 neutrino, 439
 pressure-induced
  of H_2, 101, 105
 due to solid particles, 102
 sources in late-type stars, 97-103
see also Bound-free, Free-free, Helium, Hydrogen
Optical depth, 234-39, 259, 311
Optical galaxy, 412-13, 415
Optical identification
 of radio galaxies, 413, 416
 table of, 408-10
see also Galactic nuclei, Galaxies
Optical size of galactic nuclei
see Galactic nuclei
Orion A, 170, 232, 238, 244, 246, 252, 260
Orion Nebula, 169, 225
Oscillator strength, 246, 340, 342, 347, 353
Outbursts
 of galactic nuclei, 413-15
  frequency of, 413, 415
  harmonic mean life for, 414
see also Galactic nuclei
O/C ratio
 in late-type stars, 90, 106, 109
O/H ratio
 in M15, 168
Oxygen
 forbidden-line emission, 373, 378-79, 382, 384, 392-93, 396-97, 402, 404, 408, 415, 420
see also Forbidden lines

P

Parabolic cylinder function, 48
Partial-wave analysis, 333-34, 337, 340
see also Quantal approximations
Particle
 -anti-particle pair

creation, 449
auroral, 79, 81-82
  ejection of, 414-15
  high energy, 439, 452-53
  production of aurora by charged, 73, 79-81
  relativistic, 370, 389, 392-93, 414, 430, 441, 443, 445, 452
  spectra, 428
  see also Electron, Neutron, Proton, Auroras
Partition function, 96-97
Paschen lines, 421
  see also Hydrogen
Péclet number, 47
Peculiar Galaxies
  Atlas of, 432
Perihelion of Mercury
  excess motion of, 297-99, 326
Periods
  of pulsars, 268-73
  changes in, 270-72
  see also Pulsars
Perseus cluster, 371, 397, 432
Phase
  corrections, 209
  3p
  shifts, 199
Photodissociation
  of molecules, 100
Photoelectric observations, 116, 122-23
Photoionization, 421
  of molecules, 100
  of neutral hydrogen, 164
  see also Ionization
Photometer
  iris, 116
Photometry
  photoelectric, 24
  multichannel, 124
  of subdwarfs, 163
Photospheres
  stellar
    to get He abundance, 171-74
  see also Helium
PKS sources, 171, 403, 404, 400, 409, 410
Plage emission, 2, 3
  see also Sun
Planck function, 233-34
Planetary nebula, 399
Plasma
  electron-proton, 54-55
  electron-proton-alpha particle, 54
  frequency, 266
  interstellar, 279, 281, 286-87
  waves, 424
  see also Interstellar gas,

Medium, H II regions
Plate
  automatic
    scanner, 126
  parameters, 125
  photographic, 116-17, 125-26
  resolution elements in, 116
  subtraction technique, 116
Pleiades
  rotation in, 301-2, 319-20, 322
Polar-cap absorption, 76
Polarizability, 334-35
Polarization, 276-78, 292, 392, 396-98, 402
  circular, 277-78, 294
  elliptical, 276, 292
  of fine structure
    in pulsars, 278-79
    in late-type stars, 94-95
  variability of, 94
  wavelength dependence of, 94
  linear, 276-78, 292, 423
  optical, 391
  position angle of
    wavelength dependence of, 94
  potential, 335, 346
  see also Pulsars
Polar wind, 55-58
  density of, 57
  velocity of, 57
  see also Stellar winds
Poloidal fields, 8, 10-13, 15, 18-19, 21-24, 27, 316
  see also Magnetic fields
Population
  amorphous, 373
  disk, 376
  dwarf, 373-74
  halo, 436
Population I, 93, 162, 171-72, 423, 426
Population II, 164, 172-74, 447
Positive ions, 338-40
Potential
  centrifugal, 309
  Coulomb, 337
  direct, 333
  effective, 335
  exchange, 333
  gravitational, 308, 314
  polarization, 335, 346
Power spectrum, 127-28
  of pulsars, 285-86
  see also Pulsars
Prandtl number, 48
Pressure
  broadening, 101, 261
    see also Broadening
  dengenerate electron, 440

degenerate neutron, 440
dissociation, 96
ionization, 96
  see also Degeneracy, Dissociation, Ionization
Primeval explosion
  anisotropy in, 175
Primeval fireball, 175, 434, 451
Prominences, 2-3
  see also Sun
Proper motion
  of Sco X-1, 152
  survey machine, 126-27
Protogalaxy, 436
Proton
  Fermi level, 184, 203
  superfluid, 199, 206
  threshold, 183, 186, 192-93
Protosun, 302-3
PSR 0833, 272-78, 281-82, 285, 289, 294
PSR 1749, 271, 274, 277
PSR 1749-28, 271, 281
PSR 2045, 270
Pulsars, 143, 195, 197, 201, 204-5, 265-94, 444
  clock mechanism of, 289
  as pulsation of neutron stars, 290
  as rotation of neutron stars, 289-90
  discovery of, 265
  disperson measure of, 266-67, 270, 279, 281, 284, 286-87
  histogram of, 279-80
  distance estimates of, 280-82
    from dispersion measure, 281
    from 21-cm absorption, 281
  dynamic spectrum of, 266-67, 283-84, 287
  energy considerations of, 288-91
  rotation, 289-90
  galactic distribution of, 268-69
  giant, 443
  high-latitude, 268
  irregular intensity fluctuations, 283-86
    long-term, 285
    narrowband, 284
    time scales of, 283
    wideband, 284
  oblique rotator model for, 290-93
  observations of, 266-88
  periods of, 268-73
    catalogue of, 270-71
    histogram of, 269

SUBJECT INDEX

sudden changes in, 273, 294
variations in, 272-73
period changes of
 mean rate of, 270-72
periodic intensity variations, 285-86
physical nature of source of, 288-89
 brightness temperature of, 288
 dimensions of, 288
 energy requirements of, 288
 lifetime of, 288
 space density of, 288
polarization
 of fine structure in, 278-79
polarization angle
 changes in, 276, 293
power spectra of, 285-86
pulse
 dispersion, 279-80
 duration, 273-74, 288
 emission, theories for, 291-93
 envelope, 274-76
 individual fine structure, 278-79, 285-86, 288
 mean envelope, 273-76
 mean shape, 273-75
 shape, 272-73
 shape dependence on wavelength, 275-76
 smearing, 267
 trains, 284-85
 width, 276
radiation mechanisms for, 291-94
 as rotating neutron stars, 180
search techniques, 135-36, 266-68
spectra of, 282-83
 infrared, 282-83
 optical, 282-83
 radio, 282
 X ray, 282-83
spectra index of, 284
subpulses
 structure of, 278, 286
 time shifts of, 279, 285-86
swept-frequency nature of radiation, 266, 284
theory of, 288-94
see also Dispersion, Dispersion measure, Neutron stars, Magnetic fields, Polarization, X Rays
Pulsation
 class II, 279, 285

stellar, 168
 irregular, 93
 semiregular, 93

Q

Q
quality factor, 219, 222
QSO
 see Quasistellar objects
Quadrupole lines
 of H_2, 101
Quantal approximations, 336-41, 356-57
 Bethe, 340-41, 347
 Born, 335-37, 341, 343, 350
 Born (a), 356, 360
 Born (b), 356-57, 360-61
 Born II, 337, 341-42, 361
 Born-exchange, 357, 360-61
 Born-Oppenheimer, 356-57, 360
 Born, unitarized, 337-38
 others, 357
Quantum-defect extrapolation, 348-49
Quantum theory, 353-56
 asymptotic charges, 353-54
 atomic eigenfunction expansion, 332-33
 close-coupling approximation, 334-36, 343, 348-51
 direct and exchange amplitudes, 354
 partial-wave analysis, 333-34, 337, 340
Quasars
 see Quastistellar objects
Quasistellar objects, 348, 386, 399, 407-12, 419-20, 422-25, 428, 432, 439-40, 442-45
 dead, 441
 ejection of, 432
 helium abundance in, 171

R

Racah coefficient, 347
Radial velocity
 of Arcturus, 226
 contours, 260
 of Cyg X-2, 154-55
 of Sco X-1, 145
Radii
 of late-type stars, 90
Radio emission, 413, 416-17, 424
 nonthermal, 412, 432-33
 see also Synchrotron

radiation
Radio flux
 table of, 408-10
Radio galaxies, 394, 412-17, 439-40, 442, 445, 449-52
 see also Galactic nuclei, Galaxies, Radio sources
Radioheliograph
 image-forming, 123
Radiometer, 134
Radio sources, 396-97, 402-3, 406-7, 410, 413-16, 432, 446
 see also Radio galaxies
Random motions, 381, 384
Rankine-Hugoniot relations, 38
Ratio recording, 221
Rayleigh-Jeans approximation, 234, 450
Rayleigh scattering, 94, 100, 164
 by H, H_2, 100
 by He, C, N, 100
Receivers
 array, 116
 continuum, 134
 filter, 130
 multichannel, 116, 130-31
 spectral line, 116
Recombination
 dielectronic, 242-43, 255, 257
 overpopulation by, 257
 electronic, 239
 radiative, 234, 243
 three-body, 239
 see also Ionization
Recombination line(s)
 anomalous, 231, 240, 244, 254-60
 carbon, 231, 242-43, 257, 259
 helium, 231, 253-56
 hydrogen, 231, 244-54
 in Orion Nebula, 244-45, 247-52, 255-59, 261
 radiofrequency, 169, 231-63
Reddening, 421
 towards nucleus, 377-78
Redshift
 apparent magnitude relation, 378, 411
 cosmological, 369, 411, 419
 gravitational, 442-43
 in neutron stars, 194, 206
 intrinsic, 442
Redshifts, 395, 398, 402-7, 410-13, 419, 422, 441-43, 445
 table of, 408-10

Regression
 of node on solar equator, 297, 299
Relativity
 effects of, 188, 206, 412, 441-42
Relaxation, 436, 440, 448
 encounterless, 436
 times, 436-38, 441
 violent, 436-37
Relaxed systems, 435
Resolving power, 213-14
 of Fabry-Perot etalon, 213
 of grating spectrometer, 213
 see also Fourier spectroscopy
Resonance
 fluorescence, 423
 influence
  on inelastic scattering, 343-46, 348-49
Reynolds criterion, 314
Reynolds number, 14, 47-48
Richardson criterion, 314
Riometer, 80
Rotating core
 solar, rapidly, 297-99, 303, 326-27
 evolution of, 319-23
 see also Oblateness of Sun, Rotation
Rotating disk, 441, 443
Rotation
 curves, 381-82, 386, 395, 398
 differential
  of Sun, 302, 314, 316, 319
  of galactic nucleus, 381-83, 385
 galaxies, 381, 435
 in Hyades, 301, 319-20, 322
 internal
  of Sun, 297-327
 of late-type stars, 93
 of neutron star, 197, 203, 205-6
 nonuniform
  of core of Earth, 7
 in Pleiades, 301-2, 319-20, 322
 of Sun, 4, 11, 13, 16-17, 19, 21-22, 24, 26
 surface, 301, 309, 319, 322-23, 326-27
 see also Galactic nuclei, Neutron stars, Oblateness, Pulsars
r process, 448

S

S-18, 173

S 5003, 155-56
Saha-Boltzmann equation, 233
Sampling, 209, 214
 see also Data handling, Information processing
Sanford bands
 of SiC_2, 82
Satellites
 of galaxies, 432-33
Scalar-tensor theory
 of gravitation, 297-99
 see also Brans-Dicke
Scattering
 amplitudes, 354-55
 Compton, 194-95, 370, 392, 424-25, 427-28
 continuous
  sources, 100
 elastic, 334, 344
  by alkali atoms, 334
  resonances in, 334, 340
 electron, 100, 336, 344, 391, 422-23
 inelastic, 336, 343-46
  influence of resonances on, 343-46
 multiple, 337
 resonance line, 423
 -single, 337
Schroedinger equation, 332, 338
Schwarzschild radius, 188, 206, 440, 442, 444
Scintillation
 interstellar, 276, 285-88
 techniques, 370
Sco-Cen association, 150, 152
Sco X-1, 142-57
 color-magnitude locus of, 149
 distance estimates, 149-52
 flares of, 144-45, 153
 free-free self-absorption in, 153
 old nova, 152
 optical identification of, 143-44
 optical spectrum of, 144-45
 periodicities
  in light variation of, 147
 photometric properties of, 145-47
 proper motions of, 152
 radial velocity of, 145
 radio emission from, 149-50
 simultaneous optical and X-ray observations of, 152-54
 ultraviolet observations

 of, 149-50
 variability of, 145-48, 152
 of emission lines, 144
 see also X rays
Screening
 of nuclei, 196
Seeing anisotropy, 304
Seismic activity
 in neutron stars, 198
Seyfert nuclei, 391, 407, 410-12, 415-24, 436, 441
 physical conditions in, 420-30
 see also Galactic nuclei, Galaxies
Seyfert's Sextet, 431
Shapley-Ames galaxies, 377, 395
Shear
 meridional, 310
 radial, 310
 transverse, 310
Shock transitions
 in stellar winds, 36, 38-42, 45
Shock waves
 in stellar winds, 47-49
Signal averaging, 222
Signal-to-noise gain
 from multiplexing
  in infrared, 212
  in visible and near infrared, 212
Signal-to-noise ratio, 128, 130, 134-35, 209, 212, 215-18, 222, 226-27
SiO_2 particles
 in M stars, 94
Single-particle energy states
 gap in spectrum of, 198-99, 201
Sky noise, 223-24
 angular correlation of, 223
SN 1572, 155
Solar activity, 2, 307, 323
Solar limb
 position of, 308-9
 temperature of, 309
 width of, 304
Solar wind, 3
 braking, 301-2, 315-16, 320
 heat addition, 43-46
 magnetic field in, 298
 multifluid models of, 50-55
 observations of, 31
 reviews of, 31
 structure of, 299, 317, 326
 torque, 298, 302, 317-24, 326
 see also Stellar winds, Magnetic field
Solid-angle gain, 213

SUBJECT INDEX

Space velocities
 of late-type stars, 92-93
Specific intensity, 234
Spectral classification
 of C, M, and S stars, 87-88
 of galaxies, 372
 see also Classification of galaxies
Spectral index, 387, 416, 421, 429
 of pulsars, 284
Spectral type
 of late-type stars
  C, N, R, 88-90, 92
  M, 87, 89, 92
  S, 88-89, 92
Spectral type, integrated
 of galaxies, 372, 375, 383-84
Spectrometer
 autocorrelation-function, 127-28, 130-31
 Fourier, 127
  for infrared, 136
 grating, 213, 216-17
  champions, 223
  resolving power, 213
  spectrum of Venus, 220
 scanning monochromator, 128
 source of noise
  in infrared, 127
 see also Fourier spectroscopy
Spectroscopic character, 408-10, 413, 418
Spectroscopy
 Fabry-Perot, 209
 see also Fourier spectroscopy
Spectrum
 of Alpha Orionis, 227-28
 auroral, 71-73
 of Cyg X-2, 154
 dynamic
  of pulsars, 266-67, 283-84, 287
 emission
  of Earth, 228
 high resolution, 226-27
 infrared of
  Corona, 228
  Moon, 223
  Venus, 216-20, 223-24, 227
 integrated, 372, 375, 423
 nonthermal, 441
 of pulsars, 266-67, 282-84, 287
 radio, 416
 reconstructed, 210
 of Sco X-1, 144-45

Spindown
 currents, 315
 of Sun, 312-17
Spiral arms, 376, 378, 416, 420, 432
 H II regions as tracers of, 261
Spiral galaxies, 371-74, 376-81, 385, 397, 410, 416-18, 433, 435-36
 see also Galaxies
s process, 448
 elements in late-type stars, 91-92
Star cluster
 relativistic, 439, 441-44
Star formation, 375
 rate of, 447-48
Stark broadening, 261
Starquake, 197, 205-6, 294
 see also Neutron stars
Stars
 Ap, 172
 Ba II, 91-92
 carbon, 88, 90-92, 94, 99, 172
 CH, high-velcoity, 88, 91
 χ Cygni type, 108
 degenerate
  mass of, 187
 dMe, 92-93
 helium, 172
 high-luminosity, 420
  in nucleus, 370, 376, 378, 386
 horizontal-branch, 167-68, 173-74
 long-period, 92
 low-mass, 376-77
 magnetic A, 303
 in moving groups, 164-65, 176
 neutron, 179-206, 289-91, 370, 440, 442-43
 RCBr, 172
 relativistic, 442
 RR Lyrae
  sensitivity of luminosity to He abundance, 167
  variable, 93
 very late-type, 87-110
 see also Neutron stars
Statistical equilibrium, 238-39, 242
 nuclear, 188
Steady-state cosmology, 435, 445, 451
 galactic nuclei in
  formation of, 445
 inhomogeneous
  theory, 445
 universe, 441, 451
 see also Cosmology, Universe

Stellar distribution, 437
Stellar nucleus, 371, 391, 395, 403, 406-7
 see also Galactic nuclei
Stellar population, 371-78, 384, 436, 446
 intermediate, 372
 synthesis of
  of galaxies, 373-76
 see also Population
Stellar winds, 31-59, 291
 accretion solutions, 38-40
 boundary conditions for, 37-41
 braking, 298, 319-20
 electron temperature in, 50, 52-53, 58-59
 expansion
  into interstellar medium, 35, 38
 flow
  away from comet, 37
  with heat addition, 38, 43-46
   changes of temperature in, 45
   equations of motion for, 44
  with heat conduction, 38, 46-52, 58
   equations of motion for, 46
 ionization fronts in, 40
  D-type, 41
 ion temperature of, 50, 52-53
 magnetic field
  nonradial, 58
 multicomponent flow, 32, 50
 radial flow of
  with positive energy, 34-35
  with zero or negative energy, 36-37
 recombination fronts, 40
 shock transitions in, 38-39, 41
 shock waves in, 47-49
 subsonic flow in, 36-40, 42, 45
 supersonic flow in, 36-40, 42, 45, 52
 transitions
  between subsonic and supersonic flow in, 36, 40, 42, 45
 viscosity, effect of, 46-50
  equations of motion for, 46
 see also Magnetic field, Solar wind
Stephan's quintet, 431
Stepping drive, 218, 222

489

SUBJECT INDEX

Stimulated emission, 233, 236, 238, 243, 249, 251, 260
Stoke's parameters, 133
Strömgren sphere, 40, 253
Subdwarfs
  photometry of, 163
Subsonic flow, 36-40, 42, 45
  see also Stellar winds
Sun
  angular velocity change in, 315
  density stratification of, 315
  dependence of brightness on latitude, 306-9, 312, 326
  differential rotation of, 302, 314, 316, 319
  equation of state for, 166
  extrapolated limb of, 306
  gravitational quadrupole moment of, 297-99, 307-12, 319, 326
  heavy metal content of, 166
  helium abundance, 165, 166
    from prominences, 167
  instabilites in interior, 312-17
  internal rotation of, 299-327
  iron abunance of, 166
  limb darkening of, 306-8
  mixing
    in core of, 316
  moment of inertia of, 302
  oblateness of, 299, 303-7, 319, 326
  orientation of axis of changes in, 305
  rapidly rotating core of, 297-99, 303, 319-23, 326-27
  spindown of, 312-17
  temperature gradient of, 314
  velocity fields in, 299-300, 308-10, 312, 326
  see also Core, Corona, Dynamo, Magnetic field, Oblateness of Sun, Rotation, Solar wind, Sunspots
Sun-Sirius group, 164-65, 176
Sunspots, 2, 21, 309
  cycle, 27, 71, 303
  fields, 17, 309
    forward tilt of, 27
  latitude of
    variation of, 22
  relation to aurora, 66-71

Superadiabatic zone
  in late-type stars, 103-5
Superconductivity, 199
Superfluid
  protons, 199, 206
  rotation
    effect of, 200
Superfluidity
  anisotropic, 199
  in neutron stars, 198-200, 206
Supergranulation, 25
Supergranules, 3, 5, 16, 20, 27
  eddy diffusivity of, 25
Supernovae, 179, 200-1, 439, 448, 452
Supernova remnants, 155, 281, 289, 416-17, 429, 451-52
Supersonic flow, 36-40, 42, 45, 52
  see also Stellar winds
Superstar, 440
Swan bands, 88
Synchrotron radiation, 140, 201, 292-94, 370, 388-89, 391-92, 396, 424, 426-29, 441, 444
  coherent, 370, 388
  electron, 424, 426, 430
  incoherent, 370, 424
  proton, 424
Synchrotron self-absorption, 388, 398
Synchrotron sources
  total energy of optical
    table of, 425-26, 428

T

Tape recording
  magnetic, 117, 123
Tau X-1, 140, 142-43
  see also Crab Nebula, Pulsars
Technetium
  in late-type stars, 91
Telescope
  filled-aperture, 132
  one-mile, 131-33
  solar oblateness, 300, 304-5
Temperature
  antenna, 253, 258
  brightness, 238, 253, 258
  of background radio emission, 450
  of pulsars, 288
  coronal maximum, 55
  discrepancy
    between optical and radio

data, 244
effective
  of late-type stars, 89
  -luminosity plane, 162, 165
  of photosphere, 307
electron, 244, 246-47, 249-52, 258, 379, 420
  of aurora, 80
  in stellar winds, 50, 52-53, 58-59
excess
  at equator of Sun, 311
fluctuations
  in H II regions, 244
gradient, 194
  of Sun, 314
  of H I cloud, 237
ion
  in stellar winds, 50, 52-53
line, 237-38, 241
melting
  of neutron-star crusts, 196
stratification
  in late-type stars, 106
surface
  of neutron stars, 195, 201
  see also Auroras, Galactic nuclei, Neutron stars, Pulsars, Sun, Stellar winds
Thermal absorption, 395
Thermal radiation, 411, 422-23
Thermodynamic derivatives, 97
Thermonuclear processes, 415
Thermopile, 223
Thomson scattering, 100, 194-95
Threshold
  inelastic, 343
  ionization, 355-56
  linear
    law, 352, 355, 357
  negative muon, 184, 186
  neutron, 192-93
  proton, 183, 186, 192-93
Throughput factor (luminosity), 212-13, 215-16
  see also Fourier spectroscopy
Ton 1542, 171
Toroidal field, 2, 4, 8, 10, 12-13, 15-16, 18-26, 303, 316
  see also Magnetic field
Torsional oscillation, 303
Transfer
  equation of, 95
  for gaseous nebulae, 234

SUBJECT INDEX

Transitions
  allowed, 339, 341
  collisional
    by electron impact, 239
  collisionally induced, 330
  between fine-structure levels, 346-48
  forbidden, 98, 346, 348-49, 361
  free-free, 236, 340
  impact-parameter method for allowed, 341-42
  intercombinational, 337, 339
  probability dipole, 359
  quadrupole, 342
  rate, 233, 235, 239
  spontaneous, 239
  see also Forbidden lines
Transonic region, 48
Turbulence, 314-16, 318, 320, 325, 422-23, 440
  modulation of intensity, 221
  of beamshape, 221
  of position, 221
  thermally driven, 313, 319-20, 323
  see also Microturbulence
Tycho's supernova, 155

U

UBV colors, 419
Ultraviolet
  excess, 419, 433-34
  radiation
    from galactic nuclei, 433-34
Universe
  big-bang, 449
  evolving, 434
    formation of nuclei in, 436-44
  expansion of, 378
  Friedman, 444
  mass of
    observable, 445
  steady-state, 444, 451
  see also Cosmology, Steady state
URCA process, 200, 203

V

Van Allen belts, 79
Variability
  of galactic nuclei, 402, 407, 410, 412, 423-24, 426,
428-30
  of late-type stars, 93-94
  of pulsars, 285-86
  of Sco X-1, 145-48, 152
Variable stars
  eruptive, 93
  GW Aurigae type, 93
  Mira type, 94
  T Tauri type, 93
  see also Flare stars, Flash stars, Pulsating stars
Vela X, 197, 205, 206, 265, 281
Velocity dispersion, 373, 375, 383, 435, 438
  elliptical, 382
  isotropic, 382
Velocity field, 383-84, 389, 391, 397
  in Sun, 299-300, 308-10, 321, 326
Venus
  CO_2 bands in, 220, 224, 226
  Fourier spectrum of, 220
  infrared spectra of, 216-20 223-24, 227
Vibrations
  of neutron stars, 180, 201-4
  see also Neutron stars, Pulsation
Virgo cluster, 370-71, 374, 377, 380, 393, 431-32, 437, 440
Viscosity
  in stellar winds, 46-50
  turbulent, 314, 316
Viscous forces, 313
Visual observations
  of double stars, 115
Volcanic activity
  in neutron stars, 198
von Zeipel's theorem, 307-9
V_α potential, 184-85
V_γ potential, 185, 187, 200
VV 144, 405, 418
VV 150, 406, 418
VV 159, 431
VV 172, 431

W

W3, 247
W43, 170
W51, 247
Wavefunctions, 331-33, 336, 343, 348, 354-57, 359
  Coulomb, 355
  distorted-wave approximation, 336, 348
  perturbing electron, 332
Wavenumber measurements
  accuracy of, 214
Whistlers, 80
White dwarfs, 187-88, 196, 202, 289-90, 370, 440, 442
  maximum mass of, 188-89
  vibrational stability of, 188
Wiener-Khintchine theorem, 127
Wilson-Bappu effect, 165

X

X ray(s), 426, 433
  auroral events, 73, 78-79
  background radiation, 433, 449, 451-52
  flares in Sco X-1, 153
  positions
    accuracy of, 139-40
  pulsar spectra, 196, 282-83
  pulses, 143
X-ray sources, 179, 202-3, 392-93
  angular size of, 142-43
  lunar occultation of, 140, 142
  optical identification of, 139-42
    criteria for, 139
  optical observations
    of extrasolar, 139-58
  thermal bremsstrahlung in, 140, 147, 149-51, 153
  see also Cyg X-2, Sco X-1, Tau X-1

Z

Zirconium abundance
  in late-type stars, 91
Zwicky objects, 402-3, 405-6, 418-91

CUMULATIVE INDEXES

VOLUMES 4 - 8

INDEX OF CONTRIBUTING AUTHORS

A

Axford, W. I., 8:31

B

Batten, A. H., 5:25
Bely, O., 8:329
Bok, B. J., 4:95
Boksenberg, A., 7:421
Bowen, I. S., 5:45
Brandt, J. C., 6:267
Brown, R. H., 6:13
Bullen, K. E., 7:177
Burbidge, E. M., 5:399
Burbidge, G. R., 8:369

C

Cameron, A. G. W., 8:179
Carr, T. D., 7:577
Caughlan, G. R., 5:525
Cayrel, R., 4:1
Cayrel de Strobel, G., 4:1
Chapman, S., 8:61
Christy, R. F., 4:353
Clark, B. G., 8:115
Cohen, M. H., 7:619
Connes, P., 8:209
Conti, P. S., 7:99

D

Danziger, I. J., 8:161
Dicke, R. H., 8:297
Dressler, K., 4:207
Dupree, A. K., 8:231

E

Eggen, O. J., 5:105
Evans, J. V., 7:201

F

Fazio, G. G., 5:481
Fichtel, C. E., 5:351
Field, G. B., 4:207
Findlay, J. W., 4:77
Ford, W. K., Jr., 6:1
Fowler, W. A., 5:525

G

Gardner, F. F., 4:245

Garstang, R. H., 6:449
Giacconi, R., 6:373
Ginzburg, V. L., 7:375
Goldberg, L., 5:279; 8:231
Goldreich, P., 6:287
Goody, R., 7:303
Gould, R. J., 6:195
Gulkis, S., 7:577
Gursky, H., 6:373

H

Hayashi, C., 4:171
Hewish, A., 8:265
Hiltner, W. A., 8:139
Holzer, T. E., 8:31
Howard, R., 5:1
Huang, S.-S., 4:35

I

Iben, I., Jr., 5:571

J

Johnson, H. L., 4:193

K

Kellermann, K. I., 6:417
Kerr, F. J., 7:39

L

Layzer, D., 6:449
Lebovitz, N. R., 5:465
Ledoux, P., 4:293
Lin, C. C., 5:453
Lynds, B. T., 6:215

M

Mathews, W. G., 7:67
McDonald, F. B., 5:351
McGee, R. X., 5:183
Meyer, P., 7:1
Moffet, A. T., 4:145
Mook, D. E., 8:139
Morrison, P., 5:325

N

Ness, N. F., 6:79
Neupert, W. M., 7:121
Newkirk, G., Jr., 5:213

Novikov, I. D., 5:627

O

O'Dell, C. R., 7:67
Öpik, E. J., 7:473

P

Parker, E. N., 8:1
Pauliny-Toth, I. I. K., 6:417
Peale, S. J., 6:287
Pikel'ner, S. B., 6:165
Popper, D. M., 5:85

R

Renson, P., 4:293
Robinson, B. J., 5:183
Ryle, M., 6:249

S

Sawyer, C., 6:115
Schatzman, E., 5:67
Scheuer, P. A. G., 6:321
Schmidt, M., 7:527
Smak, J. I., 4:19
Somerville, W. B., 4:207
Souffrin, P., 5:67
Spinrad, H., 7:249
Strittmatter, P. A., 7:665
Strömgren, B., 4:433
Sweet, P. A., 7:149
Swenson, G. W., Jr., 7:353
Syrovatskii, S. I., 7:375

T

ter Haar, D., 5:267

U

Underhill, A. B., 6:39

V

van de Hulst, H. C., 5:167
Van Regemorter, H., 8:329
Van Speybroeck, L. P., 6:373
Vardya, M. S., 8:87

INDEX OF CHAPTER TITLES   493

Vasilevskis, S., 4:57
Vaughan, A. H., Jr., 5:139

W

Wagoner, R. V., 7:553
Wallerstein, G., 7:99

Weidemann, V., 6:351
Wheeler, J. A., 4:393
Whiteoak, J. B., 4:245
Wickramasinghe, N. C., 6:215
Williams, P. J. S., 6:321
Wilson, R., 7:421

Wing, R. F., 7:249

Z

Zeldovič, Ya. B., 5:627
Zimmerman, B. A., 5:525
Zwaan, C., 6:135

INDEX OF CHAPTER TITLES

VOLUMES 4 - 8

THE SOLAR SYSTEM

Magnetic Field of the Sun (Observational)	R. Howard	5:1-24
Waves in the Solar Atmosphere	E. Schatzman, P. Souffrin	5:67-84
Structure of the Solar Corona	G. Newkirk, Jr.	5:213-66
Ultraviolet and X Rays from the Sun	L. Goldberg	5:279-324
Energetic Particles from the Sun	C. E. Fichtel, F. B. McDonald	5:351-98
Observed Properties of the Interplanetary Plasma	N. F. Ness	6:79-114
Statistics of Solar Active Regions	C. Sawyer	6:115-34
The Structure of Sunspots	C. Zwaan	6:135-64
The Physics of Comet Tails	J. C. Brandt	6:267-86
The Dynamics of Planetary Rotations	P. Goldreich, S. J. Peale	6:287-320
X Rays from the Sun	W. M. Neupert	7:121-48
Mechanisms of Solar Flares	P. A. Sweet	7:149-76
The Interiors of the Planets	K. E. Bullen	7:177-200
Radar Studies of Planetary Surfaces	J. V. Evans	7:201-48
Motions of Planetary Atmospheres	R. Goody	7:303-52
The Moon's Surface	E. J. Öpik	7:473-526
The Magnetosphere of Jupiter	T. D. Carr, S. Gulkis	7:577-618
The Origin of Solar Magnetic Fields	E. N. Parker	8:1-30
The Theory of Stellar Winds and Related Flows	T. E. Holzer, W. I. Axford	8:31-60
Auroral Physics	S. Chapman	8:61-86
Internal Rotation of the Sun	R. H. Dicke	8:297-328
STELLAR SPECTRA AND ATMOSPHERES		
Atmospheres of Very Late-Type Stars	M. S. Vardya	8:87-114
PHYSICS OF STARS		
Abundance Determinations from Stellar Spectra	R. Cayrel, G. Cayrel de Strobel	4:1-18
The Long-Period Variable Stars	J. I. Smak	4:19-34
Magnetic Stars	P. Ledoux, P. Renson	4:293-352
Pulsation Theory	R. F. Christy	4:353-92
Superdense Stars	J. A. Wheeler	4:393-432
Extrasolar X-Ray Sources	P. Morrison	5:325-50
The Wolf-Rayet Stars	A. B. Underhill	6:39-78
White Dwarfs	V. Weidemann	6:351-72
Lithium and Beryllium in Stars	G. Wallerstein, P. S. Conti	7:99-120
Infrared Spectra of Stars	H. Spinrad, R. F. Wing	7:249-302
Physics of Massive Objects	R. V. Wagoner	7:553-76
Stellar Rotation	P. A. Strittmatter	7:665-84

Neutron Stars	A. G. W. Cameron	8:179-208
Pulsars	A. Hewish	8:265-96

ASTROMETRY

The Accuracy of Trigonometric Parallaxes of Stars	S. Vasilevskis	4:57-76

STELLAR ASTRONOMY

Problems of Close Binary Systems That Involve Transfer of Angular Momentum	S.-S. Huang	4:35-56
On the Interpretation of Statistics of Double Stars	A. H. Batten	5:25-44
Determination of Masses of Eclipsing Binary Stars	D. M. Popper	5:85-104
Masses of Visual Binary Stars	O. J. Eggen	5:105-38

THE INTERSTELLAR MEDIUM

OH Molecules in the Interstellar Medium	B. J. Robinson, R. X. McGee	5:183-212
Structure and Dynamics of the Interstellar Medium	S. B. Pikel'ner	6:165-94
Interstellar Dust	B. T. Lynds, N. C. Wickramasinghe	6:215-48
The Large-Scale Distribution of Hydrogen in the Galaxy	F. J. Kerr	7:39-66
Evolution of Diffuse Nebulae	W. G. Mathews, C. R. O'Dell	7:67-98
Radiofrequency Recombination Lines	A. K. Dupree, L. Goldberg	8:231-64

THE GALAXY

The Polarization of Cosmic Radio Waves	F. F. Gardner, J. B. Whiteoak	4:245-92
Observing the Galactic Magnetic Field	H. C. van de Hulst	5:167-82
Radio Spectra	P. A. G. Scheuer, P. J. S. Williams	6:321-50
Variable Radio Sources	K. I. Kellermann, I. I. K. Pauliny-Toth	6:417-48
Cosmic Rays in the Galaxy	P. Meyer	7:1-38
Optical Observations of Extrasolar X-Rays Sources	W. A. Hiltner, D. E. Mook	8:139-60

EXTRAGALACTIC ASTRONOMY

Magellanic Clouds	B. J. Bok	4:95-144
The Structure of Radio Galaxies	A. T. Moffet	4:145-70
Quasistellar Objects	E. M. Burbidge	5:399-452
The Dynamics of Disk-Shaped Galaxies	C. C. Lin	5:453-64
Intergalactic Matter	R. J. Gould	6:195-214
The Counts of Radio Sources	M. Ryle	6:249-66
Quasistellar Objects	M. Schmidt	7:527-52
The Nuclei of Galaxies	G. R. Burbidge	8:369-460

COSMOLOGY AND COSMOGONY

Evolution of Protostars	C. Hayashi	4:171-92
On the Origin of the Solar System	D. ter Haar	5:267-78
Rotating Fluid Masses	N. R. Lebovitz	5:465-80
Stellar Evolution Within and Off the Main Sequence	I. Iben, Jr.	5:571-626
Cosmology	I. D. Novikov, Ya. B. Zeldovič	5:627-48
The Cosmic Abundance of Helium	I. J. Danziger	8:161-78

INSTRUMENTATION

Astronomical Optics	I. S. Bowen	5:45-66
Astronomical Fabry-Perot Interference Spectroscopy	A. H. Vaughan, Jr.	5:139-66
Electronic Image Intensification	W. K. Ford, Jr.	6:1-12
Synthetic-Aperture Radio Telescopes	G. W. Swenson, Jr.	7:353-74
Information-Processing Systems in Radio Astronomy and Astronomy	B. G. Clark	8:115-38
Astronomical Fourier Spectroscopy	P. Connes	8:209-30

OBSERVATIONAL TECHNIQUES

Absolute Intensity Calibrations in Radio Astronomy	J. W. Findlay	4:77-94
Astronomical Measurements in the Infrared	H. L. Johnson	4:193-206
Spectral Classification Through Photo-		

INDEX OF CHAPTER TITLES

electric Narrow-Band Photometry	B. Strömgren	4:433-72
Measurement of Stellar Diameters	R. H. Brown	6:13-38
Observational Techniques of X-Ray Astronomy	R. Giacconi, H. Gursky, L. P. Van Speybroeck	6:373-416
Ultraviolet Astronomy	R. Wilson, A. Boksenberg	7:421-72
High-Resolution Observations of Radio Sources	M. H. Cohen	7:619-64

PHYSICAL PROCESSES

Hydrogen Molecules in Astronomy	G. B. Field, W. B. Somerville, K. Dressler	4:207-44
Gamma Radiation from Celestial Objects	G. G. Fazio	5:481-524
Thermonulcear Reaction Rates	W. A. Fowler, G. R. Caughlan, B. A. Zimmerman	5:525-70
Theoretical Atomic Transition Probabilites	D. Layzer, R. H. Garstang	6:449-94
Developments in the Theory of Synchrotron Radiation and Its Reabsorption	V. L. Ginzburg, S. I. Syrovatskii	7:375-420
Excitation and Ionization by Electron Impact	O. Bely, H. Van Regemorter	8:329-68